Digital Sensors and Sensor Systems: Practical Design

Sergey Y. Yurish

IFSA International Frequency Sensor Association Publishing

Sergey Y. Yurish
Digital Sensors and Sensor Systems: Practical Design

e-mail (for orders and customer service enquires): ifsa.books@sensorsportal.com
Visit our Home Page on http://www.sensorsportal.com

ISBN-10: 84-616-0652-3
ISBN-13: 978-84-616-0652-8
BN-20111111-XX
BIC: TJFC

Contents

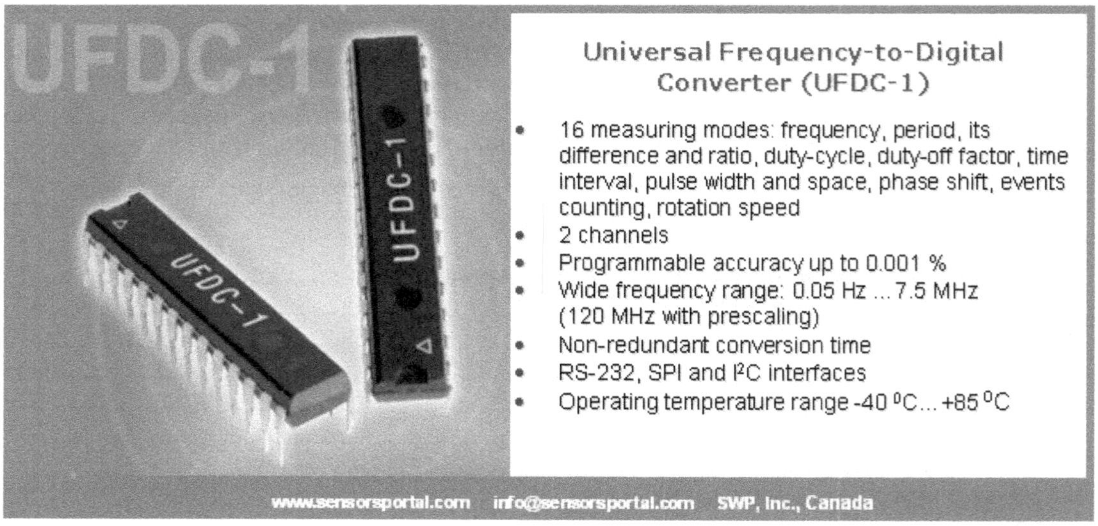
UFDC-1
Universal Frequency-to-Digital Converter (UFDC-1)
• 16 measuring modes: frequency, period, its difference and ratio, duty-cycle, duty-off factor, time interval, pulse width and space, phase shift, events counting, rotation speed
• 2 channels
• Programmable accuracy up to 0.001 %
• Wide frequency range: 0.05 Hz ... 7.5 MHz (120 MHz with prescaling)
• Non-redundant conversion time
• RS-232, SPI and I²C interfaces
• Operating temperature range -40 °C ... +85 °C
www.sensorsportal.com info@sensorsportal.com SWP, Inc., Canada

Easy and quick
sensors systems development
Evaluation Kit CD
EVAL UFDC-1/UFDC-1M-16
International Frequency Sensor Association
OPTYS Corporation
• 16 measuring modes
• Frequency range from 0.05 Hz up to 7.5 MHz (120 MHz)
• Programmable accuracy from 1 % up to 0.001 %
• RS232 (USB optional)
sales@sensorsportal.com
http://www.sensorsportal.com/HTML/E-SHOP/PRODUCTS_4/Evaluation_board.htm

Preface

A sensor is a universal and one of the most ubiquitous components being used in the simplest to the most complex application of any industry, including automotive, communications, computers, consumer electronics, medical, aerospace, etc. There is almost not one industry where a sensor is not used.

Semiconductor and integrated sensor design are heavily driven by technology scaling. Rapid advances in microelectronics and nano-technologies have brought new challenges to the digital, smart, intelligent sensors and sensor systems design.

Digital sensors and systems appeared from a necessity to input results of measurement into a computer. Traditionally, the design task of such sensors and sensor systems was solved by converting an analog quantity (voltage or current) into a digital code by an analog-to-digital converter (ADC). Most books about digital sensors and sensor systems design focus on such classical approach. However, at the *nm* technology level, the design of analog and mixed-signal circuit becomes essentially more difficult and expensive. Only a few books, book chapters, articles and papers, written mainly by the author, consider a technology driven and coming alternative to the ADC – a frequency (time)-to-digital conversion, well suited for technology scaling.

Because digital, smart, intelligent sensors and sensor systems design based on the frequency (time)-to-digital conversion has not been adequately covered in the literature before, this book aims to fill a significant gap and presents new knowledge in this emerging area of modern sensors.

Research and development results, described in this book, are relevant to the author's international research activities in the frame of European FP6 Marie Curie Chairs Excellence (EXC) project MEXT-CT-2005-023991 Smart Sensors Systems Design (SMARTSES), and International Frequency Sensor Association (IFSA) activities.

Who Should Read this Book ?

This book will greatly benefit undergraduate and at PhD students, engineers, scientists and researchers in both industry and academia. It is especially suited as a reference guide for practicians, working for Original Equipment Manufacturers (OEM) electronics market (electronics/hardware), sensor industry, and using commercial-off-the-shelf components, as well as anyone facing new challenges in technologies, and those involved in the design and creation of new digital sensors and sensor systems, including smart and/or intelligent sensors for physical or chemical, electrical or non-electrical quantities. The goal of this book is to help the practicians achieve the best metrological and technical performances at low cost, and significantly to reduce time-to-market. It should be also be useful for students, lectures and professors to provide a solid background of the novel concepts and design approach because of till now such topics have been covered adequately only in a few European and American universities.

How this Book is Organized ?

This book has been organized into 19 Chapters and one Appendix.

Chapter 1. Starting with detailed modern sensor market analyze the Chapter 1 discusses main technological challenges and two possible digital sensors and sensor systems design approaches.

Chapters 2 and 3 covers detailed information about novel integrated circuits – Universal Frequency-to-Digital Converters (UFDC-1, UFDC-1M-16) and its development board; Universal Sensors and Transducers Interface (USTI) and its future modifications - USTI-EXT for extended, high operational temperature range up to +150 ^{0}C, and USTI-WSN for wireless sensor network applications.

Chapter 4 devoted to the Frequency-to-Digital Converter (FDCP) integrated circuit with parallel, high speed interface.

Chapters 5–12 describes appropriate digital sensors, transducers and sensor systems designs: optoelectronic, temperature, pressure, humidity, acceleration, magnetic, rotational speed, chemical and biosensors. Each chapter starts from appropriate sensor market analyze and state-of-the-art review, contains practical examples, and followed by lists of references with dozens titles. These chapters can be used separately.

Chapter 13–15 deals with a direct interfacing of resistive, capacitive and resistive bridge sensing elements.

Chapter 16 describes the data acquisition systems design for both: quasi-digital and analog sensors. This chapter also discusses multifunctional and multiparametric sensors, transducers and sensor systems.

Chapter 17 devoted to intelligent sensors and sensor systems design including extension of IEEE 1451 standard for quasi-digital sensors and transducers.

Chapter 18 covers other quasi-digital sensors and transducers based designs, and applications, including wireless sensor networks.

Chapter 19 describes System-on-Chip (SoC), System-in-Package (SiP) designs based on MEMS sensors and outlines the future perspectives and modern trends.

Appendix contains useful information about engineering technique how to estimate a resulting, total error of designed sensor system in practice. An example and practical consideration are given.

How to Use this Book ?

If you are familiar with the UFDC-1, UFDC-1M-16 and USTI integrated circuits, you can pass the Chapter 2 and 3 and read appropriate chapters devoted to particular digital sensors and sensor systems designs. Each chapter in this book can be used independently and contains its own detailed list of references. For additional and up-dated information, the appropriate sensor web sections at Sensors Web Portal launched by the author are recommended to use at http://www.sensorsportal.com/HTML/Sensor.htm

Universal Frequency-to-Digital Converter
(UFDC-1 and UFDC-1M-16)
in MLF (5 x 5 x 1 mm) package
UFDC-1
UFDC-1
UFDC-1
SMALL WORLD -
BIG FEATURES
SWP, Inc., Toronto, Ontario, Canada,
Tel. + 34 696067716, fax: +34 93 4011989, e-mail: sales@sensorsportal.com
http://www.sensorsportal.com/HTML/E-SHOP/PRODUCTS_4/UFDC_1.htm

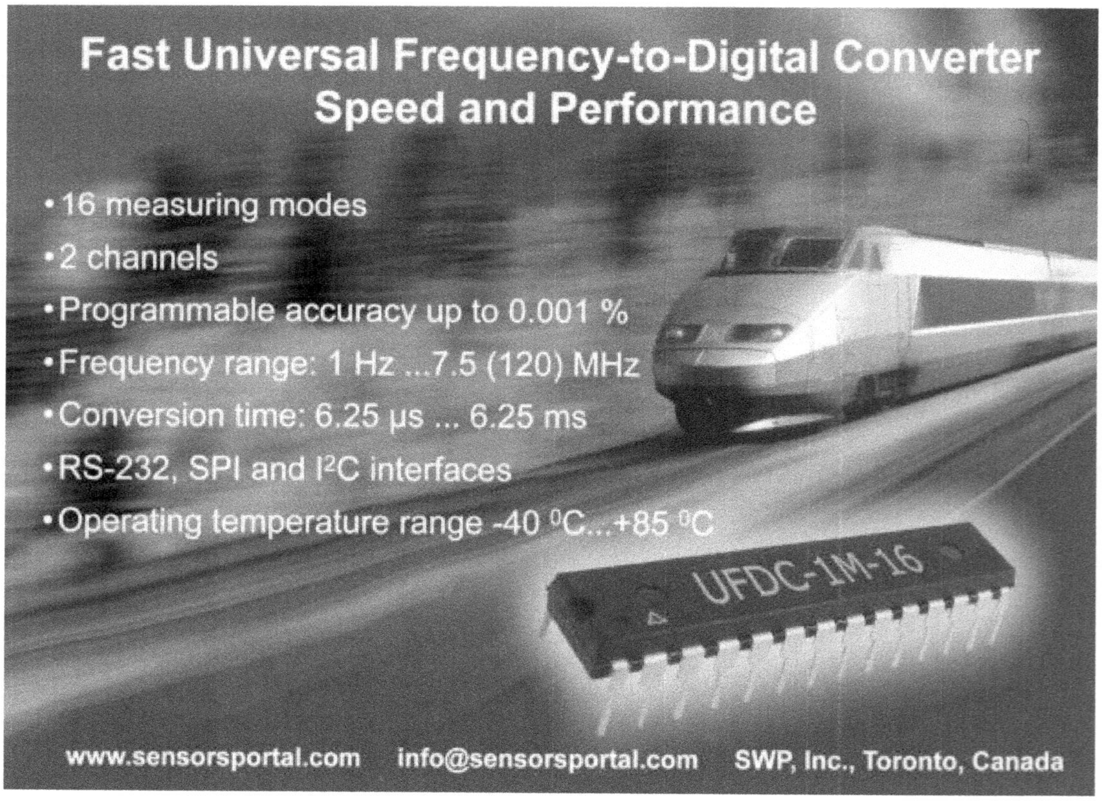
Fast Universal Frequency-to-Digital Converter
Speed and Performance
• 16 measuring modes
• 2 channels
• Programmable accuracy up to 0.001 %
• Frequency range: 1 Hz ...7.5 (120) MHz
• Conversion time: 6.25 μs ... 6.25 ms
• RS-232, SPI and I²C interfaces
• Operating temperature range -40 °C...+85 °C
UFDC-1M-16
www.sensorsportal.com info@sensorsportal.com SWP, Inc., Toronto, Canada

About the Author

Sergey Y. Yurish received MS and PhD degrees in electrical engineering and instrumentation from the Department of Automation, National University Lviv Polytechnic, Ukraine, in 1985 and 1997 respectively.

He is a president of International Frequency Sensor Association (IFSA), co-founder of Sensors Web Portal, Inc. (Canada) and co-founder and R&D Director of Technology Assistance BCNA 2010, S. L. (Spain). During his long and successful academic career he was a visiting professor of University of Applied Sciences (Furtwangen, Germany) and Technical University of Catalonia (UPC, Barcelona, Spain).

With over twenty five years of research experience he has made fundamental contributions to digital, smart and intelligent sensors and sensor systems, and advanced methods and devices for measurements of frequency-time parameters of electrical signals.

He has published more than 160 articles and papers in international peer reviewed journals and conference proceedings and got numerous best papers and articles awards. He hold nine patents and is the co-authors of three books 'Data Acquisition and Signal Processing for Smart Sensors' *(John Wiley & Sons*, 2002), 'Smart Sensors and MEMS' *(Kluwer, 2003)* and 'Smart Sensor Systems' (*John Wiley & Sons*, 2008).

Dr. Sergey Y. Yurish is an Editor-in-Chief of *Sensors & Transducers* journal (ISSN 1726-5479), *IFSA Newsletter* (ISSN 1726-6017) and a member of Editorial Board of *International Journal of Measurement Technologies and Instrumentation Engineering*.

For his continuous leadership roles and outstanding scientific research results Sergey Yurish has got an International Academy, Research and Industry Association (IARIA)'s Fellow Award in 2010.

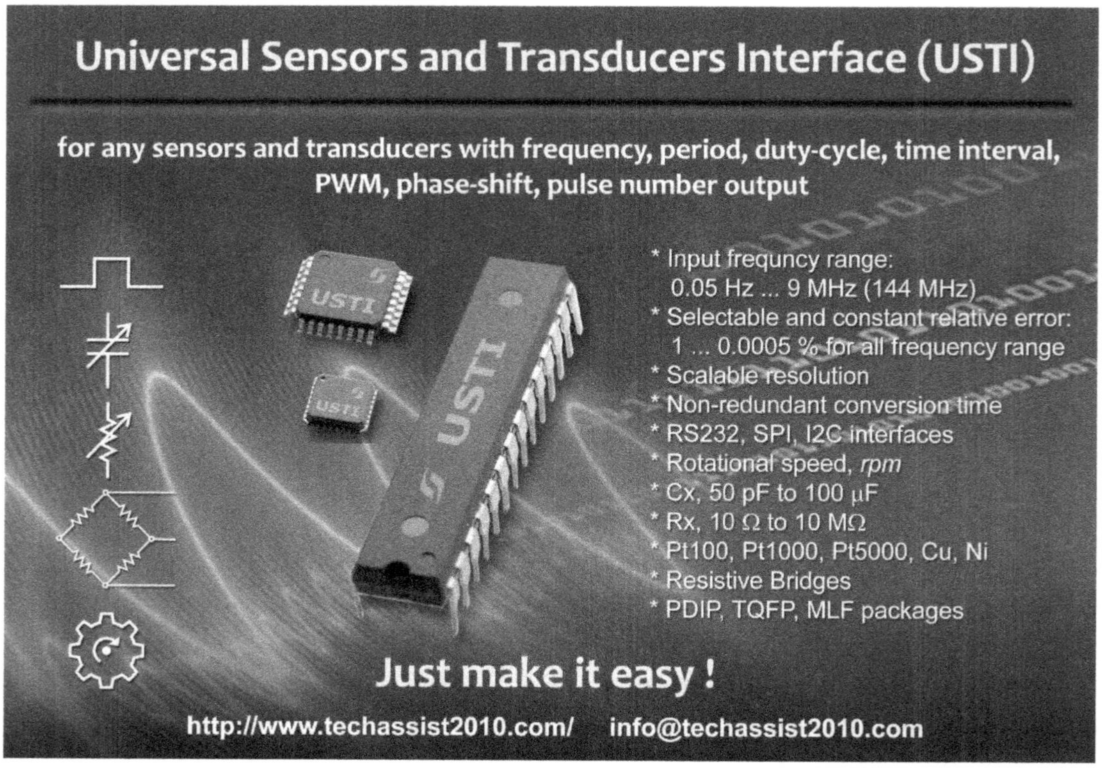
Universal Sensors and Transducers Interface (USTI)
for any sensors and transducers with frequency, period, duty-cycle, time interval, PWM, phase-shift, pulse number output
USTI
* Input frequncy range: 0.05 Hz ... 9 MHz (144 MHz)
* Selectable and constant relative error: 1 ... 0.0005 % for all frequency range
* Scalable resolution
* Non-redundant conversion time
* RS232, SPI, I2C interfaces
* Rotational speed, rpm
* Cx, 50 pF to 100 μF
* Rx, 10 Ω to 10 MΩ
* Pt100, Pt1000, Pt5000, Cu, Ni
* Resistive Bridges
* PDIP, TQFP, MLF packages
Just make it easy !
http://www.techassist2010.com/ info@techassist2010.com

SENSORS WEB PORTAL IFSA
MEMS
NEMS
NANOSENSORS
SMART SENSORS
Sensors Web Portal
All about SENSORS
http://www.sensorsportal.com

List of Abbreviations

A/D	analog-to-digital
ADC	analog-to-digital converter
ALU	arithmetic logic unit
ASIC	application specific integrated circuit
ASIP	application specific integrated processor
BAW	bulk acoustic wave
C/I	capacitance-to-current
CMOS	complementary metal oxide semiconductor
DAQ	data acquisition
D.C.	duty-cycle
DCM	duty-cycle modulated
DSC	digital signal controller
DSP	digital signal processor
FDC	frequency-to-digital converter
FDCP	frequency-to-digital converter with parallel interface
FM	frequency modulated
FS	full scale
FTF	fast Fourier transform
HVAC	heating, ventilation, and air conditioning
GUI	graphical user interface
I/F	current-to-frequency
I^2C	inter-integrated circuit
IC	integrated circuits
IFSA	International Frequency Sensor Association
IR	infrared
IRED	infrared light source
LCD	liquid crystal display
LCF	Lyapynov's characteristic functions
LFC	light-to-frequency converter
MDC	method of the dependent count
MEMS	micro-electro-mechanical system
MLF	micro lead frame

MNBS	micro-nano bio systems
MUX	digital multiplexer
NCAP	network-capable application processor
NTC	negative temperature coefficient
OA	operational amplifier
OEM	original equipment manufacturers
PBL	project based learning
PBeL	project based e-learning
PCB	printed circuit board
PDA	personal digital assistants
PDIP	plastic dual inline package
PWM	pulse-width modulated
RH	relative humidity
RTD	resistance temperature detectors
SAW	surface-acoustic wave
SMD	surface mount device
QCM	quartz crystal microbalance
SiP	system-in-package
SNR	signal-to-noise ratio
SoC	system-on-chip
SPI	serial peripheral interface
STIM	smart transducer interface module
TDC	time-to-digital converter
TEDS	transducer electronic data sheet
TII	transducer-independent interface
TQFP	thin quad flat pack
UART	universal asynchronous receiver transmitter
UFDC	universal frequency-to-digital converter
USTI	universal sensor and transducer interface
VCO	voltage-controlled oscillator
VFC	voltage-to-frequency converter
VFI	variable frequency inputs
WSN	wireless sensor network

Chapter 1

Introduction

1.1 Modern Sensor Markets and Trends

According to *Research and Markets* the Worldwide OEM Electronics Manufacturing Market 2010 divides the global electronics industry into eight main market segments [1]:

- Automotive;
- Communications;
- Computers /Peripherals;
- Consumer Electronics;
- Industrial;
- Medical;
- Aerospace;
- Other Transportation.

And there are a lot of sensors and sensor systems, which are using now and will be necessary in the nearest future in each of these segments.

Sensors are primary elements for most technology products built today. Sensors' application areas span simple switches and temperature sensing devices as a part of the most basic of home appliances to sophisticated sensors in advanced manufacturing equipment and factory automation systems. The advent of micro-technology has revolutionized the electronics industry so much so that miniaturization has now become a cue word across the globe. Further, the role played by the micro-technology is enormous and sensors, as an offshoot of this know-how, are used in most of the industrial electronic applications. Smart sensors have the digital capability and at the same time perform additional intelligent functions such as signal processing, self-diagnosing, self-identification, self-validation, self-adaptation or integral fault detection. Presently, smart sensors are having significant presence in only few select applications. However, recent advancements in sensing technologies and supporting technologies such as MEMS and Bluetooth, and growing acceptance of IEEE 1451 interface standards family, are creating new applications for smart sensors and are allowing them to penetrate further into existing applications [2].

Sensors are of various types, based on what they sense. The number of sensor types is large but the most common ones include level, pressure, flow, temperature, MEMS accelerometer, MEMS pressure, MEMS gyroscope, MEMS microphone, magnetic, proximity, displacement, image, wireless, smart, nano, touch, load cell, and gas (sensor, detector, analyzer). There are also a number of other sensors (such as pH, humidity, biosensors, immunosensors, etc.) that make this market one of the most fragmented markets. In addition, each of these sensors can be further split into a number of sensing techniques [3].

The *Yole Development Report* also marks most popular and promised MEMS devices markets: acceleration sensors, gyroscopes, pressure sensors, ink jet heads, silicon microphones, RF MEMS, and microbolometers. Main application areas of modern MEMS devices are: consumer, space/military, automotive, medicine and telecommunications.

A modern car, as usually, has more than one hundred different sensors. It is an attractive and still-growing multibillion-dollar market. It is characterized by very high production volumes of sensors that must be extremely reliable and low cost. Microsystem sensor technologies meet these needs and thus, are widely accepted for different automotive sensor applications.

As stated by the new market research report on smart sensors, Europe and the US, together account for a major share of the global market. While the US and Europe are the largest markets for smart sensors globally, growth in future will be driven by expanding Indian and Chinese economies. Sensor market in Brazil, Russia, India and China has a high growth rate 29 % [4]. Low manufacturing cost, cheap labor, presence of Original Equipment Manufacturers (OEMs) in BRIC countries has driven the BRIC sensor market. Initially started with automotive and industrial applications the sensors have made its way in other areas such as consumer electronics, medicals and others.

According to the *Global Industry Analysts, Inc.* (February 2010) - a U.S. based world leader in business intelligence and strategy support, the global smart sensors market to reach US $ 6.0 billion by 2015 [2].

The world plug and play Smart TEDS Sensors (IEEE 1451) market is to reach $40.9 million in 2013 according to the *Frost & Sullivan.*

'Smart Sensors: an International Market Report' by *BizAcumen, Inc.* declares that sensor networks and smart sensors are being used widely in industries, including automotive, medical, industrial, entertainment, security, and defense due to increased usage of process controls and sensing elements in different sectors. Use of smart sensors in counter-terrorism applications including cargo tracking, biometrics among others is also increasing. Smart systems are being readily accepted by the auto industry as automakers explore ways to save lives, deflect safety concerns and preempt costly litigation. Smart systems can sense an impending crash, its nature, the presence of occupants, their position, and determine the type of airbags to be fired and the force and speed of their deployment. Motion-tracking sensors and inertial sensors based on micro-electromechanical system (MEMS) are widely being used in medical applications.

Implantable devices that are small in size and yet possess high reliability, and handheld devices for use in diagnostics and home monitoring mainly drive the demand for such sensors.

Let consider the global sensor market (not only smart sensor market) in the past, today and in the future. According to the 'Sensor Markets 2008: Worldwide Analyses and Forecasts for the Sensor Markets until 2008' report from the *Intechno Consulting* (Switzerland) global non-military sensor market was growing from US $ billion in 2003 to US $ 50.6 billion in 2008 with the growing at an average annual rate of 4.5 %.

The year 2009 was especially challenging for telecommunications, automotive, petrochemical and industrial process industries, due to which growth in smart sensors market was marginally affected. The sharp downturn in the global economy, uncertainty over the development of end-markets, inventory levels, and pricing pressures affected the overall business environment.

Nevertheless of the past financial crisis and the decreasing in motor vehicles sector, the sensor marker's recession has not been observed, and it was achieved the predictable level in US $ 56.3 billion in 2010.

In February 2010 the *Frost & Sullivan* declares that 'Intelligence is yet another important trend gaining ground in all sensor markets. Sensors are no longer dumb components but are made intelligent to meet the growing needs of the end-user. The integration of diagnostics and verification features has made sensors smarter. These features enable sensors to inform engineers and the like about their status and the status of the instrument/device depending on the application. Almost all sensor types are being made intelligent according to the application they are used in' [3].

Another recent report from *Frost & Sullivan* '**Developments in Sensor Interfaces**', finds that data acquisition systems form an important end-user segment for sensor interfacing. The numerous applications for smart sensor systems require efficient data acquisition strategies to handle the variety of sensory data produced. However, given the sheer number of sensors currently available in the market and that various devices have adopted numerous sensing principles, the task of designing appropriate interfacing ICs to satisfy them all appears daunting.

The new report **'Sensors Market to 2020'** (January 2011) from the *Global Business Intelligence (GBI) Research* - a market-leading provider of 'best-in-class' business intelligence reports, also has confirms the rising need for intelligent sensors in automotive and consumer electronics [5].

According to the *MarketsandMarkets (M&M)* - a global market research and consulting company (USA), the global market for sensors in consumer electronics is expected to grow from $ 10,221 million in 2009 to $ 22,190 million in 2015. The market is expected to have a high CAGR of 12.44 % from 2010 to 2015 due to the increasing demand for sensors in products such as mobile phones, smartphones, game consoles, and laptops [6]. Among all segments, image sensors command the largest share in terms of value. Communication forms the largest application market due to the high demand for mobile phones and PDAs across the globe; while entertainment applications are expected to have the highest CAGR of 16.5 % from 2010 to 2015 [6].

The AMA Association for Sensor Technology's (Germany) report '**Sensor Trends 2014: Trends in Future-Oriented Sensor Technologies**' has marked that driven by increasing quantities and increasing price pressure, many sensor suppliers are moving their creation of value more and more to signal processing, energy management, self-monitoring, and miniaturization, and digital interfaces, intelligence, communication and system integration are among the major development trends in the future sensor technologies, common to almost all of the sensors markets irrespective of their types and differences. In other words, the aforementioned trends hold true for all types of sensors from the ones used in aircrafts to the ones used in oil and gas applications or iPhones [7].

Strong growth expected for sensors based on MEMS-technologies, smart sensors, sensors with bus capabilities and embedded processing. It means that the global demand for highly integrated smart systems will increase dramatically in the years to come. Sensors based on MEMS and NEMS technologies and intelligent (smart) sensors are the focus of the current sensor development. MEMS technologies allow miniaturization of sensors and, at the same time, the integration of their sensor elements with microelectronic functions in minimal space. MEMS technologies make it possible to mass-produce sensors more and more cost-effectively while improving their functionality and miniaturizing them. Thus, major sensor and electronics manufacturers expect that within a few years, there will be complete sensors integrated on one chip and having their own local intelligence for information processing.

In 2012 the forecasted turnover generated by MEMS components will be around 10.5 bn Euro; and the turnover generated by MEMS based systems is expected to reach to 77.8 bn Euro.

Europe represents the largest and fastest growing sensors segment in this market. Today, European industry still occupies a leading position in the manufacturing of products incorporating smart systems technologies.

The '**Strategic Analysis of the Sensors Market in Europe**' from *Frost & Sullivan*, June 2010, finds that the market earned revenues of $12.5 billion in 2009 and estimates this to reach $ 19.0 billion in 2016 [8]. Europe represents also the largest and fastest growing $ 2.1 billion smart sensors market by 2010, as stated *by Global Industry Analysts, Inc.*

The resent research report on global sensor market (May 2011) from the *Market Research* - the world's largest and continuously updated collection of market research, has declared that any measurement or an automation system would definitely have a sensor integrated into it. Sensors are responsible for converting physical changes into quantities that can be measured by data-acquisition systems. The rapid growth of electronic control systems has provided users with latest advancements in sensor technology such as response time, miniaturization, and accuracy and communications capability. The positive growth in sensor technology has paved way for more intense research and development and has created new fields for sensor applications [9].

According to report from the *Global Industry Analysts, Inc.* (October 2010) the global sensors market will reach US $ 74.2 billion by 2015 [2].

The most recent report from the Market Research has declared the addition of a new report entitled 'Sensors: Technologies and Global Markets' (May 2011) in accordance with it, in the year 2010, the global sensor market was projected at 56.3 billion dollars and for 2011 it is predicted to reach 62.8 billion dollars. The market is growing at a compound annual growth rate of 7.8 % [9]. Demand for sensors in the USA will rise 6.1 percent annually to 2014 [10]. Taking into account the CAGR, it is predicted that the global sensor market growth will be as it is shown in Table 1.

Table 1. Global Sensor Market Size Prediction (annual growth rate of 7.8 %).

Year	2009	2010	2011	2012	2013	2014	2015	2016
Global Sensor Market (billion $ US)	–	56.3	62.8	67.7	73.0	78.7	84.9	91.5
European Sensor Market (billion $ US)	12.5	–	–	–	–	–	–	19.0
Global Smart Sensors Market (billion $ US)	–	–	–	–	–	–	6.0	–
European Smart Sensors Market (billion $ US)	–	2.1	–	–	–	–	–	–

'–' – information not available.

However, according to new Report from *Global Industry Analysts, Inc.* (October 2011), global sensors market to reach US $ 76.7 billion in 2017. It means annual growth rate of 4 %.

New analysis from *Frost & Sullivan* (November 2011), World Emerging Sensors Markets, finds that in the next 7 to 10 years miniaturization of systems is expected to drive innovation. OEMs will increasingly incorporate smaller sensors to improve performance, reliability, and longevity as well as reduce costs. Despite the potential and benefits of sensors, many end users prefer traditional technologies, due to their inadequate knowledge of innovations and application areas. However, cutting-edge sensor technologies offer many benefits, such as greater ease of installation, accuracy, as well as energy and cost savings. Vendors therefore need to educate end users about the benefits of advanced and emerging sensor technology through literature.

The most recent sensors market reports and analyses are available at Sensors Web Portal (http://www.sensorsportal.com) in Marketplace 2 web section [11].

According to the *International Frequency Sensor Association (IFSA)* study 2011 the global sensor market divided according to sensors' outputs types as it is shown in Figure 1.1.

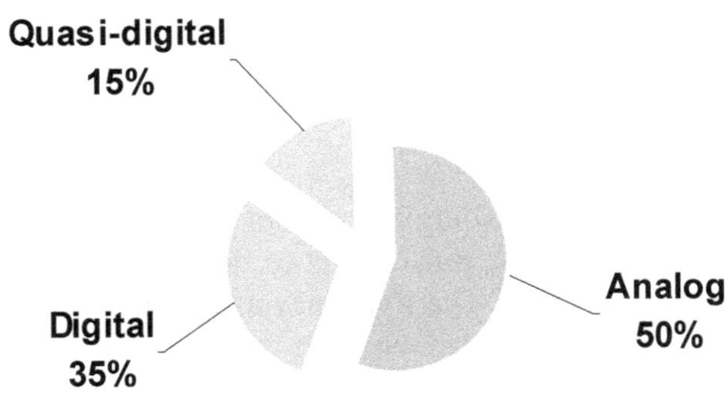

Fig. 1.1 Global Sensor Market Divided According to Sensors Outputs' Types (*IFSA*, 2011).

It is clear, that all smart and intelligent sensors are sensors with digital outputs (common serial, parallel or bus interfaces such as RS232/422/485, USB, SPI, I2C, CAN, SMBus, LIN, etc.). Today the digital sensors market segment is more dynamical growing segment. Digital, smart and intelligent sensors and sensor systems can be easy made based on so-called quasi-digital sensors (sensors with frequency, period, duty-cycle, time interval, pulse width modulated, phase-shift, pulse number outputs) [12]. A quasi-digital sensor in sensor systems hierarchy is shown in Figure 1.2.

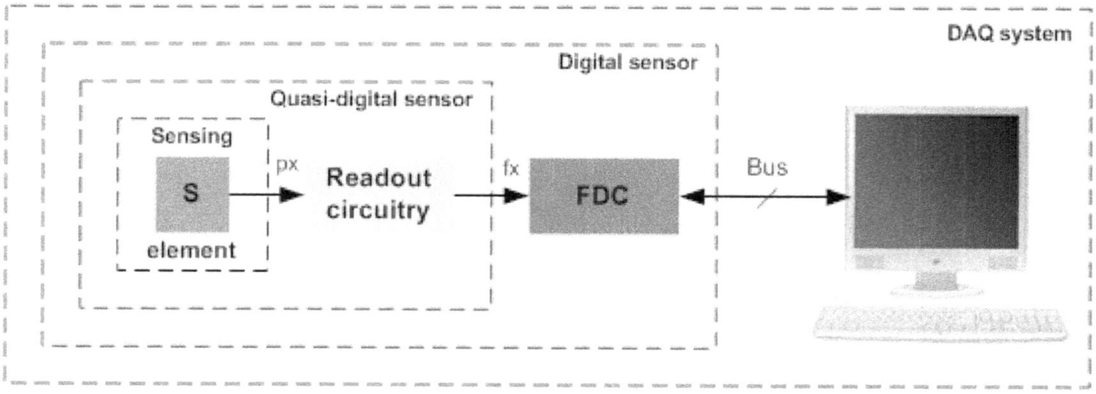

Fig. 1.2 A quasi-digital sensor in sensor systems hierarchy.

A classification of quasi-digital sensors in term of possible output signal is shown in Figure 1.3.

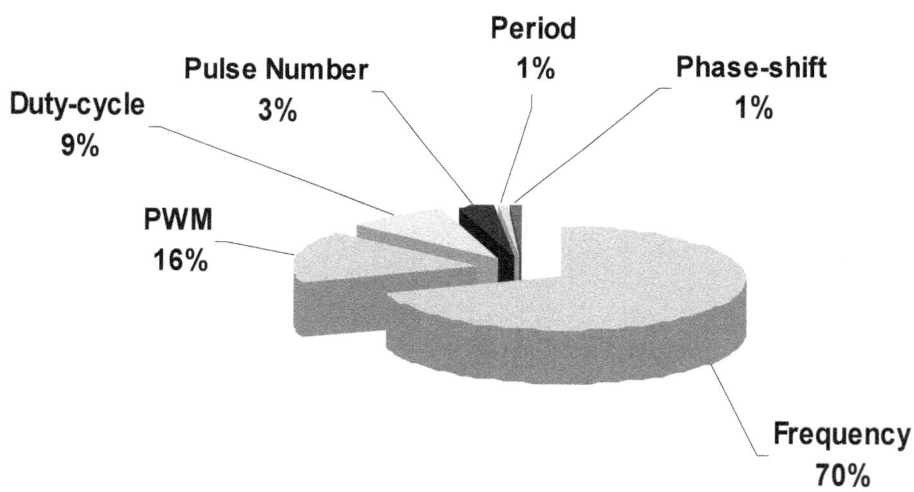

Fig. 1.3 Classification of Quasi-Digital Sensors in Term of Output Signal (*IFSA*, 2011).

But why customers should use quasi-digital sensors with frequency, period, duty-cycle, time interval, PWM, phase-shift or pulse number output instead of traditional analog (voltage or current) output sensors ? The frequency as informative parameters of sensor's output has many advantages in comparison with analog informative parameters. It is a high noise immunity, wide dynamic range, simple interfacing, multiplexing, integration and coding, robustness, multiparametricity, no sensor output standardization is necessary, availability of most accurate standards and references [12]. Let consider these main advantages of frequency in details.

- **High noise immunity.** In frequency output sensors and transducers, it is possible to reach a higher accuracy in comparison with sensors and transducers with analog-to-digital converters. This property of high noise immunity proper to a frequency modulation is apparently the principal premise of frequency output sensors in comparison with analog ones. The frequency signal can be transmitted by communication lines for a much greater distance than analog and digital signals. The frequency signal transmitted practically represents a serial digital signal. Thus, all advantages of digital systems are demonstrated. Also, only two physical wires (signal and ground) are necessary for transmission of such signal. In comparison with the usual serial data transmission, for example, according to a popular RS232 interface, it has the advantage of not requiring any synchronization, and therefore any addition physical wire. Frequency signals are ideal for electrically noisy industrial environments where common ADCs are susceptible to interference, and for remote sensing applications, including environmental monitoring. Voltage is inherently vulnerable to noise sources coupled to the desired signal, and therefore it is less convenient in those applications.

- **High output signal power**. From the power point of view, the section from a sensor output up to an amplifier input is the heaviest section in a measuring channel for transmitting signals. Here the signal is transmitted by a very low power level. The losses, originating in this section, cannot be filled any more by further signal processing. Output powers of frequency output sensors are, as a rule, considerably higher than for voltage- or current-output sensors. In this case, the power affecting the generated frequency stability is the oscillation (reactive) power of the oscillating loop circuit and due to the higher quality factor of the oscillating loop its power is higher. This power can be increased by designing an oscillating loop with a high quality factor. For analog signals, the low-level voltage output from a bridge-type sensor, for example, must be amplified to a usable level and converted to a non-amplitude-related encoding scheme in order to obtain a robust signal.

- **Wide dynamic range.** Because the information is in the form of frequency, the dynamic range is not limited by the supply voltage and voltage noise. A dynamic range of over 120 dB may be easily obtained. There are many frequency output sensors on the modern sensor market working in such wide dynamic range. First of all it is light and color sensors with frequency output. Such devices have broad frequency changes on its outputs: form part of Hz to several MHz.

- **High accuracy of frequency standards.** The frequency reference, for example, quartz crystal oscillators, can be made more stable than the voltage reference. This can be explained by the same objective differences of information-carrying properties of amplitude-modulated and frequency-modulated signals. So, a non-compensated quartz crystal oscillator has the relative error less than $\pm$ (1...50)·10-6 error and temperature-compensated quartz crystal oscillator has the relative error $\pm$ (10-8 $\div$10-10).

- **Simplicity of commutation and interfacing.** Parasitic electromotive forces (emf), transient resistances and cross-feed of channels in analog multiplexers for analog sensors are reduced to the occurrence of additional errors. These errors can be reduced by using differential signals but this often results in more complex circuit designs. The frequency-modulated signal is not sensitive to all the above listed factors. Multiplexers for frequency output sensors and transducers are simple enough and do not introduce any errors into observed results.

- **Simplicity of signal integration and coding.** The precise integration in time of frequency sensors' output signals can be realized simply enough. An incremental pulse counter is an ideal integrator with unlimited measurement time. The frequency signal can be directly processed by microcontrollers without any additional interface circuitry.

- **Multiparametricity.** The single sensor's output of quasi-digital sensors can contain two informative parameters at the same time, for example, frequency, proportional to one value, and duty-cycle, proportional to other value. Examples of such devices exist on the modern sensor market. It is multiparametric light/color and pressure/temperature sensors and transducers.

Using frequency as the output signal for sensors is both an extremely useful alternative to the conventional analog voltage output signal and is easily accomplished with relatively few components. By eliminating the need for an ADC, frequency output sensor schemes reduce the cost of sensor systems.

No output standardization is necessary as in the case of analog signal domain (4-20 mA or 0.5 … 4.5 V). It means, that frequency (time) outputs sensors manufacturers can produce sensors and transducers with very wide dynamic range. Many types of sensing elements and read-out circuitry can be merged in this way on a single chip, in System-on-Chip (SoC) or Systems-in-a-Package (SiP). In addition, such approaches will provide a great opportunity to create new self-adaptive smart sensors and systems.

A duty-cycle output signal is also widely used as an information-carrying parameter for different quasi-digital sensors. It is quite immune to interfering signals, such as voltage spikes, and the ratio between time intervals does not depend on the absolute value of any component.

A phase–shift output is used in the measurement of distances, small rotation angles, electric potentials difference, linear and angular displacements, strain and other variables.

Today there are quasi-digital sensors and transducers for practically any physical and chemical, electrical and non-electrical quantities. The obvious tendency of the relative error decreasing up to 0.003 % and below is observed. These devices are working in broad frequency ranges: from several hundredth parts of Hz up to several MHz. The extension of their "intelligent" capabilities including intelligent signal processing is traced.

The process of miniaturization boosts the creation of multichannel, multifunction (multiparametric) one-chip digital sensors and sensors arrays. Today advanced CMOS 90 nm, 45 nm, 40 nm, 32 nm and 28 nm technological processes are available in many companies and through various technology brokers such as EUROPRACTICE and CPM [13]. However, today we are at the in front of technology limitations and challenges. How far can we go with this scaling ? Where's the limit ? Let consider the coming technology changes and trends in size reduction below.

1.2 Technology Challenges

Dr. Stephan Henzler form the TU Munchen (Germany) has written in his book [14] published in 2010: "Microelectronics and integrated circuit design are heavily driven by technology scaling. The main reason for scaling is increased system performance at reduced manufacturing cost (per system). In most systems digital circuits dominate with respect to die area and functional complexity. Digital blocks take full advantage of reduced device geometries in terms of area, power per functionality, and switching speed. On the other hand, analog circuits rely not on the fast transition speed between a few discrete states but mainly on the actual shape of the transistor characteristic. Technology scaling continuously degrades these characteristics with respect to analog performance parameters, such as output resistance or intrinsic gain. **Below the 100 nm technology, the design of analog and mixed-signal circuits becomes perceptibly more difficult.** This is particularly true for low supply voltage near 1 V or below". The result is not only an increased design effort, long development time, high risk, cost, and the need for very high volumes [15], but also growing power consumption [13].

There are many reasons that analog doesn't scale as readily [16]. The area shrinks considerably less than predicted by the digital scaling factor. Obviously, both effects are contradictory to the original goal of scaling. However, digital circuits become faster, smaller, and less power hungry [14]. The fast switching transitions reduce the susceptibility to noise, e.g. flicker noise in the transistors. There are also a few drawbacks, such as the generation of power supply noise or the lack of power supply rejection. Because of these drawbacks the analog circuits do not become much smaller in area in smaller technologies.

In the saturation region, there is a much lager change in current in 40 nm process in comparison with the 130 nm process. This change leads to the fact that the transistor has much less control over the current. The design of critical analog circuits that don't scale easily, for example, amplifiers, analog front ends and ADCs, is therefore much more difficult, as the realizable gain per stage is significantly reduced. To compensate for that reduction, more sophisticated circuits are required, leading to lager area and higher consumption [16].

Another factor is the reduced supply voltage and dynamic range. As the supply voltage is reduced with the migration to lower process nodes, the available signal range shrinks. This lower signal range requires proportionately lower noise levels to maintain the same dynamic range. In mixed-signal circuits, such as ADCs using switches capacitors, the reduction in the noise level can be achieved with larger capacitors. To compensate for a 2X reduction in signal range, the capacitors must be increased 4X – making it quite difficult to scale down the area. In addition, the larger loading capacitances require larger currents for charging and discharging, invariably leading to higher consumption. In addition, the designs will become more complicated in order to yield good performance at those low voltages [16].

Their power consumption and speed performance improve significantly due to reduced parasitic capacitance. On the other hand, this results in major challenges for the design of interface electronics: matching and noise becomes a serious issue, while the dynamic range is reduced due to the supply voltage reduction [17].

The scaling for the analog/mixed ICs can be achieved using a number of different design techniques, not using design CAD tools but rather different architectures. At 28 nm, second- and third-order effects have become serious concerns. As a result, existing design and verification techniques are starting to fall short. Smarter solutions are needed to manage issues arising from tighter spacing [18].

However, digital circuits scale very well with scaling CMOS technologies. So, another trend is the transition of traditional analog functions to the quasi-digital domain, for example, to use a frequency (time)-to-digital conversion instead of analog-to-digital conversion, and implement as many system components as possible in the digital or quasi-digital domain. It lets get over technological limitation at scaling. Moving form traditional analog to the quasi-digital domain has many advantages and no disadvantages.

It was also predicted in the old technology forecast [19]: "*Many basic measurements will be related to or measured by time/frequency techniques. This situation will occur because of the increasing facility and accuracy provided through time frequency measurements... Frequency and computing counters can be expected to become smaller, lighter and cheaper, provide direct frequency readout and will be more universally used... Progress will continue in the field of time and frequency during the next five years and will contribute materially to the state of the art in electronics*".

As far back as 1961 professor P. V. Novitskiy wrote: "... *In the future we can expect, that a class of frequency output sensors will get such development, that the number of now known frequency output sensors will exceed the number of now known amplitude sensors...*" [20]. No any doubt that now we are living in time, when this prediction becomes true !

Frequency (time)-to-digital converters can be used for high volume micro-electronics in order to circumvent analog impairments in nanometer-scale CMOS technologies. Hence, only FDC concept is promising that has the ability to exploit the advantages of both: analog and digital circuits. In the nearest future the FDCs will be a trend-setting technology for ultimately scaled CMOS technologies for both a monolithic approach (System-on-Chip, System-in-Package) and multi-chip approach to drastically increase the integration level. So, it is a reason why this book is well-timed and useful for all who is involved today or will be involved in the nearest future to the modern design of digital sensors and sensor system. The next chapters of this book are devoted namely how to create such devices quickly and by most effective way.

1.3 Digital Sensors and Systems Design Approach

The number of physical phenomena, on the basis of which direct conversion sensors with digital outputs can be designed, is essentially limited. There are a very limited number of true digital sensors, for example, so-called digital output angular-position encoders (Fig. 1.4) and digital output piezoelectric accelerometers (Fig. 1.5). In the digital angular-position encoder, a special disk with figurate holes is moving between light sources and light sensors. The Gray code, proportional to an angle is generated on the light sensor output. No any ADC is necessary.

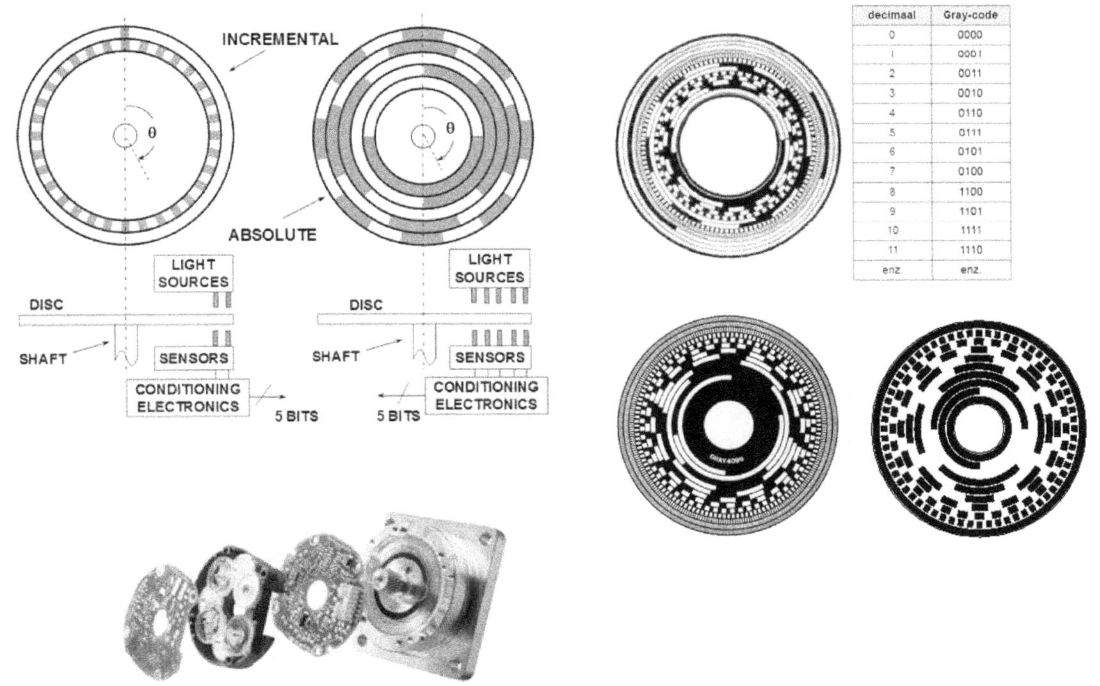

Fig. 1.4 Digital Angular-position Encoders.

The accelerometer (Fig. 1.5) is composed of an array of piezoelectric PbZrTiO3 (PZT) thin films parallel-plate capacitor structures on a Si microcantilever [21]. The PZT structures are joined together in series and connected to a CMOS switch array. Due to the piezoelectric effect, the number of on-state CMOS switches increases with increasing acceleration. This composition makes it possible to realize a digital output accelerometer in which output is the number of on-state CMOS switches [21]. The acceleration is such device can be directly converted to digital output without an ADC.

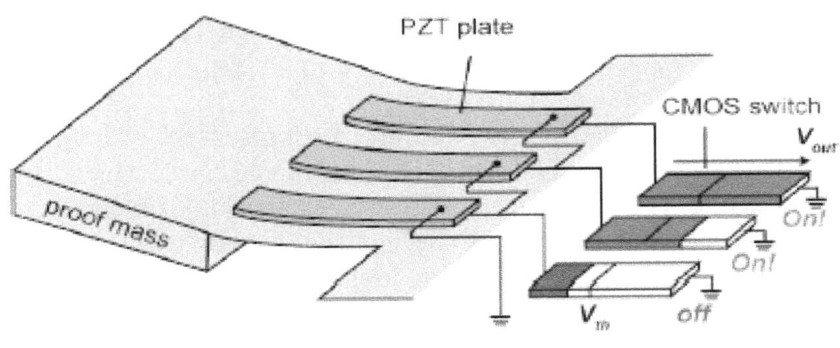

Fig. 1.5 Digital Output Piezoelectric Accelerometer [21].

However, there is no natural phenomenon with discrete performances changing under pressure, temperature, etc. In this case, an ADC and analog multiplexers (for sensor arrays, multisensing and multi-parametric systems) are used. There are three different design approaches for digital sensors, transducers and sensor systems: ADC-based, FDC-based and approach based on an intermediate voltage-to-frequency conversion. In the

ADC based classical structure, a sensor's (S) analog output (voltage or current) is converted to digital output with the help of ADC (Fig. 1.6, a). In the second FDC-based structure (Fig. 1.6, b), instead of analog sensor a quasi-digital sensor with frequency, period, duty-cycle, PWM, phase-shift or pulse number output should be used. The third approach is using an intermediate voltage-to-digital converter (VFC) to convert analog output to frequency, and then, the FDC is used to convert the frequency signal to digital (Fig. 1.6, c).

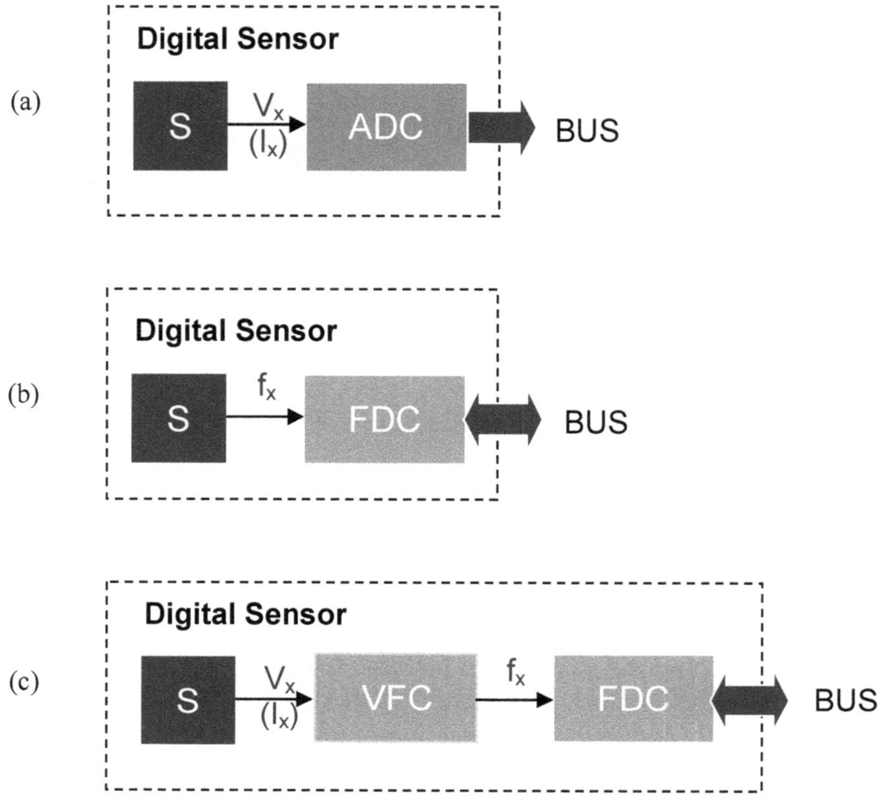

Fig. 1.6 Digital Sensors Design Approaches: ADC-based (a); FDC-based (b); and with intermediate VFC conversion (c).

The two last design approaches are covered all existing sensors with analog and quasi-digital outputs. An advanced FDC has a bidirectional bus, which give a unique opportunity to change its metrological performances and build intelligent self-adaptive sensors and sensor systems. In spite of the fact that sigma-delta ADCs contain a few number of analog component, the two last design approaches let achieve much better metrological performances and scaling due to pure digital components of FDC and VFC. When going to smaller technologies, reduced parasitic gate capacitances result in smaller gate delays. This improved frequency/time resolution makes frequency (time)-to-digital conversion an interesting and promising alternative for ADC converters. Hence, the only promising FDC concepts are those which have the ability to exploit the advantages of digital circuits. Using frequency-time parameters as the output signal for sensors is both an extremely useful alternative to the conventional analog voltage output signal and is easily accomplished with relatively few components. By eliminating the need for an ADC, frequency output sensor schemes essentially reduce the cost of digital sensors, transducers and sensor systems. Quasi-digital frequency-time domain sensors are rather interesting from a technological and fabrication compatibility point of view, due to the simplifications of the signal conditioning circuitry and measurand-to-digital converter, as well as metrology performances and the hardware for realization. The last one essentially influences the chip area [22]. Frequency and PWM outputs generation and quasi-digital sensors classification are described in [11, 23-27].

One of the main parts (a core) of FDC-based sensors, transducers and sensor systems is the frequency (period, duty-cycle, time interval or phase-shift)-to-digital converter. This unit directly influences on sensor metrological characteristics, such as accuracy and the conversion time, as well as power consumption. It is clear that such programmable, universal, high-performance FDCs must be based on novel, advanced methods of measurement for frequency-time parameters of signals. No classical frequency (time)-to-digital conversion methods nor many existing advanced methods can be used in modern digital sensors and sensor system due to their low metrological performances and many drawbacks, for example, the dependence of quantization error on converted frequency; redundant conversion time, etc. Fortunately, advanced FDCs – Universal Frequency-to-Digital Converter (UFDC-1), UFDC-1M-16 and Universal Sensors and Transducers Interface (USTI) integrated circuits (IC) are now available on the modern market [28] and can be used in digital sensors (including smart and intelligent), transducers and sensor systems design as discrete ICs for OEM or IP to be used in monolithic ASIC or IC designs.

Speaking about industrial quasi-digital sensors and transducers, continuously up-dated information about many of them are available on Sensors Web Portal web site [29] in appropriate list of manufacturers.

The next two chapters of this book will be devoted to the new electronics components – UFDC-1, UFDC-1M-16 and USTI. If a reader familiar with these components, he can pass through these chapters and continue reading from chapters about particular sensors design.

1.4 References

[1]. Randall Sherman, The Worldwide OEM Electronics Manufacturing Market - 2010 Edition, *New Venture Research Corp.*, December 2010.

[2]. Smart Sensors. A Global Strategic Business Report, *Global Industry Analysts, Inc.*, February 2011.

[3]. Global Sensors Market - Unity in Diversity!, *Frost & Sullivan*, February 2010.

[4]. Sensor Market Opportunities in BRIC Countries (2011-2016) - Brazil, Russia, India, China, March 2011.

[5]. Sensors Market to 2020 - Rising Need for Intelligent Sensors in Automotive and Consumer Electronics to Increase Demand, GBISC009MR, *GBI Research*, January 2011.

[6]. World Sensors Market in Consumer Electronics (2010-2015), SE 1079, *MarketsandMarkets*, October 2010.

[7]. Sensor Trends 2014: Trends in Future-Oriented Sensor Technologies, *AMA Association for Sensor Technology*, Germany, July 2010.

[8]. Strategic Analysis of the Sensors Market in Europe, *Frost & Sullivan*, June 2010.

[9]. Andy Choi, Research Report on Global Sensor Market, May 2011.

[10].Sensors to 2014, *Freedonia*, September 2010.
[11].Kirianaki N.V., Yurish S.Y., Shpak N.O., Deynega V.P., Data Acquisition and Signal Processing for Smart Sensors, *John Wiley & Sons*, 2002.
[12].Sensors Markets II (http://www.sensorsportal.com/HTML/Marketplace_2.htm).
[13].STMicroelectronics Makes 28 nm CMOS Process Available Through CMP, Press Release, *Sensors & Transducers Magazine (e-Digest)*, Vol. 129, Issue 7, June 2011: http://www.sensorsportal.com/HTML/DIGEST/Digest_June_2011.htm
[14].Stephan Henzler, Time-to-Digital Converters, *Springer*, 2010.
[15].Backram H.-E., Scared of mixed-signal and RF ASICS? Think twice! *EE Times Europe*, May 2010, p.50.
[16].Carlos Azeredo-Leme, Navraj Nandra, Analog Circuits Benefit from Scaling Trends, Chip Design, Spring 2011, pp. 18-23.
[17].Denneels H., Piette F., Gielen G., A Novel, PLL-based Frequency Conversion Mechanism for Sensor Interfaces, *Sensors & Actuators A: Physical*, (in press). Available online 26 May 2011.
[18].Joe Davis, The Challenge of Analog, Mixed-signal and Custom Physical Implementation at 28 nm, *Tech Design Forum*, Vol.8, Issue 3, June 2011, pp. 40-43.
[19].Stone R.R., Phillips D.H., Berg W.B., Technology Forecasts: Frequency and Time, in *Proceedings of the IEEE Conference on Technology Forecasts and Assess*, New York, USA, 1971, pp. 54-55.
[20].Novitskiy P.V. Frequency Output Sensors Design Problem for all Electrical and Non-electrical Quantities, *Measurement Technology*, No. 4, 1961, pp.16-21 (in Russian).
[21].Toshihiro Itoh, Takeshi Kobayashi, Hironao Okada, A Digital Output Piezoelectric Accelerometer for Ultra-low Power Wireless Sensor Node, in *Proceedings of IEEE Sensors 2008*, 26-29 October 2008, Lecce, Italy, pp.542-545.
[22].Sergey Y. Yurish, Sensors and Transducers: Frequency Output Versus Voltage Output, *Sensors & Transducers*, Vol.49, Issue 11, November 2004, pp.302-305.
[23].Gui Yun Tian, Sofiane Zairi, Ali Sophian, Craig A. Grimes, Frequency Output Sensors, in Encyclopedia of Sensors, Edited by C. A. Grimes, E. C. Dickey, and M. V. Pishko, Vol. 4, *American Scientific Publishers*, 2006, pp. 73-88.
[24].Mayela Zamora, Manus Henry and Christian Peter, Generation of Frequency Output for Instrumentation Applications using Digital Hardware, *Sensor Review*, Vol. 23, No. 2, 2003, pp. 143–149.
[25].Eric Jacobsen, Creating a PWM Output Sensor Using a Field-Programmable Analog Array, *Sensors Magazine*, May 1998.
[26].Frequency Output Circuit, Application Notes, APP-104, *ICSensors*, August 1999.
[27].Novitskiy P. V., Knorring V. G., Gutnikov V. S., Digital Measuring Instruments with Frequency Sensors, *Energiya*, Leningrad, 1970 (in Russian).
[28].Technology Assistance BCNA 2010, S. L. web site: http://www.techassist2010.com/
[29].Sensors Web Portal: http://www.sensorsportal.com/HTML/Sensor.htm

Chapter 2

Universal Frequency-to-Digital Converters

2.1 Background and State-of-the-Art

Many modern manufacturers today produce sensors and transducers with frequency, period, duty-cycle, time interval, frequencies difference. PWM or pulse number output signal for various physical and chemical, electrical or non-electrical quantities. Despite the presence of many highly versatile sensor interfacing chips in the market, they remain limited to analog sensor types [1].

One of the first ICs for frequency-to-digital conversion was designed at the end of 1980s in the former Soviet Union (USSR). It is the 48-pin one-chip specialized microprocessor USPS- 30 [2]. The USP works in a pipeline mode and can measure frequency, period, time interval, pulse width and count pulse number. The main disadvantages of this device are its narrow frequency range: from 0.1 Hz to 100 kHz and high power consumption.

The IC ALU for time interval measurements that was also developed in the USSR uses a standard counting method for frequency measurements and indirect counting methods for time interval measurements [3]. The absolute accuracy $\Delta T = \pm 33$ ns at the reference frequency $f_0 = 30$ MHz. This IC also has relatively high power consumption.

The universal 42-pin CMOS frequency-to-digital converter K512PS11 was manufacturing in the USSR and worked in two modes: single conversion and multiple conversions. It is based on the indirect counting method with interpolation [4]. This IC has parallel 16-bit output. The maximal converted frequency was limited by 1 MHz.

The universal sensor interface chip (USIC) can measure frequency and pulse width. It has an 80-pin pack (QFP) and a limited high frequency of 4 MHz [5].

The single-chip, FPGA-based, 84-pin interpolating time counter described in [6] is based on the classical time interpolation technique. It has the limited maximal measurable time interval of 43 s. and narrow functionality.

An ASIC of single channel frequency-to-digital converter has been designed to handle one input channel [7]. This converter is capable of measuring frequencies from 100 Hz to 100 kHz. The frequency measurement relative error is 0.1 %. This chip can be interfaced to a 16-bit bus. The hybrid technique for frequency measurements has been used in order to meet these specifications. Drawbacks of this IC are its low accuracy and the narrow converted frequency range.

A frequency-to-digital converter (FDC) from AutoTEC is implemented in a Xilinx FPGA and provides digital data for frequency signals [8]. The FDC has 16-bit and 12-bit counters and uses a 1 MHz free running clock frequency as a reference. The frequency range is from 35 Hz to 24 kHz, with the absolute error $\Delta_x = \pm$ 5 Hz and relative full-scale error $\delta_x =$ 0.2 %. Disadvantages of this IC are the same: low accuracy and narrow converted frequency range.

Known form the literature a frequency-to-digital converter that consists in pulse-shrinking elements, latches and D-flip-flops and is based on frequency comparison by the pulse-shrinking element. The frequency-to-digital converter has no analog element but its error is very high – up to 0.1 %.

Acam-messelectronic GmbH (Germany) produces the family of CMOS Time-to-Digital Converters (TDC) ICs such as TDC-GP1, TDC-GP2, TDC-GP21 and TDC-GPX (Table 2.1). Such ICs is intended for time interval and pulse width measurements, which usually necessary in various physical investigations. Geodesics, security systems, production controls and distance measurements [9].

Table 2.1 The family Time-to-Digital Converters (TDC) ICs.

IC			
TDC-GP1	**TDC-GP2**	**TDC-GP21**	**TDC-GPX**

44-lead TQFP (12×12 mm)	32-pad QFN32 (5×5 mm)	32-pad QFN32 (5×5 mm)	100-lead TQFP (16×16 mm)
Packages (Not to scale)			

The TDC-GP1 is a universal 2-channel time-to-digital converter with a 250 ps (125 ps optional) resolution and two measurement ranges: 3 ns to 7.6 μs, and 60 ns to 200 ms (with a predivider). A quartz clock in the range from 500 kHz to 35 MHz (100 MHz with internal predivider) is needed for reference. The TDC-GP1 is realized in a 0.8 μm CMOS process and packaged in 44 surface mount TQFP. The IC offers a standard 8-bit microcontroller interface and has also 4 ports for the measurement of resistors, inductors or capacitors with precision of 50 ppm in single measurement. The current consumption during measurements is about 20 mA, supply voltage 2.7 – 5.5 V [10]. The TDC-GPX can directly to measure period in two measuring ranges: 100 ms to 60 ns (10 Hz to 16.7 MHz) and 6.7 μs to 3 ns (150 kHz to 333 MHz). But the relative error is high enough: ± 0.015 % to 0.11 % respectively at sampling 1000 measurements per second, and depends on measurand.

The IC TDC-GP2 is the next generation of general-purpose 2 channel TDCs. It has typical resolution of 65 ps and two measurement ranges: 3.5 ns to 1.8 μs, and 500 ns to 4 ms. The TDC-GP2 has 4-wire SPI compatible serial digital output [9]. 2 or 4 temperature sensors such as PT500/PT1000 or higher can be connected to the IC in the temperature measurement mode. The IC is available in the QFN 32 package. Supply voltage is 1.8 – 5.5 V [11].

TDC-GP21 is next generation's upgrade for TDC-GP2. It offers pin-to-pin and functional compatibility, several feature improvements and an extended functionality. The IC has 2 channels with typical 90 ps resolution, or one channel with the double resolution mode with 45 ps and quad resolution mode with 22 ps resolution. The measurement range is 500 ns to 4 ms. The TDC-GP2 also has 4-wire SPI compatible serial digital output [9]. The same temperature measurement mode as in the TDC-GP2 is also available. The IC is available in the QFN 32 package. Supply voltage is 2.5-3.6 V [12].

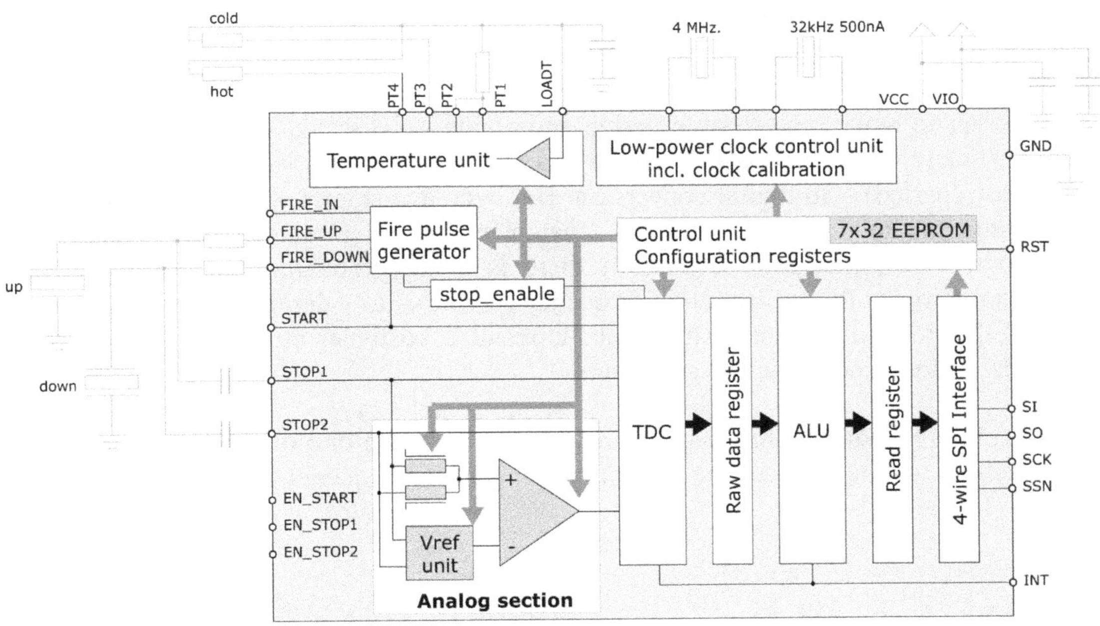

Fig. 2.1 TDC-GP21 block diagram [12].

The TDC-GPX is the most powerful member in the TDC family. It has 8 channels with 81 ps resolution and 2 channels with 41 ps, 27 ps and 10 ps rms resolution. Measuring ranges: from 0 to 9.4 μs, from 0 to 64 μs, from 0 to 40 μs and from 0 to 10 μs [9]. The TDC-GPX is available in TQFP 100 package and TFBGA 120 package. The IC supports 28- bit data asynchronous bus 28 or 2 × 16 bit buses [13]. Supply voltage is 3.0-3.6 V and the current consumption is about 45 mA.

All four ICs are using the modified method of delayed coincidences. The family of TDC ICs has a high time resolution but a narrow functionality, limited low period (frequency) range, relatively high price and can work only in the slave communication mode. It can not be used widely in the sensor industry because of very limited number of quasi-digital sensors exist on the modern market with the time interval and pulse width informative parameters on its output.

Further research and development in the modern TDCs is directed to improve a time resolution up to 20 ps and increase rage for to 43 s [14] and 73 min [15], improve time resolution 3.9 ps, 5 ps and 1 ps [16] and decrease measurable time interval to 6 ns [17].

All integrated converters and interfacing ICs considered above except the time-to-digital converters from *Acam-messelectronic GmbH* are based on classical conversion methods for frequency-time parameters to digital. Hence, they cannot be used with all existing modern frequency-time domain sensors due to low accuracy in comparison with sensor's accuracy or/and narrow frequency range. Further. They do not cover all frequency–time informative parameters of modern quasi-digital sensors and transducers such as duty-cycle, duty-off factor, phase-shift, frequency (period) ratio and difference, frequency deviation, etc.

Of course now such design problems can be solved partially by using standard microcontrollers. However, it is very high labor-intensive approach because of it requires a low level programming on assembler language. For the other side, the customer can not obtain good metrological performances because of the classical (free available) frequency (period)-to-digital conversion methods have a lot of disadvantages, such as dependent of quantization error on measurand, narrow frequency range, redundant conversion time, etc. It order to make a reasonable design and obtain good results from the metrological point of view a customer must buy a license for particular advanced methods for frequency (period) – to digital conversion. But even a customer will use such advanced conversion methods there are no any guarantee for a high-quality design and good metrological performances, because of there is a big probability to introduce many additional errors at the realization of frequency (period) – to – digital conversion methods [18-21]. In order to guarantee an acceptable result, a customer must be an expert in the methods for frequency-time measurements.

This oppressive situation has been change gradually at the end of 2004 at creation and introduction on the market the Series of Universal Frequency-to-Digital Converters (UFDC-1) ICs [22].

2.2 Universal Frequency-to-Digital Converter (UFDC-1)

2.2.1 General Description

The Universal Frequency-to-Digital Converter (UFDC-1) is a fully digital CMOS integrated circuit [23. 24] based on the novel patented methods for frequency, period, its ratio, duty-cycle and phase-shift measurements [25-27]. It provides the technologies that both reduce the cost and time of sensor application development and improve digital sensor performance. The photo of IC in 32-lead TQFP package is shown in Figure 2.2.

Fig. 2.2 IC UFDC-1 in 32-lead TQFP package.

By using this IC it is possible to build a wide variety of digital, smart sensors and intelligent sensor systems. The device represents a high-resolution conversion for data acquisition systems designed to support all range of frequency-time domain and digital sensor applications. It offers high performance and wide functionality with flexibility and requires minimum possible number of external components. Just as a standard ADC is working with any analog sensors' output, this revolution electronic component – UFDC-1 (see Table 2.2) is working with any quasi-digital sensors' output. Now customers are not involved in the methods selection and its realization by software (firmware). No any additional sensor signal conditioner is necessary. Practically any quasi-digital sensors existing on the modern sensor market can be directly connected to the UFDC-1.

Table 2.2 Revolution electronic components during one century electronics development.

Year				
1905	**1947**	**1963**	**1974**	**2004**
			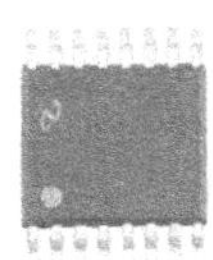	
Vacuum valve	Transistor	Operational amplifier	Analog-to-digital converter	Frequency-to-digital converter
Electronic Component				

The UFDC-1 can be easy included into digital environment, controlled by an external microcontroller (slave mode) or PC, or work independently as a separate measuring unit without an external control (master mode). The function selection can be configured in both software and hardware.

The IC is consists on the two main blocks: measuring and communication units. The block diagram is shown in Figure 2.3. Only one external component – 16 MHz quartz crystal resonator is necessary.

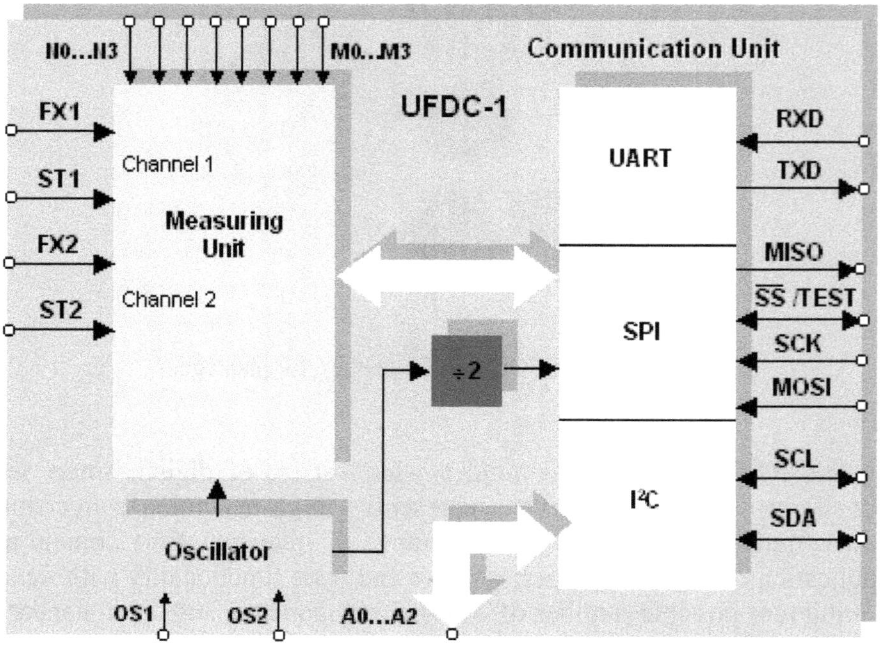

Fig. 2.3 UFDC-1 block diagram.

The IC converts frequency-time parameters of electrical signals into digital signals according to three popular serial buses and provides interface to microcontroller, DAQ or sensor system to read these digital results with scalable resolution and programmable accuracy during the non-redundant conversion time.

The UFDC-1 has 16 measuring and one generating modes. The measuring mode can be hardware-selected by using four selection external pins M0-M2 and the relative error by using four selection pins N0-N2 (RS232 master communication mode). The mode and accuracy can be also software-selected by using any of three possible interfaces (slave communication mode). A communication interface type (RS232, SPI or I2C) can be automatically selected according to the circuit connection.

2.2.2 Features, Metrological Performance and Main Electrical Characteristics

The UFDC-1 has the following key features:

- 16 measuring modes
- Provides interfacing for many types of frequency-time domain sensors and transducers

- Frequency range from 0.05 Hz up to 7.5 MHz (at 40 ÷ 50 % duty-cycle for f_{xmax}) without prescaling and 120 MHz with prescaling
- Programmable accuracy (relative error) for frequency (period) conversion from 1 % to 0.001 % for rectangular waveform input pulse signals
- Relative quantization error is constant in all specified frequency range
- Non-redundant conversion time, s: 0.0002 to 0.2
- Internal reference clock 500 kHz, 16 MHz @ 16MHz quartz oscillator
- Quartz-accurate automated calibration
- Simple 3-wire serial interface (SPI compatible)
- 2-wire interface (I^2C compatible)
- Two direction RS232/422 serial interface
- Master and slave communication modes for RS232/422
- Frequency generator mode $f_{osc}/2$ = 16 MHz / 2 = 8 MHz with quartz crystal stability
- Single power supply +4.5 ... +5.5 V
- Current consumption at 16 MHz, 5V, 25 ^{0}C below 20 mA
- Operating temperature range (- 40 ^{0}C to + 85 ^{0}C)
- Package Types: 28-lead Plastic Dual Inline Package (PDIP), compact 32-lead small Thin Profile Plastic Quad Flat Package (TQFP) and 32-pad Micro Lead Frame Package (MLF)

The UFDC-1 can measure the following frequency-time parameters of signals:

- Frequency, $f_{x1,}$ 0.05 Hz ... 7.5 MHz directly and up to 120 MHz with prescaling
- Period, $T_{x1,}$ 150 ns ... 20 s
- Phase shift, φ_x, 0 ... 3600 at f_{xmax} = 500 kHz at 50 % duty-cycle
- Time interval between Start and Stop pulses, $\tau_{x,}$ 2 μs ... 250 s
- Duty-cycle, D.C., 0 ... 1 at $f_{xmax} \leq 500$ kHz
- Duty-off factor, Q 10^{-8} ... 10^6 at $f_{xmax} \leq 500$ kHz
- Frequency difference, $f_{x1} - f_{x2}$ with sign recognition. 0 ... 7.5 MHz
- Period difference, $T_{x1} - T_{x2}$ with sign recognition. 0 ... 20 s
- Frequency ratio, f_{x1}/f_{x2} 7×10^{-9} ... 1.4×10^8
- Period ratio. T_{x1}/T_{x2} 7.5×10^{-9} ... 1.33×10^8
- Rotational speed, $n_x = (f_x \times 60)/Z$. where Z is the number of encoder teeth
- Pulse width, t_p 2 μs ... 250 s
- Space interval, t_s 2 μs ... 250 s
- Pulse number (events) counting, N_x 0 ... 4×10^9

2.2.3 Functionality

2.2.3.1 Accuracy

For quasi-digital output sensors and transducers, the UFDC relative error should be selected small enough. If the sensor error is, for example, 0.01 %, the conversion error should be ten times (or at least. five times), smaller, i.e. the relative error of UFDC-1 should be selected equals to 0.001 % or at least, 0.002 % [28]. In this case the conversion error can be neglected. The UFDC-1 has non-redundant conversion time. The higher conversion accuracy needs longer conversion time. However, it is non-redundant for any chosen accuracy and frequency. The quantization error is constant and does not depend on the measures frequencies in all frequency range for 0, 1, 6-10, 14-15 modes (Table 2.3).

Table 2.3 Measuring modes of the UFDC-1.

M3	M2	M1	M0	Mode	Name	Mode No.
0	0	0	0	Frequency (channel 1), f_{x1}	FX1	0
0	0	0	1	Period (channel 1), T_{x1}	TX1	1
0	0	1	0	Phase-shift, φ_x	Fi	2
0	0	1	1	Time interval between start- and stop-impulse, τ_x	T	3
0	1	0	0	Duty-Cycle, *D.C.*	D.C.	4
0	1	0	1	Duty-Off Factor. Q	Q	5
0	1	1	0	Frequency difference with sign determination, f_{x1}-f_{x2}	FD	6
0	1	1	1	Period difference with sign determination, T_{x1}-T_{x2}	TD	7
1	0	0	0	Frequency ratio, f_{x1}/f_{x2}	FR	8
1	0	0	1	Period ratio, T_{x1}/T_{x2}	TR	9
1	0	1	0	Rotation speed, (rpm) n_x	n	10
1	0	1	1	Pulse width, t_p	tp	11
1	1	0	0	Space interval, t_s	ts	12
1	1	0	1	Pulse number (events) count, N_x	N	13
1	1	1	0	Frequency (channel 2), f_{x2}	FX2	14
1	1	1	1	Period (channel 2), T_{x2}	TX2	15

Table 2.4 Programmable accuracy of the UFDC-1.

N3	N2	N1	N0	Relative Error. δ_x %	Note
0	0	0	0	1	-
0	0	0	1	0.5	-
0	0	1	0	0.25	-
0	0	1	1	0.1	-
0	1	0	0	0.05	-
0	1	0	1	0.025	-
0	1	1	0	0.01	-
0	1	1	1	0.005	-
1	0	0	0	0.0025	-
1	0	0	1	0.001	-
1	0	1	0	1	With prescaler, f_{x1max}, f_{x2max} = 120 MHz
1	0	1	1	0.25	With prescaler, f_{x1max}, f_{x2max} = 120 MHz
1	1	0	0	0.05	With prescaler, f_{x1max}, f_{x2max} = 120 MHz
1	1	0	1	0.01	With prescaler, f_{x1max}, f_{x2max} = 120 MHz
1	1	1	0	0.0025	With prescaler, f_{x1max}, f_{x2max} = 120 MHz
1	1	1	1	0.001	With prescaler, f_{x1max}, f_{x2max} = 120 MHz

In modes 2, 4-5, the accuracy does not depend on the period of pulse sequence. The programmable accuracy influences only the conversion error of period. Measurement results in these modes only depend on the pulse width. The conversion error for modes 3, 11-12 strictly depends on the converted values and does not depend on value programmed using selection pins N0_N3 or accuracy selection commands (A0-AF). This error can be calculated according to the following equation:

$$\delta_q = \frac{1}{16 \cdot 10^6 \cdot t_x} \cdot 100 \ \%, \tag{2.1}$$

where t_x is the appropriate measuring value of t_p. t_s or τ_x are depending on the mode. For example, at measurement of time interval between Start and Stop pulses $\tau_x = 0.1$ ms, the relative error will be $\delta_{\tau x} = \pm 0.0625$ %. The absolute error is constant: $\Delta_{\tau x} = \pm$ 62.5 ns. Programmed converted accuracy also does not influence on a conversion result in the mode 13.

In the master mode, the relative error and measurement mode should be selected by external switches. In this mode the UFDC-1 works through the RS232 interface in a unidirectional mode. In the slave mode (with an external microcontroller), all measuring modes and programmed accuracy can be set up by a data bus or through communication I/O ports of master microcontroller. This allows us to use the UFDC-1 as a simple peripheral circuit for a microcontroller. A bidirectional data exchange is possible with any of three interfaces RS232, SPI or I2C. The UFDC-1 works as a standard element in different systems with I2C bus architecture, or can be controlled through the RS232 or SPI interfaces. The choice of interface can be performed automatically depend on the circuit connections.

2.2.3.2 Communication Interfaces

2.2.3.2.1 RS232 Interface

The converter can work in an autonomous operating mode (master mode. unidirectional data transfer) as well as in systems (slave operating mode. bidirectional data transfer). If during five seconds after being powered-on the UFDC-1 has not received any command through the RS232, the converter reads the external switches and is placed in the unidirectional mode of data transfer through the RS232 in BCD ASCII format. By default, the transmission speed is 2400, mode - 0 (frequency measurement in the first channel), accuracy – A0 (1 %). The data transfer rate can be chosen from the following range: 300, 600, 1 200, 2 400, 4 800, 9 600, 19 200 and 38 400 (for slave mode).

The brief communication command set description for RS232 interface, appropriate modes and programmed accuracies are adduced in Table 2.5. Circuit diagrams for master mode and slave mode are shown in Figures 2.4 and 2.5 receptively. Pin numbers in all circuit diagrams correspond to the converter in 28-lead Plastic Dual Inline Package (PDIP).

The connector XP1 (Figure 2.5) is intended for in-circuit automated calibration. In the working mode jumper J1 must be in ON position and in the calibration mode it must be in OFF position.

Table 2.5 Communication command set for RS232 interface.

Interface RS232	Description	Parameter
A	Get current accuracy number	–
An	Set accuracy number 'n' (0 ≤ n ≤ 9) without an external divider; (A ≤ n ≤ F) – with an external divider by 16	0 – 1 % 1 – 0.5 % 2 – 0.25 % 3 – 0.1 % 4 – 0.05% 5 – 0.025% 6 – 0.01% 7 – 0.005% 8 – 0.0025% 9 – 0.001% A – 1 % B– 0.25 % C – 0.05% D – 0.01% E – 0.0025% F – 0.001%
B	Get current baud rate number	–
Bn	Set com port baud rate number 'n' (0 ≤ n ≤ 7)	0 – 300 1 – 600 2 – 1 200 3 – 2 400 4 – 4 800 5 – 9 600 6 – 19 200 7 – 38 400
C	Check the measurement status. Returns "r" if result is ready and "b" if measurement is in progress	–
F	Get current calibration constant	–
Fsnnn	Set calibration constant. "s" is the sign ('+' or '-') "nnn" - three hex digits (000 ≤ nnn ≤ FFF)	Integer hex digit from 000 up to FFF
M	Get current measuring mode number	–
Mn	Set measurement mode number 'n' (0 ≤ n ≤ F)	0 – frequency, FX1 1 – period, TX1 2 - phase shift, Fi 3 - time interval, T 4 – duty-cycle, D.C. 5 - duty-off factor, Q 6 - frequencies difference, FD 7 - periods difference, TD 8 - frequencies ratio, FR 9 - periods ratio, TR A - rotation speed, n B - pulse interval, tp C - space interval, ts D - pulse counter, N E – frequency 2, FX2 F - period 2, TX2
R	The same as R0	–
Rn	Get measurement result in format number 'n' (0 ≤ n ≤3)	0 - Decimal ASCII 1 - BCD HEX 2 - BIN HEX 3 - BIN
S	Start measurement	–
T	Test mode. After this command signal $f_{osc}/2$ is generated on pin 16.	–
Z	Get current number of modulation rotor gradations	–
Znn	Set the number "nn" (01 ≤ nn ≤ FF) of modulation rotor gradations for rotation speed measurement mode ("nn" - hex digits)	1–255

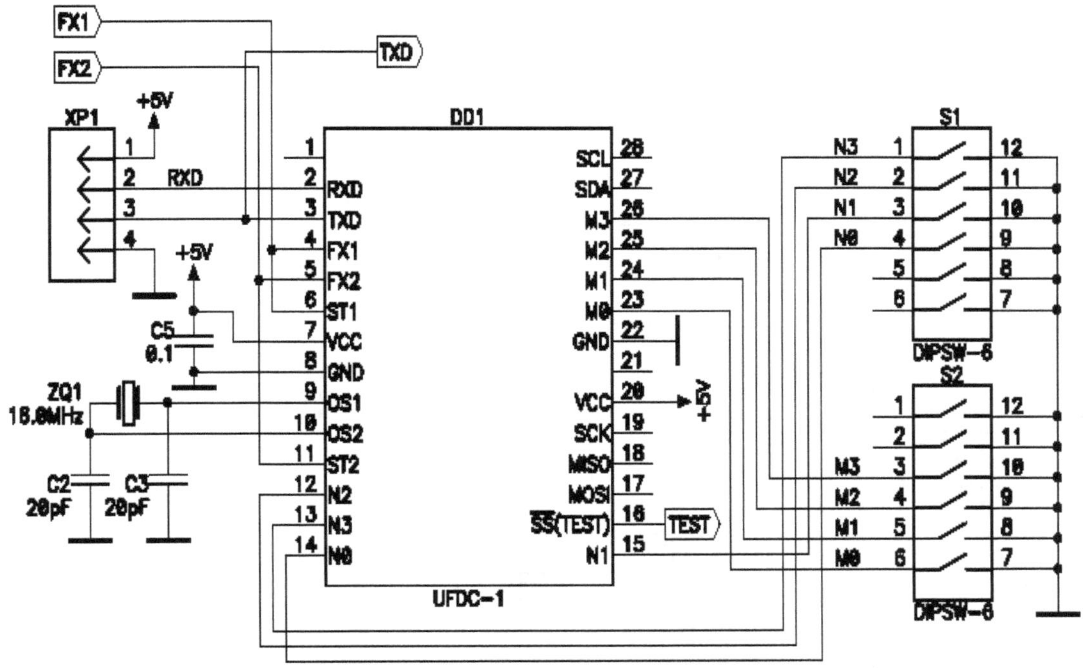

Fig. 2.4 Circuit diagram for the UFDC-1 working in autonomous (Master) mode, RS232 interface.

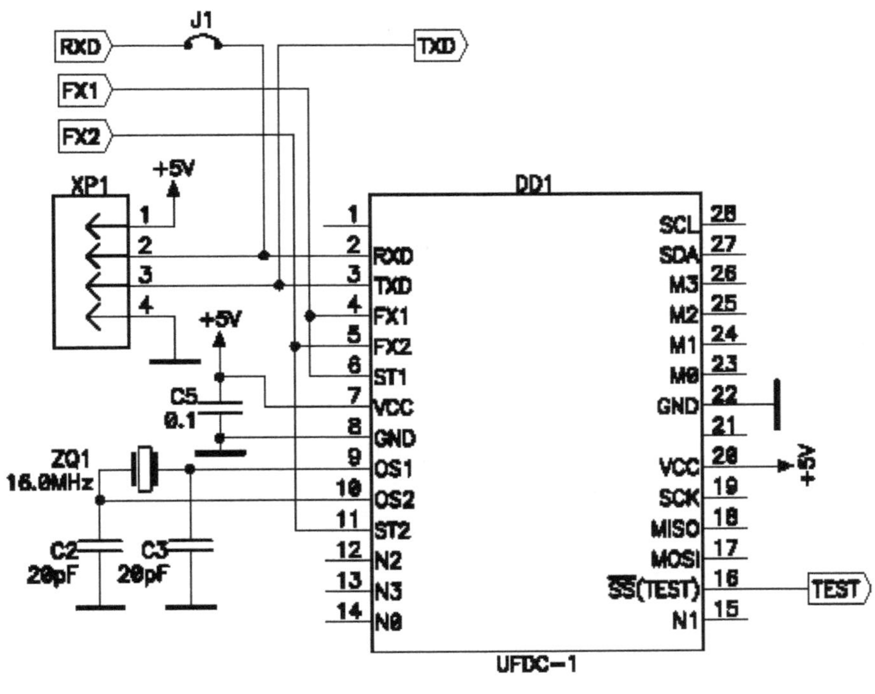

Fig. 2.5 Circuit diagram for the UFDC-1 working embedded in sensor or DAQ board (Slave Mode).

For connection the UFDC-1 to a PC through the RS232 communication serial interface the additional circuit (Figure 2.6) for interfacing function is recommended.

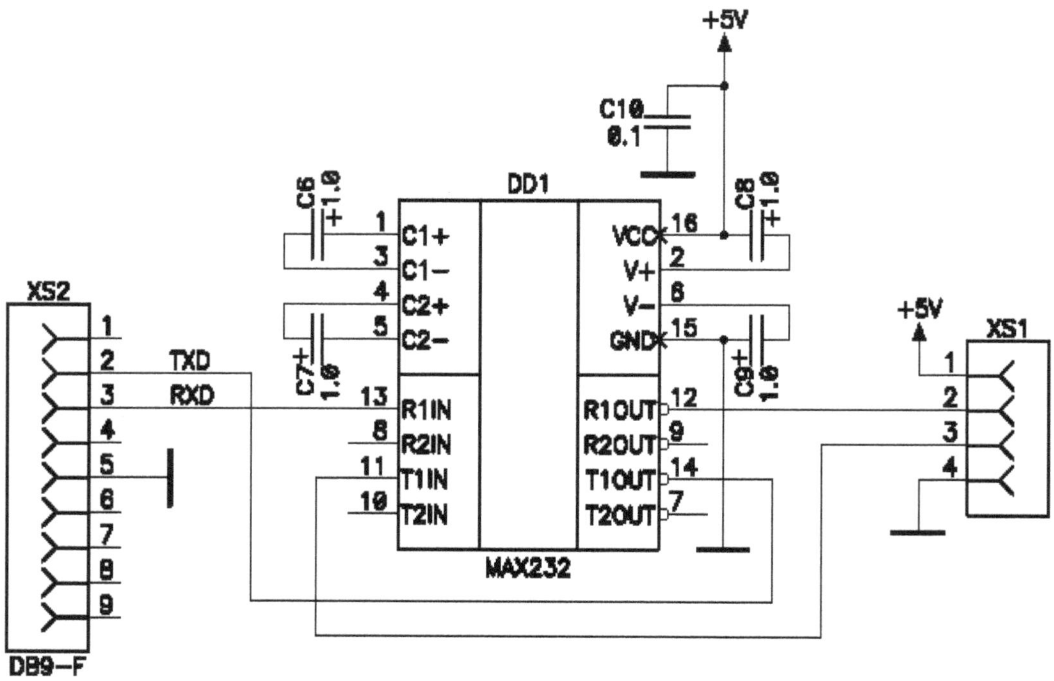

Fig. 2.6 Interfacing circuit for connection the UFDC-1 to PC.

2.2.3.2.2 Three-wire Serial Interface (SPI Compatible)

The brief description of communication command set for SPI interface. Appropriate modes and programmed converted accuracies are adduced in Table 2.6. The time diagrams for SPI interface are shown in Figure 2.7. The circuit diagram for the 3-wire Serial Interface communication mode is shown in Figure 2.8. The connector XP1 is intended for in-circuit automated calibration. Jumpers J1 and J2 must be in the ON position in the working mode and in the OFF position in the calibration mode. Maximum frequency of SPI data-rate should be chosen from 100 kHz to 500 kHz to guarantee a reliable operation.

2.2.3.2.3 Two-wire Interface (I^2C Compatible)

The brief communication command set description for I^2C interface, appropriate measuring modes and programmed converted accuracies are adduced in Table 2.8. The UFDC-1 can have one of 8 possible addresses that can be setup by A0-A2 pins. The absolute address of a device has the following format 1110 <A2><A1><A0>.

The circuit diagram for the 2-wire I^2C interface is shown in Figure 2.9. The connector XP1 is used for in-circuit automated calibration. In the working mode the jumper J1 must be set in the ON position and in the calibration mode it should be switched to the OFF position. Maximum possible communication speed in this mode is 100 kHz.

Table 2.6 Communication command set for SPI interface.

Interface		Description	Parameter
SPI	RS232 command analog		
<01><FF><FF>	A	Get current accuracy number	–
<02><nn>	An	Set accuracy number 'n' (00 ≤ nn ≤ 09) without an external divider; 0A ≤ nn ≤ 0F– with an external divider by 16	00 – 1 % 01 – 0.5 % 02 – 0.25 % 03 – 0.1 % 04 – 0.05% 05 – 0.025% 06 – 0.01% 07 – 0.005% 08 – 0.0025% 09 – 0.001% 0A – 1 % 0B– 0.25 % 0C – 0.05% 0D – 0.01% 0E – 0.0025% 0F – 0.001%
<03><FF><FF>	C	Check measurement status. Returns 0 if result is ready and not 0 if measurement is in progress.	–
<05><FF><FF>	M	Get current measuring mode number	–
<06><nn>	Mn	Set measurement mode number 'n' (00 ≤ nn ≤ 0F)	00 – frequency, FX1 01 – period, TX1 02 - phase shift, Fi 03 - time interval, T 04 – duty-cycle, D.C. 05 - duty-off factor, Q 06 - frequencies difference, FD 07 - periods difference, TD 08 - frequencies ratio, FR 09 - periods ratio, TR 0A - rotation speed, n 0B - pulse interval, tp 0C - space interval, ts 0D - pulse counter, N 0E - frequency 2, FX2 0F - period 2, TX2
<07><FF><FF>	R1	Get measurement result in BCD format. Returns sign byte and 12 bytes of result in BCD code: <Sign><I5><I4><I3><I2><I1><I0><F0><F1><F2><F3><F4><F5>	–
<08><FF><FF>	R2	Get measurement result in BINARY format. Returns 12 bytes of result in binary complementary code: <I5><I4><I3><I2><I1><I0><F0><F1><F2><F3><F4><F5>	–
<09>	S	Start measurement	–
<0B><FF><FF>	Z	Get current number of modulation rotor gradations	–
<0C><nn>	Znn	Set number of modulation rotor gradations "nn" = (01...FF) - set number "nn" of modulation rotor gradations	1–255

Legend:

<sign> - sign of result: 0x20 (space char) for positive result. 0x2D (minus char) for negative result
<In> - byte n of integer part of result
<Fn> - byte n of fractional part of result
<FF> - dummy byte

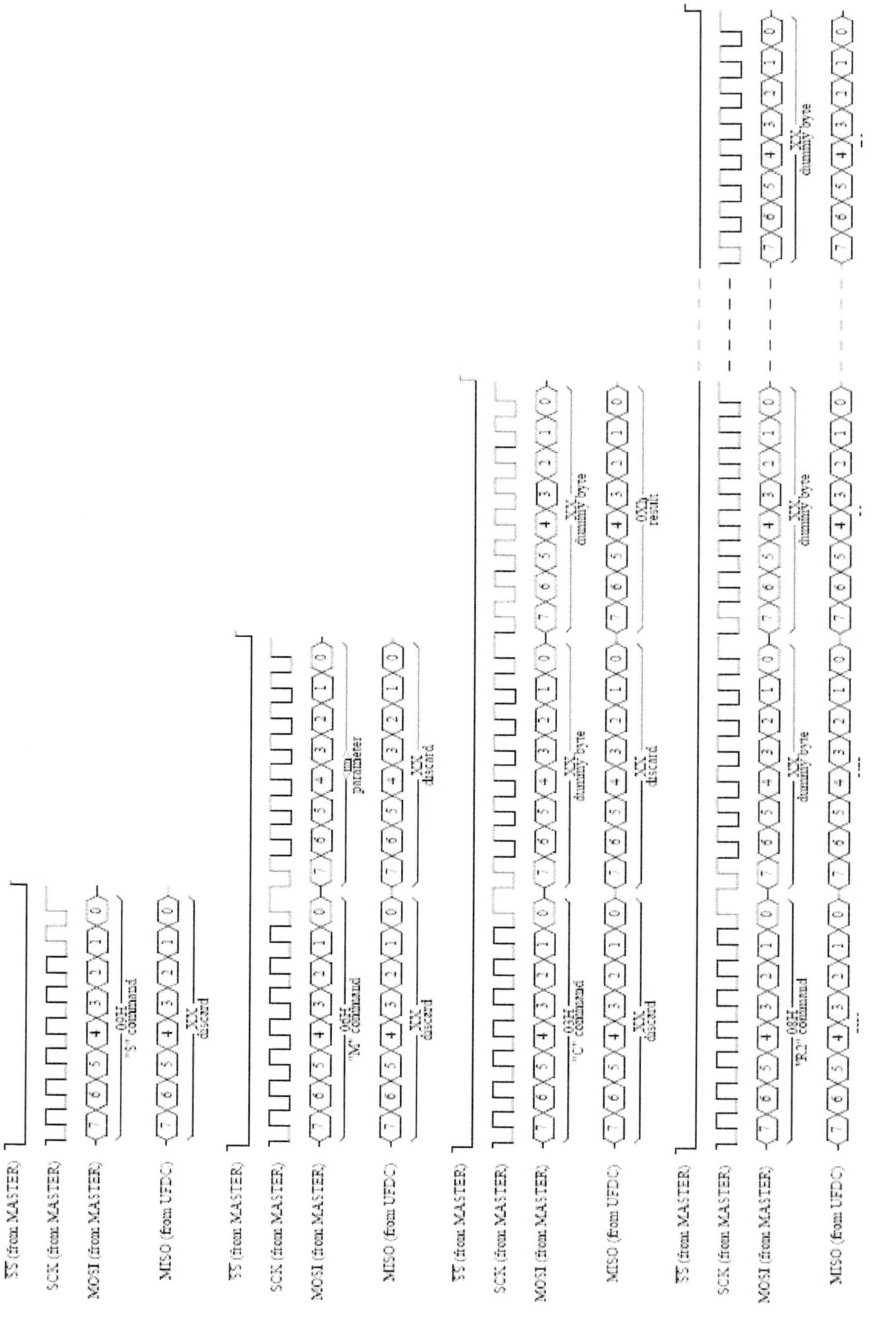

Fig. 2.7 Time diagrams for SPI interface.

Table 2.7 Communication command set for I^2C interface.

Interface		**Description**	**Parameter**
I^2C	**RS232 command analog**		
<01>	A	Get current accuracy number	–
<02><nn>	An	Set accuracy number 'n' (00 ≤ nn ≤ 09) without an external divider; 0A ≤ nn ≤ 0F– with an external divider by 16	00 – 1 % 01 – 0.5 % 02 – 0.25 % 03 – 0.1 % 04 – 0.05% 05 – 0.025% 06 – 0.01% 07 – 0.005% 08 – 0.0025% 09 – 0.001% 0A – 1 % 0B– 0.25 % 0C – 0.05% 0D – 0.01% 0E – 0.0025% 0F – 0.001%
<03>	C	Check measurement status. Returns 0 if result is ready and not 0 if measurement is in progress.	–
<05>	M	Get current measuring mode number	–
<06><nn>	Mn	Set measurement mode number 'n' (00 ≤ nn ≤ 0F)	00 – frequency, FX1 01 – period, TX1 02 - phase shift, Fi 03 - time interval, T 04 – duty-cycle, D.C. 05 - duty-off factor, Q 06 - frequencies difference, FD 07 - periods difference, TD 08 - frequencies ratio, FR 09 - periods ratio, TR 0A - rotation speed, n 0B - pulse interval, tp 0C - space interval, ts 0D - pulse counter, N 0E - frequency 2, FX2 0F - period 2, TX2
<07>	R1	Get measurement result in BCD format Returns sign byte and 12 bytes of result in BCD code: <Sign> <I5> <I4> <I3> <I2> <I1> <I0> <F0> <F1><F2><F3><F4><F5>	–
<08>	R2	Get measurement result in BINARY format Returns 12 bytes of result in binary complementary code: <I5> <I4> <I3> <I2> <I1> <I0> <F0> <F1> <F2> <F3> <F4> <F5>	–
<09>	S	Start measurement	–
<0B>	Z	Get current number of modulation rotor gradations	–
<0C><nn>	Znn	Set number of modulation rotor gradations "nn" = (01...FF) - set number "nn" of modulation rotor gradations	1–255

Legend:

<sign> - sign of result: 0x20 (space char) for positive result. 0x2D (minus char) for negative result
<In> - byte n of integer part of result
<Fn> - byte n of fractional part of result

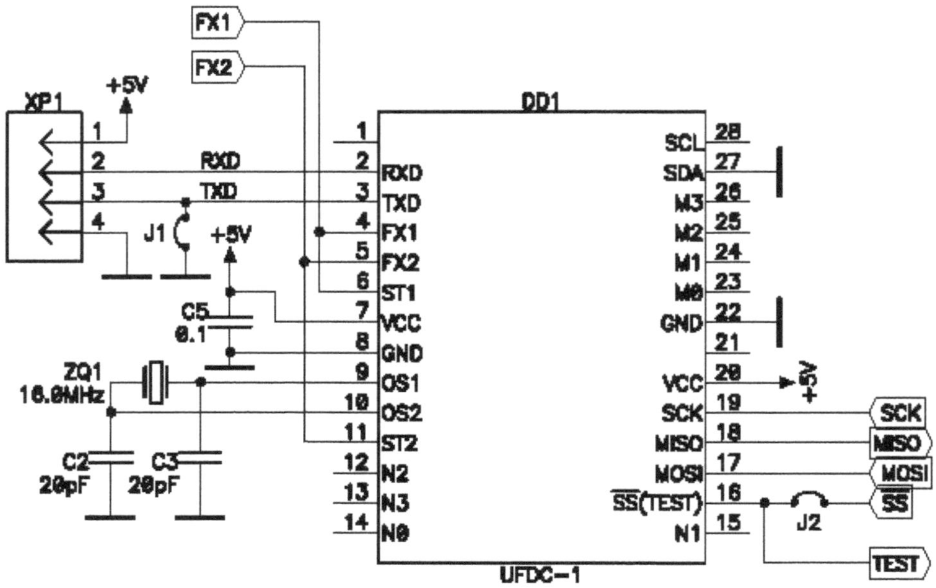

Fig. 2.8 The UFDC-1 circuit diagram for the 3-wire serial interface (SPI).

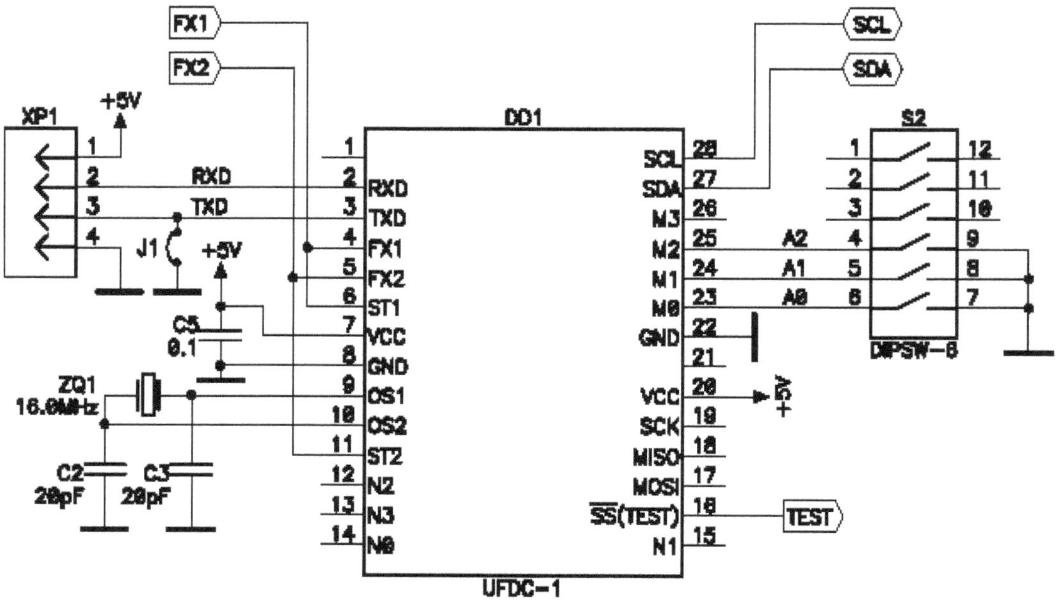

Fig. 2.9 The UFDC-1 circuit diagram for the 2-wire I^2C interface.

2.2.3.3 Quartz-Accurate Calibration Procedure

Taking into account a high programmable accuracy (up to relative error ± 0.001 %) of the Universal Frequency-to-Digital Converter (UFDC-1) it needs a very accurate frequency standard at least ≤ ±0.0001 %. Such relative error has a temperature compensated crystal oscillator (TCXO) or oven controlled crystal oscillator. However, these oscillators are relatively expensive devices and its price is higher than the UFDC-1 price. In addition, the oven controlled crystal oscillator can not be used in embedded applications as digital or smart microelectronic sensors.

A crystal oscillator is a relatively inexpensive method of obtaining an accurate frequency standard if it will be periodically calibrated [29]. The frequency or rate of vibration of quartz oscillator is determined by the cut. size, shape of the resonator, etc. Most commonly used are slices cut under an angle of $35^0 15'$ with the Z axis (the AT cut). These AT-cut slices can be used for frequencies from 0.5 MHz to 350 MHz. The primary frequency determining factor for AT cut is thickness since they vibrate in the thickness shear mode. The precision with which the thickness is controlled determines the variation from crystal to crystal from a nominal center frequency.

The temperature characteristics of AT-cuts crystals are described by a 3rd order parabola. It is then possible to describe the relative change of frequency:

$$\frac{\Delta f}{f} = A_i \Delta T + C_i \Delta T^3, \qquad (2.2)$$

where $\Delta T = T - T_i$ and T_i is the inflection temperature; A_i is the coefficient, which is depend on the quartz crystal construction and determined by the angle of cut; C_i is the constant; T_i is varied in the range from +25 to +35 ^{0}C for AT-cut depend on the crystal size [30. 31].

The frequency variation due to adjustment inaccuracy for a room temperature crystal oscillator (RTXO) can exceed 30 ppm (or ± 0.003 %) and has a systematic character. For example, the low cost 16 MHz crystal oscillators from *Siward* with 30 ppm determined tolerance has the real frequency 16 001 400 Hz that corresponds to 90 ppm (± 0.009 %) of reference error.

In order to use the UFDC-1 with low cost crystal oscillators for conversions with quantization error better than ±0.01 % it is expediently to make a calibration with the aim to compensate the adjustment or trimming inaccuracy. In other case, the calibration is not necessary.

The calibration should be made in real working conditions with the use of recommended 16 MHz crystal oscillator. At this calibration procedure, the converter should be connected to a PC through the serial interface RS232. Any terminal software can be used for this purpose, for example, Terminal V1.9b for Windows.

The calibration sequence for the UFDC-1 is the following:

1) The test command "T" toggles the UFDC-1 in the calibration mode. The sequence of rectangular pulses with frequency $f_{osc}/2$ is generated on its TEST output.
2) The frequency of this pulse sequence should be measured with the help of any external frequency counter with the relative error not worse than ± 0.0001 % or at least ± 0.0005 %.
3) The fractional part of measuring result is rejected and the number 8 000 000 (Hz) should be subtracted from the integer part.
4) The received correction factor Δ needs to be converted into the hexadecimal number.
5) Use the command "F ± Δ" (with taking into account the result sign) to put the correction value into the UFDC-1. The command "F" without parameters indicates the internal value of correction factor and can be used for checking the loaded number.

The calibration is performed only once. The correction factor Δ is stored in the nonvolatile memory of the UFDC-1, which holds this value after power-off.

Example:

Toggle the UFDC-1 in the calibration mode by the "T" command entered in any terminal software, working under Windows operation systems:

```
>T
```

Let the measured frequency on the TEST output is 8 000 694.257865 Hz. for example. We reject a fractional part and received the integer number is 8 000 694 Hz. Then we calculate the correction factor 8 000 694 - 8 000 000 = 694 Hz and convert the result into the hexadecimal number $(694)_{10} = (2B6)_{16}$. Finally, we send the correction command with taking into account the correction factor's sign:

```
>F+2B6
```

Then to check programmed value we send the command:

```
>F
2B6
```

The returned result 2B6 testifies about the properly stored value of correction factor for the given quartz crystal. After that, the UFDC-1 is ready for conversion in the frame of programmable accuracy ranges from ±1 to ±0.001 %.

If the primary frequency of crystal oscillator is 16000937.5 Hz and the converted frequency f_x = 25 000 Hz. the result without calibration (Δ = 0) and with specified quantization error $\delta_{q\ setup} = \pm 0.001$ % would be f_{xi} = 24998.0002 Hz. This corresponds to the conversion error $\delta_q = \pm 0.008 \% > \delta_{q\ setup}$. After calibration, the measured frequency is f_{xi} = 25000.1781 Hz. It corresponds to the relative error $\delta_q = \pm 0.0007 \% < \delta_{q\ setup}$.

Besides the trimming inaccuracy, the primary frequency of crystal oscillator is also influenced by different drifts due to temperature changes, aging, supply-voltage changes and other environmental changes. The periodical calibration is also useful for elimination of aging influence as well as temperature changes.

Taking into account the industrial working temperature range for the UFDC-1 (-40 ^{0}C ...+85 ^{0}C) it is also expediently to investigate the conversion error in this temperature range, a possibility for temperature drift calibration and temperature margins in which the calibration will be necessary.

The experimental research can be carry out in a climatic chamber. A data acquisition, result indication and communication between the UFDC-1 and a PC can be made in the LabView environment. An example of appropriate virtual instrument's front panel is shown in Figure 2.10. It consists of four main modules: serial port configuration, measuring modes and conversion error setup, results indication and statistical calculations, and calibration modules. The last one lets input the correction factor Δ and its sign. The virtual instrument's block diagram (measuring mode setup. indication and calculation modules) is shown in Figure 2.11, and port configuration module – in Figure 2.12. The experimental setup is shown in Figure 2.13.

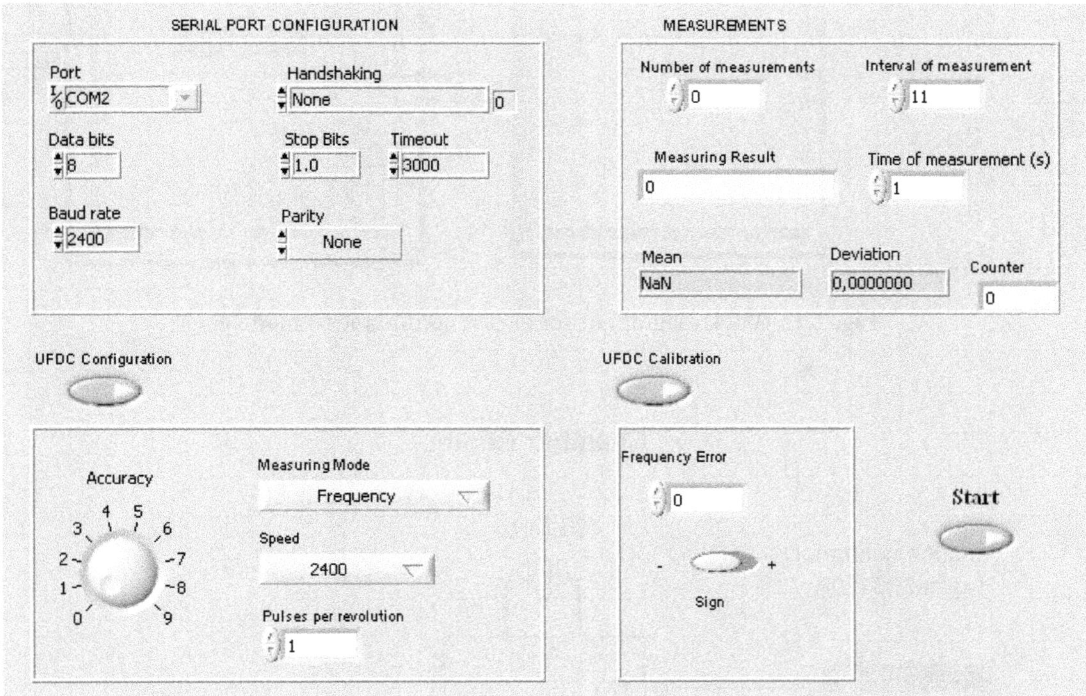

Fig. 2.10 Virtual instrument's front panel for the UFDC-1 experimental research.

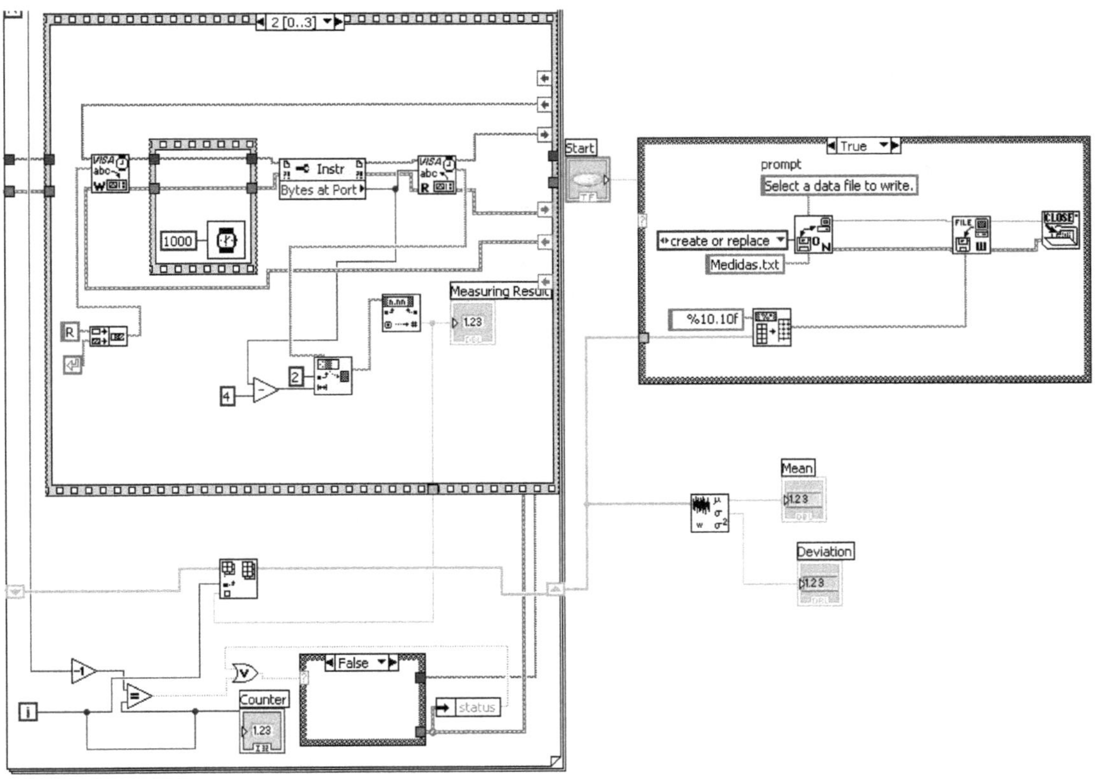

Fig. 2.11 Block diagram of Measuring Mode Setup, Calculation and Results Indication modules.

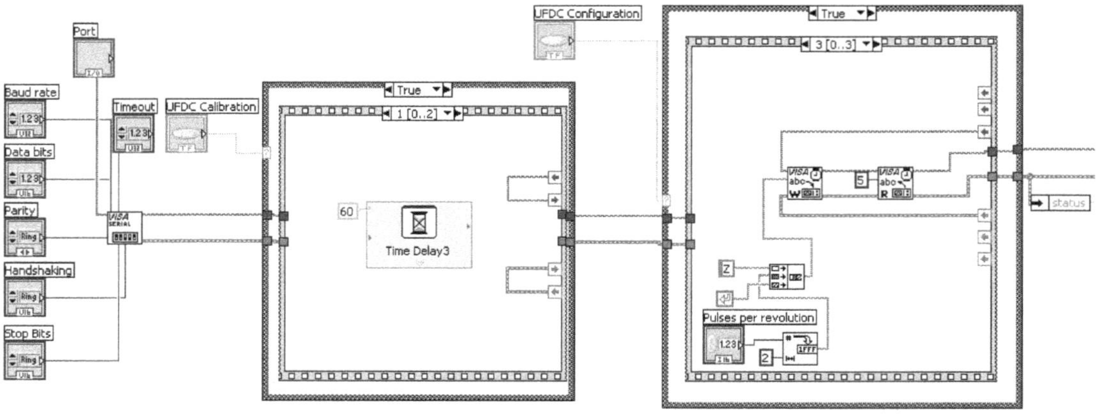

Fig. 2.12 Block diagram of serial port configuration module.

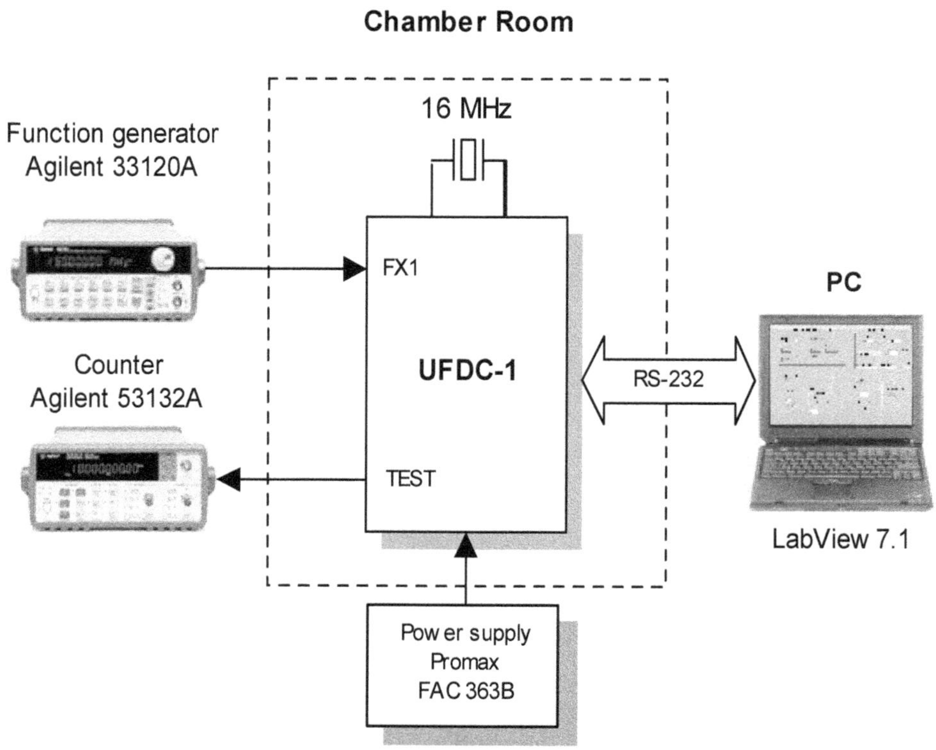

Fig. 2.13 Experimental setup.

After 45 minutes temperature stabilization the UFDC-1 was calibrated at + 21 ^{0}C. The quantization error was chosen ± 0.001 % and frequencies 1 Hz, 1 kHz and 1 MHz were measured one hundred times of each through 2 s time interval. The investigated temperature range was –40 ^{0}C ... +80 ^{0}C with 10 ^{0}C step at 50 % relative humidity. The input signal of appropriate frequency was generated by the function generator Agilent 33120A. The frequency counter Agilent 53132A was used for frequency measurement at the UFDC's TEST output. Frequency vs. temperature characteristics for 1 Hz, 1 kHz and 1 MHz are shown in Table 2.8 and in Figure 2.14 (a-c) accordingly.

Table 2.8 Experimental results: frequency vs. temperature (data in frames correspond to the ranges where correction does not need).

	1 Hz		1 kHz		1 MHz	
t, °C	$f_{x\ mean}$, Hz	|Δf|, Hz	$f_{x\ mean}$, Hz	|Δf|, Hz	$f_{x\ mean}$, Hz	|Δf|, Hz
-40	0.99998840381	0.00001159619	999.989301511	0.010698488	999989.883188	10.1168118
-30	0.99997040622	0.00002959378	999.980002999	0.019997000	999979.581952	20.4180470
-20	0.99997516445	0.00002483555	999.979203143	0.020796856	999979.881988	20.1180110
-10	0.99997804490	0.0000219551	999.980002999	0.019997000	999984.582552	15.4174478
0	0.99998572280	0.0000142772	999.986014637	0.013985362	999989.183104	10.8168958
+10	0.99998984286	0.00001015714	999.998800167	0.001199832	999998.484220	1.5157798
+20	0.99998966262	0.00001033738	1000.000000000	0.000000000	1000002.634718	2.6347180
+30	1.00000342059	0.00000342059	1000.009998800	0.009998800	1000009.185504	9.1855041
+40	1.00001093949	0.00001093949	1000.009998800	0.009998800	1000016.636398	16.63639813
+50	1.00001253972	0.00001253972	1000.019197880	0.019197880	1000024.087292	24.08729212
+60	1.00002213831	0.00002213831	1000.029497040	0.029497040	1000029.287916	29.28791612
+70	1.00002811700	0.00002811700	1000.029997000	0.029997000	1000032.038246	32.03824611
+80	1.00002355818	0.00002355818	1000.029997000	0.029997000	1000032.288276	32.28827611

Due to temperature changes, the conversion error at 1 Hz frequency measurement can be increased up to ± 0.003 % (or 30 ppm). It is true for temperature ranges –30 ^{0}C … 0 ^{0}C and +50 ^{0}C … +80 ^{0}C. In the rest part of temperature range, the conversion error does not exceed the programmed value ± 0.001 % (or 10 ppm).

At 1 kHz frequency measurement the conversion error also can be 30 ppm and its value exceeds the programmed error in the temperature ranges –30 ^{0}C …0 ^{0}C and +40...+80 ^{0}C.

Analogically, at 1 MHz frequency measurement the conversion error is 32 ppm in the temperature ranges –30 ^{0}C ...+10 ^{0}C and +30 ^{0}C ...+80 ^{0}C.

Obtained experimental results coincide with the temperature-frequency characteristic for standard quartz oscillator of RTXO type with the AT-cut, described by a 3rd order parabola. Such quartz resonators have minimal temperature changes in the temperature range of 0...+50 ^{0}C. At frequency increasing, the temperature range is narrowed, probably, because of strong influence of short-term instability at high frequency measurements.

At frequency measurement in temperature range near –40 ^{0}C no any correction is necessary. It is explained by the temperature-frequency characteristic of this type of quartz oscillators.

If the UFDC-1 is working outside of experimental determined temperature ranges with the quantization error better then ± 0.01 %, the correction procedure described above should be used. If due to some reasons the correction is not possible, it is expediently to use a TXCO type quartz crystal with the frequency variation not more than 5 ppm in the working temperature range –20 ^{0}C ...+70 ^{0}C. Besides, the UFDC-1 can work well also with an external precise quartz generator.

Figure 2.15 shows the temperature intervals in which it is necessary to calibrate the UFDC-1.

The main contribution to the frequency reference error contributes the trimming inaccuracy (or calibration tolerance). This component has a systematic character. The correction procedure lets to the UFDC-1 work well with a low-cost RTXO quartz oscillator. It is a very important property for different embedded applications, for example, digital and smart sensors and transducers.

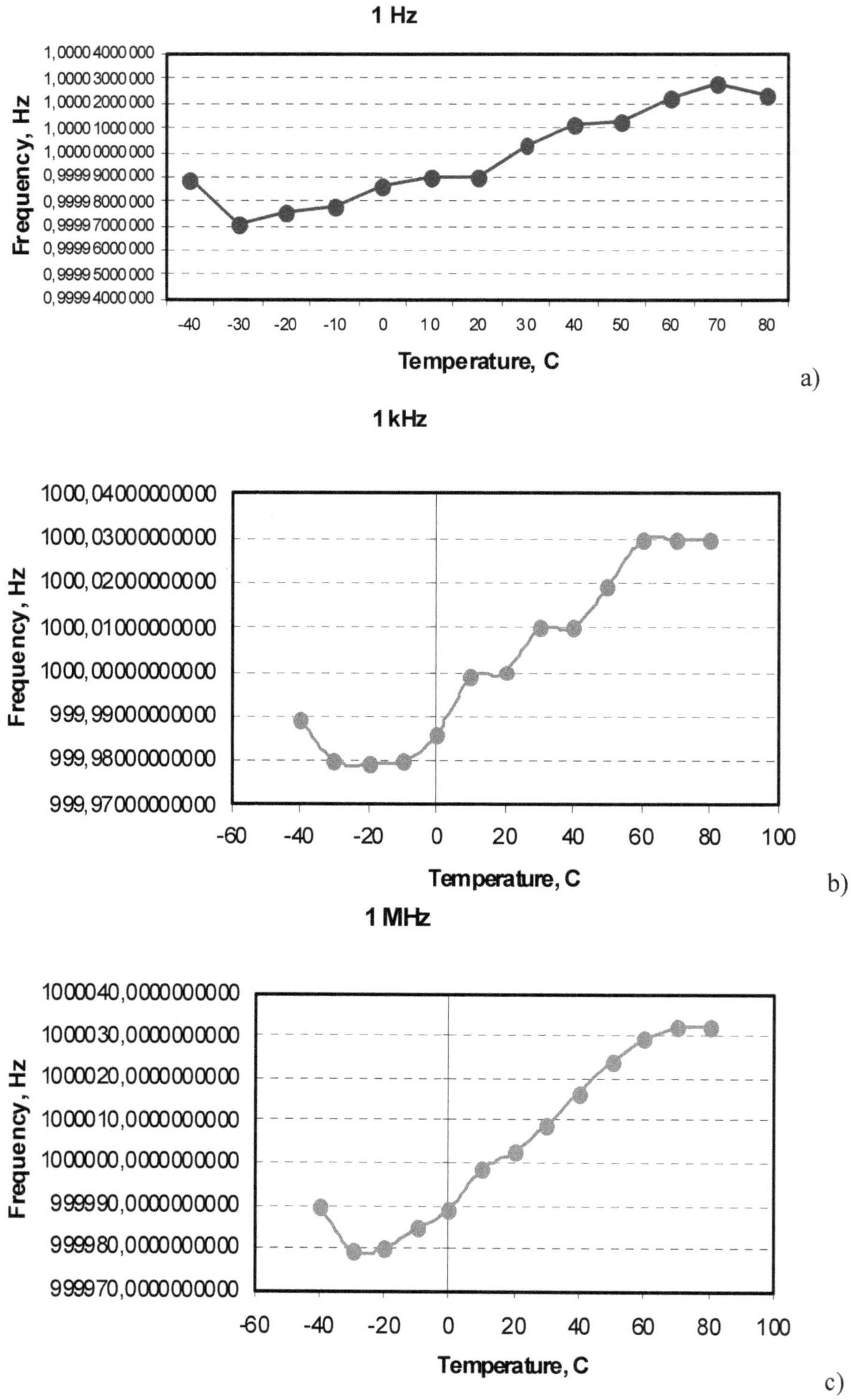

Fig. 2.14 Frequency vs. temperature characteristics for 1 Hz (a), 1 kHz (b), and 1 MHz (c).

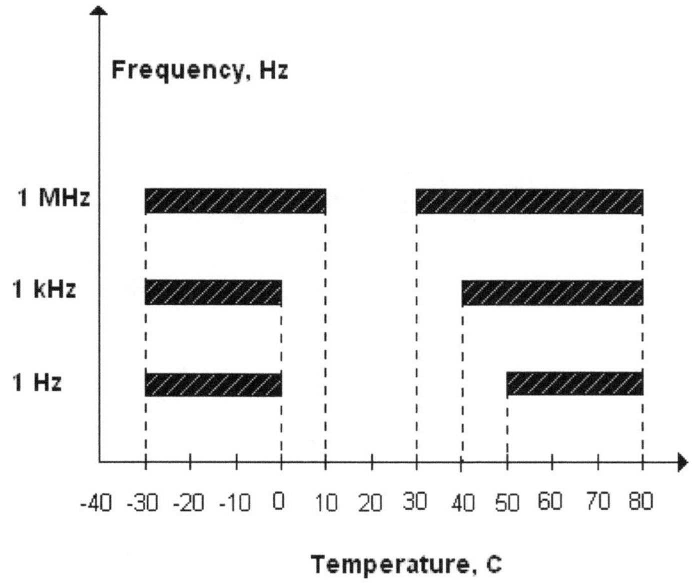

Fig. 2.15 Temperature intervals in which it is necessary to calibrate the UFDC-1.

In the case if the calibration is not possible for any reason, it is expediently to use a TCXO type quartz resonator of with the error not more than 5 ppm in the range of operating temperatures from -20 ^{0}C to +70 ^{0}C. In addition, the UFDC-1 can work not only with the built-in crystal oscillator, but also with precision external crystal generator. In this case ask for the custom designed UFDC-1 option.

The similar calibration also is useful at work of the converter outside normal conditions (for example, at the increased or decreased ambient temperature) with the purpose of correction of an additional temperature error. After a long term UFDC-1 usage, such periodical calibration will also reduce the quartz crystal error due to the crystal aging.

Experimental investigation of UFDC-1 with the low cost 16 MHz, SDE quartz crystal oscillator (Figure 2.16) have been carry out in the climatic chamber CTS T-40/50 (Figure 2.17) in the industrial temperature range from -40 ^{0}C to +85 ^{0}C.

The aim of experiments was to determine the temperature instability for crystal oscillator's frequency of this type. The PCB containing UFDC-1 and quartz resonator (Figure 2.16) has been placed into the climatic chamber, where temperature was decreased to the lowest possible from the working temperature range, and after some retention interval the frequency on the test output of TEST of UFDC-1 was measured using the precision frequency counter Agilent 53132A with the high accurate temperature-compensated reference. The retention interval depends on the thermal characteristics of the resonator and was determined experimentally by methods of experimental investigation of quartz oscillators [32]. This frequency was measured at regular intervals. This interval equals to a time, from which the difference between adjacent frequency measurements did not exceed the given value caused by the required accuracy of measurement. Taking into account the non-linear temperature-frequency characteristics of quartz oscillator, the measurement frequency at the TEST output of UFDC-1 were carried out every 10 ^{0}C throughout the operating temperature range from tmin to tmax. Preliminary UFDC-1 calibration was not conducted in order to determine the temperature range in which calibration is not required.

Fig. 2.16 Printed circuit board with UFDC-1 IC and *SDE* quartz resonator.

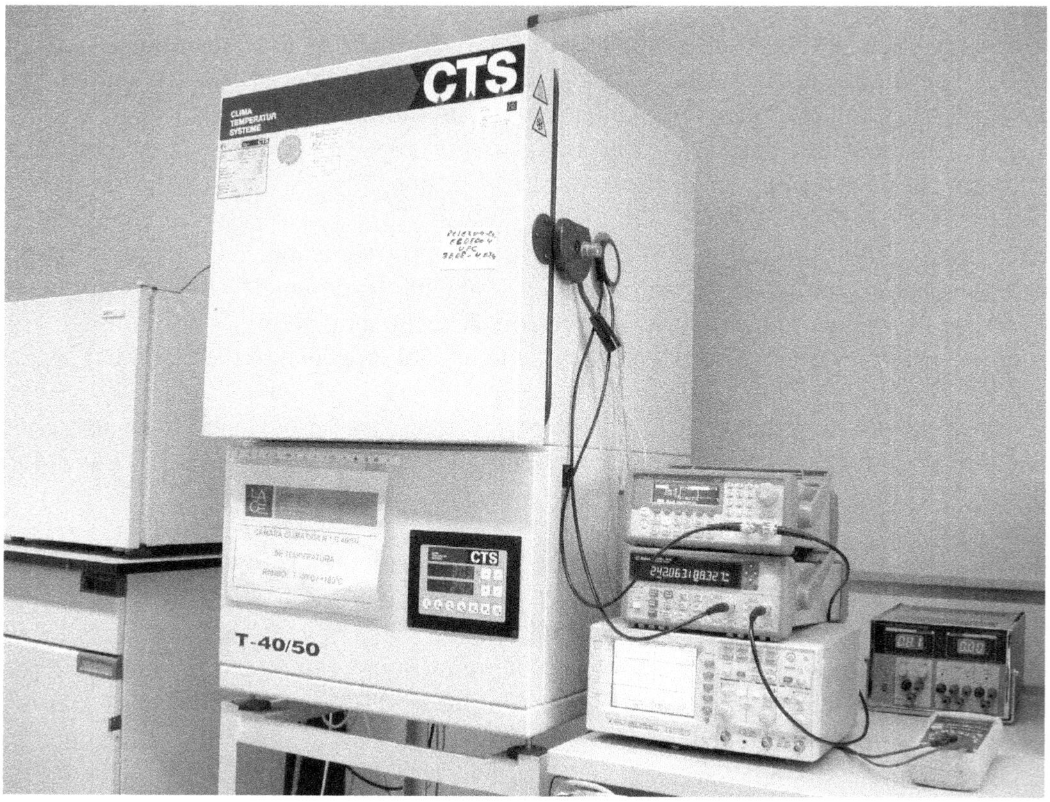

Fig. 2.17 Climatic chamber CTS T-40/50 and experimental setup.

Error for temperature setup and maintaining it in the climatic chamber CTS T-40/50 was 0.5 ^{0}C in the same gradient value within the chamber. The relative error of UFDC-1 reference frequency was determined as

$$\delta_{f_0} = \frac{f_{osc} - 2 \cdot f_{TEST}}{f_{osc}} \times 100 \text{, \%,} \tag{2.3}$$

where f_{osc} = 16 MHz is the nominal frequency of quartz resonator; f_{TEST} is the frequency has been measured on the TEST output. The experimental dependents of $\delta_{fo} = f(t\ ^0C)$, is shown in Figure 2.18.

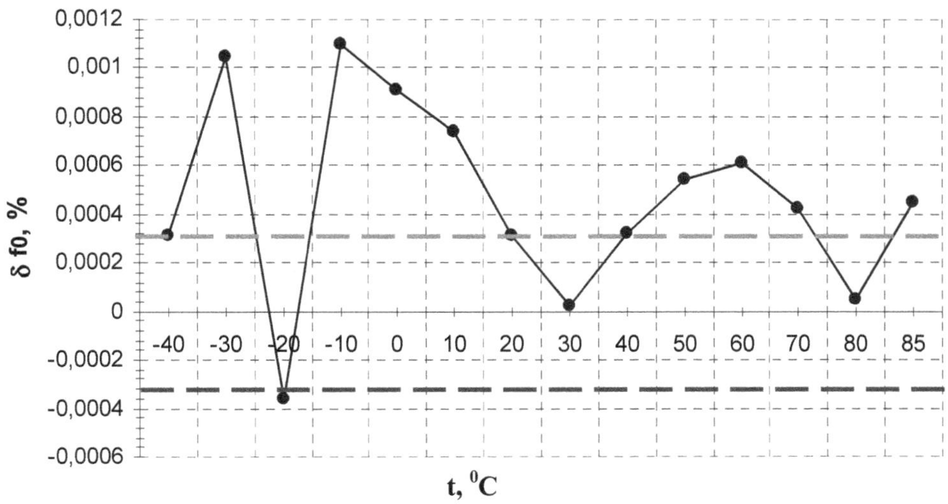

Fig. 2.18 Experimental dependence $\delta_{fo} = f(t\ ^0C)$.

As it is visible form the Figure 2.18, the calibration is not necessary, when the relative error does not exceed $|\delta_{f0}| \leq \pm$ 0.00033 %, i.e. when the reference frequency error in three times less than the minimal possible relative error of UFDC-1 $\delta = \pm$ 0.001 %. In this case the reference frequency error can be neglected [28] in comparison with the IC's relative error. In other words, the UFDC-1 calibration is necessary for the following temperature subranges: from - 30 ^{0}C to +10 ^{0}C; from +50 ^{0}C to +70 ^{0}C and from +80 ^{0}C to +85 ^{0}C. In other temperature subranges the calibration is not necessary.

The experimental results are well coincide with the temperature-frequency characteristic of modern RTXO type quartz crystal resonators with the AT-cut, which are described by the third order parabola.

It is clear; customers do not need to carry out such experiments for calibration purpose. The calibration procedure is simple enough. The aim of experiments descriptions in this book is to demonstrate high metrological performance of the UFDC-1 IC. However, based on detailed explanation of these investigation and measuring setups, readers can easy to repeat it in a laboratory with educational or/and research purposes. It should be also useful if the IC is used in other designs as silicon IP.

In the Table 2.9 there is a short summary of calibration procedure for the UFDC-1. The calibration example has been discussed above, see p. 48.

Table 2.9 Summary of calibration procedure for the UFDC-1.

<table>
<tr><th>① What calibrate?</th><th>② Why calibrate ?</th></tr>
<tr><td>• Systematic quartz-crystal error to reduce the adjustment or trimming inaccuracy
• Temperature drift
• Quartz-crystal aging error</td><td>• Taking into account a high UFDC-1 accuracy (up to 0.001 %) it needs a very accurate reference at least ≤ 0.0001 %
• Low cost crystal oscillators does not have a good stability due to systematic error</td></tr>
<tr><th>③ When calibrate ?</th><th>④ How to calibrate ?</th></tr>
<tr><td>• In order to use the UFDC-1 with any low cost crystal oscillators for conversions with the relative error less than 0.01 % it is necessary to calibrate it with the aim to compensate the adjustment or trimming inaccuracy
• If application needs relative error ≥ 0.01 % no calibration is necessary
• If the UFDC-1 is working in specified temperature range with high accuracy</td><td>• Should be made in real working conditions with the 16 MHz crystal oscillator
• Connect the UFDC-1 to a PC through the serial interface RS232
• Use the test command "T“
• Measure the frequency at the TEST pin by any external frequency counter with accuracy not worse than 0.0001 % or at least 0.0005 %
• Calculate the correction factor Δ
• Input it into the UFDC-1</td></tr>
<tr><th colspan="2">⑤No calibrate if:</th></tr>
<tr><td colspan="2">• Relative error > 0.01 %
• Use a precision temperature-compensated external integrated generator ± 3 ppm frequency stability over the -40 °C to +85 °C</td></tr>
<tr><td colspan="2">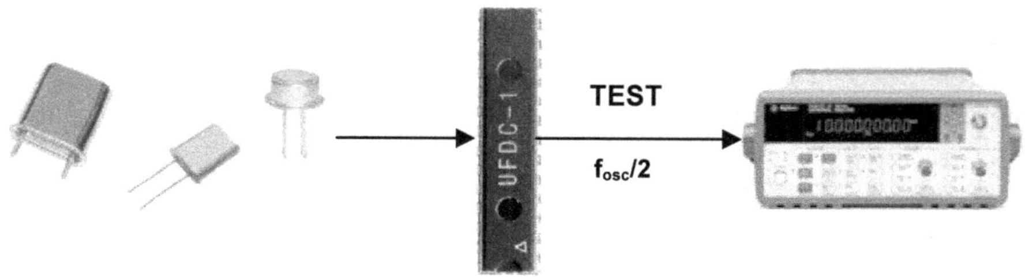
</td></tr>
</table>

2.2.3.4 Measuring Modes

In this section, the UFDC-1's properties for all measuring modes will be specified. Names of these modes are the same as used in Table 2.3.

2.2.3.4.1 Mode 0. FX1: Frequency, Channel I, 0.05 Hz – 7.5 MHz without Prescaling

In this mode a frequency signal f_{x1} (should be TTL rectangular pulses) must be connected in parallel to inputs FX1 and ST1. If the pulse shape differs from the square wave, the usage of an external input shaper (Schmitt-trigger signal conditioner) is recommended. At low and infralow frequency measurements it is recommended to use a low-pass filter.

If the maximum converted frequency exceeds 7.5 MHz, it is necessary to use an external prescaler (external divider by 16) and to choose the appropriate mode and accuracy by means of commands or external switches N0-N3. By this way the frequency range will be expanded up to 120 MHz.

The conversion error in this mode is constant and does not exceed the programmed in all frequency range. The conversion result is represented in Hz. All conversion accuracies specified in Table 2.4 are available in this mode.

2.2.3.4.2 Mode 1. TX1: Period, Channel I, 150 ns – 20 s

This mode of measurement is similar to the Mode 1. The conversion result is represented in microseconds. All conversion accuracies specified in Table 2.4 are available in this mode. The mode with prescaling (external divider by 16) is also possible.

2.2.3.4.3 Mode 2. Fi: Phase-Shift, 0- 360^0, f_{xmax} = 500 kHz at 50 % duty-cycle

The pulse sequences with the same frequency, a phase-shift between which need to be converted into a digital are connected to inputs FX1, ST1 and FX2, ST2 accordingly. Units for conversion result are degrees. The programmed accuracy influences on the period conversion. The conversion error of phase-shift does not depend on the pulse sequences period, but depends on the pulse width.

2.2.3.4.4 Mode 3. T: Time Interval between Start- and Stop-Impulses, 2 μs - 250 s

Start- and stop-pulse are connected to FX1, ST1 and FX2, ST2 inputs accordingly. The UFDC-1 converts the time interval between these pulses into digital. In this mode the conversion error does not depend on the value on inputs N0-N3 or given in a command mode through one of possible interfaces and can be calculated according to the equation (2.1) for each value of converted time interval. Units of measurements for conversion result are microseconds.

2.2.3.4.5 Mode 4. D.C.: Duty-Cycle: 0 – 1 at $f_{xmax} \leq$ 500 kHz for 40-60 % D.C.

The pulse signal is coming on inputs FX1, ST1. The programmed accuracy influences on the period conversion. The conversion error does not depend on the period of pulse signal, but depends on the pulse width.

2.2.3.4.6 Mode 5.Q: Duty-Off Factor, $10^{-8} - 8\times10^{6}$, at ≤ 500 kHz for 40-60 % D.C.

Similarly to the Mode 4, the pulse sequence is coming on FX1, ST1 inputs. The programmed accuracy influences on the period conversion. The conversion result depends on the pulse width.

2.2.3.4.7 Mode 6. FD: Frequency Difference with Sign Determination f_{x1}-f_{x2}, 0 - 7.5 MHz

The frequency signal f_{x1} is coming on FX1, ST1 inputs and signal f_{x2} - on FX2, ST2 inputs. The sign "minus" before the result of measurement testifies that $f_{x1} < f_{x2}$. All conversion accuracies specified in Table 2.4 are available in this mode. The mode with prescaling (external divider by 16) is also possible.

2.2.3.4.8 Mode 7. PD: Period Difference with Sign Determination T_{x1}-T_{x2}, 0 – 20 s

The pulse signal with period Tx1 is coming on FX1, ST1 inputs and signal with period Tx2 - on FX2, ST2 inputs. The sign "minus" before the result of measurement testifies that $T_{x1} < T_{x2}$. All conversion accuracies specified in Table 2.4 are available in this mode. The mode with prescaling (external divider by 16) is also possible.

2.2.3.4.9 Mode 8. FR: Frequency Ratio, f_{x1}/f_{x2}, $7\times10^{-9} - 1.4 \times 10^{8}$

The frequency signal f_{x1} is coming on FX1, ST1 inputs and signal f_{x2} - on FX2, ST2 inputs. All conversion accuracies specified in Table 2.4 are available in this mode. The mode with prescaling (external divider by 16) is also possible.

2.2.3.4.10 Mode 9. PR: Period Ratio, T_{x1}/T_{x2}, $7.5 \times 10^{-9} - 1.33 \times 10^{8}$

The pulse sequence with period T_{x1} is coming on FX1, ST1 inputs, and pulse sequence with period T_{x2} – on FX2, ST2 inputs. All conversion accuracies specified in Table 2.4 are available in this mode. The mode with prescaling (external divider by 16) is possible.

2.2.3.4.11 Mode 10. n: Rotation Speed (rpm), n_x

The UFDC-1 can measure a rotation speed and present the results in *rpm* units. The number of encoder's teeth Z (1...255) must be set up in advance. In the Master Mode (interface RS232, unidirectional mode), by default *Z*=1.

In a system the UFDC-1 can also measure rotation acceleration. An external microcontroller or PC calculates acceleration of rotation based on two measurements n1 and n2 of rotation speed and time interval for the second measurement t_2:

$$\varepsilon = \frac{n_2 - n_1}{t_2}, \text{ r/min2} \qquad (2.4)$$

The sign "-" indicates deceleration of rotation, and "+" indicates acceleration of rotation.

2.2.3.4.12 Mode 11. t_p: Pulse Width, t_p 2 μs – 250 s

Pulse signal is coming on FX1, ST1 inputs. In this mode the conversion error does not depend on the value on N0-N3 inputs or given through one of possible interfaces and can be calculated according to the equation (2.1) for each value of converted pulse width. Units of measurements for result are microseconds.

2.2.3.4.13 Mode 12. t_s: Space Interval, t_s 2 μs – 250 s

This mode of measurement is similar to the Mode 11.

2.2.3.4.14 Mode 13. N: Pulse Number (Events) Count, N_x 0 – 4×10^9

The UFDC-1 can count the pulse number (events) on FX1, ST1 inputs. Each reinitialization of the mode resets the counter to zero.

2.2.3.4.15 Mode 14. FX2: Frequency, Channel II, 0.05 Hz – 7.5 MHz without prescaling

In this mode a frequency signal f_{x2} (should be TTL rectangular pulses) must be connected in parallel to inputs FX1 and ST1. If the pulse shape differs from rectangular, the usage of an external input shaper is recommended. At low and infralow frequency measurements it is recommended to use the low-pass filter.

If the maximum converted frequency exceeds 7.5 MHz, it is necessary to use an external prescaler (external divider by 16) and to choose the appropriate mode and accuracy by means of commands or external switches N0-N3. By this way the frequency range will be expanded up to 120 MHz.

The conversion error in this mode is constant and does not exceed the programmed in all frequency range. The conversion result is represented in Hz. All conversion accuracies specified in Table 2.4 are available in this mode.

2.2.3.4.16 Mode 15. TX2: Period, Channel II, 150 ns – 20 s

This mode of measurement is similar to the Mode 14. The conversion result is represented in microseconds. All conversion accuracies specified in Table 2.4 are available in this mode. The mode with prescaling (external divider by 16) is also possible.

2.2.4 Software

There are many ways to display the measurement results. Any terminal software for Windows can be used with the UFDC-1 using the serial port of a PC. The configuration of the Terminal must be the following: 2400 baud, 8 data bit, 1 stop bit and No parity error check. Following the RS232 communication commands from Table 2.5 it is easy to manage and display measuring results. There are four possible data format in the Terminal: BCD ASCII, BCD Hex, BIN Hex and BIN. One of the possible choices is Terminal V1.9b [33] freeware software. Its screen is shown in Figure 2.19 Here is a calibration procedure example and conversion result.

In order to have correct measurement results the following options must be selected in the main terminal window: an appropriate COM port; baud rate 2400; data bits – 8; parity – non; stop bits – 1; handshaking – non. Commands must be printed in the Transmit window, the results will be shown in the Receive window

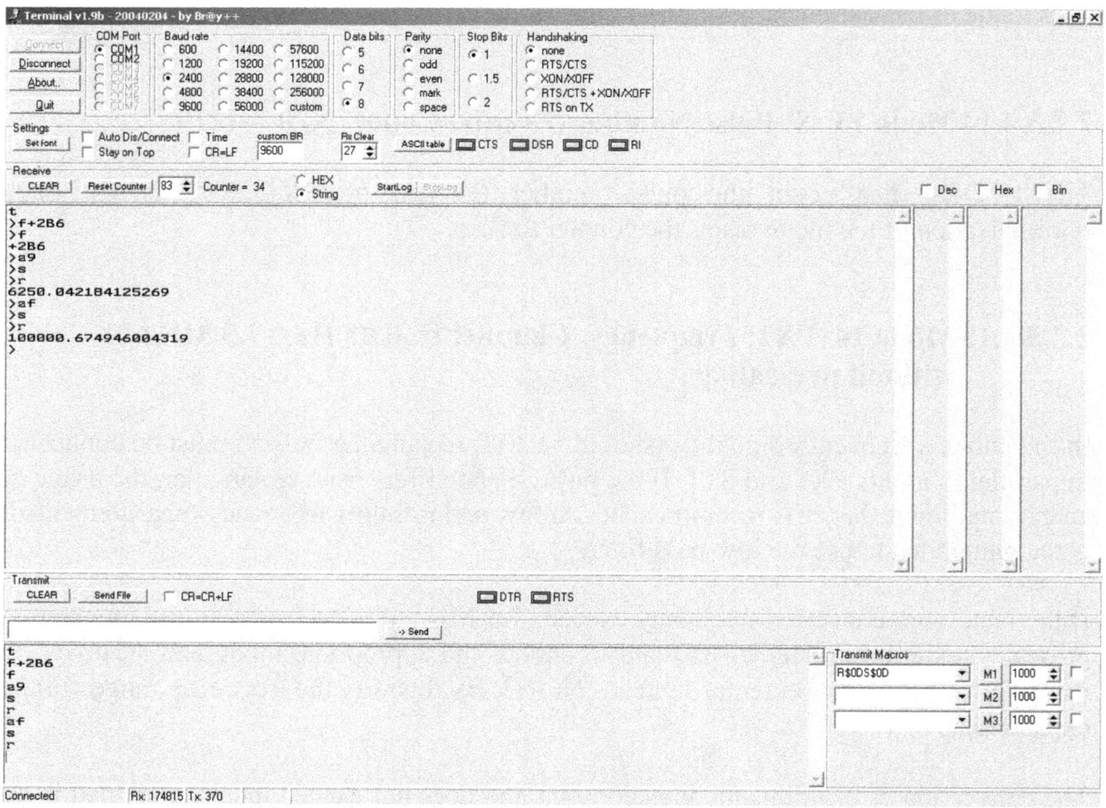

Fig. 2.19 Terminal V1.9b window.

Example:

```
>A9
>S
>R
12000.089992800575
```

- These commands set up the accuracy 0.001 %, start measurement (S), ask for BCD ASCII format (R) and return the result of measurement in Hz.

NOTE: For low and infralow frequencies, low rotation speed measurements as well as for long time intervals, periods, pulse widths and spaces measurements it is strongly recommended to use the command "C" between "S" and "R" commands (see Table 2.5) in order to check if the result is ready. In opposite side it is a risk to get a result of previous measurement.

It is clear, high level software for sensor and DAQ systems based on the UFDC-1 IC can be easily realized in the LabView environments. See Figure 2.10-2.12, for example.

2.2.5 Applications

The UFDC-1 has many applications in different industries of OEM markets (Figure 2.20), comparable or even exceed the ADC market.

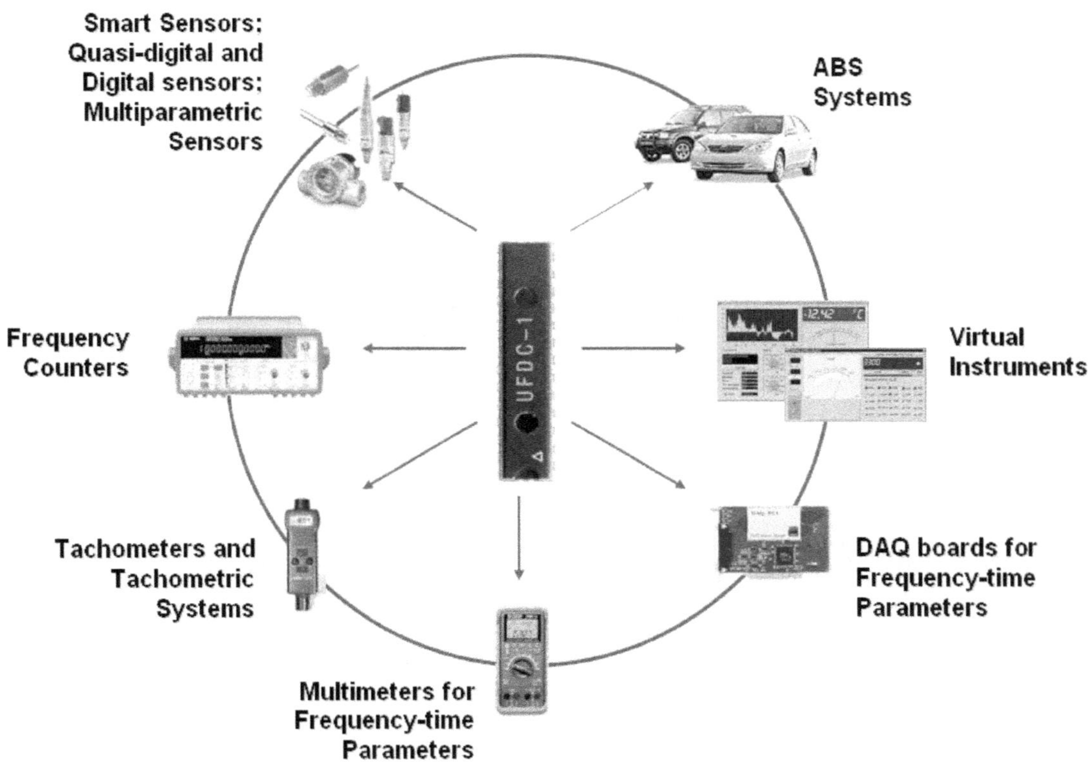

Fig. 2.20 UFDC-1 applications.

Some typical applications are following:

- Any physical, chemical or biosensor and sensor systems;
- Any frequency, period, duty-cycle, time interval, phase-shift, pulse number output sensors in order to produce a digital output;
- Digital sensors design, SoC and SiP;
- Smart (self-adaptive) sensors due to programmable accuracy and non-redundant conversion time - adaptive possibilities of UFDC-1 automatically choice the conversion time depending on the given error of measurement and opposite;

- Multifunctional and multiparamters sensors for simultaneous detection various parameters proportional to frequency and duty-cycle carrying the information provided by sensing elements;
- Data acquisition (DAQ) boars and systems for frequency-time parameters;
- Virtual instruments;
- Desktop multifunctional frequency counters;
- Tachometers and tachometric systems;
- Handheld multimeters for frequency-time parameters of electric signals;
- High-end, mid- and low-range ABS;
- High accuracy programmable frequency-time parameters – to – digital converters for different measuring and communication applications.

The UFDC-1 can also work with voltage and current output sensors. It this case, any voltage–to-frequency converter (VFC) can be used (Figure 2.21).

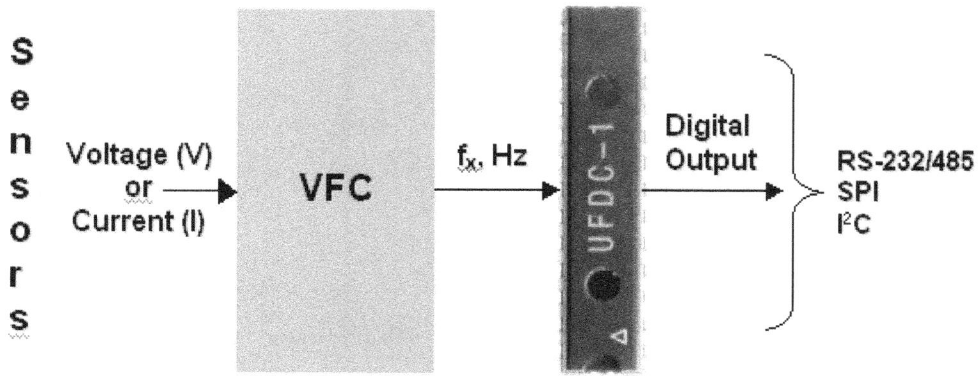

Fig. 2.21 UFDC-1 and analog domain signals.

The accuracy of such system will be determined by sensor accuracy as well as by the VFC's accuracy.

2.2.6 Pin Out and Housing

The UFDC-1 is available in a 28-lead Plastic Dual Inline Package (PDIP), 32-lead small Thin Profile Plastic Quad Flat Package (TQFP) and 32-pad Micro Lead Frame Package (MLF). Such packages are very suitable for both: prototyping and mass production of various digital sensors, sensor system and measuring instruments (from desktop measuring instruments to hand-held and pen-sized devices, Figure 2.22).

Figure 2.23 (a) and (b) show the pin configuration of PDIP and TQFP/MLF accordingly. The function of the pins is listed in Table 2.10. The detailed packaging information is available in [24]. This device can also be supplied in wafer form and/or as an IP [34, 35]. ICs' footprints for ECAD tools are shown in Figure 2.24 (a-c).

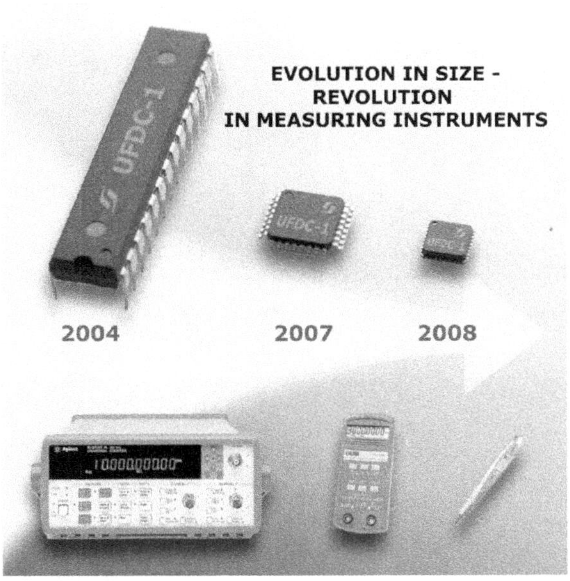

Fig. 2.22 UFDC-1 in different packages.

Pin Number	Function	Pin Number	Function
1	FX2	17	SCK
2	ST1	18	VCC
3	GND	19	-
4	VCC	20	-
5	GND	21	GND
6	VCC	22	-
7	OS1	23	A0 (M0)
8	OS2	24	A1 (M1)
9	ST2	25	A2 (M2)
10	N2	26	M3
11	N3	27	SDA
12	N0	28	SCL
13	N1	29	-
14	SS/TEST	30	RXD
15	MOSI	31	TDX
16	MISO	32	FX1

(a)

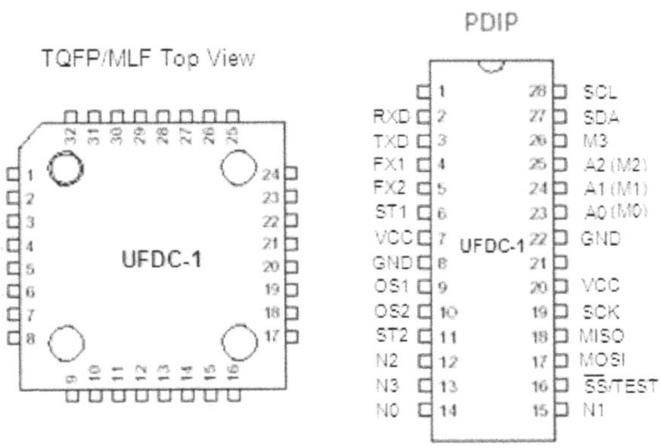

(b)

Fig. 2.23 Pin configuration and housing (Not to scale): (a) for TQFP and MLF packages; (b) for PDIP package; NOTE: Bottom pad for MLF package should be soldered to ground.

Table 2.10 Pin description.

Name	Pin Descriptions	Name	Pin Descriptions
V_{DD}	Supply voltage	SCK	SPI Bus Master clock Input
GND	Ground	MISO	SPI Bus Master Input/Slave Output
FX1, ST1	Input channel 1	MOSI	SPI Bus Master Output/Slave Input
FX2, ST2	Input channel 2	SS	SPI Bus Master Slave select
OS1, OS2	Oscillator	SCL	Two-wire (I^2C compatible) Serial Bus Clock Line
M0…M3	Mode selection (see table 2)	SDA	Two-wire (I^2C compatible) Serial Bus Data Input/Output Line
N0…N3	Accuracy selection (see table 3)	RXD	UART Input Pin
A0…A2	Addresses (for I^2C bus communication)	TXD	UART Output Pin
TEST	Calibration signal		

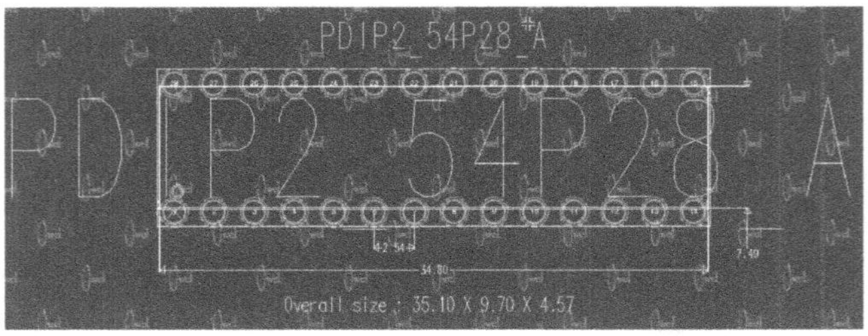

(a)

(b)

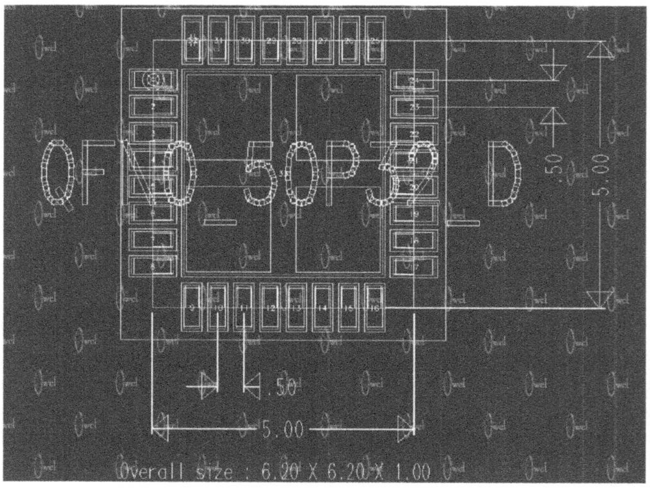

(c)

Fig. 2.24 UFDC-1 footprints for ECAD tools: (a) PDIP package; (b) TQFP package; and (c) MLF package.

2.2.7 Experimental Investigations

In order to experimentally determine the limiting performance of the UFDC-1, some metrological experiments for the maximal possible measured frequency f_x and duty-cycle values for the maximal frequency range were carried out [36]. Fig. 2.25 shows the measurement set up.

The square waveform pulse signal whose frequency must be measured, was fed from a function generator Agilent 33120A to inputs FX1 and ST1 (the first channel) of the UFDC-1 running on a 16 MHz clock. The UFDC-1 converted the frequency of the input signal to digital according to the modified method of the dependent count and was programmed to measure frequency with the minimal possible relative error ± 0.001 %. The supply voltage of the UFDC-1 was +5 V_{dc}, provided by a Promax FAC-363B power supply.

The frequency generated by the Agilent 33120A was in the range from 0.03 Hz up to 7.5 MHz. It was monitored by a frequency counter Agilent 53132A. A Promax OD-512 oscilloscope monitored the signal waveform.

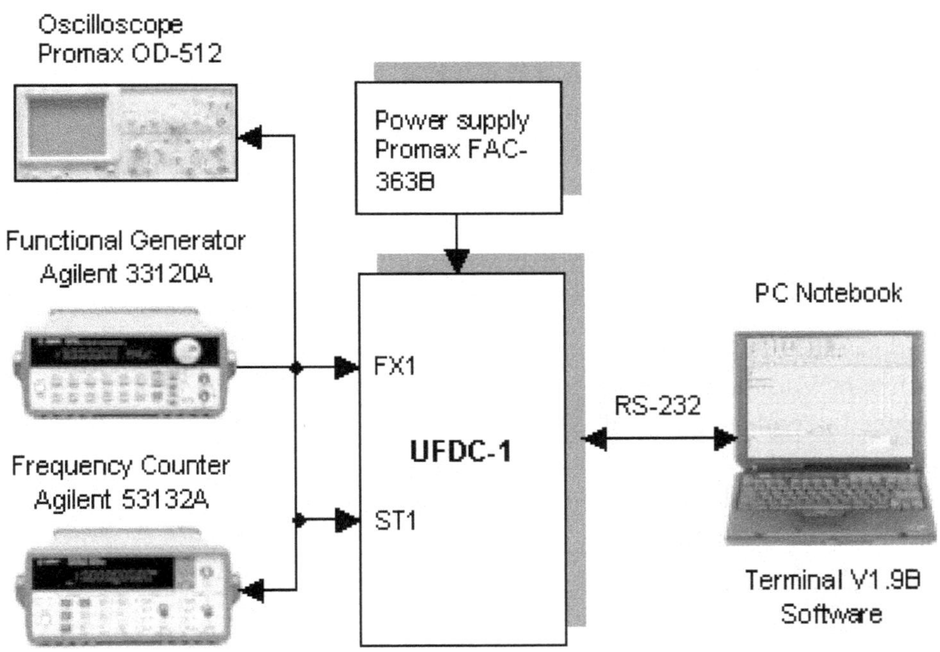

Fig. 2.25 Measurement set up to determine the maximum possible frequency f_x.

The measured frequency value was sent to a PC via an RS232 interface implemented with the MAX232 IC. There are several ways to display the measurement results. The user interface was implemented with the help of terminal software Terminal V1.9b for Windows. Frequency measurements were taken every second until totaling 60 measurements. Three of them were later excluded from the universe as gross errors, determined with the help of the censored sample method [28]. The measurement error was evaluated from appropriate statistics. The results for a 7.5 MHz square waveform pulse signal are shown in Figure 2.26. The statistical characteristics are adduced in Table 2.11.

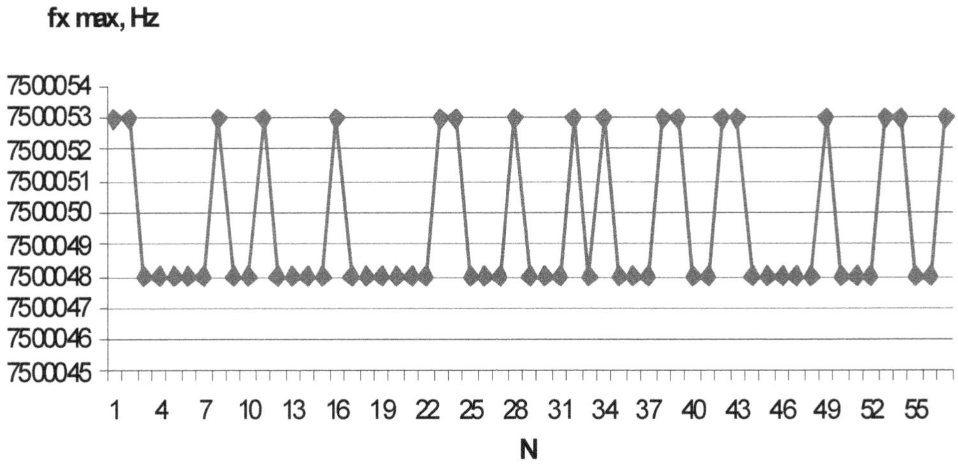

Fig. 2.26 Measured results for square waveform signal with frequency 7.5 MHz.

Table 2.11 Statistical characteristics of measurement results.

Parameter	Value
Number of measurements, N	60-3=57
Minimum f_x (min), Hz	7 500 048
Maximum f_x (max), Hz	7 500 053
Sampling Range, f_x (max)- f_x (min), Hz	5
Median	0
Arithmetic Mean, Hz	7 500 050
Variance	5.5
Standard Deviation	2.35
Coefficient of Variation	3 198 578.51
Confidence Interval at probability $P = 97$ %	$f_x \in [7\,500\,048.90 \div 7\,500\,050.25]$
$\chi 2$ –test at $P = 97$ %: - Gaussian (normal) distribution: - Uniform distribution:	 No No
Relative error, %	0.0007 < 0.001 (programmable)

The maximal possible frequency of a square waveform pulse signal for the UFDC-1 without prescaling was 7.5 MHz; the minimal possible frequency was 0.03 Hz. With an external prescaler (predivider based on a 16-bit counter), the maximal possible frequency $f_{x,max}$ was 120 MHz. The maximal frequency for non-square waveform signals (for example, triangular or sine waveforms) can be increased by an external Schmitt-trigger.

During the experiments, the minimal possible pulse width (t_p), pulse space (t_s) and time interval (t_x) were also determined. The minimal possible value for these time parameters was 2 μs. Their range was set up by the function generator Agilent 33120A and was monitored by the frequency counter Agilent 53132A set in the appropriate measuring mode.

Another critical parameter at high frequency measurements for a square waveform pulse signal was the duty-cycle. The duty-cycle values determined for the maximal frequency ranges are adduced in Table 2.12 and shown in Fig. 2.27.

Table 2.12 Limit duty-cycle for the input signal depending on the maximal frequency.

Duty-cycle, %	Frequency, MHz
40...50	7
40...57	6
30...64	5
23...71	4
20...79	3
20...80	2 and less

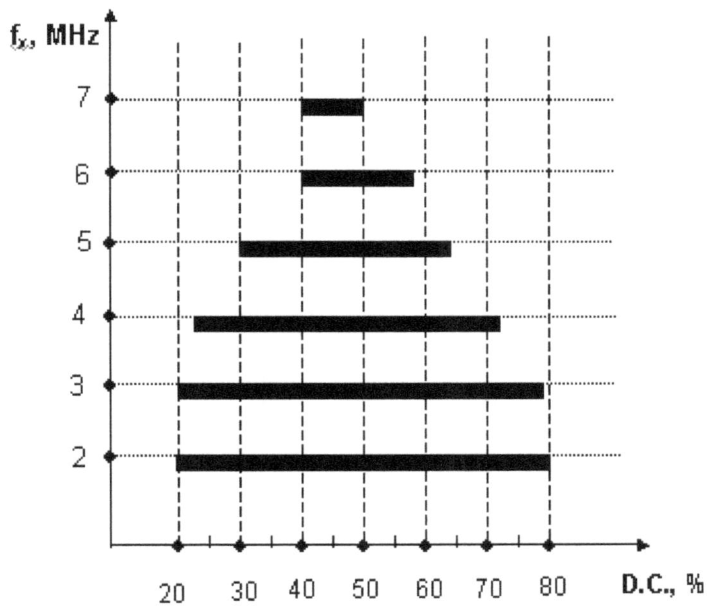

Fig. 2.27 Dependence between $f_{x,max}$ and duty-cycle (D.C.).

Experimental results have validated the high metrological performance of the UFDC-1, including the maximal possible measuring frequency and duty-cycle requirement for a square waveform pulse signal. The relative error for square waveform pulse signal measurement at the maximal possible frequency was below ± 0.0007 % (worst case condition). This was less than the programmable relative error (± 0.001 %).

2.3 Universal Frequency-to-Digital Converter (UFDC-1M-16)

The Universal Frequency-to-Digital Converter (UFDC-1M-16) is a high-speed version of the UFDC-1 [37] (Figure 2.28), which has been introduced on the market in 2007. The UFDC-1M-16 has the same functionality, pin out, housing, set of commands, measuring and communication modes as the UFDC-1 IC. The differences between the UFDC-1M-16 and UFDC-1 are shown in Table. 2.13.

Fig. 2.28 High-speed Universal Frequency-to-Digital Converter (UFDC-1M-16) in 28-lead PDIP package.

Table 2.13 Differences between UFDC-1M-16 and UFDC-1 ICs.

Performance	IC	
	UFDC-1	**UFDC-1M-16**
Internal reference frequency, MHz	0.5	16
Frequency range	0.05 Hz ... 7.5 (120) MHz	1 Hz ... 7.5 (120) MHz
Conversion time, s	0.0002 ... 0.2	0.00000625 ... 0.00625

Let determine a measurement time for the UFDC-1M-16 IC. The measurement time T_{meas} for this converter includes three main components: conversion rate (t_{conv}), communication (t_{comm}) time and calculations (t_{calc}) time [38]:

$$T_{meas} = t_{conv} + t_{comm} + t_{calc} \qquad (2.5)$$

At the same chosen relative error, in order to reduce the conversion rate t_{conv} the reference frequency f_0 should be increased. So, in comparison with the UFDC-1 [24], which has the internal reference frequency f_0=500 kHz, the UFDC-1M-16 has internal reference frequency f_0=16 MHz.

According to the modified method of the dependent count the conversion rate is $t_{conv}=1/f_x$ only for the frequency range up to 160 Hz at maximum possible relative error $\delta = \pm 1$ % and for the frequency range up to 160 kHz at the minimum possible relative error $\delta = \pm 0.001$ %.

The communication time for a slave communication mode (RS232 interface) can be calculated according to the following equation:

$$t_{comm} = 10 \cdot n \cdot t_{bit}, \qquad (2.6)$$

where t_{bit} = 1/300, 1/600, 1/1200, 1/2400, 1/4800, 1/9600, 1/14400, 1/19200, 1/28800 or 1/38400 is the time for one bit transmitting; n is the number of bytes (n=13÷24 for ASCII format).

As usually, at the right chosen of baud rate (maximum possible for a certain application) the $t_{comm} \leq t_{conv.}$ For example, the communication time at 38400 baud rate will be $t_{comm} = (0.0034 \div 0.00625)$ s.

The communication time for SPI interface should be calculated as:

$$t_{comm} = 8 \cdot n \cdot \frac{1}{f_{SCLK}}, \tag{2.7}$$

where f_{SCLK} is the serial clock frequency, which should be chosen for the UFDC-1M-16 in the range from 100 to 500 kHz; n=12÷13 is the number of bytes. The number n is dependent on measurement result format: BCD (n=13) or binary (n=12).

The communication standard mode speed for the I^2C interface can be determined according to the following equation:

$$t_{comm} = 8 \cdot n \cdot \frac{1}{f_{SCL}}, \tag{2.8}$$

where f_{SCL} is the serial clock frequency, which should be equals to 100 kHz for the UFDC-1M-16; n=12÷13 is the number of bytes for measurement result: BCD (n=13) or binary (n=12).

The calculation time depends on operands and is as usually $t_{calc} \leq 4.5$ ms.

The dependence of conversion rates t_{conv} from the relative error δ for the UFDC-1 and IC UFDC-1M-16 are shown in Figures 2.29 (a) and (b) respectively.

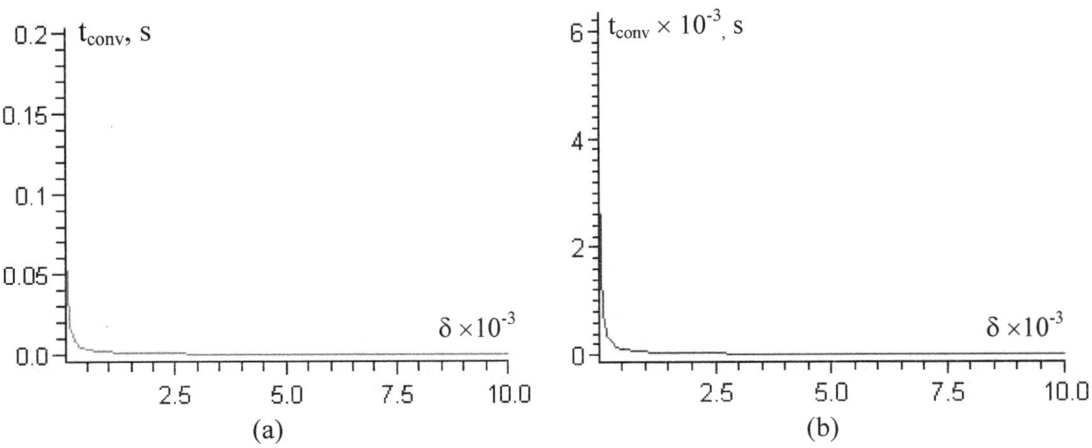

Fig. 2.29 Conversion rate vs. relative error for UFDC-1 (a), and UFDC-1M-16 (b) at $N_\delta/f_0 \geq T_x$, where $N_\delta = 1/\delta$; δ is the relative error; f_0 is the internal reference frequency; T_x is the period of measured frequency f_x

The UFDC-1M-16 has programmable relative error from 1 to ± 0.001 % at appropriate conversion rate from 6.25 μs to 6.25 ms. Due to the programmable relative error, the conversion time and accuracy can be optimized for specific applications. Relative errors and appropriate conversion times are shown in Table 2.14.

The described above methodology for the measurement time T_{meas} calculation can be used also for the UFDC-1 IC. The difference will be only in the conversion rate (t_{conv}) due to decreased internal reference frequency f_0.

Let's compare the conversion time of UFDC-1M-16 based on the modified method of the dependent count with conversion times of other converters based on any classical or other advanced frequency-to-digital conversion methods. For the indirect counting method, with the reference frequency f_0=16 MHz, relative error ± 0.001 % and conversion time that does not exceed one period T_x the frequency range will be essentially limited by ≤ 160 Hz. In case of the direct counting technique, for the same relative error and the gate time 6.25 ms it will be possible to convert only frequencies from the range $f_x \in$ [16 kHz ÷ 16 MHz] and for all relative error except the δ_x = ± 0.001 % the conversion time will be redundant. For advanced conversion methods, for example, reciprocal, ratiometric, M/T, constant elapse time, single- and double buffered, DMA method [26], at the reference frequency f_0=16 MHz and gate time T_0= 6.25 ms the conversion time also will be redundant for all relative error from the Table 2.14 except the δ_x = ± 0.001 %. In other words, the conversion time will be the same - 6.25 ms for the relative error ± 0.001 % and relative error ± 1 %.

Table 2.14 Relative errors and appropriate conversion times.

Relative error, δ_x %	$N_\delta=1/\delta_x$	**UFDC-1** (at f_0=500 kHz)	**UFDC-1M-16** (at f_0=16 MHz)
		T_{conv}, s	
1	100	0.0002	0.00000625
0.5	200	0.0004	0.0000125
0.25	400	0.0008	0.000025
0.1	1000	0.002	0.0000625
0.05	2000	0.004	0.00125
0.025	4000	0.008	0.0025
0.01	10000	0.02	0.00625
0.005	20000	0.04	0.00125
0.0025	40000	0.08	0.0025
0.001	100000	0.2	0.00625

Experimental and modeling results confirm high metrological performances of high-speed universal converter UFDC-1M-16. Its conversion time is non-redundant, adaptive, minimum possible and in 32 times less than in the IC UFDC-1.

Due to the modified methods of the dependent count for a frequency-to-digital conversion the conversion rate (6.25 μs to 6.25 ms) is programmable, non-redundant, shorter than for the pulse counting technique and comparable with the conversion time for successive-approximation and Σ-Δ ADC.

The ICs UFDC-1 and UFDC-1M-16 is available for purchase at Sensors Web Portal and Technology Assistance BCNA 2010, S. L. web sites (online stores):
http://www.sensorsportal.com/HTML/E-SHOP/PRODUCTS_4/UFDC_1.htm
http://www.techassist2010.com/en/orders-buy-products/place-orders.html

2.4 Evaluation Board EVAL UFDC-1/1M-16

The EVAL-UFDC-1/UFDC-1M-16 is a simple full featured evaluation and development board that allows users to quickly evaluate the performance of the Universal Frequency-to-Digital Converters UFDC-1 and UFDC-1M-16 without the need for external components and design DAQ systems on its basis [39]. This 2-channel Evaluation Board lets use such ICs in 16 frequency-time parameters measuring modes, one generation (for a calibration purpose) mode and build various digital sensors and sensor systems including smart, self-adaptation and self-identification sensor systems. All existing frequency, period, duty-cycle, time interval, pulse-width modulated, pulse number and phase-shift output sensors and transducers can be directly interfaced to this evaluation board. The user can connect TTL-compatible sensors' outputs to the Evaluation Board, measure any output frequency-time parameters, and test out the sensor systems functions.

Using the accompanying software, the EVAL-UFDC-1/UFDC-1M-16 evaluation board can be interfaced to any personal computer running Windows XP/Vista/Windows 7 or later, via one of the serial computer's com port RS232 or USB port (with an additional USB to Serial (9-pin) DB9 RS232 adapter cable, for example, EMINENT USB to Serial Converter).

Four holes are provided for mechanical attachment of the Evaluation Board to users' applications.

The component placement drawing, a photograph of the EVAL-UFDC-1/UFDC-1M-16 evaluation board and circuit diagram are shown in Figures 2.30 (a, b) and 2.31 respectively.

The RS232 interfacing chip MAX232CSE offers a serial communication between the UFDC-1/UFDC-1M-16 and, for instance, a personal computer or external (master) microcontroller. A 16 MHz crystal completes the on-chip oscillator of the UFDC. The board is working from 8 to 14 V AC/DC external power supply. The LM7805CT voltage regulator provides the +5 V, DC supply voltages for the integrated circuit.

The evaluation board can be used for all functions of the UFDC integrated circuits. The setting of one from 16 measuring mode and accuracy is performed by the PC or microcontroller in a slave communication mode or by external switches SW2 and SW3 respectively in the master communication mode. The I/O configuration selector (SW1) lets to choose +5 V for powering external sensors (1 & 3 switches) and 10 K pull-up resistors for open collector sensor outputs (2 & 4 switches). The pin 2 in each of channel input connector should be used for more "tolerant" but "slower" input signals; the pin 3 is for faster signals.

The software allows the EVAL-UFDC-1/UFDC-1M-16 Evaluation and Development board's functions to be controlled from a PC via an easy-to-use interface operating under

Windows XP/Vista/Windows 7 or higher. There are many ways to display evaluated results, such as the terminal programs for Windows, LabView and special software supplied with the Evaluation Board.

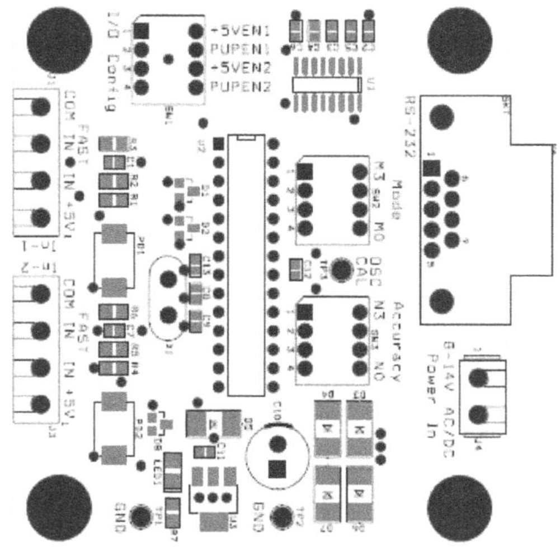

(a)

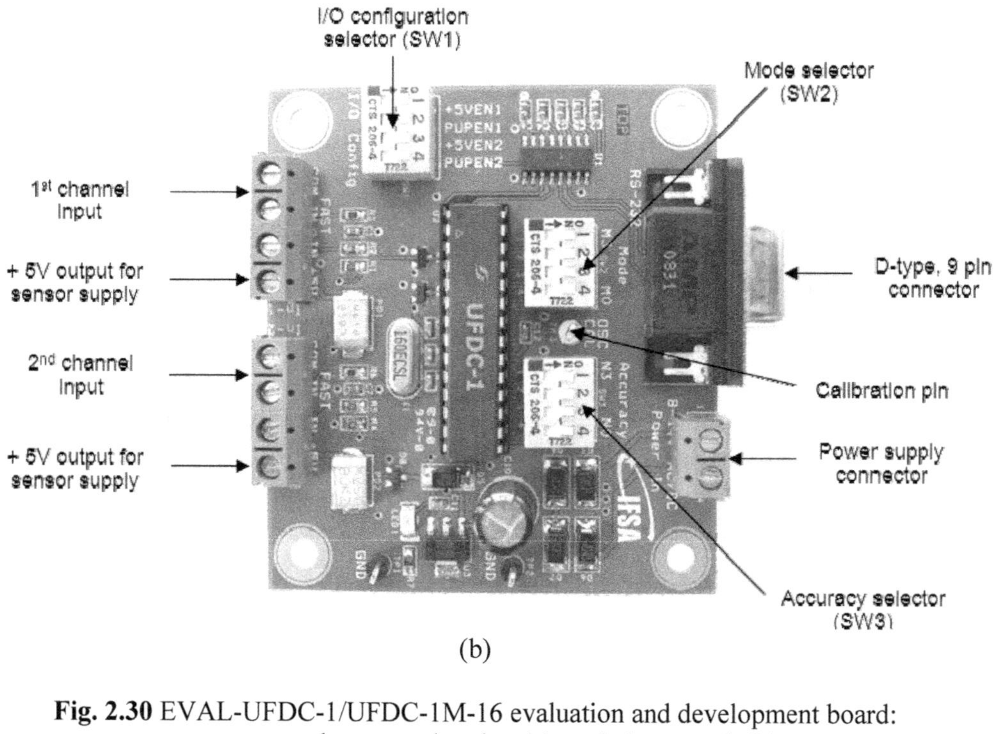

(b)

Fig. 2.30 EVAL-UFDC-1/UFDC-1M-16 evaluation and development board: components placement drawing (a), and photograph (b).

Any terminal program for Windows, for example, Terminal V1.9b, can easily display the measurement results and accept the programming commands for measuring modes, accuracy selections, etc. via the serial port of the PC.

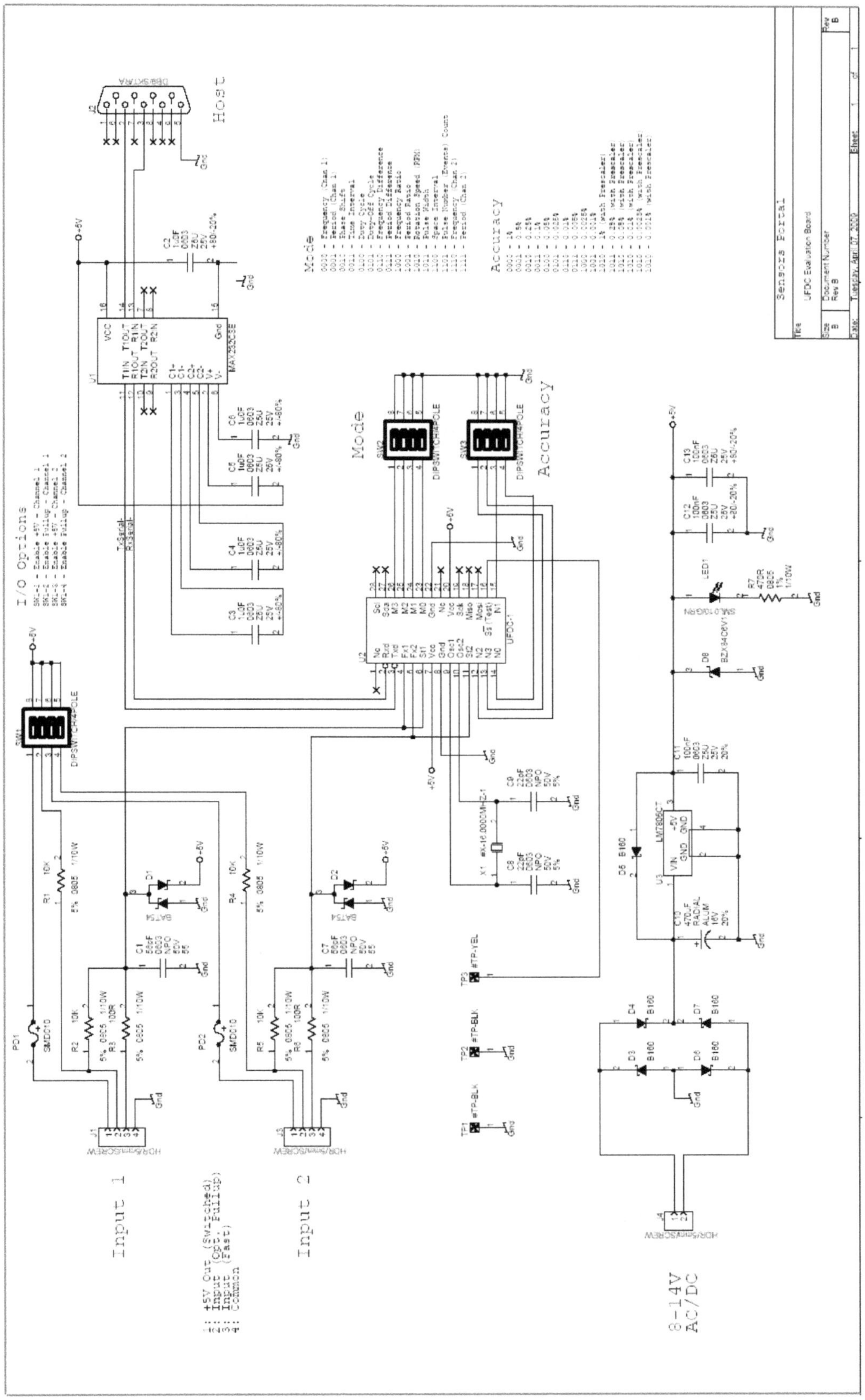

Fig. 2.31 Circuit diagram of evaluation and development board EVAL-UFDC-1/UFDC-1M-16.

The initial configuration of Terminal V1.9b must be the following: Baud rate 2400; 8 data bit, 1 stop bit and No parity error check and handshaking. Once the communication between the UFDC and the PC is actuated you will see the invitation sing “>” in the Receive window of Terminal V1.9b after press the <Enter> key. The baudrate can be changed by the appropriate UFDC-1 command and by the option in the Terminal V1.9b software window at the same time.

The Terminal V1.9b software does not need any installation. It is supplied on the EVAL-UDFC-1/UFDC-1M-16 CD.

The original UFDC EVAL Software V1.14 also does not need any installation. Please copy the UFDC_Eval.exe file from the CD into an appropriate folder in your PC’s hard disc and run it. The main window of the software is shown in Figure 2.32. At first, click on the Communication Menu and chose the appropriate COM Port from the pull down menu. The communication dialog window will appear (Figure 2.33).

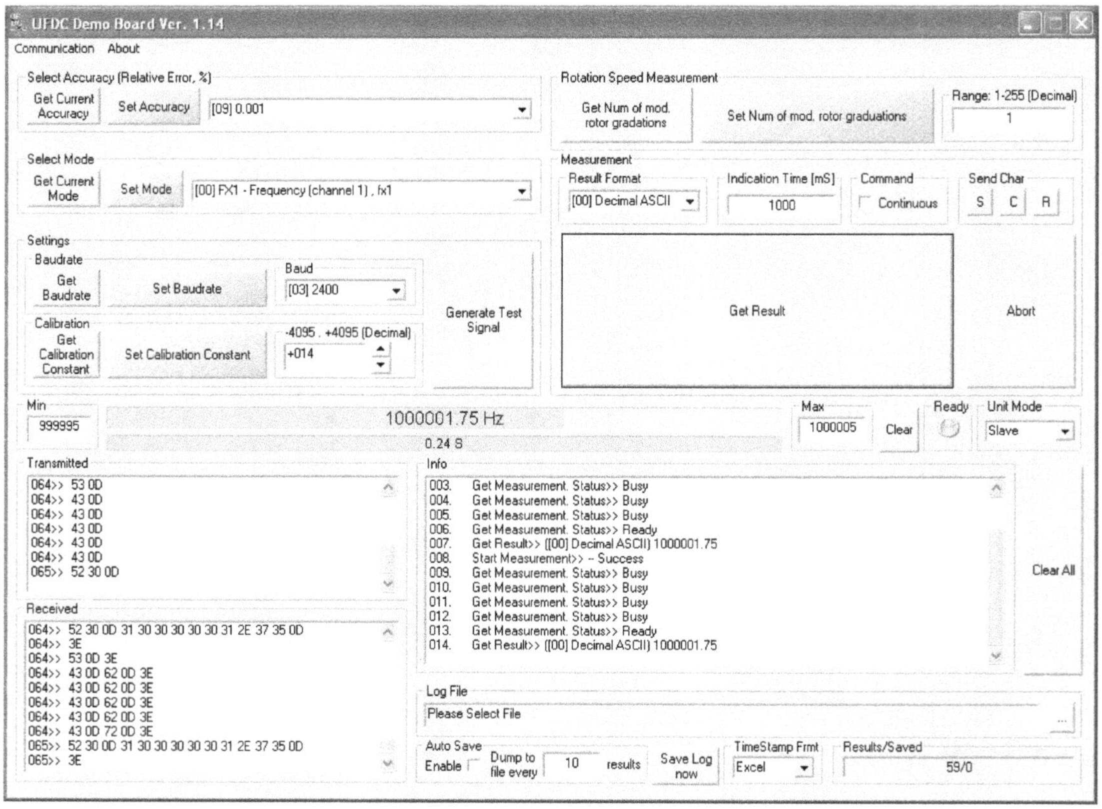

Fig. 2.32 UFDC EVAL software main window.

Select an appropriate COM port number available on your PC and confirm your choice by OK button. Then switch power on for the evaluation board. The evaluation board is ready to operate in the Slave communication mode at 2400 baudrate. You can now select an appropriate measuring mode and relative error by choosing the appropriate values from the pull down menus and confirm it by pressing proper ‘Set’ button. A ‘Get’ button returns current value.

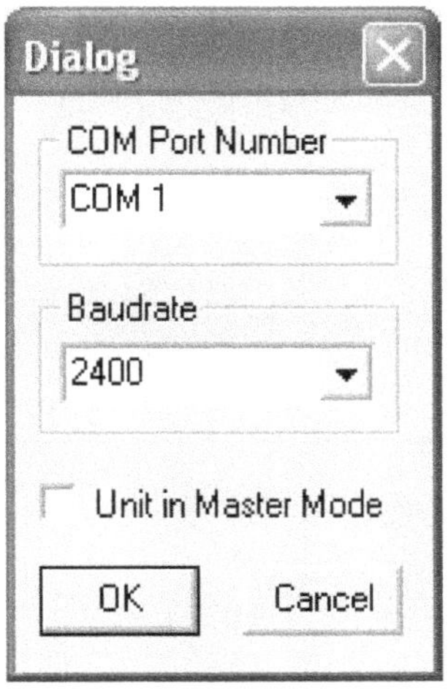

Fig. 2.33 Communication dialog window.

To change the current 2400 baudrate please make the following:

1) Select new baudrate from the pull down menu in the main window;
2) Click on "Set Baudrate". In the Info Box you should get the following message: "Set Baudrate -- Success";
3) Click on Communication Menu, and select the same baudrate to let application know you are now communicating with different COM port settings.

The software also lets to calibrate the converters and put a calibration constant into the converter's memory (see Calibration Procedure, described above, p.48).

The Min and Max values in appropriate fields of the main window are relevant to how the result progress bar behind the displayed result would "zoom in" to those limits, and give you a better visual indication of the measuring signal. For example, if you are monitoring a signal that is always between 100 and 105 Hz, setting the Min to 100, and Max to 105 will display a progress bar between these 2 points.

When monitoring either the UFDC-1 or the UFDC-1M-16 converters, you can log the measuring results to a log file. It is possible to choose the *.txt or *.csv log file extension in the Log File dialog window (Figure 2.34).

The log file contains the following fields: Time Stamp in the Unix or Excel format (It can be selected in the TimeStampFrmt pull down menu in the main window); RxDelay (s); Accuracy; Mode; Baudrate; Calibration Constant; Result Format (can be selected in the appropriate pull down menu in the main window); Indication Time (ms); Modulation Rotor Gradation Number (for rotation speed measurement mode only); Result of measurement and Units (Hz, μs, rpm, %, etc.). The RxDelay represents the time between start the measurement command 'S' and received result of measurement after the read command 'R'. The 'About' menu item shows the current version of the software.

The Evaluation Board can be working also in the Master communication mode. In this case the appropriate measuring mode and accuracy should be chosen with the help of external switches SW2 and SW3 respectively on the evaluation board. The Master communication mode option 'Unit in Master Mode' can be selected in the communication dialog window and from the Unit Mode pull down menu in the main window. At communication modes changing the power supply for Evaluation Board must be switch off.

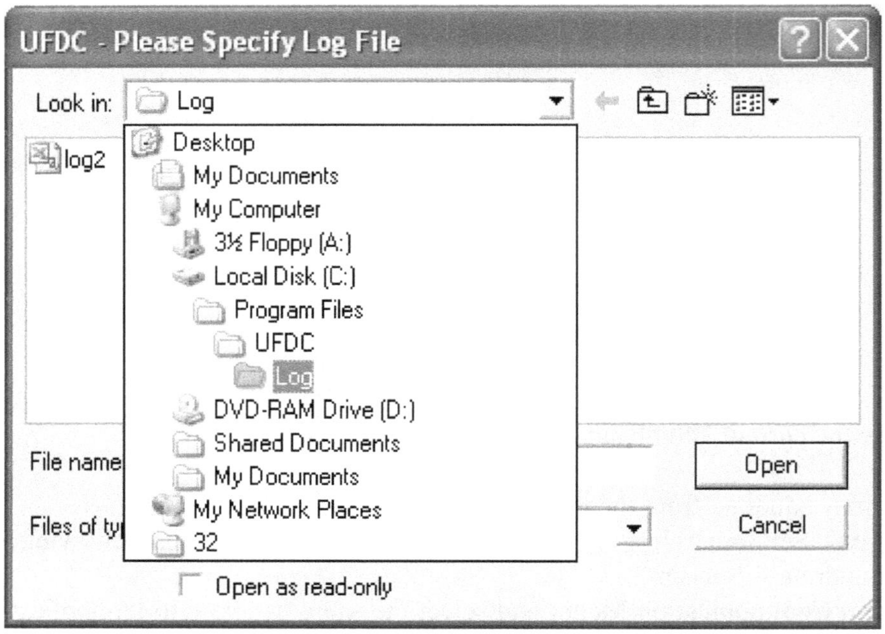

Fig. 2.34 Log File dialog window.

In cases when measuring signal is disconnected from the evaluation board, a shaft is stopped in rotation speed measurement mode; a wrong port number selected or input signal is out of range, for example, less than 0.05 Hz, the appropriate message after 40 s delay will be indicated under the result indicator.

The Evaluation Board can work in single and continuous measuring modes. At the first mode the 'Get Result' button must be pressed every time to obtain measuring result. In the continuous measuring mode the 'Continuous' option in the Command field in the main window must be checked. For continuous measurement, the indication time can be chosen from 1 ms to 10000 ms.

The EVAL-UFDC-1/UFDC-1M-16 is designed for operation in the industrial temperature range from -40 ^{0}C to +85 ^{0}C.

The possible applications of the board (as it is) are numerous:

- Digital sensor systems
- Smart and intelligent sensor systems
- Data Acquisition systems for frequency-time parameters of electric signals
- Frequency counters
- Tachometers and tachometric systems
- Various virtual instruments
- Educational process in sensors and measurements
- Remote laboratories and distance education

The evaluation and development board is available for purchase from the Sensors Web Portal, Inc. and Technology Assistance BCNA 2010, S. L. web sites (online stores):
http://www.sensorsportal.com/HTML/E-SHOP/PRODUCTS_4/UFDC_1.htm
http://www.techassist2010.com/en/orders-buy-products/place-orders.html

2.5 References

[1]. Yurish S. Y., Novel Universal Frequency-to-Digital Converters and Sensors Interface Integrated Circuits, *Sensor Electronics and Microsystem Technologies*, 3, 2008, pp.81-90.

[2]. Gruman L. N., Z. Ya. Leitan, V. A. Murzin and Yu. V. Frolov, Specialized Measuring Microprocessor USP-30, *Measuring Instruments and Control Systems*, 7, 1987, pp. 18-20 (in Russian).

[3]. Balashov S. M., L. A. Nesterova and Yu. P. Rodionov, ALU of Digital Measuring Instruments for Time Intervals, *Electronic Industry*, 3, 1989, pp.10-12 (in Russian).

[4]. Belous A. I., V. K. Kovalevskiy, V. A. Kosobryukhov, A. A. Parhomchuk and B. N. Chernukha, Universal Time Converter based on K512PS11 IC, *Microprocessor Devices and Systems*, 2, 1990, pp. 15-17 (in Russian).

[5]. Wilson P. D., R. S. Spraggs and S. P. Hopkins, Universal Sensor Interface Chip (USIC): Specification and Application Outline, *Sensor Review*, Vol. 16. No. 1, 1996, pp. 18-21.

[6]. Kalisz J., R. Szplet, R. Pelka and A. Poniecki, Single-chip Interpolating Time Counter with 200-ps Resolution and 43-s Range, *IEEE Transaction on Instrumentation and Measurement*. Vol. 46, No. 4, 1997, pp. 851-856.

[7]. Ramalingam N., V. K. Varadan and V. V. Varadan, Innovative Frequency Measurement Technique used in the Design of a Single Channel Frequency to Digital Converter ASIC, *Smart Materials and Structures*, No.8, 1999, pp. 243-251.

[8]. AutoTEC web site: http://www.autotecsystems.com

[9]. Acam mess electronic Product Overview, *Acam mess electronic GmbH,* November 2010.

[10].TDC-GP1 General Purpose TDC, Functional Description, Datasheet, *Acam mess electronic GmbH*, February 2001.

[11].TDC-GP2 2-channel Universal Time-to-Digital Converter, Target Specification. DB_GP2_e V2.0, *Acam mess electronic GmbH*, May 2010.

[12].TDC-GP21 Universal 2-Channel Time-to-Digital Converter, Preliminary Datasheet, DB_GP21_e V1.2, *Acam mess electronic GmbH*, March 2011.

[13].TDC-GPX Ultra-high Performance 8 Channel Time-to-Digital Converter, Datasheet, *Acam messelectronic GmbH.* January 2007.

[14].Jozef Kalisz, Review of Methods for Time Interval Measurements with Picoseconds Resolution, *Mertologia*, 41, 2004, pp. 17-32.

[15].R. Szplet. J. Kalisz and Z. Jachna, A 45 ps Time Digitizer with a Two-phase Clock and Dual-edge Two-stage Interpolation in a Filed Programmable Gate Array Device, *Measurement Science and Technology*, 20, 2009. pp. 1-11.

[16].Stephan Henzler, Time-to-Digital Converters, *Springer*, 2010.

[17].Krzysztof Chmielewski, Multi-vernier Time-to-digital Converter Implemented in a Field-programmable Gate Array, *Measurement Science and Technology*, Vol. 22, Issue 1, 2011.

[18].Reverter F., Jordana J., Pallàs-Areny R., Program-Dependent Uncertainty in Period-to-Code Converters Based on Counters Embedded in Microcontrollers, in

Proceedings of the 20th IEEE Instrumentation and Measurement Technology Conference (IMTC' 03), Vail, Colorado, USA, 20-22 May 2003, 2, pp. 977-980.

[19].F. Reverter, O. Casas, J. Jordana, R. Pallàs-Areny, Trigger Uncertainty in Period-to-Code Converters Based on Counters Embedded in Microcontrollers, *Sensors and Actuators A*., Vol. 110, 2004, pp.439-446.

[20].F. Reverter, M. Gasulla, R. Pallàs-Areny, Analysis of Interference Rejection in Period-to-Code Converters, in *Proceedings of the 21st IEEE Instrumentation and Measurement Technology Conference (IMTC' 2004)*, Como, Italy, 18-20 May 2004, pp.839-842.

[21].F. Reverter, R. Pallàs-Areny, Uncertainty Reduction Techniques in Microcontroller-based Time Measurement, *Sensors and Actuators A*., Vol. 127, 2006, pp.74-79.

[22].Sergey Y. Yurish, Nikolay V. Kirianaki and Ramon Pallas-Areny, Universal Frequency-to-Digital Converter for Quasi-Digital and Smart Sensors: Specifications and Applications, *Sensor Review*, Vol. 25, No.2, April 2005, pp.92-99.

[23].Universal Frequency-to-Digital Converter (UFDC-1), Datasheets, *Technology Assistance BCNA 2010, S. L.,* 2011.

[24].Universal Frequency-to-Digital Converter (UFDC-1), Specification and Application Note, *Technology Assistance BCNA 2010, S.L.,* 2011.

[25].Krianaki N. V., Yurish S. Y., Shpak N. O., Methods of Dependent Count for Frequency Measurements, *Measurement*, Vol. 29, Issue 1, January 2001, pp.31-50.

[26].Kirianaki N. V., Yurish S. Y., Shpak N. O., Deynega V. P., Data Acquisition and Signal Processing for Smart Sensors, *John Wiley & Sons*, Chichester, UK, 2002.

[27].Yurish S. Y., Data Acquisition for Frequency- and Time Domain Sensors, in Smart Sensor Systems (Ed. by Gerard C. M. Meijer), *John Wiley & Sons*, Chichester, UK 2008, pp.313-341.

[28].Novitskiy P. V., Zograf I. A., Errors Estimation of Measurement Results, *Energoatomizdat*, 1991 (in Russian).

[29].Sergey Y. Yurish, Nikolay V. Kirianaki, Automated Calibration Technique for Programmable Universal Frequency-to-Digital Converter, in *Proceedings of the IFAC Workshop on Programmable Devices and Embedded Systems (PDeS'2006)*, Czech Republic, Brno, 14-16 February 2006, pp.72-77.

[30].Al'tshuller G. B., Frequency Control in Quartz Generators, *Svyaz'*, Moscow, 1975 (in Russian).

[31].Plons'kiy O. P., Quartz Generators on Transistors, *Tekhnika*, Kyjv, 1966 (in Ukrainian).

[32].Al'tshuller G. B., Elfimov N. N., Shakulin V. G., Economy, Miniaturized Quartz Generators, *Svyaz'*, Moscow, 1979 (in Russian).

[33].Terminal V1.9b: http://www.sensorsportal.com/DOWNLOADS/Terminal.exe

[34].Worldwide Component Library: http://www.2wcl.com/

[35].Design & Reuse: http://www.design-reuse.com/

[36].Sergey Y. Yurish, Nikolay V. Kirianaki and Ramon Pallas-Areny, Low-Cost Precision Universal Programmable Frequency-to-Digital Converter, in *Proceedings of the IFAC Workshop on Programmable Devices and Systems (PDS'2004),* Poland, Cracow, 18-19 November 2004, pp.70-75.

[37].High-Speed Universal Frequency-to-Digital Converter UFDC-1M-16, Datasheets, *Technology Assistance BCNA 2010, S.L.,* 2011.

[38].Yurish S. Y., High-Speed Universal Frequency-to-Digital Converter for Quasi-Digital Sensors and Transducers, *Sensors & Transducers*, Vol.80, Issue 6, June 2007, pp.1225-1229.

[39].Evaluation Board for Universal Frequency-to-Digital Converters UFDC-1 and UFDC-1M-16, EVAL-UFDC-1/UFDC-1M-16, *IFSA*, 2011.

Chapter 3

Universal Sensors and Transducers Interface (USTI)

3.1 General Description

The Universal Sensors and Transducers Interface (USTI) is the next generation of integrated frequency-to-digital converters and sensor signal conditioners [1-3]. It offers metrological improvements and an extended functionality. The USTI is a fully digital CMOS integrated circuit of a universal, 2-channel, high precision multifunctional converter based on novel frequency-time conversion methods. It is perfectly suited to any applications where frequency-time parameters but also sensors output signal, have to be measured with highest resolution and programmable accuracy during the non-redundant conversion time and self-adaptive possibilities. The USTI covers all specified converted range of frequencies and accuracy for modern quasi-digital (frequency-time domain) sensors. In addition, the USTI provides interfacing for resistive (including platinum and cuprum resistors and thermistors), capacitive sensors and resistive bridges [4, 5]. The photos of IC in different possible packages are shown in Figure 3.1.

(a) (b) (c)

Fig. 3.1 IC USTI in 32-lead TQFP (a); and 32-pad MLF packages (b); and 28-lead PDIP packages.

By using this IC it is possible to build a wide variety of digital, smart sensors and intelligent sensor systems. It can be used in various SoC, SiP, BioMEMS and Lab-on-Chip. The main differences between USTI and UFDC series of ICs are shown in Table 3.1.

Table 3.1 Comparative performances of USTI and UFDC series of ICs.

Parameters \ IC	UFDC-1	UFDC-1M-16	USTI
Programmable relative error, %	1 ... 0.001	1 ... 0.001	1 ... 0.0005
Minimal frequency, Hz	0.05	1	0.05
Maximal frequency, MHz	7.5 (120)*	7.5 (120)*	9 (144)*
Internal reference frequency, MHz	0.5	16	0.625
Conversion time, s	0.0002 … 0.2	0.00000625 … 0.00625	0.00016 … 0.32
Frequency generator mode, MHz	8	8	10
Number of channels	2 different	2 different	2 identical
Number of measuring modes	16	16	29
TEDS Support	-	-	●
Frequency	●	●	●
Period	●	●	●
Phase shift	●	●	●
Time interval	●	●	●
Duty-cycle	●	●	●
Duty-off factor	●	●	●
Frequency/ period difference	●	●	●
Frequency/ period ratio	●	●	●
Rotation speed	●	●	●
Pulse width	●	●	●
Space interval	●	●	●
Pulse number (events) counting	●	●	●
Frequency deviation relative/ absolute	-	-	●
Resistive	-	-	●
Capacitance	-	-	●
Resistive bridge	-	-	●

* - with prescalling

The IC is consists on the three main blocks: time-to-digital converter (TDC), measuring communication units. The TDC is used in parameter-to-digital conversion for a direct interfacing of capacitive, resistive and bridge sensing elements to the USTI. The block diagram is shown in Figure 3.2. Only one external component – 20 MHz quartz crystal resonator is necessary.

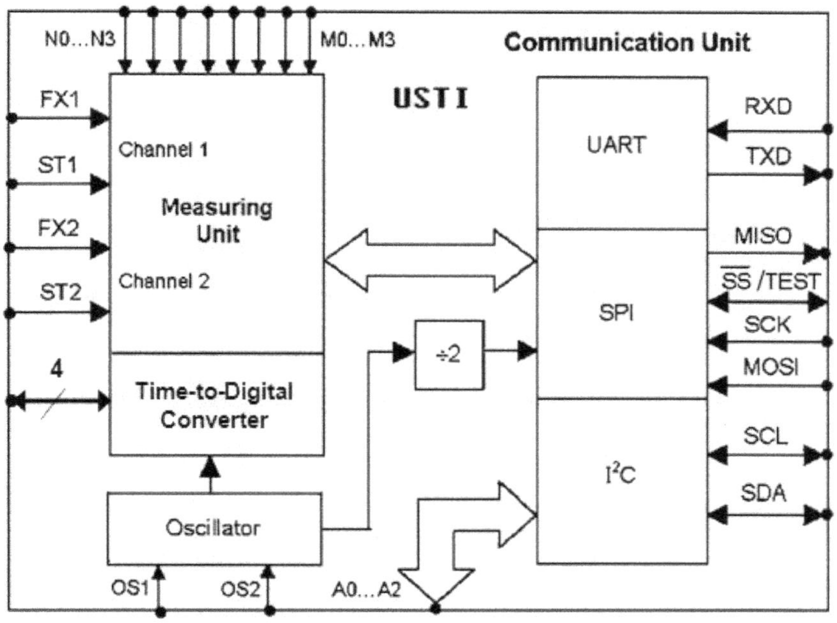

Fig. 3.2 USTI block diagram.

Similarly to UFDC-1/UFDC-1M-16 the USTI can be easy included into digital environment controlled by an external microcontroller (slave mode) or PC, or work independently as a separate measuring unit without an external control (master mode). The function selection can be configured in both software and hardware.

3.2 Features, Metrological Performance and Main Electrical Characteristics

The USTI has the following key features:

- 29 measuring modes and one generator mode;
- 2 identical channels for each of frequency-time parameters;
- Provides frequency (time)-to-digital conversion for many types of sensors and transducers
- Provides interfacing for resistive (including platinum (Pt), cuprum (Cu) resistors and thermistors (Ni)), capacitive sensing elements and passive resistive bridges;
- Frequency range from 0.05 Hz up to 9 MHz without prescaling and 144 MHz with prescaling;
- Programmable accuracy (relative error) for frequency (period) conversion from 1 % to 0.0005 %;

- Relative quantization error is constant in all specified frequency range;
- Non-redundant conversion time;
- Scalable resolution;
- Internal reference clock 625 kHz, 20 MHz @ 20 MHz quartz oscillator;
- Quartz-accurate automated calibration;
- 3-wire serial interface (SPI compatible);
- 2-wire interface (I^2C compatible);
- RS232/422 serial interface;
- Master and slave communication modes;
- Frequency generator mode 10 MHz with quartz crystal stability;
- Single power supply +4.5 … +5.5 V;
- Active supply current at 20 MHz, +4.5 V, 25 °C below 9.5 mA;
- Operating temperature range (- 40 ^{0}C to + 85 ^{0}C);
- Package Types: 28-lead Plastic Dual Inline Package (PDIP), compact 32-lead small Thin Profile Plastic Quad Flat Package (TQFP) and 32-pad Micro Lead Frame Package (MLF).

The USTI can measure the following frequency-time parameters of electrical signals and sensing elements:

- Frequency, f_{X1}, 0.05 Hz … 9 MHz directly and up to 144 MHz with prescaling;
- Frequency, f_{X2}, 0.05 Hz … 9 MHz directly and up to 144 MHz with prescaling;
- Period, T_{X1}, 110 ns … 20 s;
- Period, T_{X2}, 110 ns … 20 s;
- Phase shift, φ_X, 0 - 360° at f_{xmax} = 625 kHz at 50 % duty-cycle;
- Time interval between Start – and Stop-pulse, τ_X, 1.5 μs … 250 s;
- Duty-cycle, DC_1, 0 … 1 at $f_{xmax} \leq 625$ kHz;
- Duty-cycle, DC_2, 0 … 1 at $f_{xmax} \leq 625$ kHz;
- Duty-off factor, Q_1, 10^{-8} … $8 \cdot 10^6$ at $f_{xmax} \leq 625$ kHz;
- Duty-off factor, Q2, 10^{-8} … $8 \cdot 10^6$ at $f_{xmax} \leq 625$ kHz;
- Frequency difference, $f_{X2} - f_{X2}$, with sign recognition, 0 … 9 MHz;
- Period difference, $T_{X1} - T_{X2}$, with sign recognition, 0 … 20 s;
- Frequency ratio, f_{X1}/f_{X2}, $5.5 \cdot 10^{-9}$ … $1.8 \cdot 10^8$
- Period ratio, T_{X1}/T_{X2}, $4.0 \cdot 10^{-9}$ … $2 \cdot 10^8$
- Rotational speed, $n_{X1} = (f_X \cdot 60)/Z$, where Z is the number of encoder teeth;
- Rotational speed, $n_{X2} = (f_X \cdot 60)/Z$, where Z is the number of encoder teeth;
- Pulse width, t_{P1}, 1.5 μs … 250 s;
- Pulse width, t_{P2}, 1.5 μs … 250 s;
- Space interval, t_{S1}, 1.5 μs … 250 s;
- Space interval, t_{S2}, 1.5 μs … 250 s;
- Pulse number (events) counting, N_{X1}, $0 - 4 \cdot 10^9$
- Pulse number (events) counting, N_{X2}, $0 - 4 \cdot 10^9$
- Frequency deviation absolute, D_{A1}, 0 … 9 MHz;
- Frequency deviation relative, D_{R1}, 0 … 100 %;
- Frequency deviation absolute, D_{A2},, 0 … 9 MHz;
- Frequency deviation relative D_{R2}, 0 … 100 %;
- Resistance, R_x, 10 Ω to 10 MΩ with average relative error ± 0.47 % and ± 0.01 % at splitting of the range into sub ranges;

- Capacitance, C_x 50 pF to 100 μF with average relative error ± 0.036 % and ± 0.7 % the worst case relative error;
- Resistive bridges, B_x, must not include any internal components other than 4 arms forming a bridge.

3.3 Functionality

3.3.1 Accuracy

For frequency-time output sensors and transducers, the USTI relative error should be selected small enough. If the sensor's error is, for example, ± 0.01 %, the conversion error should be ten times (or at least, five times), smaller, i.e. accuracy of USTI should be ± 0.001 % or at least, ± 0.002 %. In this case the conversion error can be neglected. The USTI has non-redundant conversion time. The higher conversion accuracy needs longer conversion time. However, it is non-redundant for any selected accuracy and frequency. The quantization error is constant and does not depend on the measures frequencies in all frequency range for 0, 1, 6 -10, 14 -15, 19, 22 and 26 modes (Table 3.2).

In modes 2, 4-5, 20-21 the accuracy does not depend on the period of pulse sequence. The programmable accuracy influences only the conversion error of period. Measurement results in these modes only depend on the pulse width. The conversion error for modes 3, 11-12, 23-24 strictly depends on the converted values and does not depend on value programmed using selection pins N0-N2 or accuracy selection commands (A00-A0F). This error can be calculated according to the following equation:

$$\delta_q = \frac{1}{20 \cdot 10^6 \cdot t_x} \times 100\%, \tag{3.1}$$

where t_x is the appropriate measuring value of t_p, t_s or τ_x depending on the mode. Programmed converted accuracy also does not influence on a conversion result in the mode 13.

The USTI can work as an independent measuring unit (master mode), controlled by any external microcontroller, or embedded into a sensor, DAQ board or PC (slave mode). In the master mode, the conversion accuracy and measurement mode should be selected by external switches. In this mode the USTI works only through the RS232 interface in a unidirectional mode. In the slave mode (with an external microcontroller), all measuring modes and programmed accuracy can be set up by a data bus or through communication I/O ports of master microcontroller. This allows us use the USTI as a simple peripheral circuit for a microcontroller or DSP. A bidirectional data exchange is possible with any of three interfaces RS232, SPI or I^2C. The USTI works as a standard element in different systems with I^2C bus architecture, or can be controlled through the RS232 or SPI interfaces. The choice of interface can be performed automatically depend on the circuit connections.

Table 3.2 Measurement mode number, accuracy number and charge time number to ternary jumpers' state conversion.

Dec	Hex	Measurement Mode	Accuracy	Time	J2	J1	J0
0	00	Frequency 1	1	100 μs	O	O	O
1	01	Period 1	0.5	200 μs	O	O	L
2	02	Phase Shift	0.25	300 μs	O	O	H
3	03	Time Interval	0.1	400 μs	O	L	O
4	04	Duty Cycle 1	0.05	500 μs	O	L	L
5	05	Duty Off 1	0.025	600 μs	O	L	H
6	06	Frequency Difference	0.01	700 μs	O	H	O
7	07	Period Difference	0.005	800 μs	O	H	L
8	08	Frequency Ratio	0.0025	900 μs	O	H	H
9	09	Period Ratio	0.001	1 ms	L	O	O
10	0A	Rotational Speed 1	0.0005	2 ms	L	O	L
11	0B	Pulse Interval 1	1 Presc.	3 ms	L	O	H
12	0C	Space Interval 1	0.5 Presc.	4 ms	L	L	O
13	0D	Pulse Counter 1	0.25 Presc.	5 ms	L	L	L
14	0E	Frequency 2	0.1 Presc.	6 ms	L	L	H
15	0F	Period 2	0.05 Presc.	7 ms	L	H	O
16	10	Resistance	0.025 Presc.	8 ms	L	H	L
17	11	Capacitance	0.01 Presc.	9 ms	L	H	H
18	12	Bridge	0.005 Presc.	10 ms	H	O	O
19	13	Frequency Deviation 1*	0.0025 Presc.	20 ms	H	O	L
20	14	Duty Cycle 2	0.001 Presc.	30 ms	H	O	H
21	15	Duty Off 2	0.0005 Presc.	40 ms	H	L	O
22	16	Rotation Speed 2	N/A	50 ms	H	L	L
23	17	Pulse Interval 2	N/A	60 ms	H	L	H
24	18	Space Interval 2	N/A	70 ms	H	H	O
25	19	Pulse Counter 2	N/A	80 ms	H	H	L
26	1A	Frequency Deviation 2*	N/A	90 ms	H	H	H

* - measurement mode not available in master RS232 interface configuration.

3.3.2 Communication Interfaces

The selection of communication interface is performed automatically based on the state of IS0 and IS1 inputs at the power switch on (Table 3.3).

Table 3.3 Communication interface selection.

ISO	IS1	Communication Interface
0	0	SPI slave
0	1	I2C slave
1	0	RS232 master
1	1	RS232 slave

3.3.2.1 RS232 Interface

The converter can work in an autonomous operating mode (master mode, unidirectional data transfer) as well as in systems (slave operating mode, bidirectional data transfer). If during five seconds after being powered-on the USTI has not received any command through the RS232, the converter reads the external switches and is placed into the unidirectional mode of data transfer through the RS232 in BCD ASCII format. By default, the transmission speed is 2400, mode - 0 (frequency measurement in the first channel), accuracy – A00 (1 %). The data transfer rate can be chosen from the following range: 300, 600, 1200, 2400, 4800, 9600, 14400, 19200, 28800 and 38400 (for slave mode).

The brief communication command set description for RS232 interface, appropriate modes and programmed accuracies are adduced in Table 3.4. Circuit diagrams for master mode and slave mode are shown in Figure 3.3 and 3.4 accordingly. Pin numbers in all circuit diagrams correspond to the converter in 28-lead Plastic Dual Inline Package (PDIP).

The connector XP1 is intended for in-circuit automated calibration. In the working mode jumper J1 must be in ON position and in the calibration mode it must be in OFF position.

For connection the USTI to a PC through the RS232 interface the additional circuit (Figure 3.5) for interfacing function is recommended.

3.3.2.2 Three-wire Serial Interface (SPI Compatible)

The brief description of communication command set for SPI interface, appropriate modes and programmed converted accuracies are adduced in Table 3.5. The circuit diagram for the 3-wire Serial Interface communication mode is shown in Figure 3.6. The connector XP1 is intended for in-circuit automated calibration. Jumpers J1 and J2 must be in the ON position in the working mode and in the OFF position in the calibration mode.

3.3.2.3 Two-wire Interface (I^2C Compatible)

The brief communication command set description for I^2C interface, appropriate measuring modes and programmed converted accuracies are adduced in Table 3.6. The USTI can have one of 8 possible addresses that can be setup by A0-A2 pins. The absolute address of a device has the following format 1110<A2><A1><A0>.

Table 3.4 Communication command set for RS232 interface.

Command	Description	Parameter
A	Get current accuracy number	–
Ann	Set accuracy number "nn" "nn" – two hex digits (00 ≤ nn ≤ 15) (00 ≤ nn ≤ 0A) – without an external divider (0B ≤ nn ≤ 15) – with an external divider by 16	00 – 1 % 01 – 0.5 % 02 – 0.25 % 03 – 0.1 % 04 – 0.05 % 05 – 0.025 % 06 – 0.01 % 07 – 0.005 % 08 – 0.0025 % 09 – 0.001 % 0A – 0.0005 % 0B – 1 % 0C – 0.5 % 0D – 0.25 % 0E – 0.1 % 0F – 0.05 % 10 – 0.025 11 – 0.01 % 12 – 0.005% 13 – 0.0025 14 – 0.001 % 15 – 0.0005 %
B	Get current baud rate number	–
Bn	Set com port baud rate number "n" (0 ≤ n ≤ 9)	0 – 300 1 – 600 2 – 1200 3 – 2400 4 – 4800 5 – 9600 6 – 19200 7 – 38400 8 – 14400 9 – 28800
C	Check the measurement status Returns "r" if result is ready and "b" if measurement is in progress	–
E	Get current reference value	–
Ei.f	Set reference value	"i.f" – fixed point decimal number "I" – integer part (1 to 12 decimal digits) "." – decimal point "f" – fractional part (0 to 12 decimal digits)
F	Get current calibration constant	–
Fi.f	Set calibration constant	"i.f" – fixed point decimal number "I" – integer part (1 to 12 decimal digits) "." – decimal point "f" – fractional part (0 to 12 decimal digits)
M	Get current measurement mode number	–
Mnn	Set measurement mode number "nn" (00 ≤ nn ≤ 1A)	00 – Frequency 1, F_x1 01 – Period 1, T_x1 02 – Phase Shift, φ 03 – Time Interval, T 04 – Duty Cycle 1, DC1 05 – Duty-Off Factor 1, Q1

		06 – Frequencies Difference, FD 07 – Periods Difference, TD 08 – Frequencies Ratio, FR 09 – Periods Ratio, TR 0A – Rotation Speed 1, V1 0B – Pulse Interval 1, T_p1 0C – Space Interval 1, T_s1 0D – Pulse Counter 1, N1 0E – Frequency 2, F_x2 0F – Period 2, T_x2 10 – Resistance, R_x 11 – Capacitance, C_x 12 – Bridge, B_x 13 – Frequency Deviation 1, D1 14 – Duty Cycle 2, DC2 15 – Duty-Off 2, Q2 16 – Rotation Speed 2, V2 17 – Pulse Interval 2, T_p2 18 – Space Interval 2, T_s2 19 – Pulse Counter 2, N2 1A – Frequency Deviation 2, D2
R	The same as R0	–
Rn	Get measurement result in format number 'n' (0 ≤ n ≤3)	0 – BCD ASCII 1 – BCD HEX 2 – BIN HEX 3 – BIN
S	Start measurement	–
T	Test mode. After this command signal fosc/2 is generated on pin 16.	–
W	Get current charging time constant	–
Wnn	Set charging time constant number (00 < n < 1B)	00 – 100 μs 01 – 200 μs 02 – 300 μs 03 – 400 μs 04 – 500 μs 05 – 600 μs 06 – 700 μs 07 – 800 μs 08 – 900 μs 09 – 1 ms 0A – 2 ms 0B – 3 ms 0C – 4 ms 0D – 5 ms 0E – 6 ms 0F – 7 ms 10 – 8 ms 11 – 9 ms 12 – 10 ms 13 – 20 ms 14 – 30 ms 15 – 40 ms 16 – 50 ms 17 – 60 ms 18 – 70 ms 19 – 80 ms 1A – 90 ms 1B – 100 ms
Z	Get current number of modulation rotor teeth number	–
Znn	Set the number "nn" of modulation rotor teeth number for rotation speed measurement mode (01 ≤ nn ≤ FF)	1–255

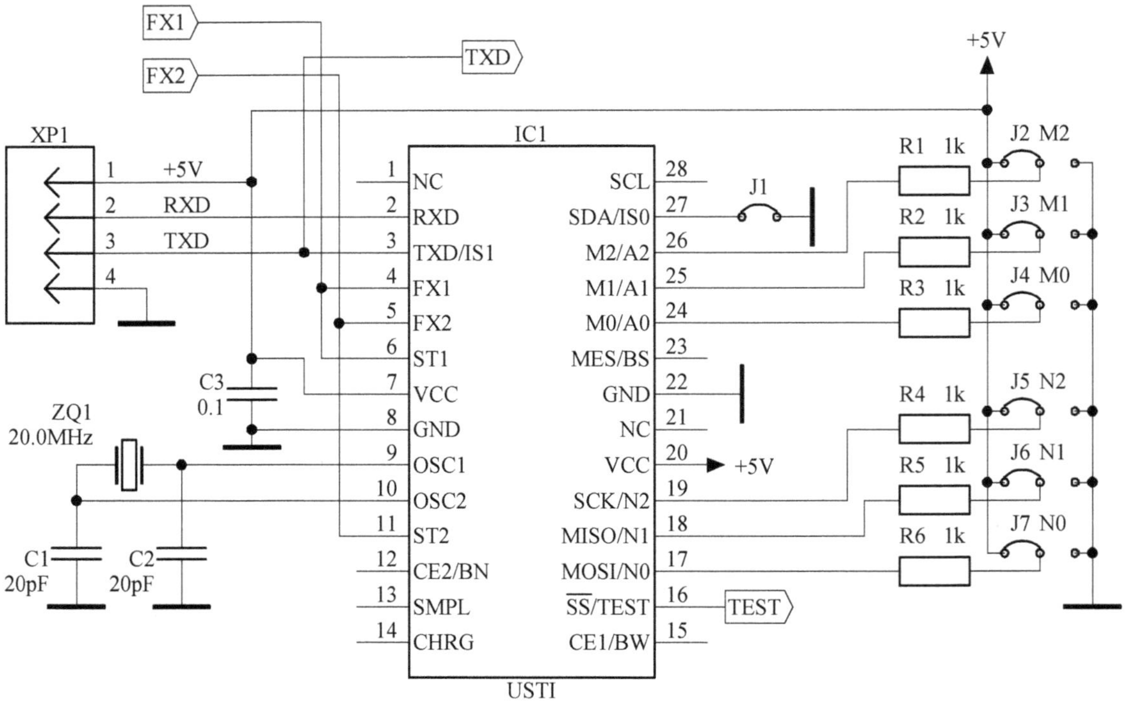

Fig. 3.3 Circuit diagram for the USTI working in autonomous (master) mode, RS232 interface.

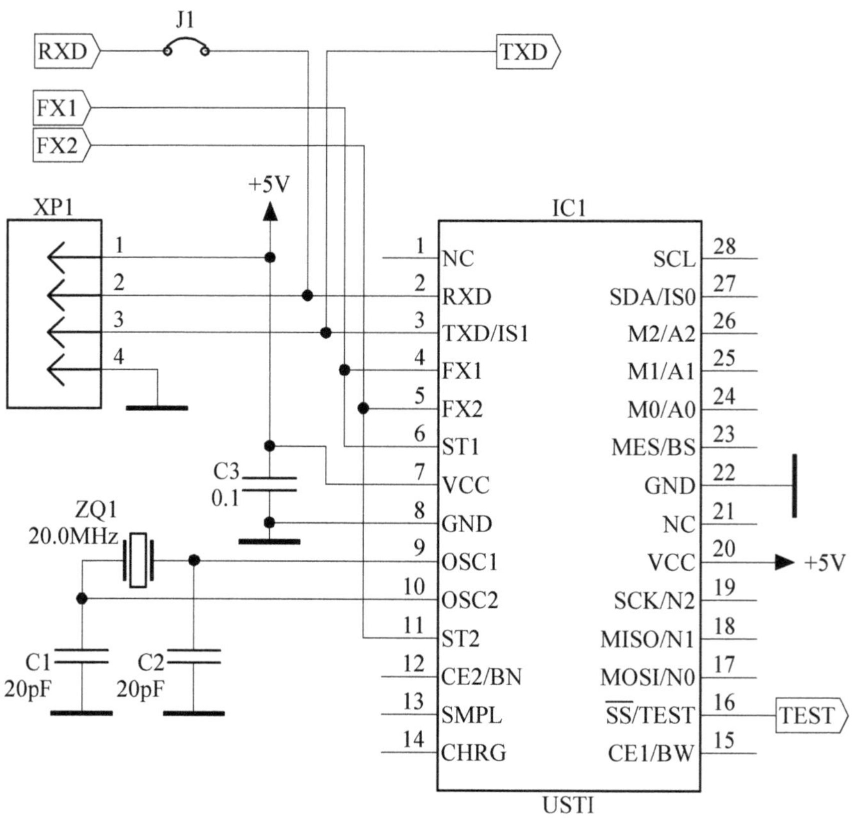

Fig. 3.4 Circuit Diagram for the USTI Working Embedded in a Sensor or DAQ Board (Slave Mode).

Table 3.5 Communication command set for SPI interface.

Interface		Description	Parameter
SPI	RS232 command analog		
<01><FF><FF>	A	Get current accuracy number	–
<02><nn>	An	Set accuracy number "nn" (00 ≤ nn ≤ 0A) – without an external divider (0B ≤ nn ≤ 15) – with an external divider by 16	00 – 1 % 01 – 0.5 % 02 – 0.25 % 03 – 0.1 % 04 – 0.05 % 05 – 0.025 % 06 – 0.01 % 07 – 0.005 % 08 – 0.0025 % 09 – 0.001 % 0A – 0.0005 % 0B – 1 % 0C – 0.5 % 0D – 0.25 % 0E – 0.1 % 0F – 0.05 % 10 – 0.025 11 – 0.01 % 12 – 0.005% 13 – 0.0025 14 – 0.001 % 15 – 0.0005 %
<03><FF><FF>	C	Check measurement status Returns 0 if result is ready and not 0 if measurement is in progress	–
<05><FF><FF>	M	Get current measurement mode number	–
<06><nn>	Mnn	Set measurement mode number "nn" (00 ≤ nn ≤ 1A)	00 – Frequency 1, $F_{x}1$ 01 – Period 1, $T_{x}1$ 02 – Phase Shift, φ 03 – Time Interval, T 04 – Duty Cycle 1, DC1 05 – Duty-Off Factor 1, Q1 06 – Frequencies Difference, FD 07 – Periods Difference, TD 08 – Frequencies Ratio, FR 09 – Periods Ratio, TR 0A – Rotation Speed 1, V1 0B – Pulse Interval 1, $T_{p}1$ 0C – Space Interval 1, $T_{s}1$ 0D – Pulse Counter 1, N1 0E – Frequency 2, $F_{x}2$ 0F – Period 2, $T_{x}2$ 10 – Resistance, R_x 11 – Capacitance, C_x 12 – Bridge, B_x 13 – Frequency Deviation 1, D1 14 – Duty Cycle 2, DC2 15 – Duty-Off 2, Q2 16 – Rotation Speed 2, V2 17 – Pulse Interval 2, $T_{p}2$ 18 – Space Interval 2, $T_{s}2$ 19 – Pulse Counter 2, N2 1A – Frequency Deviation 2, D2

<07><FF><FF>	R1	Get measurement result in BCD format Returns sign byte and 12 bytes of result in BCD code: <Sign><I5><I4><I3><I2><I1><I0> <F0><F1> <F2><F3><F4><F5>	–
<08><FF><FF>	R2	Get measurement result in BINARY format Returns 12 bytes of result in binary complementary code: <I5><I4><I3><I2><I1><I0> <F0><F1><F2><F3><F4><F5>	–
<09>	S	Start measurement	–
<0B><FF><FF>	Z	Get current number of modulation rotor teeth number	–
<0C><nn>	Znn	Set the number "nn" of modulation rotor teeth number for rotation speed measurement mode (01 ≤ nn ≤ FF)	1–255
<0D><FF><FF>	E	Get current reference value	
<0E><i><f>	Ei.f	Set reference value <I5><I4><I3><I2><I1><I0> <F0><F1><F2><F3><F4><F5>	<I5>…<I0> integer part of BCD number <F0>…<F5> fractional part of BCD number
<0F><FF><FF>	W	Get current charging time constant	–
<10><nn>	Wnn	Set charging time constant number (00 < n < 1B)	00 – 100 µs 01 – 200 µs 02 – 300 µs 03 – 400 µs 04 – 500 µs 05 – 600 µs 06 – 700 µs 07 – 800 µs 08 – 900 µs 09 – 1 ms 0A – 2 ms 0B – 3 ms 0C – 4 ms 0D – 5 ms 0E – 6 ms 0F – 7 ms 10 – 8 ms 11 – 9 ms 12 – 10 ms 13 – 20 ms 14 – 30 ms 15 – 40 ms 16 – 50 ms 17 – 60 ms 18 – 70 ms 19 – 80 ms 1A – 90 ms 1B – 100 ms

Legend:

<sign> – sign of result: 0x20 (space char) for positive result; 0x2D (minus char) for negative result

<In> – byte n of integer part of result

<Fn> – byte n of fractional part of result

<FF> – dummy byte

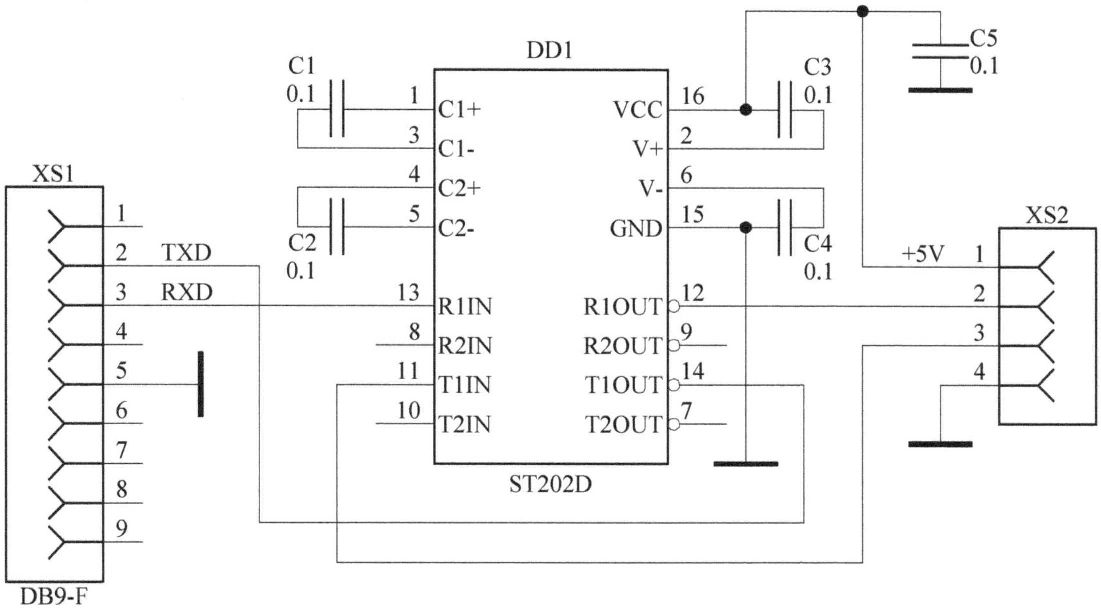

Fig. 3.5 Interfacing circuit for connection the USTI to PC.

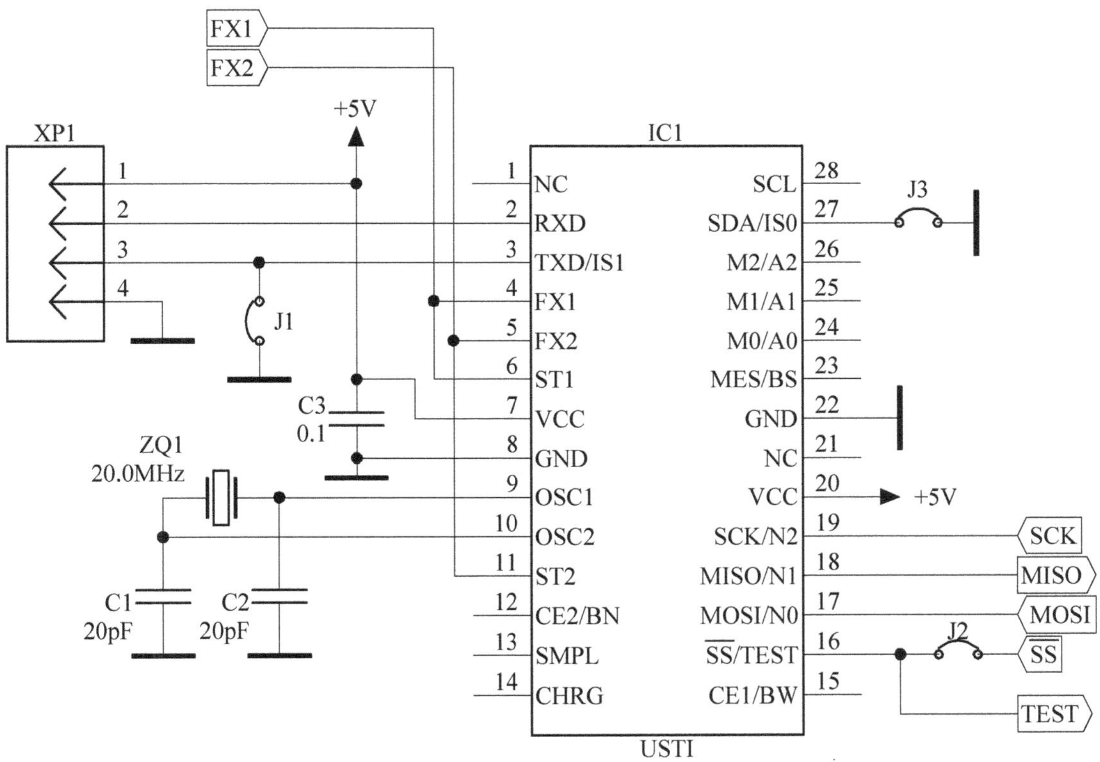

Fig. 3.6 USTI circuit diagram for the 3-wire Serial Interface (SPI).

The circuit diagram for the 2-wire I^2C interface is shown in Figure 37. The connector XP1 is used for in-circuit automated calibration. In the working mode the jumper J1 must be set in the ON position and in the calibration mode it should be switched to the OFF position.

Table 3.6 Communication command set for I^{2}C interface.

Interface		**Description**	**Parameter**
I^{2}C	**RS232 command analog**		
<01>	A	Get current accuracy number	–
<02><nn>	An	Set accuracy number "nn" (00 ≤ nn ≤ 0A) – without an external divider (0B ≤ nn ≤ 15) – with an external divider by 16	00 – 1 % 01 – 0.5 % 02 – 0.25 % 03 – 0.1 % 04 – 0.05 % 05 – 0.025 % 06 – 0.01 % 07 – 0.005 % 08 – 0.0025 % 09 – 0.001 % 0A – 0.0005 % 0B – 1 % 0C – 0.5 % 0D – 0.25 % 0E – 0.1 % 0F – 0.05 % 10 – 0.025 11 – 0.01 % 12 – 0.005% 13 – 0.0025 14 – 0.001 % 15 – 0.0005 %
<03>	C	Check measurement status Returns 0 if result is ready and not 0 if measurement is in progress	–
<05>	M	Get current measuring mode number	–
<06><nn>	Mn	Set measurement mode number "nn" (00 ≤ nn ≤ 1A)	00 – Frequency 1, F_x1 01 – Period 1, T_x1 02 – Phase Shift, φ 03 – Time Interval, T 04 – Duty Cycle 1, DC1 05 – Duty-Off Factor 1, Q1 06 – Frequencies Difference, FD 07 – Periods Difference, TD 08 – Frequencies Ratio, FR 09 – Periods Ratio, TR 0A – Rotation Speed 1, V1 0B – Pulse Interval 1, T_p1 0C – Space Interval 1, T_s1 0D – Pulse Counter 1, N1 0E – Frequency 2, F_x2 0F – Period 2, T_x2 10 – Resistance, R_x 11 – Capacitance, C_x 12 – Bridge, B_x 13 – Frequency Deviation 1, D1 14 – Duty Cycle 2, DC2 15 – Duty-Off 2, Q2 16 – Rotation Speed 2, V2 17 – Pulse Interval 2, T_p2 18 – Space Interval 2, T_s2 19 – Pulse Counter 2, N2 1A – Frequency Deviation 2, D2

<07>	R1	Get measurement result in BCD format Returns sign byte and 12 bytes of result in BCD code: <Sign><I5><I4><I3><I2><I1><I0> <F0><F1> <F2><F3><F4><F5>	–
<08>	R2	Get measurement result in BINARY format Returns 12 bytes of result in binary complementary code: <I5><I4><I3><I2><I1><I0> <F0><F1><F2><F3><F4><F5>	–
<09>	S	Start measurement	–
<0B>	Z	Get current number of modulation rotor teeth number	–
<0C><nn>	Znn	Set the number "nn" of modulation rotor teeth number for rotation speed measurement mode (01 ≤ nn ≤ FF)	1–255
<0D>	E	Get current reference value	
<0E><i><f>	Ei.f	Set reference value <I5><I4><I3><I2><I1><I0> <F0><F1><F2><F3><F4><F5>	<I5>…<I0> integer part of BCD number <F0>…<F5> fractional part of BCD number
<0F>	W	Get current charging time constant	–
<10><nn>	Wnn	Set charging time constant number (00 < n < 1B)	00 – 100 μs 01 – 200 μs 02 – 300 μs 03 – 400 μs 04 – 500 μs 05 – 600 μs 06 – 700 μs 07 – 800 μs 08 – 900 μs 09 – 1 ms 0A – 2 ms 0B – 3 ms 0C – 4 ms 0D – 5 ms 0E – 6 ms 0F – 7 ms 10 – 8 ms 11 – 9 ms 12 – 10 ms 13 – 20 ms 14 – 30 ms 15 – 40 ms 16 – 50 ms 17 – 60 ms 18 – 70 ms 19 – 80 ms 1A – 90 ms 1B – 100 ms

Legend:

<sign> – sign of result: 0x20 (space char) for positive result; 0x2D (minus char) for negative result

<In> – byte n of integer part of result

<Fn> – byte n of fractional part of result

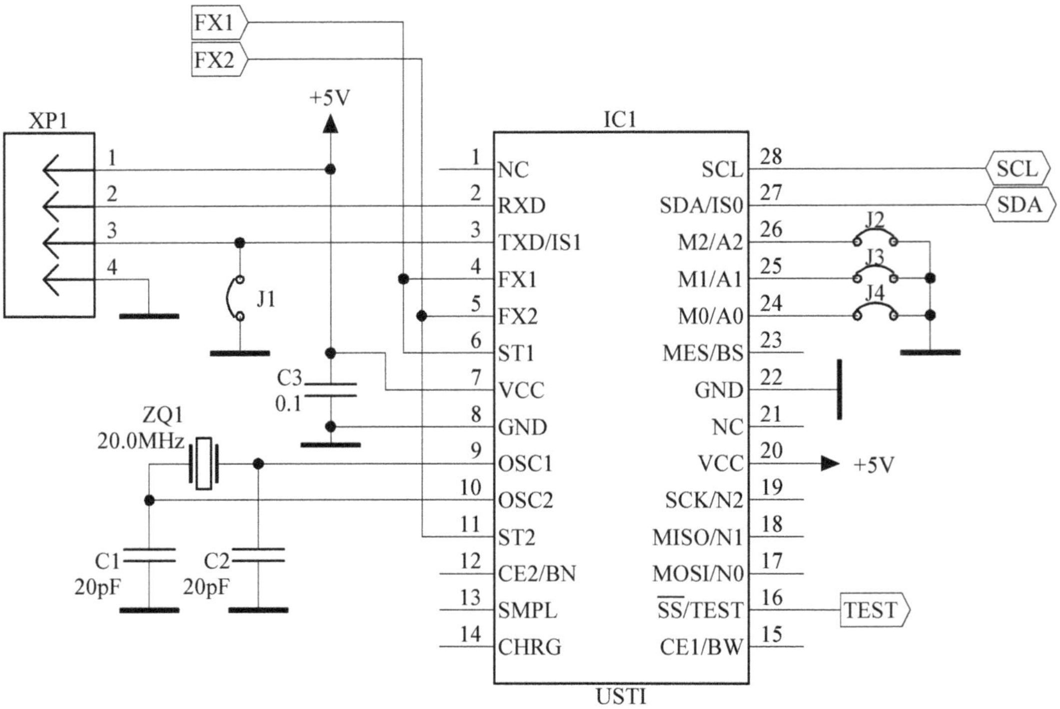

Fig. 3.7 USTI circuit diagram for the 2-wire I^2C interface.

3.3.3 Quartz-Accurate Calibration Procedure

The Universal Sensors and Transducers Interface (USTI) integrated circuit was designed for precision conversion of different frequency, time or phase-shift parameters of electric signals to digital. Taking into account a high accuracy (up to ± 0.0005 % constant relative error) of the USTI [1] it needs a very accurate frequency reference at least ± 0.00005 (or ± 0.0001 %) error. Such relative error can be obtained at the use of temperature compensated quartz crystal oscillators (TCXO) [6] or oven controlled quartz crystal oscillators [7]. For example, precision, temperature-compensated, 10 MHz to 51.84 MHz integrated generator from *Maxim* for highly accurate timing applications provides ±1 ppm frequency stability over the -40 ^{0}C to +85 ^{0}C industrial temperature range [6], Figure 3.8. However, these oscillators are very expensive devices and its price is in twice (or even more) higher than the USTI price. In addition, the oven controlled quartz crystal oscillator can not be used in different embedded applications as system-on-chip, digital or smart microelectronic sensors.

For inexpensive quartz crystal oscillators, for example, from *Siward Crystal Technology, Co. Ltd*, the frequency variation due to adjustment inaccuracy can exceed 20 - 50 ppm (or ±0.002 … ± 0.005 %) and has a systematic character. For example, 20 MHz crystal oscillators from *Siward Crystal Technology, Co. Ltd*, has the real frequency 20 004 984 Hz that corresponds to ± 0.025 % relative reference error.

In order to use the USTI with low cost crystal oscillators for conversions with quantization error better than $\delta_q < \pm 0.025$ % it is expediently to make a calibration with the aim to compensate the adjustment or trimming inaccuracy. In other cases if the quantization error should be $\delta_q \geq 0.025$ %, the calibration is not necessary. Temperature changes in the wide working industrial temperature range -40 ^{0}C to +85 ^{0}C as well as a quartz crystal aging also require a calibration

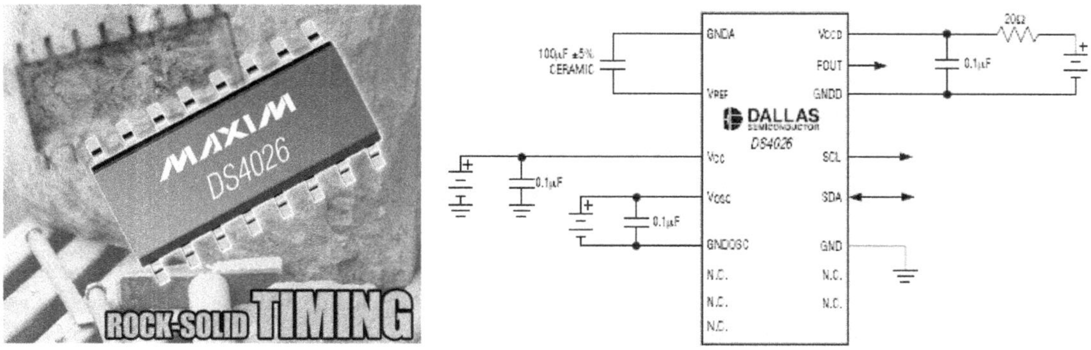

Fig. 3.8 Precision temperature-compensated integrated generator Maxim DS4026.

The calibration should be made in real working conditions with the use of recommended 20 MHz crystal oscillator. Any terminal software, for example, Terminal V1.9b for Windows [8], or LabView based software can be used for this communication purpose.

The USTI provides so-called in-circuit quartz-accurate automated calibration. The calibration is made by means of the communication interface RS232. The USTI is connected to a PC by connector XP1 (Figures 3.3-3.4, 3.6-3.7). In the calibration mode all jumpers in Figures 3.4, 3.6 and 3.7 must be in the OFF position.

The calibration sequence for the USTI is the following.

1) The test command "T" toggles the USTI to the calibration mode. A sequence of rectangular pulses with frequency $f_{osc/2}$ is generated at the TEST output.
2) The frequency of this pulse sequence should be measured with the help of an external frequency counter with accuracy not worse than ± 0.0001 % or at least ± 0.0002 %.
3) Use the command "Fi.f" to put the clock constant value into the USTI. The command "F" without parameters indicates the internal value of the clock constant and can be used for checking the number loaded.

The calibration is performed only once. The correction factor Δ is stored in the nonvolatile memory of the USTI, which holds this value after power-off.

Example:

Toggle the USTI in the calibration mode by "T" command entered in any terminal software, working under Windows operation systems:

```
>T
```

Let the measured frequency on the TEST output is 10 000 694.257865 Hz, for example. So we send the clock constant command:

```
>F10000694.257865
```

Then to check programmed value we send the command

```
>F

10000694.257865
```

The returned result 10 000 694.257865 testifies about the properly stored value of correction factor for the given quartz crystal. After that the USTI is ready for conversion in the frame of accuracy ranges specified in this document.

In comparison with the automated calibration procedure for the integrated Universal Frequency-to-Digital Converter (UFDC-1) [9], the advanced calibration procedure for the USTI takes into account not only the integer i but also the fractional part f of clock constant (correction factor Δ) measured by a frequency counter. The proposed advanced calibration technique is easier, because of it is not necessary to subtract the fractional part f of correction factor Δ from the integer part i and convert the correction factor from a decimal number into a hexadecimal number.

Besides the trimming inaccuracy, the primary frequency of crystal oscillator is also influenced by different drifts due to temperature changes, aging, supply-voltage changes and other environmental changes. The periodical calibration is also useful for elimination of aging influence as well as temperature changes.

Taking into account the industrial working temperature range for the USTI (–40 ^{0}C ... +85 ^{0}C), it is also expediently to investigate the conversion error in this temperature range and a possibility for temperature drift calibration.

To investigate the temperature dependence of conversion error in all working industry temperature range (– 40 ^{0}C...+ 85 ^{0}C) the measuring setup shown in Figure 3.9 has been used [10].

The experimental research was carry out in the climatic chamber CTS T-40/50 (Figure 2.16). A data acquisition, result indication and communication between the USTI and PC were made with the help of terminal software Terminal V1.9b [8].

The USTI's relative error of measurement was chosen 0.0005 % and frequencies 10 Hz, 1 kHz and 1 MHz were measured 60 times of each through the equidistant time intervals (~ 1 s). The investigated temperature range was –40 ^{0}C ... +80 ^{0}C with the 10 ^{0}C step at 53.7 % relative humidity. The absolute temperature error in the climatic chamber was ± 0.5 ^{0}C. The input signal of appropriate frequency was generated by the function/ arbitrary waveform generator Agilent 33220A. The high precision calibrated universal counter Agilent 53132A was used for frequency measurement at the USTI's TEST output. The digital oscilloscope Promax OD-571 was used for monitoring of impulse sequence's shape.

The frequency measurement in each of point from the investigated temperature range was made at fist without the USTI calibration (the calibration factor Δ= 10 000 000 Hz) and then after appropriate calibration. Statistical characteristics have been calculated. In addition, the calibration factors for all industrial temperature range with the step 0.5 ^{0}C where determined. The dependence of value of calibration factors on temperature is shown in Figure 3.10 and comparative logarithmic dependences of relative error on temperature at calibration and without calibration for 10 Hz, 1 kHz and 1 MHz frequencies are shown in Figures 3.11 (a-c) respectively. Dependences of relative errors on temperatures before and after calibration at 10 Hz, 1 kHz and 1 MHz are shown in Figures 3.12-3.14 respectively.

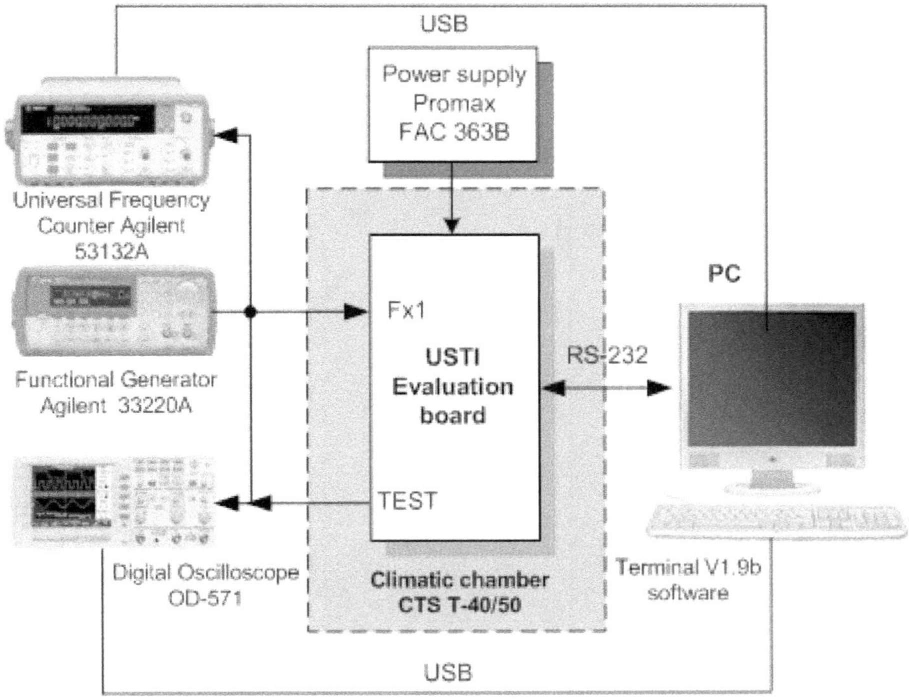

Fig. 3.9 Experimental setup for temperature dependence of conversion error investigation in all working temperature range.

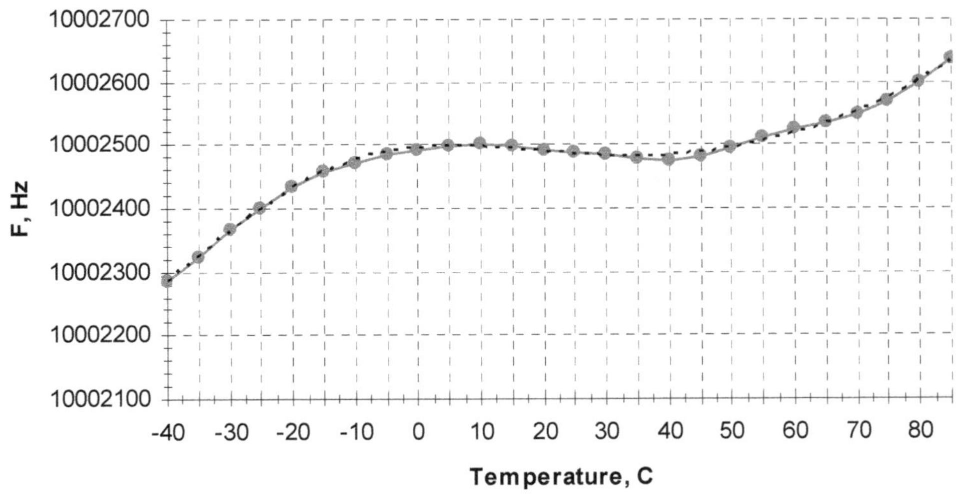

Fig. 3.10 Dependence of value of calibration factors on temperature.
($y = 3 \cdot 10^{-5} x^6 - 0.0025x^5 + 0.0871x^4 - 1.3113x^3 + 6.3954x^2 + 26.852x + 10^7$;
$R_2 = 17.991$ – squared value).

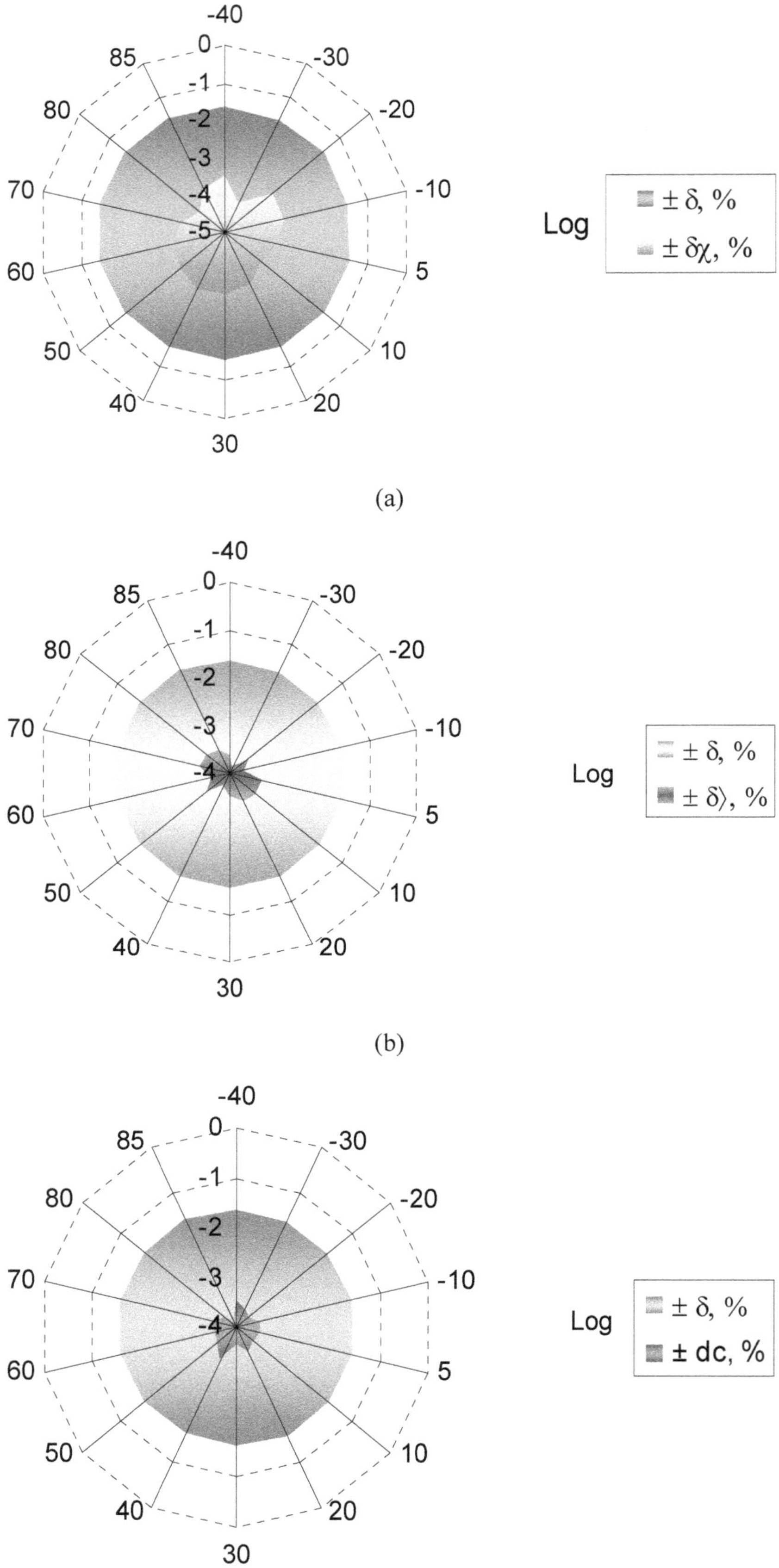

Fig. 3.11 Comparative logarithmic dependences of relative error on temperature at calibration (δ) and without calibration (δ_c) for 10 Hz (a), 1 kHz (b), and 1 MHz (c) frequencies.

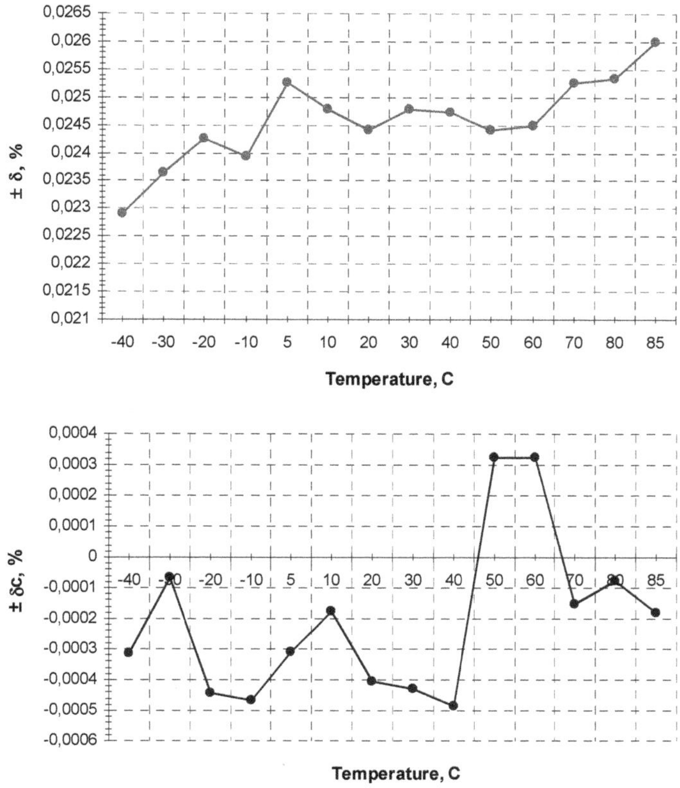

a)

b)

Fig. 3.12 Dependence of relative error on temperatures before (a), and after (b) calibration at $f_x = 10$ Hz.

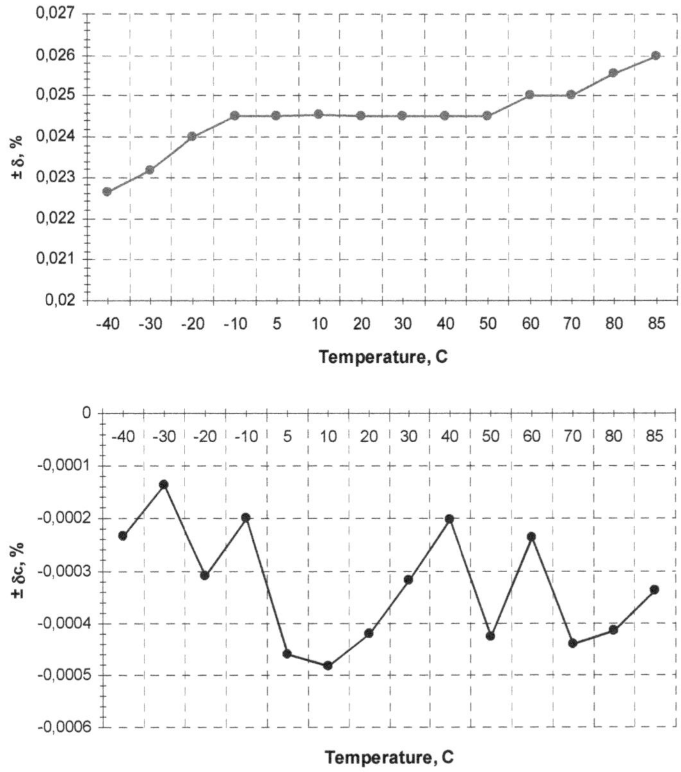

a)

b)

Fig. 3.13 Dependence of relative error on temperatures before (a), and after (b) calibration at $f_x = 1$ kHz.

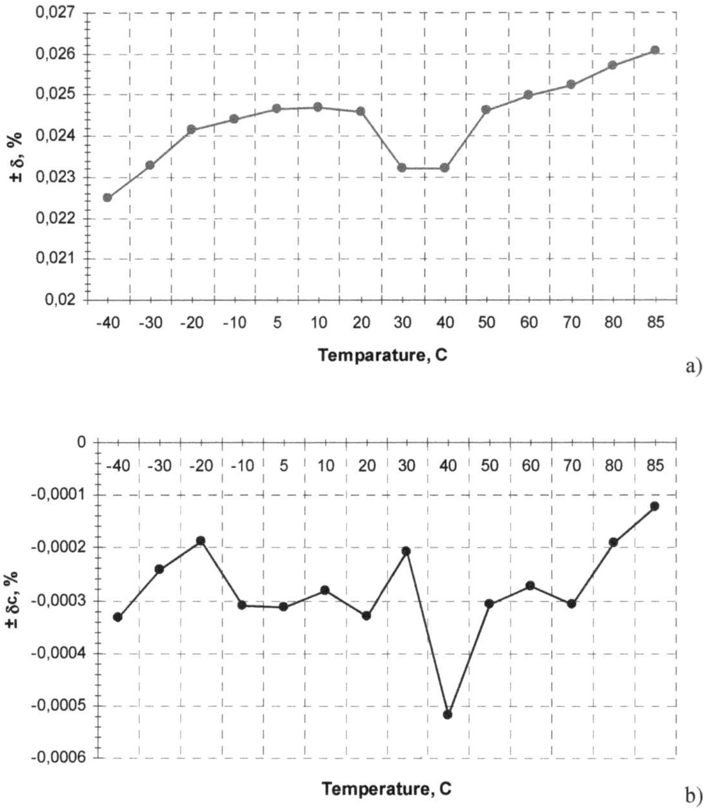

Fig. 3.14 Dependence of relative error on temperatures before (a), and after (b) calibration at f_x = 1 MHz.

As the experiments show the advanced calibration procedure lets decrease the relative error of measurement in 50-80 times. After calibration, the relative error did not exceed the programmable relative error δ = 0.0005 % in all frequency and temperature ranges. Such calibration technique can be used also for reducing an error due to a quartz crystal aging and a temperature drift at the same time.

3.3.4 Measuring Modes

In this section, we specify the USTI properties for all measuring modes. Names of these modes are the same as used in Table 3.2.

3.3.4.1 Mode 0. F_{X1}: Frequency, channel 1, 0.05 Hz … 9 MHz without prescaling

In this mode a frequency signal fx1 (should be TTL rectangular pulses) must be connected in parallel to inputs FX1 and ST1. If the pulse shape differs from square wave, the usage of an external input shaper (Schmitt-trigger signal conditioner) is recommended. At low and infralow frequency measurements it is recommended to use a low-pass filter.

If the maximum converted frequency exceeds 9 MHz, it is necessary to use an external prescaler (external divider by 16) and to choose the appropriate mode and accuracy by means of commands or external switches N0-N2. By this way the frequency range will be extended up to 144 MHz.

The conversion error in this mode is constant and does not exceed the programmed in all frequency range. The conversion result is represented in Hz. All conversion accuracies specified in the Table 3.2 are available in this mode.

3.3.4.2 Mode 1. T_{X1}: Period, channel 1, 110 ns ... 20 s

This mode of measurement is similar to the Mode 1. The conversion result is represented in microseconds. All conversion accuracies specified in Table 3.2 are available in this mode. The mode with prescaling (external divider by 16) is also possible.

3.3.4.3 Mode 2. φ: Phase-Shift, 0 - 360°, f_{xmax} = 625 kHz at 50 % duty-cycle

The pulse sequences with the same frequency, a phase-shift between which need to be converted into a digital are connected to inputs FX1, ST1 and FX2, ST2 accordingly. Units for conversion result are degrees. The programmed accuracy influences on the period conversion. The conversion error of phase-shift does not depend on the pulse sequences period, but depends on the pulse width.

3.3.4.4 Mode 3. T: Time Interval Between Start- and Stop-Impulse, 1.5 μs ... 250 s

Start- and stop-pulse are connected to FX1, ST1 and FX2, ST2 inputs accordingly. The USTI converts the time interval between these pulses into digital. In this mode the conversion error does not depend on the value on inputs N0-N3 or given in a command mode through one of possible interfaces and can be calculated according to the equation (3.1) for each value of converted time interval. Units of measurements for conversion result are microseconds.

3.3.4.5 Mode 4. D.C.: Duty-Cycle 1: 0 ... 1 at $f_{xmax} \leq$ 625 kHz for 40-60 % D.C.

The pulse signal is coming on inputs FX1, ST1. The programmed accuracy influences on the period conversion. The conversion error does not depend on the period of pulse signal, but depends on the pulse width.

3.3.4.6 Mode 5. Q: Duty-Off Factor (1/D.C), $10^{-8} – 8{\cdot}10^{6}$, at ≤ 625 kHz for 40-60 % D.C.

Similarly to the Mode 4, the pulse sequence is coming on FX1, ST1 inputs. The programmed accuracy influences on the period conversion. The conversion result depends on the pulse width.

3.3.4.7 Mode 6. FD: Frequency Difference with Sign Determination f_{x1} - f_{x2}, 0 … 9 MHz

The frequency signal f_{x1} is coming on FX1, ST1 inputs and signal f_{x2} - on FX2, ST2 inputs. The sign "minus" before the result of measurement testifies that $f_{x1} < f_{x2}$. All conversion accuracies specified in Table 3.2 are available in this mode. The mode with prescaling (external divider by 16) is also possible.

3.3.4.8 Mode 7. PD: Period Difference with Sign Determination T_{x1} - T_{x2}, 0 … 20 s

The pulse signal with period T_{x1} is coming on FX1, ST1 inputs and signal with period T_{x2} - on FX2, ST2 inputs. The sign "minus" before the result of measurement testifies that $T_{x1} < T_{x2}$. All conversion accuracies specified in Table 3.2 are available in this mode. The mode with prescaling (external divider by 16) is also possible.

3.3.4.9 Mode 8. FR: Frequency Ratio, f_{x1}/f_{x2}, $5.5{\cdot}10^{-9}$ … $1.8{\cdot}10^{8}$

The frequency signal f_{x1} is coming on FX1, ST1 inputs and signal f_{x2} - on FX2, ST2 inputs. All conversion accuracies specified in Table 3.2 are available in this mode. The mode with prescaling (external divider by 16) is also possible.

3.3.4.10 Mode 9. PR: Period Ratio, T_{x1}/T_{x2}, $4{\cdot}10^{-9} – 2{\cdot}10^{8}$

The pulse sequence with period T_{x1} is coming on FX1, ST1 inputs, and pulse sequence with period T_{x2} – on FX2, ST2 inputs. All conversion accuracies specified in Table 3.2 are available in this mode. The mode with prescaling (external divider by 16) is possible.

3.3.4.11 Mode 10. n: Rotation Speed (rpm), n_x

The USTI can measure a rotation speed and present the results in rpm units. The number of encoder's teeth Z (1…255) must be set up in advance. In the Master Mode (interface RS232, unidirectional mode), by default Z=1.

In a sensor system the USTI can also measure rotation acceleration. An external microcontroller or PC calculates acceleration of rotation based on two measurements n_1

and n2 of rotation speed and time interval for the second measurement t2. See equation (2.4).

3.3.4.12 Mode 11. t_p: Pulse Width, t_p 1.5 μs … 250 s

Pulse signal is coming on FX1, ST1 inputs. In this mode the conversion error does not depend on the value on N0-N2 inputs or given through one of possible interfaces and can be calculated according to the formula (3.1) for each value of converted pulse width. Units of measurements for result are microseconds.

3.3.4.13 Mode 12. ts: Space Interval, ts 1.5 μs … 250 s

This mode of measurement is similar to the Mode 11. Pulse signal is coming on FX1, ST1 inputs. In this mode the conversion error does not depend on the value on N0-N2 inputs or given through one of possible interfaces and can be calculated according to the formula (3.1) for each value of converted pulse width. Units of measurements for result are microseconds.

3.3.4.14 Mode 13. N: Pulse Number (Events) Count, N_X 0 … $4 \cdot 10^9$

The USTI can count the pulse number (events) on FX1, ST1 inputs. Each reinitialization of the mode resets the counter to zero.

3.3.4.15 Mode 14. F_{x2}: Frequency, channel 2, 0.05 Hz … 9 MHz without prescaling

In this mode a frequency signal f_{x2} (should be TTL rectangular pulses) must be connected in parallel to inputs FX2 and ST2. If the pulse shape differs from rectangular, the usage of an external input shaper is recommended. At low and infralow frequency measurements it is recommended to use the low-pass filter.

If the maximum converted frequency exceeds 9 MHz, it is necessary to use an external prescaler (external divider by 16) and to choose the appropriate mode and accuracy by means of commands or external switches N0-N2. By this way the frequency range will be expanded up to 144 MHz.

The conversion error in this mode is constant and does not exceed the programmed in all frequency range. The conversion result is represented in Hz. All conversion accuracies specified in Table 3.2 are available in this mode.

3.3.4.16 Mode 15. T_{x2}: Period, channel 2, 110 ns … 20 s

This mode of measurement is similar to the Mode 14. The conversion result is represented in microseconds. All conversion accuracies specified in Table 3.2 are available in this mode. The mode with prescaling (external divider by 16) is also possible.

3.3.4.17 Mode 16. Resistance, R_x, 10 Ω to 10 MΩ

In this mode unknown resistor R_x (or sensing elements as platinum or cuprum resistor or thermistor), reference resistor R_c, calibrating capacitor C and limiting resistor R_o must be connected to the USTI as it is show in Figure 3.15.

The capacitance (in Farads) of calibrating capacitor should be calculated according to the following equation:

$$C \geq \frac{0.002}{R_c}, \tag{3.2}$$

and $R_c \leq R_x$. The limiting resistor must be $R_0 \approx 300 \ldots 600\ \Omega$.

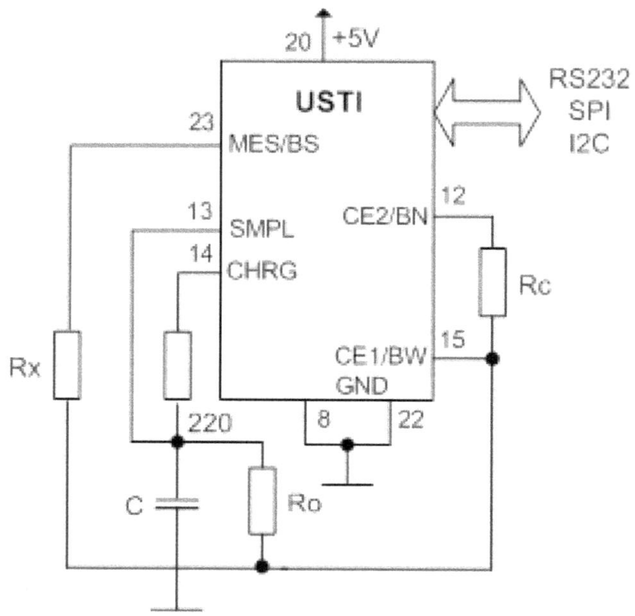

Fig. 3.15 USTI circuit diagram for resistance measurement.

The value of reference resistor R_c must be set up with the help of "Ei.f" command (see Table 3.2), where "i.f" is the fixed point decimal number: "I" is the integer part (1 to 12 decimal digits) and "f" is the fractional part (0 to 12 decimal digits).

The charging time (in seconds) should be calculated as

$$T = 2200 \times C \tag{3.3}$$

The charging time can be set up with the help of USTI's command "Wnn". All possible values for "Wnn" command are shown in Table 3.2. It is not necessary to set up accuracy in this mode.

In the master communication mode the charging time should be set up with the help of external switches N0, N1, N2 (see Table 3.2). In this mode the result at the converter's output will be presented as R_x/R_c. In order to get the actual capacitance, such result must be multiplied by R_c.

The USTI has the average relative error ± 0.47 % in a wide measuring range from 10 Ω to 10 MΩ. A conditional splitting of the range of measurement into sub ranges with a choice of appropriate values and types of R_c and C has allowed to receive also the relative error ± 0.01 % or better by using of precision external electronic components.

3.3.4.18 Mode 17. Capacitance, C_x, 50 pF to 100 μF

In this mode unknown capacitor C_x (or sensing element), reference capacitor C_{ref} and limiting resistor R must be connected to the USTI as it is show in Figure 3.16.

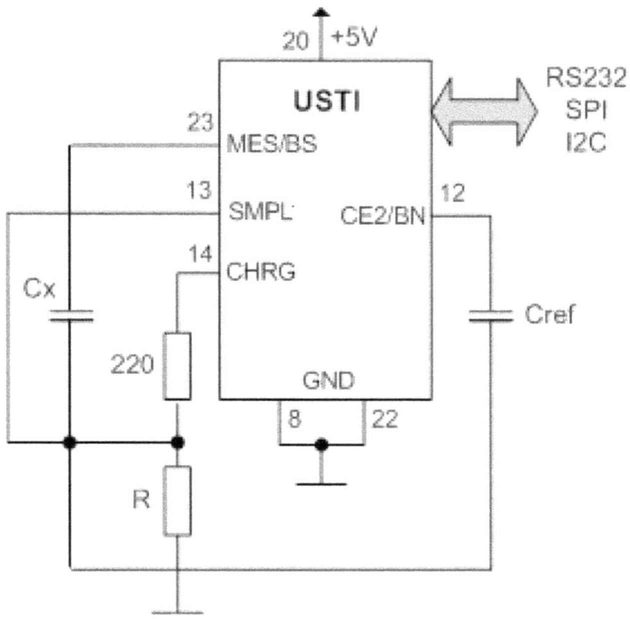

Fig. 3.16. USTI circuit diagram for capacitance measurement.

The resistance (in Ohm) should be calculated according to the following equation:

$$R \geq \frac{0.002}{C_{ref}} \tag{3.4}$$

The charging time (in seconds) should be calculated according to equation (3.3) and set up with the same command "Wnn". The value of C_{ref} should be set up with the help of appropriate USTI's command Ei.f, where "i.f" is the fixed point decimal number: "I" is the integer part (1 to 12 decimal digits) and "f" is the fractional part (0 to 12 decimal digits).

In the master communication mode the charging time should be set up with the help of external switches N0, N1, N2 (see Table 3.2). In this mode the result at the converter's output will be presented as C_x/C_{ref}. In order to get the actual capacitance, such result must be multiplied by C_{ref}.

The average relative error in this mode is ± 0.036 % over all wide measuring range and does not exceed ± 0.7 % the worst case. It can be reduced by the using of precision external components.

3.3.4.19 Mode 18. Resistive bridge

In this mode a passive resistive bridge, a limiting resistor *R* and capacitor *C* must be connected to the USTI as in shown in Figure 3.17.

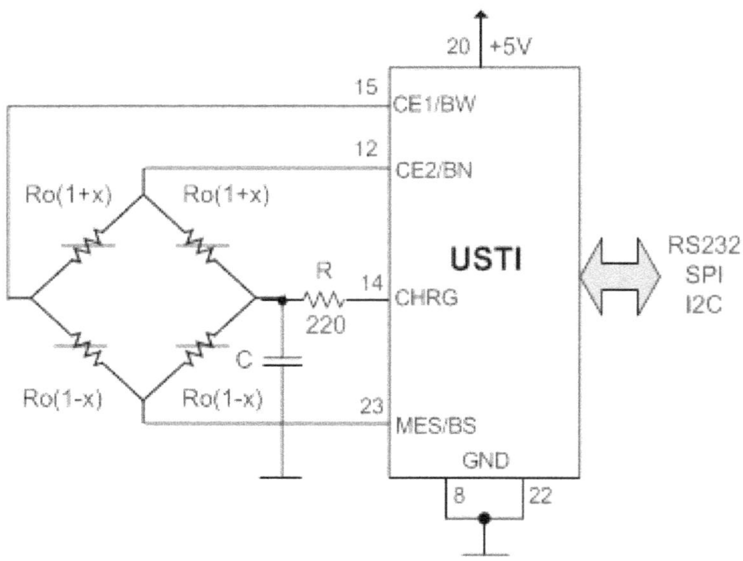

Fig. 3.17 USTI circuit diagram for resistive bridge measurement.

The capacitance (in Farads) of capacitor *C* should be calculated according to the following equation:

$$C \geq \frac{0.002}{R_a}, \tag{3.5}$$

where R_a is the resistance of a bridge arm. The charging time (in seconds) should be calculated according to equation (3.3). It must be set up with the help of command "Wnn". In the master communication mode the charging time should be set up with the help of external switches N0, N1, N2.

3.3.4.20 Mode 19. Frequency Deviation 1, DA1, 0 … 9 MHz

In this mode an absolute (in Hz) or relative (%) frequency deviation from the value, which is set-up with the help of "Ei.f" command (see Table 3.2) will be measured. The absolute or relative frequency deviation mode is selected by "Ann" command (see

Table 3.2): at A00 and A0B (a mode with prescaller) the absolute deviation will be measured; at other parameters – the relative deviation will be measured. Accuracy in this mode(s) does not depend on parameters of "Ann" command and is selected maximal possible (± 0.0005 % relative error).

3.3.4.21 Mode 20. D.C.2: Duty-Cycle 2: 0 … 1 at $f_{xmax} \leq 625$ kHz for 40-60 % D.C.

This mode is the same as the Mode 4 (see section 3.3.4.5) but measurements will be made in the 2nd channel of USTI.

3.3.4.22 Mode 21. Q2: Duty-Off Factor 2: (1/D.C), 10^{-8} … $8 \cdot 10^{6}$, at ≤ 625 kHz for 40-60 % D.C.

This mode is the same as the Mode 5 (see section 3.3.4.6) but measurements will be made in the 2nd channel of USTI.

3.3.4.23 Mode 22. n2: Rotation Speed 2 (rpm), n_x

This mode is the same as the Mode 10 (see section 3.3.4.11) but measurements will be made in the 2nd channel of USTI.

3.3.4.24 Mode 23. t_{p2}: Pulse Width 2, t_p 1.5 µs … 250 s

This mode is the same as the Mode 11 (see section 3.3.4.12) but measurements will be made in the 2nd channel of USTI.

3.3.4.25 Mode 24. t_{s2}: Space Interval 2, t_s 1.5 µs … 250 s

This mode is the same as the Mode 12 (see section 3.3.4.13) but measurements will be made in the 2nd channel of USTI.

3.3.4.26 Mode 25. N2: Pulse Number (Events) Count 2, N_X 0 … $4 \cdot 10^{9}$

This mode is the same as the Mode 13 (see section 3.3.4.14) but measurements will be made in the 2nd channel of USTI.

3.3.4.27 Mode 26. Frequency Deviation 2, DA2, 0 … 9 MHz

This mode is the same as the Mode 19 (see section 3.3.4.20) but measurements will be made in the 2nd channel of USTI.

3.4 Evaluation and Development Board

3.4.1 Hardware

For actual development purposes and non-in-circuit calibration a development kit is recommended. The development board is equipped with an RS232 interfacing chip offering a serial communication between the USTI and a personal computer. A 20 MHz crystal completes the on-chip oscillator of the USTI. A voltage regulator (7805) supplies a DC power supply of 5 V for the converter and the RS232 interface. The external power supply required for the development board amounts to 7 … 14 V DC or AC.

Connection to the USTI development board is implemented with appropriate connectors/connecting pads. XT1 and XT2 are used to connect the external frequency-time domain signal.

The development board can be used for all functions of the USTI. The setting of the operating modes and accuracy is respectively performed by external switches S1 and S2, or by appropriate commands through the RS232 interface.

The USTI evaluation and development board prototype and full circuit diagram are shown in Figure 3.18 and Figure 3.19 respectively. The silkscreen of development board is shown in Figure 3.20, and sensing elements connection diagrams are shown in Figure 3.21.

The Schmidt trigger 74HC14D is optional and can be use in separate cases to improve an input signal shape.

Fig. 3.18 USTI evaluation and development board prototype.

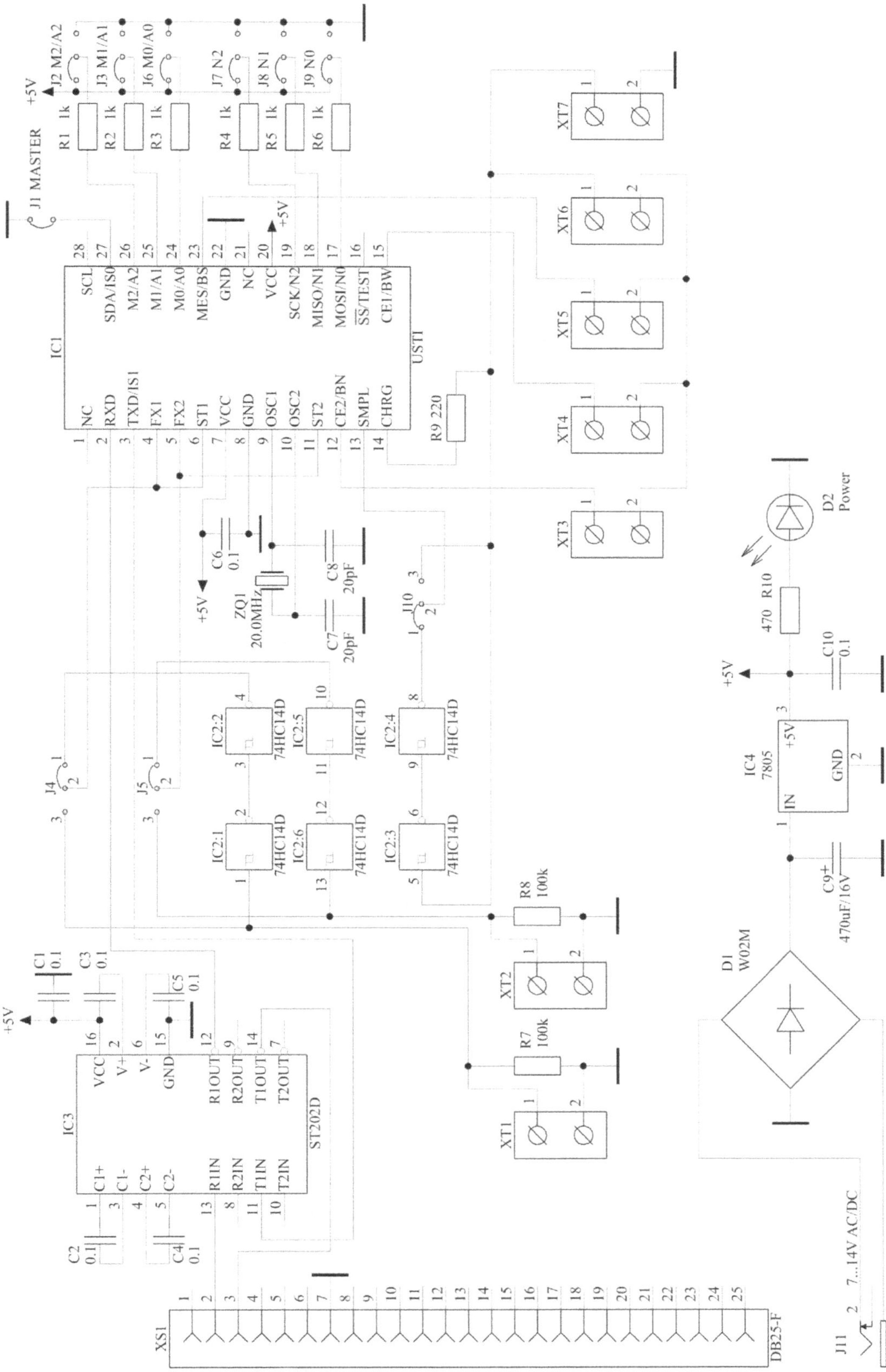

Fig. 3.19 Evaluation and development board circuit diagram.

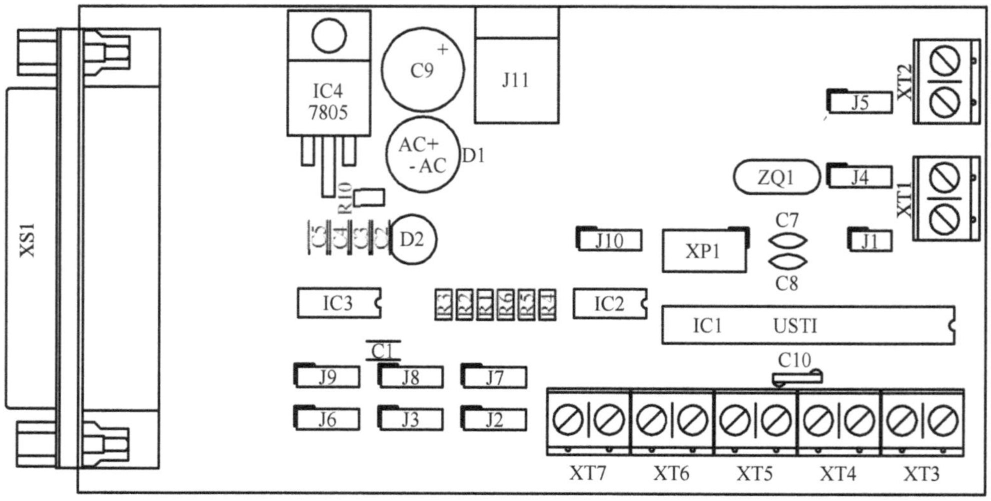

Fig. 3.20 USTI development board silkscreen (Not to scale).

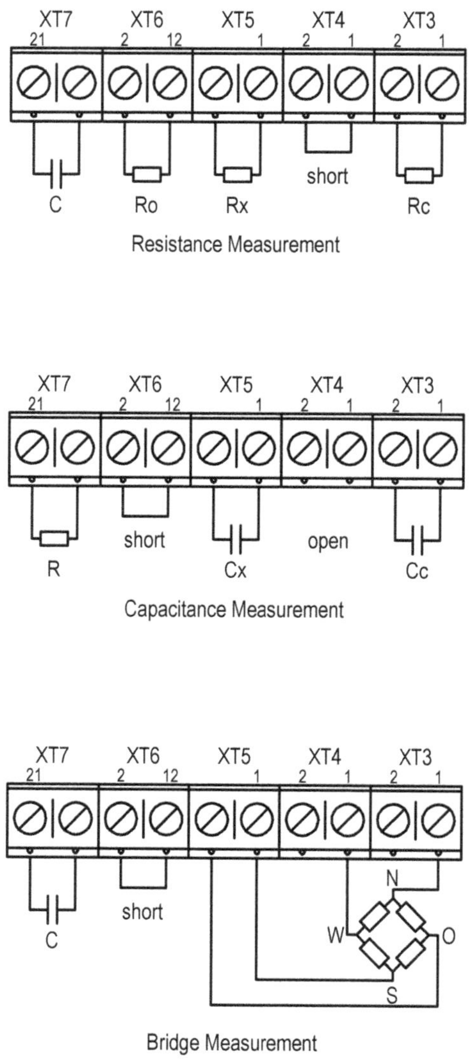

Fig. 3.21 Sensing elements connection diagrams.

3.4.2 Software

There are many ways to display the measurement results. For this purpose the LabView or custom designed software can be used. Any terminal software for Windows also can be used with the USTI development board using the serial port of a PC. One of the possible choices is the Terminal V1.9b free software [8].

In order to have correct measurement results the following options must be selected in the main terminal window: an appropriate COM port; baud rate 2400; data bits – 8; parity – non; stop bits – 1; handshaking – non.

Following the RS232 communication commands from the Table 3.4 it is easy to manage the development board and display measuring results. There are four possible data format in the Terminal: BCD ASCII, BCD Hex, BIN Hex and BIN.

Commands must be printed in the Transmit window, the results will be shown in the Receive window.

Example:

`>M00`	Select a frequency measurement mode in the 1st channel.

`>A09`
`>S`
`>R`
	These commands set up the accuracy 0.001 %, start measurement (S), ask for BCD ASCII format (R) and returns the result of measurement in Hz.

```
12000.089992800575
```

NOTE: For low and infralow frequencies, low rotation speed measurements as well as for long time intervals, periods, pulse widths and spaces measurements it is strongly recommended to use the command "C" between "S" and "R" commands (see Table 3.4) in order to check if the result is ready. In opposite side it is a risk to get a result of previous measurement.

3.5 Applications

The USTI has many applications. Some typical applications are the following:

- Sensor systems with any frequency, period, duty-cycle, time interval, phase-shift, pulse number output sensors and transducers;
- Physical, chemical, bio- and immunosensors; quartz crystal microbalance (QCM) based sensors, BioMEMS, Lab-on-Chip;
- Digital sensors design, SoC, SiP;
- ADCs based on intermediate voltage-to-frequency conversion;
- Smart, intelligent (self-adaptive) sensors and sensor systems due to programmable accuracy and non-redundant conversion time - adaptive possibilities of USTI automatically choice the conversion time depending on the given error of measurement and opposite;

- Multifunctional and multiparameters sensors for simultaneous detection various parameters proportional to frequency and duty-cycle carrying the information provided by sensing elements;
- Data acquisition (DAQ) boars and measuring systems for frequency-time parameters of electric signal;
- Virtual instruments;
- Desktop multifunctional frequency counters;
- Tachometers and tachometric systems;
- Handheld multimeters for frequency-time parameters of electric signals;
- High-end, mid- and low-range ABS;
- High accuracy programmable frequency-time parameters – to – digital converters for various measuring and communication applications.

The USTI can be used in all the same applications as the UFDC Series ICs. In addition, it can be used for analytical chemical and biochemical applications (due to frequency deviation measuring modes 13 and 1A) and Quartz Crystal Microbalance (QCM) sensor systems. As silicon IP it can be used also in various BioMEMS, Lab-on-Chip and Micro-Nano Bio Systems (MNBS).

Additionally, the USTI can directly convert to digital a resistance, capacitance and resistive bridge parameters of different sensing elements. In this case the number of external components are minimum possible. The conversion methods for these parameters are based on modified methods for sensing elements described in [13] and [14].

The chip can be embedded into any existing frequency-time domain output sensors or transducers (with the help of integrated, SoC, SiP or hybrid technologies) to produce IEEE 1451.4 compatible sensors as well as into IEEE 1451.4 compatible data acquisition systems. Design and development of IEEE 1451 compatible sensors and transducers will be quick, low-cost and effective. Such approach will improve the level of commercial adaptation of IEEE 1451 standards family in industry.

Together with all IEEE 1451 standards family benefits (including self-identification, simple installation, upgrade and maintenance) customers will receive the self-adaptation capability and digital plug-and-play sensors, using all advantages of frequency-time parameters as an informative parameters of sensors' outputs.

NOTE: Similarly to the UFDC-1 and UFDC-1M-16 ICs, the USTI can also work with any voltage and current output sensors. It this case, any voltage–to-frequency converter (VFC) can be used. The accuracy of such system will be determined by sensor accuracy as well as by VFC accuracy.

3.6 Pin Out and Housing

The USTI is available in the same packages as the UFDC Series of ICs: a 28-lead (0.300"/7.62 mm Wide) Plastic Dual Inline Package (PDIP); 32-lead 7 × 7 mm Body Size, 1.0 mm Body Thickness, 0.8 mm Lead Pitch small Thin Profile Plastic Quad Flat Package (TQFP); and 32-pad, 5 × 5 × 1.0 mm Body, Lead Pitch 0.50 mm, 3.10 mm Exposed Pad, Micro Lead Frame Package (MLF). Pin Configurations for PDIP, TQFP and MLF packages are shown in Figures 3.22 (a, b) respectively. The function of the pins is listed in Table 2.10.

The detailed packaging information and drawing is available in [5]. This device can also be supplied in wafer form and/or as an IP [11, 12]. ICs' footprints for ECAD tools are shown in Figure 3.23 (a-c).

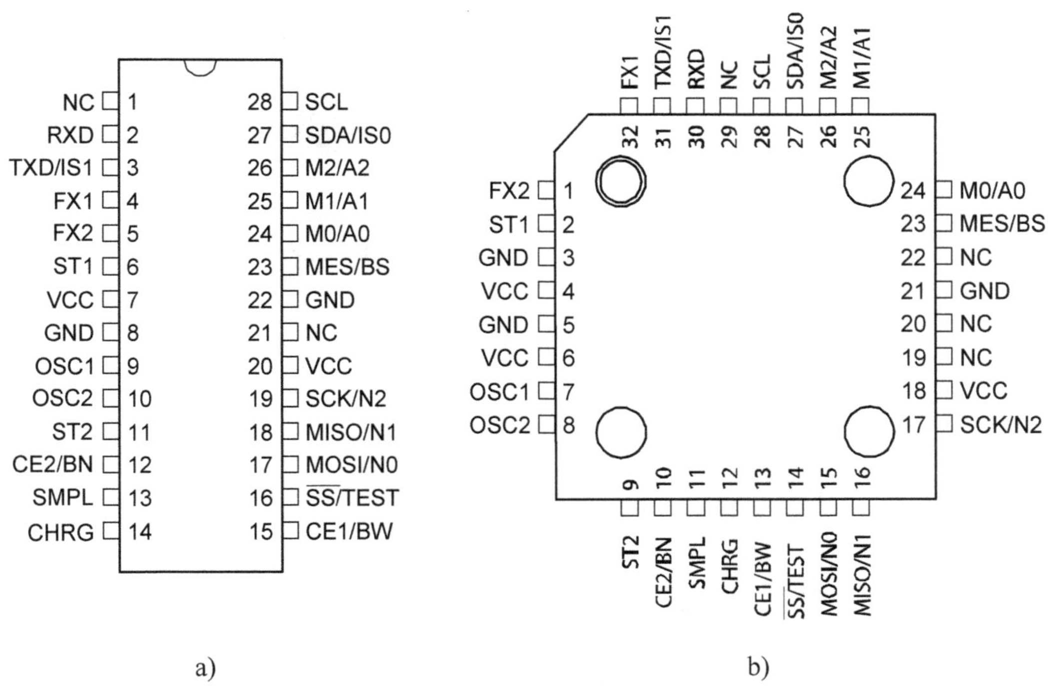

Fig. 3.22 Pin Configuration and housing (not to scale). NOTE: Bottom pad for MLF package should be soldered to ground.

Table 3.7 Pin description.

Name	Pin Descriptions	Name	Pin Descriptions
VCC	Supply voltage	SS	SPI Bus Master Slave select
GND	Ground	SCL	Two-wire (I^2C compatible) Serial Bus Clock Line
FX1, ST1	Input channel 1	SDA	Two-wire (I^2C compatible) Serial Bus Data Input/Output Line
FX2, ST2	Input channel 2	RXD	UART Input
OSC1, OSC2	Oscillator	TXD	UART Output
M0, M1, M2	Mode selection (see table)	IS0,IS1	Interface mode selection
N0, N1, N2	Accuracy selection (see table)	BN	North port of bridge
A0, A1, A2	Address of I^2C slave (for I^2C bus communication)	BW	West port of bridge
TEST	Calibration signal	BS	South port of bridge
SCK	SPI Bus Master clock Input	CE1	Calibration element 1
MISO	SPI Bus Master Input/Slave Output	CE2	Calibration element 2
MOSI	SPI Bus Master Output/Slave Input	CHRG, MES, SMPL	R_x and C_x measurement pins

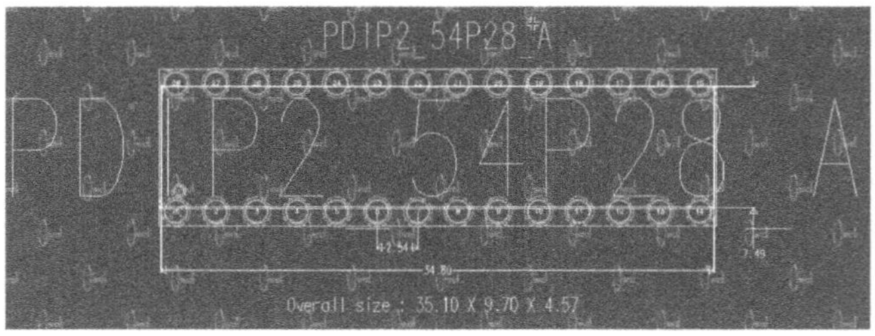

(a)

(b)

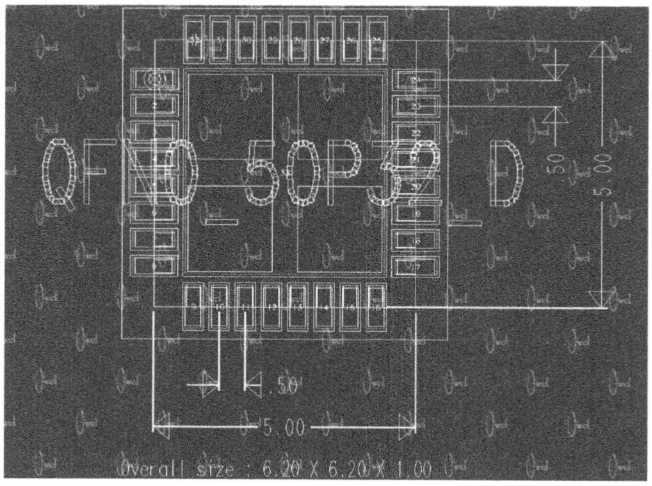

(c)

Fig. 3.23 UFDC-1 footprints for ECAD tools: (a) PDIP package; (b) TQFP package; and (c) MLF package.

3.7 Experimental Results

The objective of experimental investigations was to justify metrological performances of the USTI IC and their limits [1-3]. Such experimental investigation can be easy repeated by readers. The measurement set up is shown in Figure 3.24.

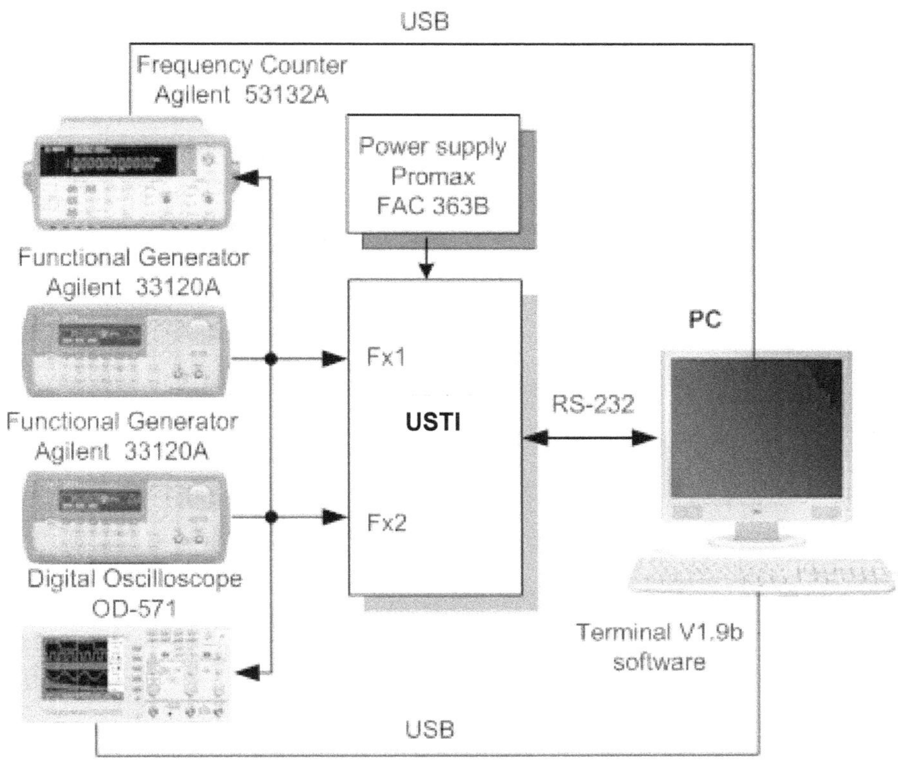

Fig. 3.24 Measurement set up for metrological performance determination.

Two TTL-level square waveform pulse signals whose frequency-time parameters must be measured, were fed from two function generators Agilent 33220A to inputs FX1 and FX2 (the 1st and 2nd channel) of the USTI evaluation board. The IC was programmed to measure frequency with the minimal possible relative error ±0.0005 %. The supply voltage of the evaluation board was +5V dc, provided by the Promax FAC-363B power supply. The frequencies generated by the generators were in the range from 0.04 Hz up to 9 MHz. It was measured by a frequency counter Agilent 53132A with ultra high oven stability internal time base. The two-channel digital oscilloscope Promax OD-571 monitored the signals waveforms as well as pulse rise and fall tail times. Before measurements, the USTI was calibrated in the working temperature range: +23.5 °C … + 25.4 °C with the purpose to compensate the quartz oscillator's systematic error.

The measured values were sent to a PC via an RS232 interface implemented with the ST202D IC. The user interface was realized with the help of terminal software Terminal V1.9b. Every measurement consisted of 60 values. The measurement errors were evaluated from appropriate statistics. The results for a 9 MHz, 1 MHz and 100 kHz square waveform pulse signal are shown in Figure 3.25 (a-c). The statistical characteristics are presented in Table 3.8. As it is visible from the table, the maximal relative error does not exceed the programmable $\delta_x < \pm\ 0.0005$ % in all frequency range including low and infralow frequencies.

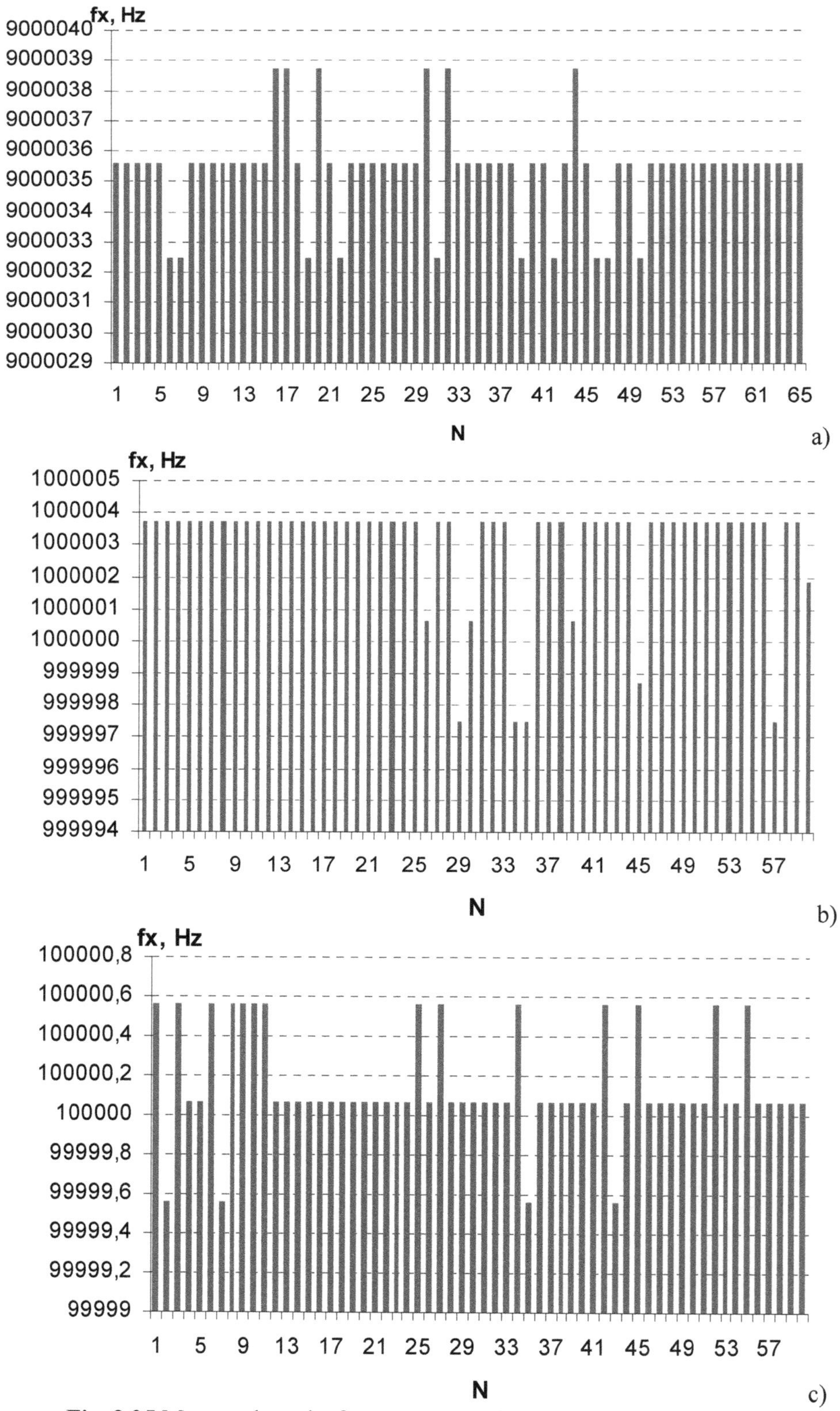

Fig. 3.25 Measured results for square waveform signal with frequency 9 MHz (a), 1 MHz (b) and 100 kHz (c).

Table 3.8 Statistical characteristics of experimental results.

Parameter	Value		
	9 MHz	1 MHz	100 kHz
Number of measurements, N	65	60	60
Minimum f_x (min), Hz	9000032.48	999997.488	99999.5635
Maximum f_x (max), Hz	9000038.73	1000003.74	100000.563
Sampling Range, f_x (max)- f_x (min), Hz	6.2515	6.2515	1
Median	0	0	0
Arithmetic Mean, Hz	9000035.42	1000003.05	100000.146
Variance	2.405	3.1488	0.0692
Standard Deviation	1.5508	1.7745	0.2631
Coefficient of Variation	5803428.66	563543.777	380129.039
Confidence Interval at probability P = 97 %	f_x ϵ [9000035 … 9000035.83]	f_x ϵ [1000002.55 …1000003.54]	f_x ϵ [100000.073 … 100000.22]
Relative error, %	0.00039 < 0.00050	0.00030 < 0.00050	0.00014 < 0.00050*

The maximal possible frequency of a square waveform pulse signal for the USTI was 9.1 MHz without prescaling, the minimal possible frequency was 0.04 Hz.

For some applications of frequency measurements, for example, special underwater communication (< 100 Hz), electrochemical converters (0.0005 … 100 Hz), bioelectrical processes (0.001 … 0.6 Hz), vibrations (7 … 400 Hz), low rotational speed measurements (a few *rpm*) in industrial process and astronomy [15], the infralow frequency range is very critical from the required accuracy point of view. Existing measuring instruments can not supply reliable measurements and low relative error in such measuring range. The USTI is based on the patented, most advanced modified method of the dependent count for precision frequency and period measurements. The results of measurements for square waveform pulse signal of infralow frequency 0.05 Hz are shown in Figures 3.26 (a-c). As it is visible form the plots, the absolute error does not exceed $\Delta_f = \pm 4.5 \times 10^{-8}$ Hz, and relative error does not exceed $\delta_f = \pm 0.00009\ \% < 0.0005\ \%$ (minimum possible programmable relative error for the USTI).

Frequency and period of non-rectangular waveform signals that have a slow slew rate - such as triangular, sine, exponential rise and fall waveform signals - must often be measured. Low frequency measurements (50, 100 and 200 Hz) involving signals with a slow slew rate are common, for example, in the conversion-to-digital of resistance, capacitance, inductance or resistive-sensor–bridge signals based on direct connection to a microcontroller [17, 18]. Such an approach offers a simple low-cost interface between microcontrollers with embedded timer/counter and quasi-digital sensors with a period, time interval or frequency output. The indirect (period) counting method for slow slew rate signals is susceptible to errors because of changes in the input trigger threshold due to several external and internal noise sources. In addition to thermal noise, there are other trigger noise sources that are program dependent [19] or result from interference in the microcontroller power supply [20, 21]. Very often, trigger errors predominate over the inherent quantization error, so that precise frequency or period measurements require the use of an external input signal-forming device, such as an amplifier-limiter with a Schmitt trigger circuit, which yields rectangular waveform impulses from the slow slew rate signal [19]. However, that solution increases the hardware cost and reduces reliability, because the probability of failure is directly proportional to the number of components being used.

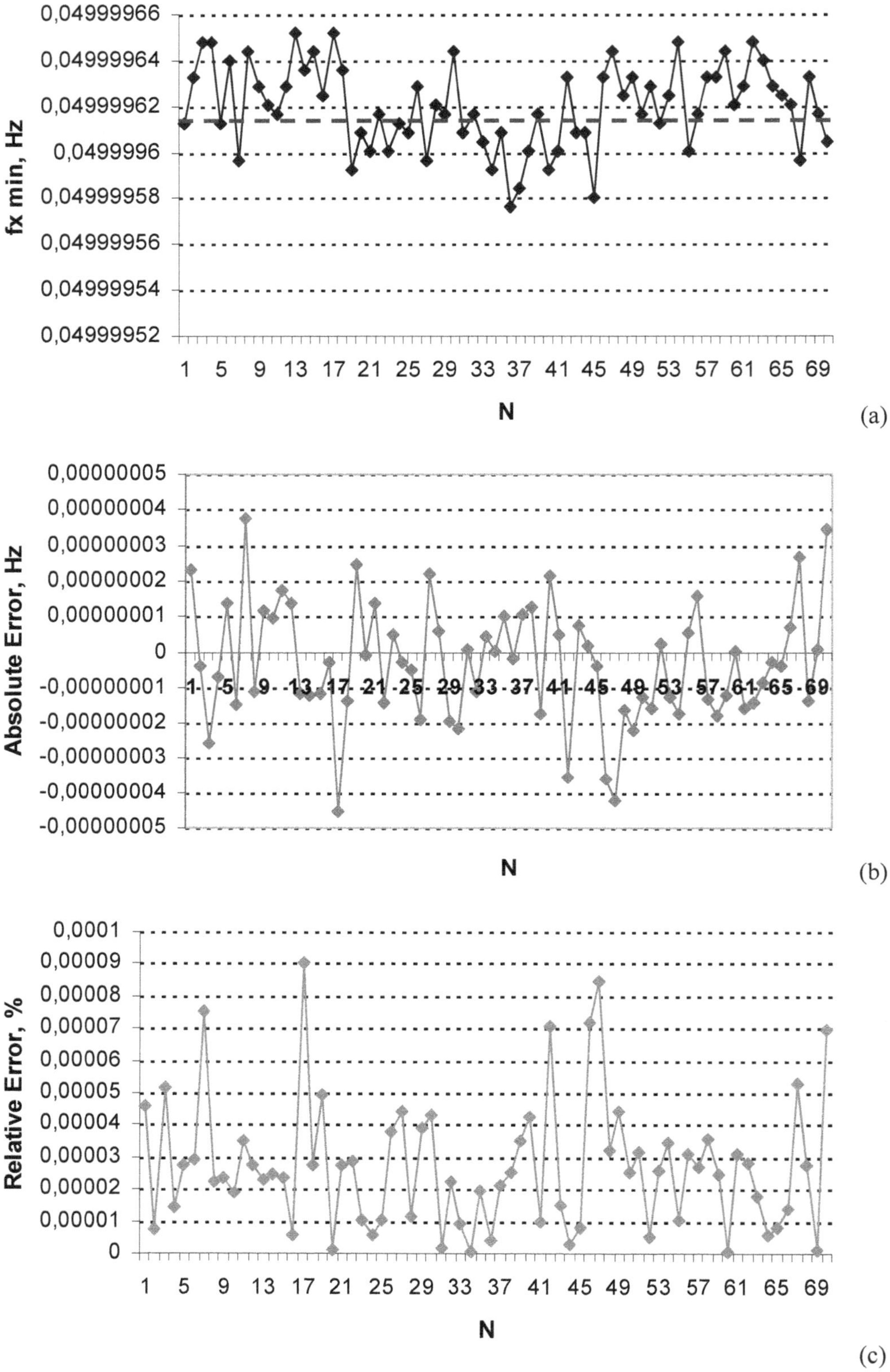

Fig. 3.26 Measuring results for infralow frequency 0.05 Hz: (a) results of 70 samples measurements; (b) absolute error; and (c) relative error.

Figure 3.27 shows a triangular signal whose period is determined by measuring the time interval between consecutive voltage crossings of its leading ramp. In a noiseless system (V_{th} undistorted), the period measured would be T_x and the input trigger points would be at S_1 and S_2. However, noise and interference added to V_{th} shift the trigger points, so that the measured period will be $T_x+\tau_1+\tau_2$. Usually, $\tau_1 \neq \tau_2$ because $\Delta V_1 \neq \Delta V_2$.

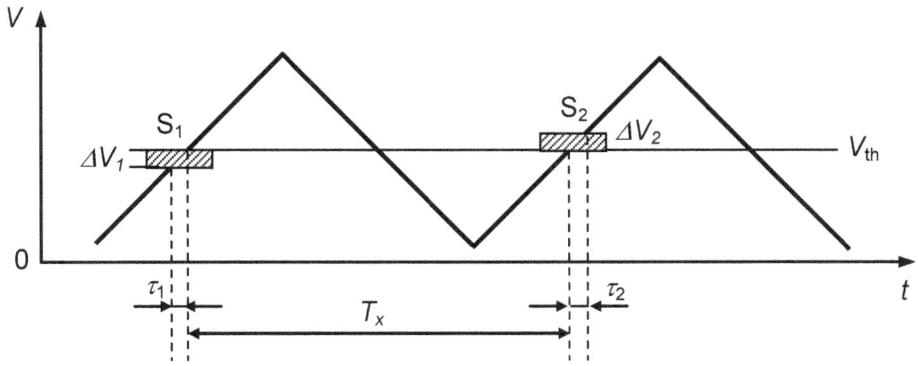

Fig. 3.27 Error level related to trigger point uncertainty when measuring the period of a triangular signal.

Frequency-to-digital conversions are immune to that error component, because it is not necessary to extract a separate period from the pulse train and $\tau_1 = \tau_2 = 0$, so that we can measure f_x and then calculate the period as $T_x=1/f_x$ [16]. But the standard direct counting method for frequency measurement applied to low frequencies has either a very high quantization error or a long measurement time. For example, in order to have a quantization error below 0.01 % at f_x = 100 Hz, the conversion time must be larger than 100 s. Advanced methods for frequency measurement, such as ratiometric, reciprocal, M/T, constant elapsed time (CET), single- and double-buffered or direct memory access (DMA) methods, and methods using a non-redundant reference frequency [16] all have a constant quantization error regardless of signal frequency; however, their conversion time is redundant, which increases the dynamic error and the measurement time for all frequencies except the nominal one. It would be preferable to use an alternative method related to frequency (period) conversion, one that would have a constant quantization error over the entire frequency range measured and a non-redundant conversion time.

The USTI uses a novel approach to the measurement of slow slew rate signal frequency (period) based on the modification of method of the dependent count. For the experimental investigation of slow slew rate signals with different waveforms the experimental set up shown in Figure 3.24 has been used. Different waveform (rectangular, triangular, sine, exponential rise and exponential fall) signals whose periods were to be measured were fed from a function generator (Agilent 33220A) to inputs FX1 and ST1 of the USTI. The internal reference frequency was f_0 = 625 kHz with the relative error δ_0 = ± 0.0001 %. The programmable relative error of period measurements was chosen as δ = ± 0.0005 %. The USTI was supplied at +5 V dc by a Promax FAC-363B power supply. The frequency generated by the Agilent 33120A was 100 Hz; its respective period was 10 ms and signals were monitored by a high precision calibrated frequency counter (Agilent 53132A). This frequency was chosen to be the same as in experiments described in [13, 19, 20, 22, 23] in order to compare the experimental results. Each slow slew rate signal was measured by the USTI directly as well as through an external

Schmitt trigger integrated circuit 74HC14D (Figure 3.28, a-e). The waveform signal parameters are shown in Table 3.9. The peak-to-peak amplitude was approximately 5 V. The Promax OD-571 oscilloscope monitored the signal waveforms.

The measured period value was sent to a PC via an RS232 interface implemented with the ST202D integrated circuit. The user interface was implemented with the help of terminal software (Terminal V1.9b for Windows). Period measurements were taken every second until a total of 60 measurements had been recorded. The measurement error was evaluated from histograms and appropriate statistical characteristics.

Table 3.9 The waveform signal parameters.

	Triangular	**Sine**	**Exponential Rise**	**Exponential Fall**
Rise Time:				
1. Input Signal	3.275 ms	2.82 ms	3.733 ms	78.49 μs
2. Schmitt Trigger Output	78.73 μs	78.73 μs	80.00 μs	81.29 μs
Fall Time				
1. Input Signal	3.3 ms	2.653 ms	75.47 μs	3.273 ms
2. Schmitt Trigger Output	78.73 μs	78.73 μs	81.31 μs	80.00 μs

The measured periods and the probability densities for triangular waveform signals of 100 Hz with 50 % symmetry are shown in Fig. 3.29 (a, b). The statistical characteristics are shown in Table 3.10 The number of equidistant classes k in the probability density function chart for a sampling of 60 measurements was calculated according to [24] and is $k = 5$. The relative error can be reduced 1.53 times by using an external Schmitt trigger.

The χ^2 test for goodness of fit test was applied to investigate the significance of the differences between observed data in the histograms and the theoretical frequency distribution for data from a normal, or uniform, population. For the triangular signal, the period of which was measured with the use of a Schmitt trigger as an input device, for five equidistant classes (Fig. 3.29, b) and a probability $P = 97$ %, according to the χ^2 test, $S < \chi^2_{max}$, where $S = 8.6667$ is the sum of deviations between the dataset and the assumed distribution; $\chi^2_{max} = 10$ is the maximum possible allowable deviation in the χ^2 distribution. Hence, the hypothesis of uniform distribution can be accepted.

The experimental results for sine wave signal period measurements and the probability densities are shown in Fig. 3.30 (a, b). In this case the use of an external Schmitt trigger does not improve the relative error (Table 3.10). There is a small error increase (from 0.000127 to 0.000159 %), but in neither cases does the relative error does exceed the programmable relative error of $\delta = \pm\ 0.0005$ %. For the sine waveform signal, at five equidistant classes (Fig. 3.30, a) and a probability $P = 97$ %, according to the χ^2 –Test, $S < \chi^2_{max}$, where $S = 3.6821$ is the sum of deviations between the dataset and the assumed distribution; $\chi^2_{max} = 7.0$ is the maximum possible deviation in the χ^2 distribution. Hence, the hypothesis of Gaussian (normal) distribution can be accepted. In a case in which a Schmitt trigger is used (Fig. 3.30, b), at five equidistant classes and the same probability $P = 97$ %, according to the χ^2 test, $S < \chi^2_{max}$, where $S = 4.2002$ is the sum of deviations between the dataset and the assumed distribution; $\chi^2_{max} = 7.0$ is the maximum possible argument of the χ^2 distribution. Hence, the hypothesis of Gaussian distribution can be also accepted in this case.

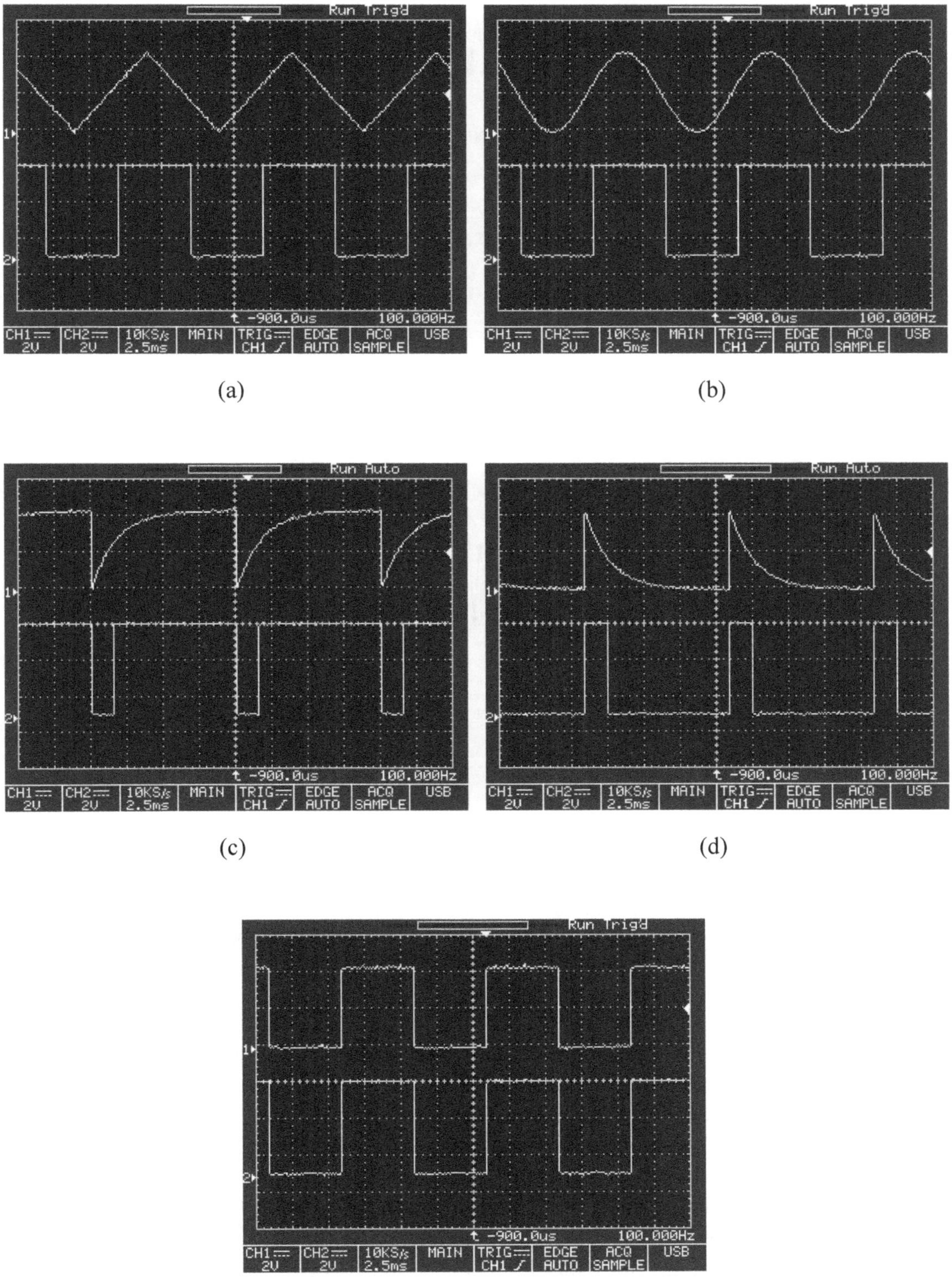

Fig. 3.28 Input forming device time diagrams: input (1), and output (2) signals for different investigated waveforms.

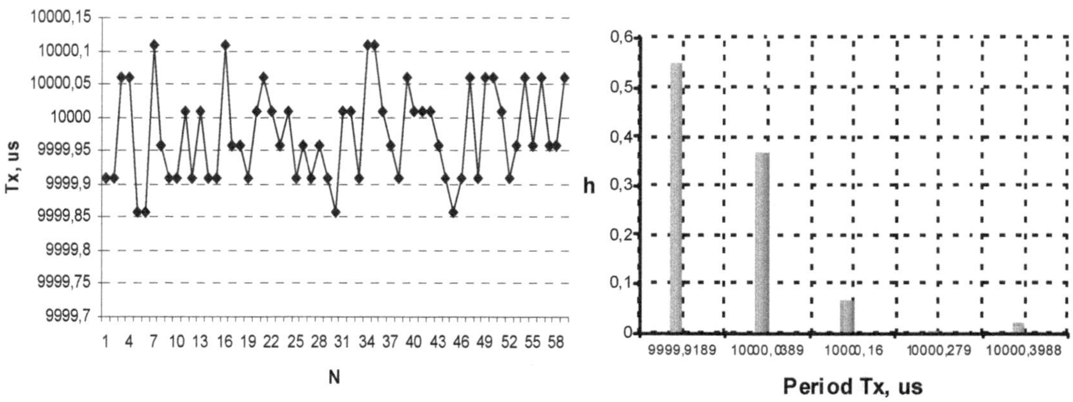

a)

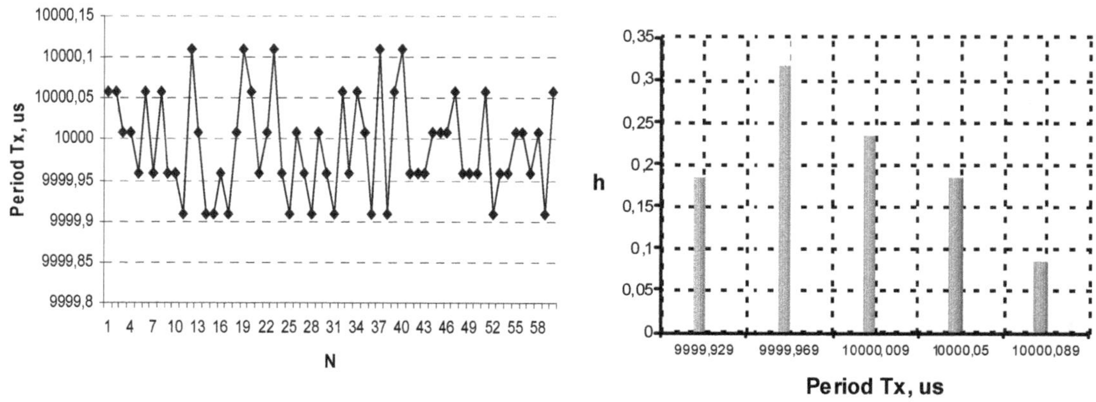

(b)

Fig. 3.29 Experimental results of direct period measurement for a triangular waveform signal and its probability density (a); experimental results and probability density with the use of an external Schmitt trigger (b).

The experimental results for exponential rise waveform signal period measurements and the probability densities are shown in Fig. 3.31 (a, b). The similar signal waveform is often used in resistance, capacitance and resistive bridge-to-time interval conversion at the direct sensor-to-microcontroller interfaces based on a charging or discharging capacitor. In case in which an external Schmitt trigger is used (Fig. 31, b) a small decrease in relative error is observed (Table 3.10). In this case at five equidistant classes and the probability P = 97 %, according to the χ^2 test, $S < \chi^2_{max}$, where $S = 1.9434$ is the sum of deviations between the dataset and the assumed distribution; $\chi^2_{max} = 7.0$ is the maximum possible deviation of the $\chi 2$ distribution. Hence, the hypothesis of Gaussian distribution can be accepted.

The experimental results for exponential fall waveform signal period measurements and the probability densities are shown in Fig. 3.32 (a, b). The use of an external Schmitt trigger as a forming device does not give any relative error improvement (Table 3.10).

Table 3.10 Statistical characteristics of the results for period measurements (10 ms) of rectangular waveform signal and slow slew rate signals of different waveforms.

Signal Wave Form / Parameter	Rectangular		Triangular		Sine		Exponent Rise		Exponent Fall	
	Directly	With Schmitt trigger	Directly	With Schmitt trigger	Directly	With Schmitt trigger	Directly	With Schmitt trigger	Directly	With Schmitt trigger
Minimal T_x (min), µs	9999.9605	9999.9605	9999.859	9999.90	9999.8605	9999.8605	9999.8582	9999.8582	9999.9587	9999.9587
Maximal T_x (max), µs	10000.0105	10000.0105	10000.4588	10000.1089	10000.0605	10000.0605	10000.1082	10000.0582	10000.0587	10000.0587
Sampling Range, T_x (max) - T_x (min), µs	0.05	0.05	0.5999	0.2	0.2	0.2	0.2499	0.2	0.1	0.1
Median	0	0	0	0	0	0	0	0	0	0
Arithmetic Mean, µs	9999.963	9999.9647	9999.9831	9999.9923	9999.9755	9999.9688	9999.9882	9999.9749	9999.9937	9999.9854
Variance, µs	0.0001	0.0002	0.0088	0.0037	0.0021	0.0032	0.0034	0.0029	0.0009	0.0009
Standard Deviation, µs	0.011	0.0139	0.0936	0.0608	0.0454	0.0569	0.0584	0.0534	0.0295	0.0298
Coefficient of Variation	910206.786	717747.766	106806.024	164427.259	220416.908	175875.618	171370.344	187261.309	338555.832	335849.227
Confidence Interval at probability P=97 %	$T_x \in$ [9999.9599÷ 9999.9661]	$T_x \in$ [9999.9607÷ 9999.9686]	$T_x \in$ [9999.9569÷ 10000.0093]	$T_x \in$ [9999.9752÷ 10000.0093]	$T_x \in$ [9999.9628÷ 9999.9882]	$T_x \in$ [9999.9529÷ 9999.9847]	$T_x \in$ [9999.9719÷ 10000.0046]	$T_x \in$ [9999.9599÷ 9999.9898]	$T_x \in$ [9999.9854÷ 10000.002]	$T_x \in$ [9999.977÷ 9999.9937]
Maximal Relative Error, %	≤ ±0.000031	≤ ±0.00004	≤ ±0.000262	≤ ±0.000171	≤ ±0.000127	≤ ±0.000159	≤ ±0.000164	≤ ±0.00015	≤ ±0.000083	≤ ±0.000084
Distribution low	-	-	-	Uniform	Gaussian	Gaussian	-	Gaussian	-	-

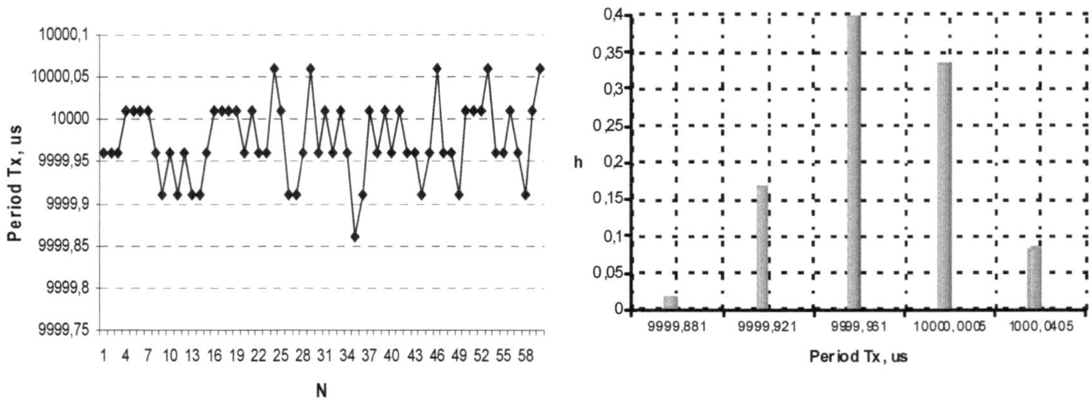

(a)

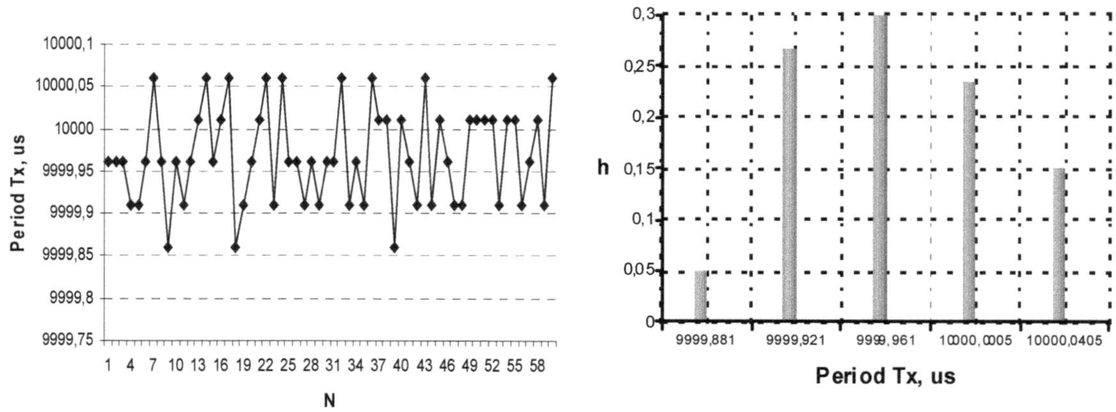

(b)

Fig. 3.30 Experimental results of direct period measurement for sine waveform signal and its probability density (a); experimental results and probability density with the use of an external Schmitt trigger circuit (b).

In the case of the rectangular waveform signal, the relative error of period measurements is minimal in comparison with non-rectangular slow slew rate signals and less in 2/8.5 times (Table 3.10). Nevertheless the use of an external Schmitt trigger decreases the pulse rise time from 24.25 ns to 21.75 ns (Fig. 3.33), and the use of an external Schmitt-trigger circuit in this case is not recommended (due to the increase in relative error) because the level of internal noises (thermal noise, interference in the IC power supply, etc.) is greater than the microcontroller's internal noises. Most significant deviations from the arithmetic mean in experimental period measurement results for square wave signals are determined by so-called program-related quantization effects, such as an interrupt response shift in time and interrupt response delay because the interrupt response time depends on the instructions executed by the microcontroller [16]. However, the total relative error does not exceed the programmable one.

These experimental results were compared with those obtained in [13, 19, 22, 23] for a 100 Hz triangular signal when measuring the period using the indirect counting method.

From the counting results of 10 003 μs and 10 006 μs reported in [13, 19, 23] and 10 003 μs and 10 004 μs reported in [22], the maximum quantization errors were ± (0.03 ... 0.06) % and ± (0.03 … 0.04) % respectively. When measuring the same frequency and waveform with the USTI IC, the maximum relative error was below ± 0.000262 % (Table 3.10). In the case of a 100 Hz square wave signal, the relative error reported in [22] was ± (0.01 … 0.03) % in comparison with the maximum relative error obtained in described experiments, which did not exceed ± 0.000031 % (Table 3.10).

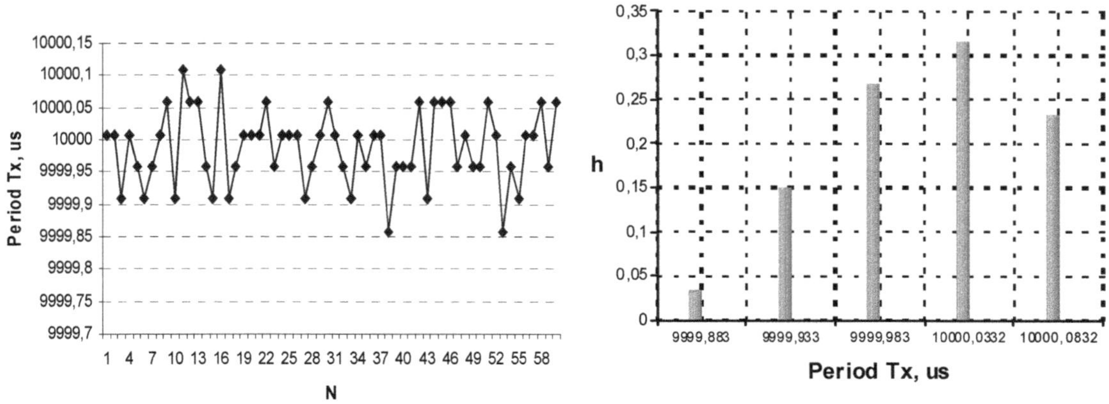

(a)

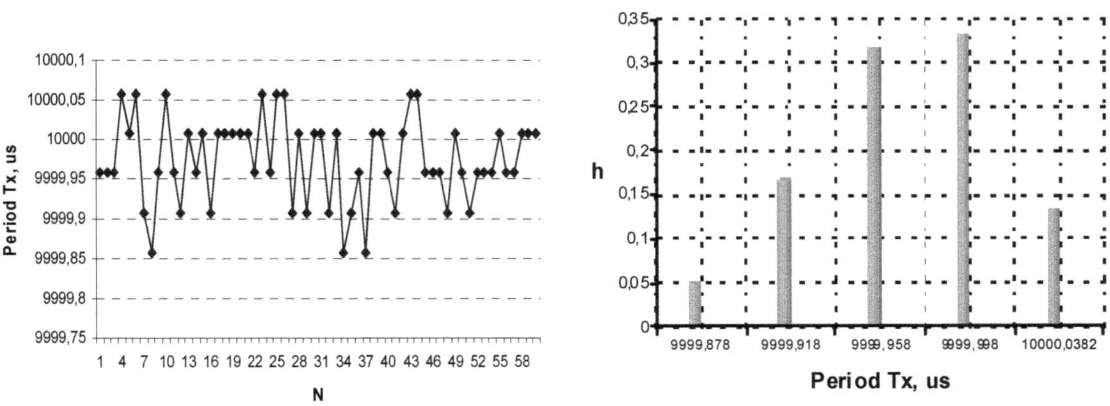

(b)

Fig. 3.31 Experimental results of direct period measurement for an exponential rise waveform signal and its probability density (a); experimental results and probability density with the use of an external Schmitt trigger circuit (b).

With regard to the conversion speed, for the ± 0.016 % relative error, neither of the two methods exceeds one period T_x. Of course, increased accuracy in the case of the modified method of the dependent count will require increasing conversion time, but, in any case, this time will be non-redundant for the proposed modified method. The reference frequency increase of up to f_0 = 20 MHz will significantly reduce the conversion time in the USTI IC. So, for the relative error δ = ± 0.0005 % the conversion time will not exceed one period T_x = 0.01 s, which corresponds to the 100 Hz input signal.

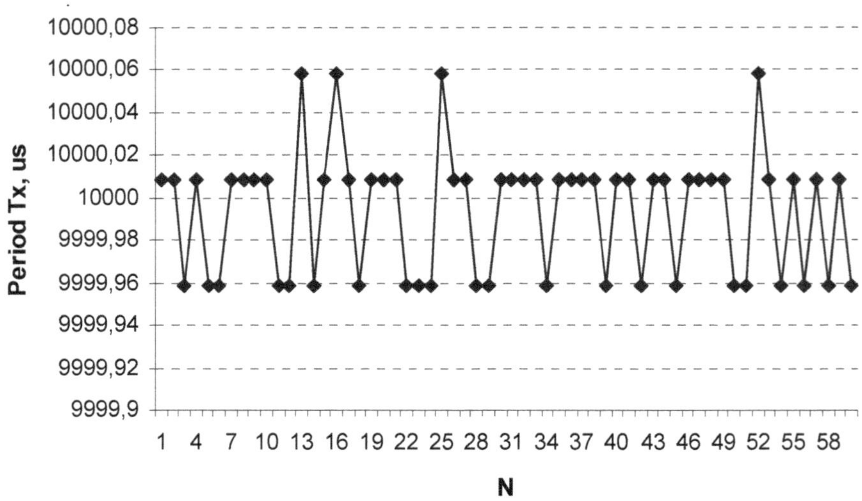

(a)

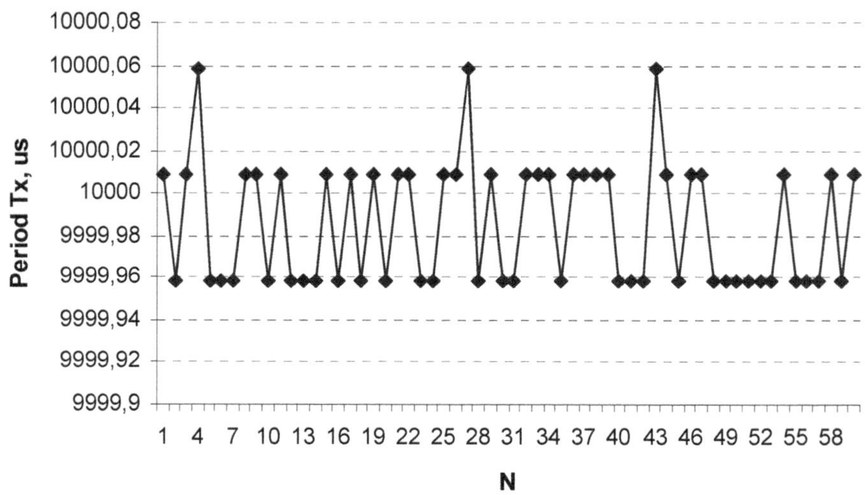

(b)

Fig. 3.32 Experimental results of direct period measurement for an exponential fall waveform signal (a); experimental results with the use of an external Schmitt trigger circuit (b).

Even though the relative error can also be reduced slightly by using an external Schmitt trigger, it is not expedient to use this device at a programmable relative error $\delta = \pm 0.0005$ %, because, as is evident from the Table 3.10, for any investigated signals the relative error without the Schmitt trigger does not exceed the programmable relative error of $\delta = \pm 0.0005$ %.

During the experiments, the minimal possible pulse width (t_p), pulse space (t_s) and time interval (t_x) were also determined. The minimal possible value for these time parameters was 1.5 μs.

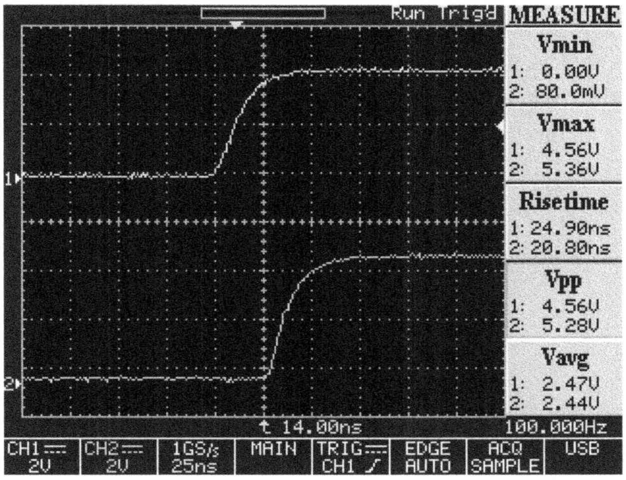

Fig. 3.33 Rectangular pulse wave front without a Schmitt trigger circuit (1) and with a Schmitt trigger circuit (2).

Another important parameter for a square and rectangular waveform pulse signal is the duty-cycle. The duty-cycle values determined for the maximal frequency ranges are adduced in Table 3.11 and shown in Figure 3.34 in comparison with the Series of UFDC ICs.

Table 3.11 Limit duty-cycle for the input signal depending on the maximal frequency.

Duty-cycle, %	Frequency f_x, MHz
47.5 … 57.0	9
42.0 … 62.0	8
36.5 … 66.0	7
32.0 … 71.5	6
26.0 … 76.5	5
20.5 … 80.0	4
any	< 3

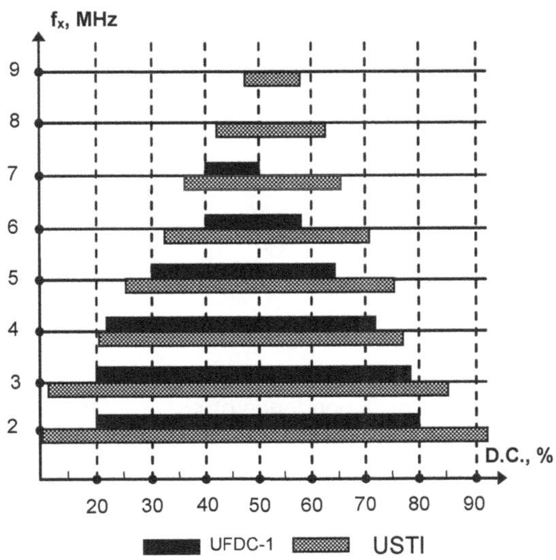

Fig. 3.34 Dependence between $f_{x,max}$ and duty-cycle for the UFDC and USTI ICs.

One of main USTI's parameter is the conversion time. According to the modified method of the dependent this time is non-redundant, adaptive (determined only by the programmable relative error δ_x) and minimum possible. The conversion time can be calculated according to the methodology, described in 2.3.

Simulation results of conversion time for the USTI prototype chip are shown in Fig. 3.35 The conversion time is most significant at precision low and infralow frequency-to-digital conversions but does not exceed the period T_x in this frequency range.

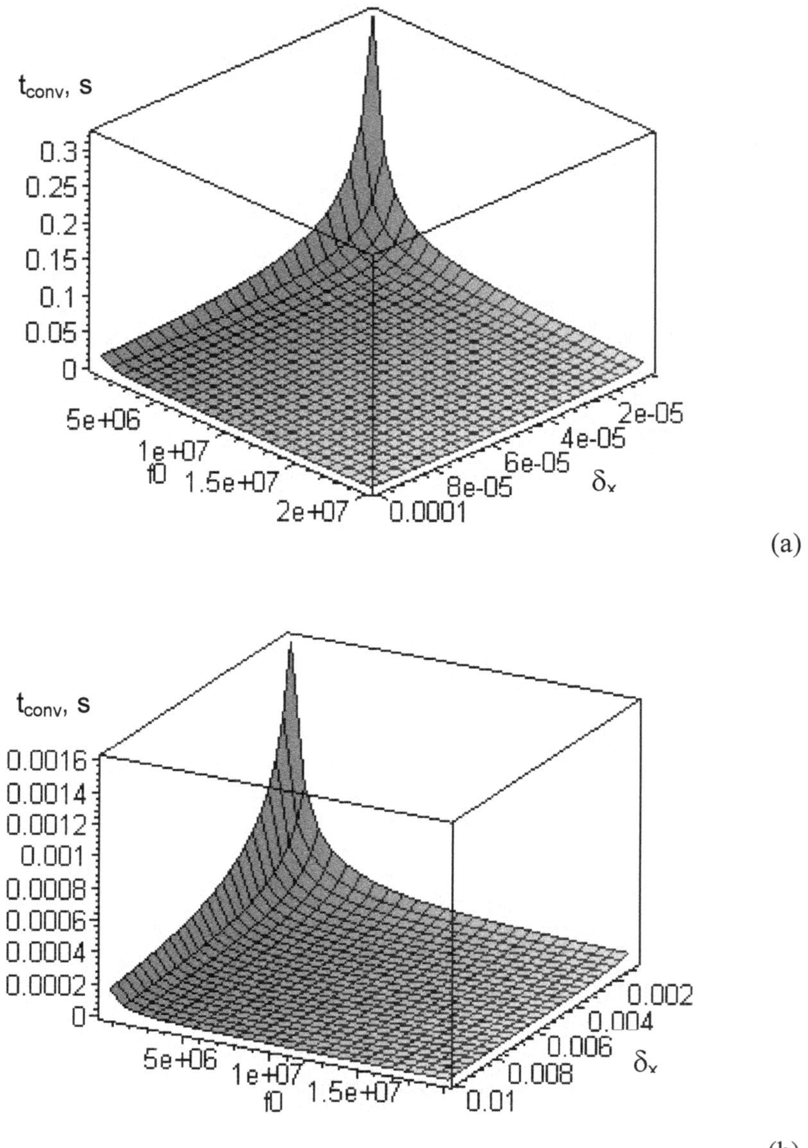

Fig. 3.35 Modeling results for dependence of $t_{conv} = \varphi (f_0, \delta)$ at range of variables f_0=625 kHz ...20 MHz, δ_x = 0.001...0.000005 (a), and δ_x = 0.01...0.001 (b).

Let's compare the conversion time of the USTI that is based on the modified method of the dependent count with conversion times, which can be achieved by using any classical or other advanced conversion methods described in [18]. For the indirect counting method with the reference frequency f_0=20 MHz, the relative error $\delta_x = \pm 0.0005$ % and conversion time that does not exceed one period T_x the frequency range will be essentially limited to 0.05...100 Hz. In case of direct counting technique, for the same relative error and gate time 0.01 s it will be possible to convert only frequencies from the range $f_x \in$ [~ 44.7 kHz ... 20 MHz] and for all relative error except the $\delta_x = \pm 0.0005$ %, the conversion time will be redundant. For advanced conversion methods, for example, reciprocal, ratiometric, M/T, constant elapse time, single- and double buffered, DMA [16], at the reference frequency f_0=20 MHz and gate time T_0=0.01 s the conversion time also will be redundant for all relative error except the $\delta_x = \pm 0.0005$ %. In other words, the conversion time will be the same - 0.01 s for the relative error ± 0.0005 % and relative error ± 1 %.

Experimental results confirm high metrological performance of USTI IC. Its conversion time is non-redundant, adaptive and minimum possible. The relative error of USTI was below ± 0.00043 % (in the worst case) in the whole frequency range including low and infralow frequencies. This was less than the programmable relative error (± 0.0005 %). Due to scalable resolution, programmable relative error and non-redundant conversion time this chip can work with any existing frequency-time domain sensor to produce a digital output or create multiparametric smart and intelligent sensors and sensor systems.

Such ICs will simplify significantly a digital sensors and smart sensor systems design process, reduce development time and production price of different applications based on innovative chip while improving sensor system performances. In comparison with the direct microcontroller interfacing the designed ICs also let to eliminate many design problems connected with the use of advanced conversion methods for frequency-time parameters, embedded microcontroller choice, its programming and additional error components due to so-called program-dependent or software-related effects.

In addition the USTI ICs can store the TEDS in its flash memory. It can be embedded into any existing frequency-time domain output sensors or transducers (with the help of integrated or hybrid technologies) to produce IEEE 1451.4 compatible sensors as well as into IEEE 1451.4 compatible data acquisition systems.

Digital transducers OEMs will receive a unique opportunity to produce low-cost IEEE 1451 compatible sensors with minimum possible hardware. They should not think now about the frequency-to-digital conversion accuracy. Only one component can cover three main functions of smart transducers: frequency (time)-to-digital conversion with high metrological performances, TEDS storage in the flash memory and communications.

3.8 USTI for Extended Temperature Range (USTI-EXT)

Some applications, for example, automotive, military and other needs ICs, which are able to work in an extended temperature ranges, for example, from -40 ^{0}C to +150 ^{0}C. The Universal Sensors and Transducers Interface (USTI-EXT) IC is working in the mentioned temperature range. In comparison with the basic USTI, the USTI-EXT has the reduced external clock frequency 16 MHz, but increased internal reference frequency from

625 kHz to 16 MHz and as consequence, the metrological performance as for the UFDC-1M-16 IC [25], but functionality as for the USTI IC [5]. Supply current does not exceed 10 mA. The IC has also increased baud rate for the RS232 interface: up to 76 800 (command B9).

The IC is available in two packages: 32-lead, 7×7 mm, Thin Profile Plastic Quad Flat Package (TQFP) and 32-pad, 5×5 mm, Quad Flat No-Lead/Micro Lead Frame Package (QFN/MLF), Figure 3.36.

Fig. 3.36 USTI-EXT IC is TQFP and MLF packages.

Both ICs: the USTI and USTI-EXT are available for purchase from the Sensors Web Portal, Inc. and Technology Assistance BCNA 2010, S. L. web sites (online stores):
http://www.sensorsportal.com/HTML/E-SHOP/PRODUCTS_4/USTI.htm
http://www.techassist2010.com/en/orders-buy-products/place-orders.html

3.9 USTI for Wireless Sensor Networks Applications (USTI-WSN)

This IC has the same metrological performance as for UFDC-1M-16 [25] and functionality as for the USTI [5] ICs, but in addition to three serial interfaces and buses (RS232, I2C and SPI) the IC has integrated low power transceiver for 2.4 GHz band. Supported data rates are 250 kb/s and 500 kb/s, 1 Mb/s, 2 Mb/s. The device is available in the 64-pad 9.0 × 9.0 mm QFN package (Figure 3.37).

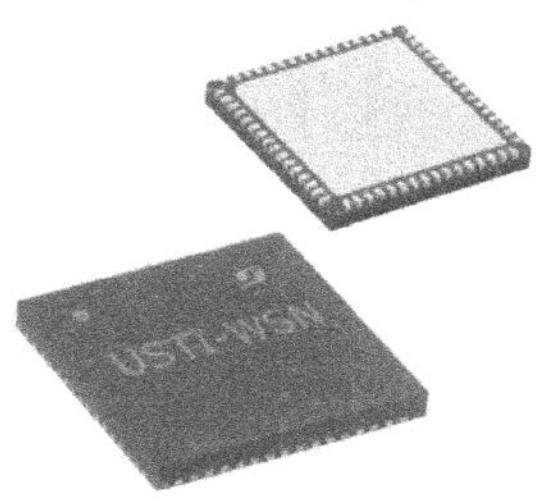

Fig. 3.37 USTI-WSN IC in 64-pad 9.0 × 9.0 mm QFN package (not for scale).

Supply voltage range is 3.6 V, and power consumption < 18.6 mA. The USTI-WSN provides frame handling, outstanding receiver sensitivity and high transmit output power enabling a very robust wireless communication. The device is fully compatible with IEEE802.15.4-2006/2003 and ZigBee standards.

3.10 References

[1]. Yurish S. Y. Smart Universal Sensor and Transducer Interface, in *Proceedings of SENSOR+TEST Conference 2007*, Nurnberg, Germany, 22-24 May 2007, Vol. 1, pp. 307-312.

[2]. Yurish S. Y., Universal Smart Sensors Interface and Signal Conditioner, in *Proceedings of 6th IEEE Conference Sensors 2007*, Atlanta, USA, 28-31 October, 2007, pp.24-27.

[3]. Yurish S. Y., Novel Universal Frequency-to-Digital Converters and Sensors Interface Integrated Circuits, *Sensor Electronics and Microsystem Technologies*, 3, 2008, pp.81-90.

[4]. Universal Sensors and Transducers Interface (USTI), Datasheets, *Technology Assistance BCNA 2010, S. L.*, 2011.

[5]. Universal Sensors and Transducers Interface (USTI), Specification and Application Note, *Technology Assistance BCNA 2010, S. L,*. 2011.

[6]. DS4026 Digital TCXO Provides Extremely High Accuracy over Wide Temperature Range, Maxim Integrated Products, *Dallas Semiconductor*, 2008.

[7]. M. E. Frerking, Fifty Years of Progress in Quartz Crystal Frequency Standards, in *Proceedings of the 1996 IEEE International Frequency Control Symposium*, June 1996, pp. 33 – 46.

[8]. Terminal V1.9b: http://www.sensorsportal.com/DOWNLOADS/Terminal.exe

[9]. Sergey Y. Yurish, Nikolay V. Kirianaki, Automated Calibration Technique for Programmable Universal Frequency-to-Digital Converter, in *Proceedings of the IFAC Workshop on Programmable Devices and Embedded Systems (PDeS'2006)*, Czech Republic, Brno, 14-16 February 2006, pp.72-77.

[10].Yurish S. Y., Advanced Automated Calibration Technique for Universal Sensors and Transducers Interface IC, in *Proceedings of IEEE International Instrumentation and Measurement Technology Conference (I²MTC 2009)*, Singapore, May 5-7, 2009, pp.402-405.

[11].Worldwide Component Library: http://www.2wcl.com/

[12].Design & Reuse: http://www.design-reuse.com/

[13].F. Reverter, R. Pallàs-Areny, Direct Sensor-to-Microcontroller Interface Circuits. Design and Characterization, Barcelona, *Marcombo*, 2005.

[14].A. Custodio, R. Bragos, R. Pallas-Areny, A Novel Sensor-Bridge-to-Microcontroller Interface, in *Proceedings of the IEEE Instrumentation and Measurement Technology Conference*, Budapest, Hungary, 21-23 May 2001.

[15].Kirianaki N. V., Dudykevich V.B., Methods and Devices for Low and Ifralow Frequency Measurements, *Vyscha Shkola*, Lvov, 1975.

[16].Kirianaki N. V., Yurish S. Y., Shpak N. O., Deynega V. P. Data Acquisition and Signal Processing for Smart Sensors, *John Willey & Sons*, Chichester, UK, 2001.

[17].R. Pallàs-Areny, J. G. Webster, Sensors and Signal Conditioning, 2nd ed., *John Wiley & Sons*, New York, USA, 2001.

[18].M. A. Atmanand, V. Jagadeesh Kumar, and Vempati G. K. Murti, A Microcontroller-Based Quasi-Balanced Bridge for the Measurement of L, G and R, *IEEE Transactions on Instrumentation and Measurement*, Vol. 45, No. 3, June 1996, pp.757-761.

[19].F. Reverter, J. Jordana, R. Pallàs-Areny, Program-Dependent Uncertainty in Period-to-Code Converters Based on Counters Embedded in Microcontrollers, in *Proceedings of the 20th IEEE Instrumentation and Measurement Technology Conference (IMTC' 03)*, Vail, Colorado, USA, 20-22 May 2003, Vol.2, pp.977-980.

[20].F. Reverter, O. Casas, J. Jordana, R. Pallàs-Areny, Trigger Uncertainty in Period-to-Code Converters Based on Counters Embedded in Microcontrollers, *Sensors and Actuators A*, Vol. 110, 2004, pp.439-446.

[21].F. Reverter, M. Gasulla, R. Pallàs-Areny, Analysis of Interference Rejection in Period-to-Code Converters, in *Proceedings of the 21st IEEE Instrumentation and Measurement Technology Conference (IMTC' 2004),* Como, Italy, 18-20 May 2004, pp.839-842.

[22].F. Reverter, R. Pallàs-Areny, Uncertainty Reduction Techniques in Microcontroller-based Time Measurement, *Sensors and Actuators A*, Vol. 127, 2006, pp.74-79.

[23].F. Reverter, Microcontroller-Based Interfaces for Quasi-Digital Sensors, PhD Thesis, *Universitat Politecnica de Catalunya (UPC-Barcelona),* 2004.

[24].]Novitskiy P. V., Zograf I. A., Errors Estimation of Measurement Results, *Energoatomizdat*, 1991 (in Russian).

[25].Universal Frequency-to-Digital Converter (UFDC-1), Datasheets, *Technology Assistance BCNA 2010, S. L.,* 2011.

Chapter 4

Frequency-to-Digital Converter with Parallel Interface (FDCP)

4.1 General Description

Some applications of frequency-to-digital converters and digital sensors on its basis, for example, control systems, together with a high conversion rate (tconv) need a high communication speed (tcomm) to reduce the measurement time (Tmeas). The last one can be achieved due to the usage of parallel communication interface (Table 4.1). In order to meet such market requirements the IC of Frequency-to-Digital Converter (FDCP) with a microcontroller (DSP microprocessor) compatible parallel interface has been introduced on the modern market in 2011 by *Technology Assistance BCNA 2010, S. L.* [1].

Table 4.1 Interface speed comparison.

Protocol	Speed
RS232	115 kbps
I^2C	400 kbps
SPI	15 Mbps
Parallel	480 Mbps

The FDCP is a fully digital, low-power CMOS integrated circuit based on the novel patented methods for frequency measurements [2-4]. It provides the technologies that both reduce the cost and time of sensor application development and improve digital sensors and systems performances. The photo of IC in 64-lead TQFP package is shown in Figure 4.1. The main metrological performances and electrical characteristics are adduced in Table 4.2. The conversion time is non-redundant from 6.7 μs to 1.6.ms depending on the programmable relative error: the smaller the relative error, the greater the conversion time.

Fig. 4.1 IC FDCP in 64-lead TQFP package.

Table. 4.2 Main metrological performances and electrical characteristics of FDCP IC.

Parameter	Value
Minimal converted frequency, Hz	≥ 500
Maximal converted frequency, MHz	≤ 16
Programmable relative errors, %	1; 0.1; 0.01; 0.002 %
Microcontroller (DSP microprocessor) compatible parallel interface, bit	32
External clock oscillator frequency, MHz	16
Internal clock frequency, MHz	32
Supply current, mA	12
Power supply, V	3.3
Operating temperature range, ^{0}C	- 40 to +85

4.2 Pin Out and Housing

The FDCP IC is available in two packages: the 64-lead, 14 × 14 mm body size, 1.0 mm body thickness, 0.8 mm lead pitch, Thin Profile Plastic Quad Flat Package (TQFP) and 64-pad, 9 × 9 × 1.0 mm body, lead pitch 0.50 mm 7.65 mm exposed pad, Micro Lead Frame Package (MLF). The pin-out is shown in Figure 4.2, and pin description in adduced in Table 4.3.

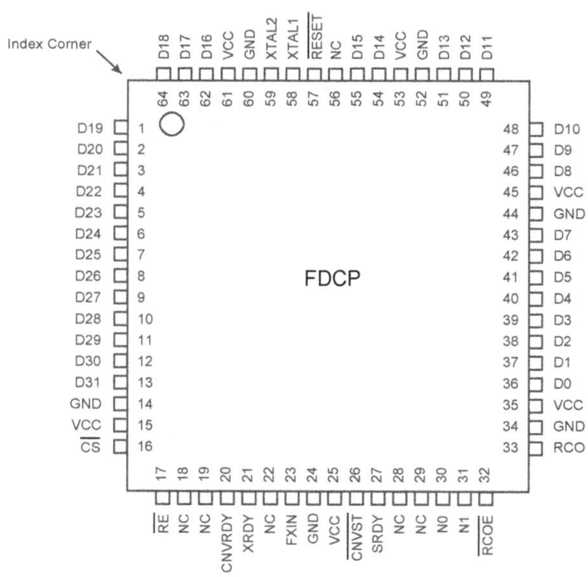

Fig. 4.2 FDCP pin-out.

Table 4.3 FDCP pin description.

Name/ Pin Number	Description
FXIN (23)	Input. Rectangular pulses with frequency range up to10 MHz.
$\overline{\text{CNVST}}$ (26)	Input. Star conversion. A wave tail on this input starts the process of frequency measurement for the input signal.
CNVRDY (20)	Output. Conversion is finished and measuring result is ready A wave front on this output indicates the presence of the result of measurement.
D[31 : 0] (13÷1, 64÷62, 55, 54, 51÷46, 43÷36)	Three-state outputs. Data bus of the converter. Read for results of measurements. Format of the output word: D[31:16]→Nr[15:0], D[15:0]→Nx[15:0]
$\overline{\text{CS}}$ (16)	Input. Chip selection. A low level on this input lets results reading.
$\overline{\text{RE}}$ (17)	Input. Read. A low level on this input together with the low level on $\overline{\text{CS}}$ setup D [31:0] in active state.
XRDY (21)	Output. Data availability on the bus. A high level on this output informs about data presence on the D[31:0].
SRDY (27)	Output. Converter is ready for working. After power-up or reset the converter needs some time for initialization. The high level on this output informs that converter is ready for working.
$\overline{\text{RCOE}}$ (32)	Output. Permits a reference frequency output. Low level on this output during power-on or reset lets a reference frequency output at the RCO output.
RCO (33)	Output. The reference frequency of converter.
$\overline{\text{RESET}}$ (57)	Input. Reseat of converter. Low level resets the converter.
N0 (30)	Input. Relative error set-up. LSB code of the conversion relative error.

<table>
<tr><th>Name/
Pin Number</th><th>Description</th></tr>
<tr><td>N1
(31)</td><td>Input. Relative error set-up.
MSB code of the conversion error.
Code on $N[1:0]$ inputs at wavetail of $\overline{CNVST}$ signal set-up the relative error.
<table><tr><th>$N[1:0]$</th><th>Error</th><th>Conversion time</th></tr><tr><td>00</td><td>1 %</td><td>6.7 μs</td></tr><tr><td>01</td><td>0.1 %</td><td>35 μs</td></tr><tr><td>10</td><td>0.01 %</td><td>320 μs</td></tr><tr><td>11</td><td>0.002 %</td><td>1.6 ms</td></tr></table></td></tr>
<tr><td>XTAL1
(58)</td><td>Output for external quartz oscillator.</td></tr>
<tr><td>XTAL2
(59)</td><td>Output for external quartz oscillator.</td></tr>
<tr><td>VCC
(15, 25, 35, 45, 53, 61)</td><td>Input. Power supply +3.3 V.</td></tr>
<tr><td>GND
(14, 24, 34, 44, 52, 60)</td><td>Ground.</td></tr>
</table>

4.3 Interfacing with Digital Signal Controller

An interfacing of the FDCP with the digital signal controller (DSC) TMS320F28335 from *Texas Instruments*, which has a 32 bit data bus [5] is shown in Figure 4.3 and time diagrams - in Figure 4.4.

The time between falling edge of CNVST signal and rise edge of the CNVRDY equals to the conversion time t_{conv} (6.7 μs for the ± 1 % programmable relative error of measurement). The signal CNVRDY confirms that result is ready for reading. The reading start at RE =0 when CS=0. In 600 ns after the RE signal's fall edge the signal XRDY will inform, that data on the 32-bit data bus are ready for reading. Data appear 8-bit by 8-bit with the100 ns delay, but all are ready at the XRDY signal appearance. All 32 bits should be read simultaneously. After set up the signals CS and RE to logical "1", the data bus returns to the high impedance state.

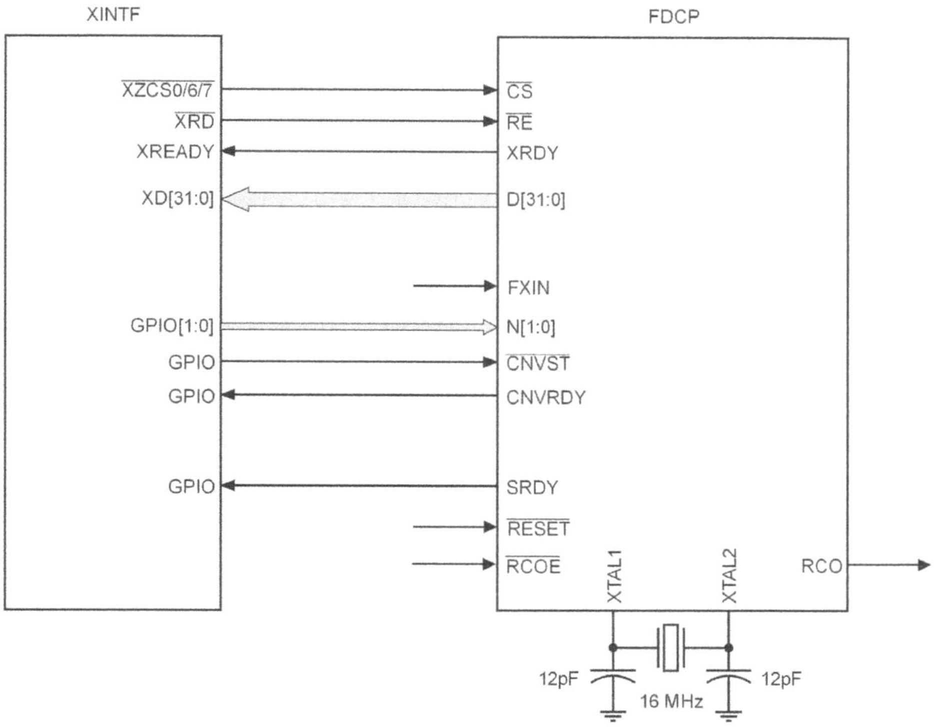

Fig. 4.3 FDCP interfacing with the DSC TMS320F28335.

In order to get measuring result the DSC must calculate the frequency or period according to the following equations:

$$f_x = \frac{N_x}{N_r} \times f_0 \tag{4.1}$$

and

$$T_x = \frac{N_r}{N_x \times f_0}, \tag{4.2}$$

where f_0 is the internal reference frequency (f_0 = 32 MHz); N_x and N_r are the numbers on lower byte and upper byte of data bus D [0:31] respectively. The format of result is the following: D[31:16]=N_r[15:0], D[15:0]=N_x[15:0].

In order to get a precision results with a relative error batter than ± 0.01 % it is necessary to measure the reference frequency f_0 at the FDCP's RCO output with an external frequency counter with the relative error better than ± 0.0007 % and use this value at the result calculaton according to equations (4.1) and (4.2).

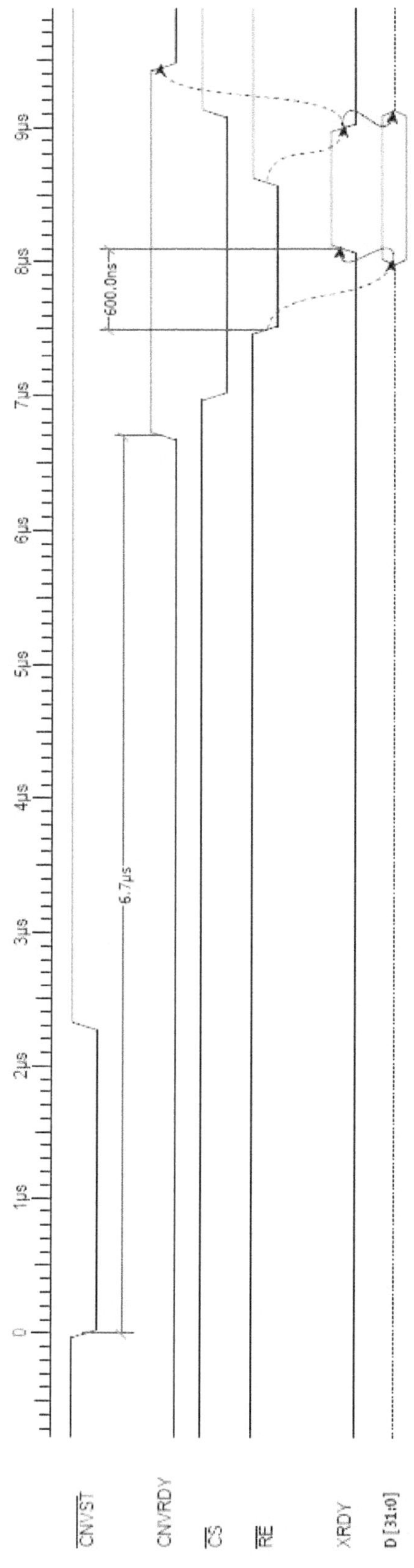

Fig. 4.4 FDCP time diagrams (conversion time 6.7 µs at ±1 % programmable relative error).

4.4 Applications

Applications of FDCP IC are numerous. From digital, smart and intelligent sensors and sensor systems to real-time control systems.

4.5 References

[1]. Technology Assistance BCNA 2010, S. L. web site: http://www.techassist2010.com/
[2]. Krianaki N. V., Yurish S. Y., Shpak N. O., Methods of Dependent Count for Frequency Measurements, *Measurement*, Vol. 29, Issue 1, January 2001, pp.31-50.
[3]. Kirianaki N. V., Yurish S. Y., Shpak N. O., Deynega V. P., Data Acquisition and Signal Processing for Smart Sensors, *John Wiley & Sons*, Chichester, UK, 2002.
[4]. Yurish S. Y., Data Acquisition for Frequency- and Time Domain Sensors, in Smart Sensor Systems (Ed. by Gerard C. M. Meijer), *John Wiley & Sons*, Chichester, UK 2008, pp.313-341.
[5]. TMS320F28335 Digital Signal Controllers (DSCs), Data Manual, SPRS439I, *Texas Instruments*, June 2007 – Revised March 2011.

Sensors Industry
News
ALL about
SENSORS
FREE Monthly
IFSA Newsletter
ISSN 1726-6017
SUBSCRIBE NOW
subscribe@sensorsportal.com

Sensors Web Portal -
world's source for sensors information
TURN
OUR VISITORS
INTO
YOUR CUSTOMERS
BY THE SHORTEST WAY
Sensors Web Portal
SENSORS
SENSORS TRANSDUCERS
Advertise in
Sensors Web Portal and its media:
sales@sensorsportal.com
http://www.sensorsportal.com/DOWNLOADS/Media_Kit_2012.pdf

Chapter 5

Optoelectronic Digital Sensors and Sensor Systems

5.1 Quasi-digital Optical Sensors State-of-the-Art

Optoelectronic sensors are widely used in various applications such as medical, automobile, environmental, bio-chemical, etc. Many of them are based on integrated light-to-frequency converters, which convert a light intensity to quasi-digital (frequency or duty-cycle) format for direct connection to a microcontroller, DSP or interfacing with a PC [1].

According to the *IC Insights* report, '2011 Optoelectronics, Sensors, and Discretes', the optoelectronics market will grow 11 % to $ 26.4 billion in 2011, with sensor/actuator sales increasing 15 % to $ 8.5 bn and discretes rising 8 % to $ 23.4 bn this year.

Today, the *Texas Advanced Optoelectronic Solutions (TAOS), Inc.* manufactures light-to-frequency converters TSL230, TSL235, TSL237, TSL238 and TSL245, and color-to-frequency converters TCS2300 and TCS2310 [2-5]. These sensors combine a configurable silicon photodiode and a current-to-frequency converter on single monolithic CMOS integrated circuits. The output can be either a pulse train or a square wave (50 % duty cycle) with frequency directly proportional to light intensity or irradiance and features a wide dynamic range of 120 ÷ 160 dB. The light levels of 0.001 to 100 000 $\mu W/am^2$ can be accommodated directly without filters. These sensors can be directly coupled to a microcontroller or logic control and its output can be handled by a counter or timer [6]. Main performances of all these sensors are shown in Table 5.1.

The *Hamamatsu Corp.,* produces a CMOS photo IC S9705 combining a current-to-frequency converter and photodiode and outputs an oscillating frequency (duty ratio 50 %) proportional to input light intensity incident in the photodiode [2, 7].

The *Melexis* manufactures the MLX75304 Light-to-Frequency SensorEyeC light sensor for brightness dimming of mobile phone and notebook screens [2, 8]. The automotive version of the MLX75304 Light-to-Frequency SensorEyeCTM is specifically designed for instrument panel dimming and headlight control. The MLX75304 enables excellent screen readability under all circumstances. In a deep dark night, it will control display brightness for perfect reading comfort. During a bright sunny day the sensor will maximize screen lighting for best possible visibility. Its huge dynamical range of more than 100 dB makes the sensor fit for all light situations.

Table 5.1 Main performances of quasi-digital light sensors.

Sensor (LFC) (Images not to scale)		**Performances**		
		Output Frequency Range	**Spectral Response, nm**	**Non-linear FS Error, %**
TAOS (USA)				
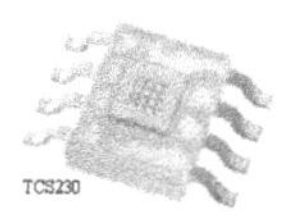	TSL230	0.4 Hz … 1.1 MHz	350 … 1000	0.2
	TSL235	0.4 Hz … 500 kHz	350 … 1000	0.2
	TSL237	2 Hz … 600 kHz	350 … 1000	1
	TSL238D	2 Hz … 600 kHz	320 … 1050	1
	TSL245	0.4 Hz … 500 kHz	850 … 1000	0.2
Hamamatsu (Japan)				
	S9705	0 Hz … 1 MHz	300 … 1000	3
Melexis (Belgium)				
	MLX75304	1 Hz … 1.6 MHz	500 … 1000	N/a

N/a – not available.

This sensor is a monolithic integrated sensor where photodiode, transimpedance amplifier and output stage are integrated on chip to minimize manufacturing cost. In addition, this approach results in a highly linear response and a low temperature dependency. A second cost saving item is the choice for a frequency output, which assures high precision control in all light conditions while simplifying the sensor interface. In fact, it only takes a digital input on a small microcontroller or even a counter input to get a precise readout of the ambient light level [8].

Because of its small size 3 mm × 3 mm SMD package and its solder reflows 260 ^{0}C compatibility, the sensor is perfectly fit for high-volume PCB assembly lines [8].

Thé frequency at the output pin (OUT) for TCL230RD is given by:

$$f_O = f_D + (Re)\ (Ee), \tag{5.1}$$

where f_O is the output frequency; D is the output frequency for dark condition (Ee = 0); Re is the device responsivity for a given wavelength of light given in kHz/(mW/cm2); Ee is the incident irradiance in mW/cm2; f_D is an output frequency resulting from leakage currents. As shown in the equation above, this frequency represents a light-independent term in the total output frequency f_O. At very low light levels, this dark frequency can be a significant portion of f_O. The dark frequency is temperature dependent [9].

Since the signal is in the form of a frequency, dynamic range is not limited by supply voltage and noise. Sensitivity of the device is maximized by a large detector area and precision input circuitry. Once converted to a frequency, the signal is virtually noise immune and may be transmitted over cables from remote sensors to other parts of system. Isolation is easily accomplished with optical couplers or transformers. Being a pulse data form, the signals of several sensors may be easily multiplexed into one microcontroller or counter using digital logic. Integration of the frequency signal is easily performed in order to eliminate low frequency interference or to measure long-term light exposure rates [6].

When a very high resolution light measurement is required, and speed is not critical, frequency or pulse counting technique would be used (the standard counting technique). This can be accomplished by counting pulses for a defined gate time. Relatively fast light measurements can be made by measuring the output period (indirect counting technique), or alternatively, pulse width [10].

One of possible light-to-digital converter based on the discrete logical components is shown in Figure 5.1 [6]. But such solution is relatively complex and needs more hardware.

The output of the light- or color-to-frequency converters may be also directly interfaced with low-cost microcontrollers. The particular choice of microcontroller will depend upon required performance and operating range of the light sensing function, as well as other system requirements. Performance factors such as accuracy and measurement range depend upon available embedded peripherals and operation speed of the microcontroller. Microcontrollers can provide a convenient direct sensor interface for frequency output light sensors in order to produce low cost digital output sensors. Such solution needs minimum hardware and external connections. The direct use of microcontrollers in the measurement chain offers simple, low-cost solutions yet introduces additional error components due to so-called program-dependent or software-related effects [11, 12] and is not a simple time-window counting task. From the other side taking into account very wide output frequency range of light-to-frequency converters (from part of Hz to 1.6 MHz, Table 5.1) no one from classical methods can be used in this full range because of dependence of quantization error on frequency and its very high value at the beginning (for frequency measurement) and at the end (for period measurement) of range. In order to get reasonable or high metrological performances of designed optical sensor systems, the frequency-to-digital conversion should be based on advanced methods for frequency measurements. Such methods must have a constant quantization relative error in a whole broad frequency range, scalable resolution, non-redundant conversion time and a possibility to measure frequency, which exceeds a reference frequency: $f_x > f_0$ in order to design a sensor systems with a reasonable power consumption. However, in case of the use any from advanced measurement method for frequency (period) including the combined conversion method, a customer must buy a license.

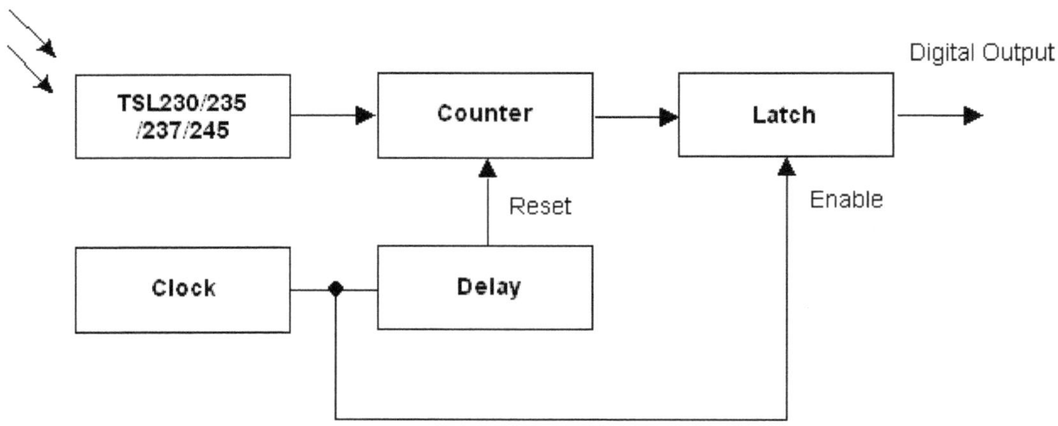

Fig. 5.1 Light-to-digital converter.

Along with quasi-digital light and color sensors the mentioned manufacturers as *TAOS, Hamamatsy, Melexis* and many others such as *Intersil, Maxim*, etc. produces integrated ADC-based light and color digital sensor. Let consider if such light and colour digital sensors satisfy to the modern market requirements. Many existing on the modern sensor market digital light sensors with embedded ADCs as usually have a slow conversion time, for example, the embedded 16-bit ADC of light sensor from *Intersil* (ISL29015), Figure 5.2 (a) has the integration time 45-90 ms [13]; the ADC from *Maxim* MAX9635 [14], Figure 5.2 (b) has the conversion time 97-107 ms [3]. Both smart sensors have a popular sensor bus I^2C, however such sensors can be used for proximity or ambient light applications, but it can not be used for light sensing applications, in which a conversion speed is a critical parameter.

Due to Series of UFDC and USTI ICs it is possible to make a universal design solution for all existing light-to-frequency converters in order to eliminate all mentioned above design problems, and introduce intelligent and smart features for various sensor systems, which can be realized in different technologies: hybrid, standard CMOS technology, System-on-Chip (SoC) or/and System-in-Package (SiP). The next sections will describe in details a design approach and examples of designed digital optoelectronic sensors and sensor systems.

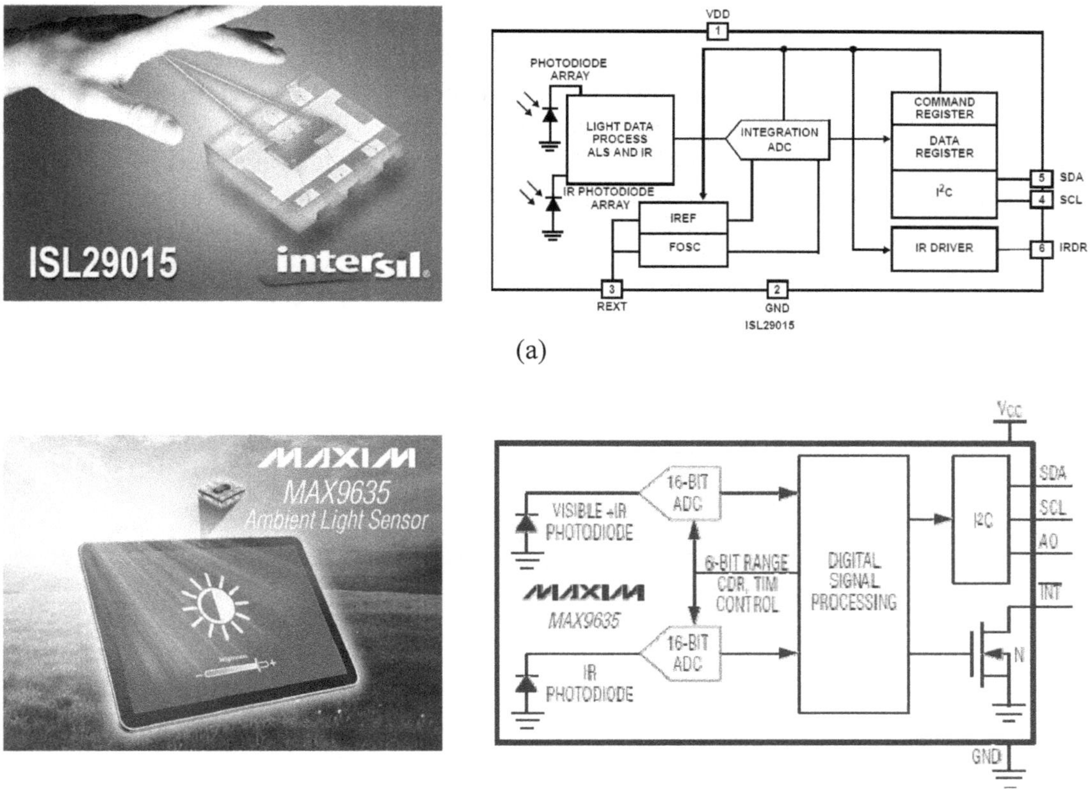

(a)

(b)

Fig. 5.2 (a) ADC – based digital light sensor ISL29015 (*Intersil*); (b) ambient light sensor MAX9635 with ADC (*Maxim*).

5.2 Design Approaches

The ADC-based sensors design has two alternative design approaches: digital sensors design with an intermediate conversion of analog output of sensing element to frequency with the help of a voltage-to-frequency converter; and digital sensor design based on the quasi-digital sensing elements (light- and color-to-frequency converters). A digital sensor diagram example according to the first design approach is shown in Figure 5.3.

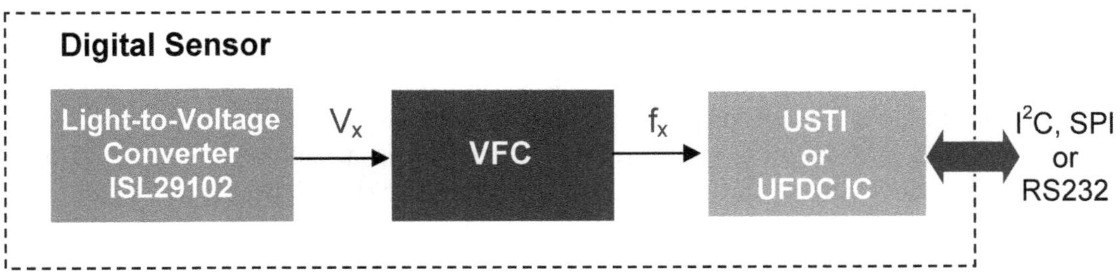

Fig. 5.3 Digital light sensor based on intermediate voltage-to-frequency conversion.

In this case any voltage output light sensor can be used, for example, the light-to-voltage converter ISL29102 from the *Intersil* [15]. Then a voltage signal should be converted to a frequency with the help of any voltage-to-frequency converters (VFC). Today, there are many low-cost VFC ICs [16] on the modern market, as well as discrete components-based solutions with high metrological performances [17]. In order to neglect by the frequency-to-digital conversion error, the programmable UFDC or USTI error must be selected 0.1% or 0.01 % depending on the sensor's error (in ten times less or at the least, it 5 times less then the sensor's error). For this relative error the conversion time will be from 0.5 to 16 ms (much better than in ADC-based light-to-digital converters).

The digital sensor diagram example according to the second design approach is shown in Figure 5.4.

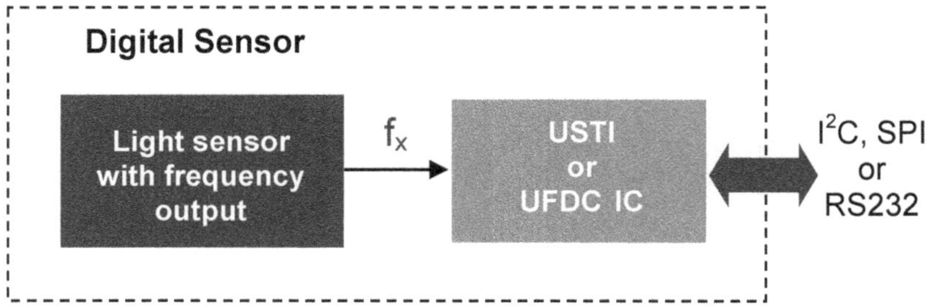

Fig. 5.4 Digital light sensor based on direct frequency-to-digital conversion.

In this case, any frequency output light sensor (light-to-digital converter) should be directly interfacing to the UFDC or USTI ICs. The hardware is minimal possible. It gives a good opportunity to design relatively low cost sensor systems. The conversion time is the same as in the first design approach.

In the following sections in this chapter are devoted namely to the second, most advanced design approach.

5.3 Digital Color and Light Sensors Based on Series of UFDC-1 ICs

5.3.1 Color-to-Digital Converter

The TCS3200 and TCS3210 programmable color light-to-frequency converters that combine configurable silicon photodiodes and a current-to-frequency converter on a single monolithic CMOS integrated circuit. The output is a square wave (50 % duty cycle) with frequency directly proportional to light intensity (irradiance).

The TCS2300's direct interfacing to the UFDC-1 IC is shown in Figure 5.5. This circuit diagram is corresponds to the 100 % scaling mode for TCS3200 (S0, S1 =1) and clear photodiode type (no filter, S2=1, S3=0). Power-supply lines must be decoupled by a 0.01 μF to 0.1 μF capacitor with short leads mounted close to the device package.

The converter's inputs S0, S1, S2, S3 and OE can be also controlled by the UFDC-1 but in this case a custom designed IC should be ordered.

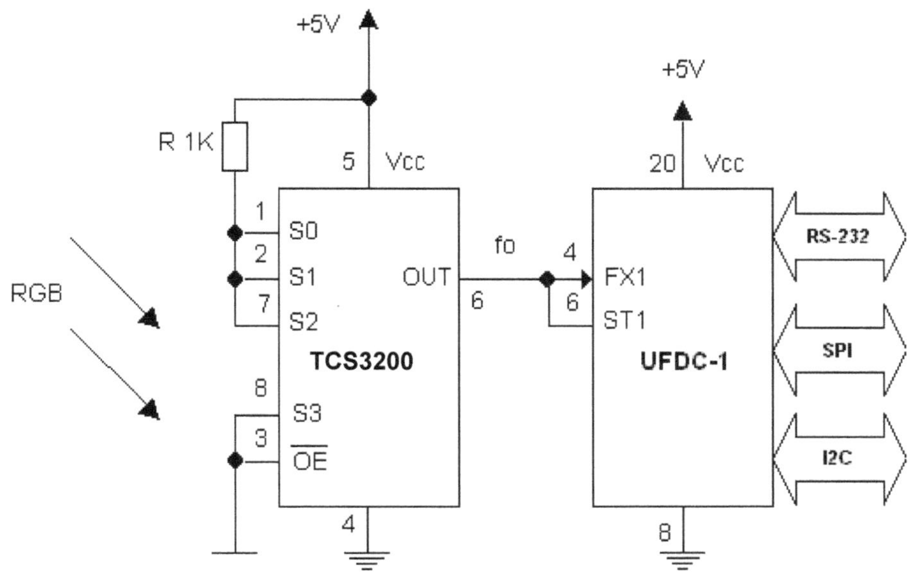

Fig. 5.5 Color detector TCS3200 with direct interface to UFDC-1.

The UFDC-1 can work in one of three possible communication modes: RS232 (slave and master), SPI and I^{2}C. A communication interface type can be automatically selected according to the circuit connection [18]. Working in the master mode (unidirectional data transfer) the UDFC-1 can be used in digital sensors design. In the master mode, the conversion accuracy and measurement mode should be selected by external jumpers. In the slave mode (with an external microcontroller), all measuring modes and programmed accuracy can be set up by a data bus or through communication I/O ports of master microcontroller. This allows us use the UFDC-1 as a simple peripheral circuit for a microcontroller. The UFDC-1 can be also controlled by PC through a COM port. It is useful for data acquisition systems design. Two different frequency output sensors can be connected to one UFDC-1. A bidirectional data exchange is possible with any of three interfaces RS232, SPI or I^{2}C.

The interfacing of UFDC-1 IC with the TCS3210 is the same as with the TCS3200. The difference between these two light-to-frequency converters is the following [19]. In the TCS3200, the light-to-frequency converter reads an 8 × 8 array of photodiodes. Sixteen photodiodes have blue filters, 16 photodiodes have green filters, 16 photodiodes have red filters, and 16 photodiodes are clear with no filters. In the TCS3210, the light-to-frequency converter reads a 4 × 6 array of photodiodes. Six photodiodes have blue filters, 6 photodiodes have green filters, 6 photodiodes have red filters, and 6 photodiodes are clear with no filters.

5.3.2 Light-to-Digital Converter

The interfacing of programmable light-to-digital converters as TSL230 to the UFDC-1 is the same as it is shown in Figure 5.5. The difference is only in inputs assignment. So, the inputs S0 and S1 are sensitivity-select inputs and S2, S3 - f_o scaling-select inputs. The devices are available with absolute output frequency tolerances of ±10 % (TSL230BRD), ±15 % (TSL230ARD), and ±20 % (TSL230RD) [9].

The light-to-frequency converters TSL235R, TSL237 and TSL245R with the direct interface to the UFDC-1 IC is shown in Figure 5.6 and the light-to-frequency converters TSL 237T, TSL 238T and TSL 238 interfacing is shown in Figure 5.7. Power-supply lines must be decoupled by a 0.01 μF to 0.1 μF capacitor with short leads placed close to the devices. The difference for the TSL 238 is only in pin number [20].

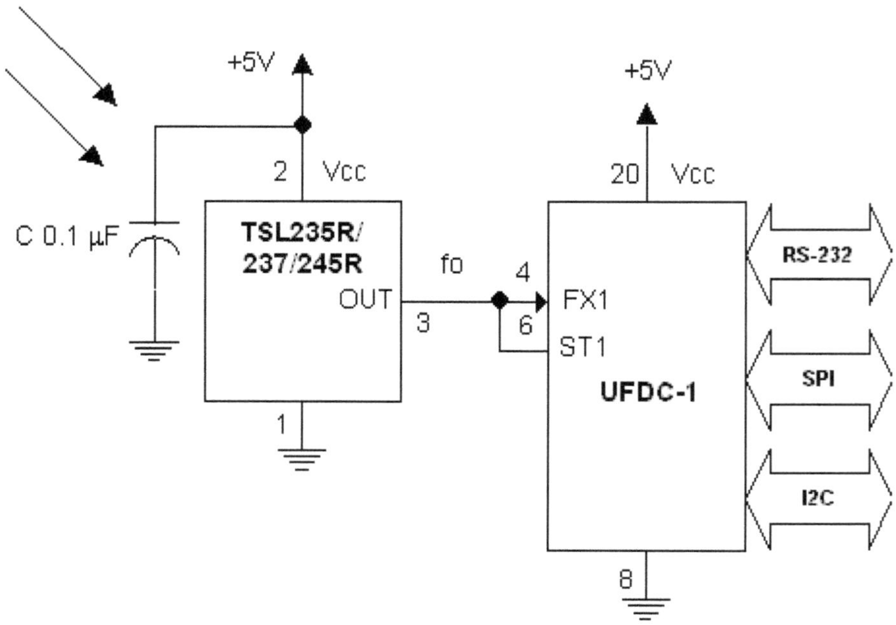

Fig. 5.6 Light detectors TSL235R, TSL237 and TSL245R with direct interface to UFDC-1.

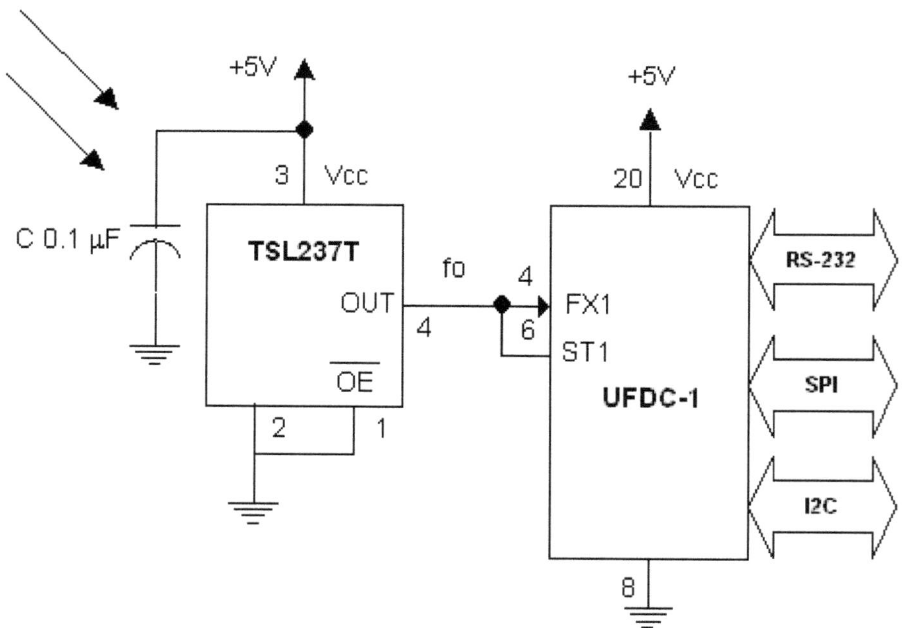

Fig. 5.7 Light detector TSL237T with direct interface to UFDC-1.

The comparative system performance summary with some popular microcontrollers and the UFDC-1 is shown in Table 5.2.

Table 5.2 The comparative system performance summary with some popular microcontrollers and the UFDC-1.

	Period Measurement			
Device	**Resolution**	**Min. Frequency**	**Max. Frequency**	**Measuring Time**
TMC370	16 bit	76 Hz	92 kHz	29.3 µs … 13.1 ms
MC68HC11	16-bit	46 Hz	111 kHz	15.3 µs … 21.8 ms
PIC16C5x	8-bit	4.8 kHz	1 MHz	1 µs … 210 µs
UFDC-1	16-bit	0.05 Hz	7.5 MHz	0.2 ms + (0÷T_x)

Taking into account the converters' nonlinearity error and absolute output frequency tolerance, the UFDC-1 conversion error should be chosen 1 %. Appropriate commands (RS232 interface) for the UFDC-1 operation are shown in Figure 5.8.

Example:

```
>M0                     ; Frequency measurement initialization
>A0                     ; 1 % conversion error set up
>S                      ; Start a measurement
>C                      ; Check a result status (returns 'r'-if ready or 'b'- if busy)
>R                      ; Read a result
1000.674946004319       ; Measurement result indication
>
```

Fig. 5.8 Appropriate commands (RS232 interface) for the UFDC-1 operation with TAOS light- and color-to-frequency converters.

5.3.3 Bus Capabilities

Due to the UFDC-1 bus capabilities, digital sensors with 2-wire I2C and 3-wire SPI bus can be easily created. Working with the I2C the UFDC-1 can have one of 8 possible addresses that can be setup by its A0-A2 pins. The absolute address of a device has the following format 1110 <A2><A1><A0> [18]. The circuit diagram for the 2-wire I2C interface to different TAOS optoelectronics sensors (for example, light, infrared and color) is shown in Figure 5.9. Appropriate commands (analog to example above) are shown in Figure 5.10.

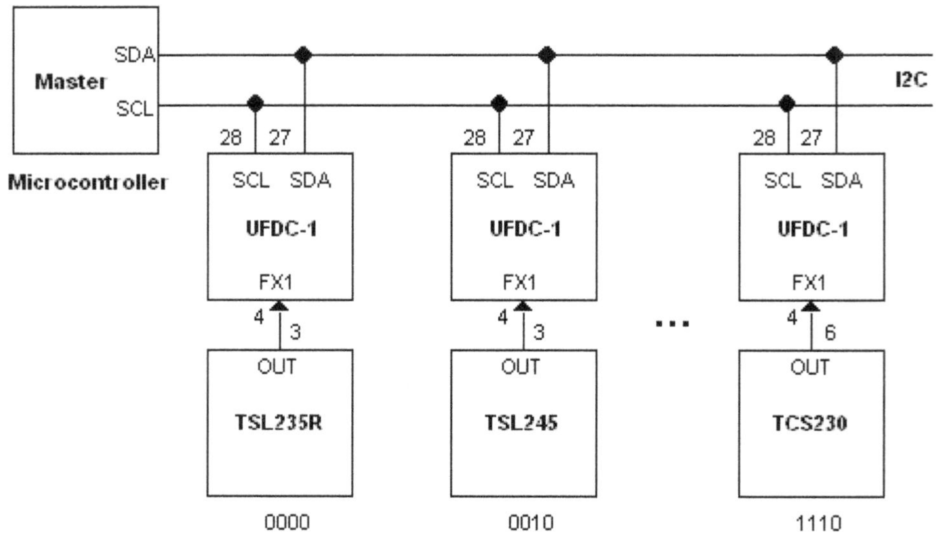

Fig. 5.9 Two-wire I^2C interface to different TAOS optoelectronics sensors.

Example:

<06><00>	; Frequency measurement initialization
<02><00>	; 1 % conversion error set up
<09>	; Start a measurement
<03>	; Check a result status (returns 0 if ready and not 0 if in progress
<07>	; Get measurement result in the BCD format.

Returns sign byte and 12 bytes of result in BCD code:
<Sign><I5><I4><I3><I2><I1><I0><F0><F1><F2><F3><F4><F5>,
where <sign> - sign of result 0x20 (space char) for positive result 0x2D (minus char) for negative result; <In> - byte n of integer part of result; <Fn> - byte n of fractional part of result.

Fig. 5.10 Appropriate commands (I^2C - protocol) for the UFDC-1 operation with TAOS sensors.

The I2C is a 2-wire, half-duplex, serial bus, as shown in Figure 5.9. The two I2C signals are serial clock (SCL) and serial data (SDA). Both lines are bidirectional and must be connected to V_{cc} via pull-up resistors. The maximal possible clock frequency is 100 kHz.

The SPI is full-duplex, (3+*n*)-wire serial busses. (*n* is the number of slaves). The circuit diagram for the 3-wire SPI is shown in Figure 5.11. In a single master, multiple slave configurations, as shown in Figure 5.11, each slave device requires a dedicated Slave Select (SS) signal, which is created by the master using standard I/O pins.

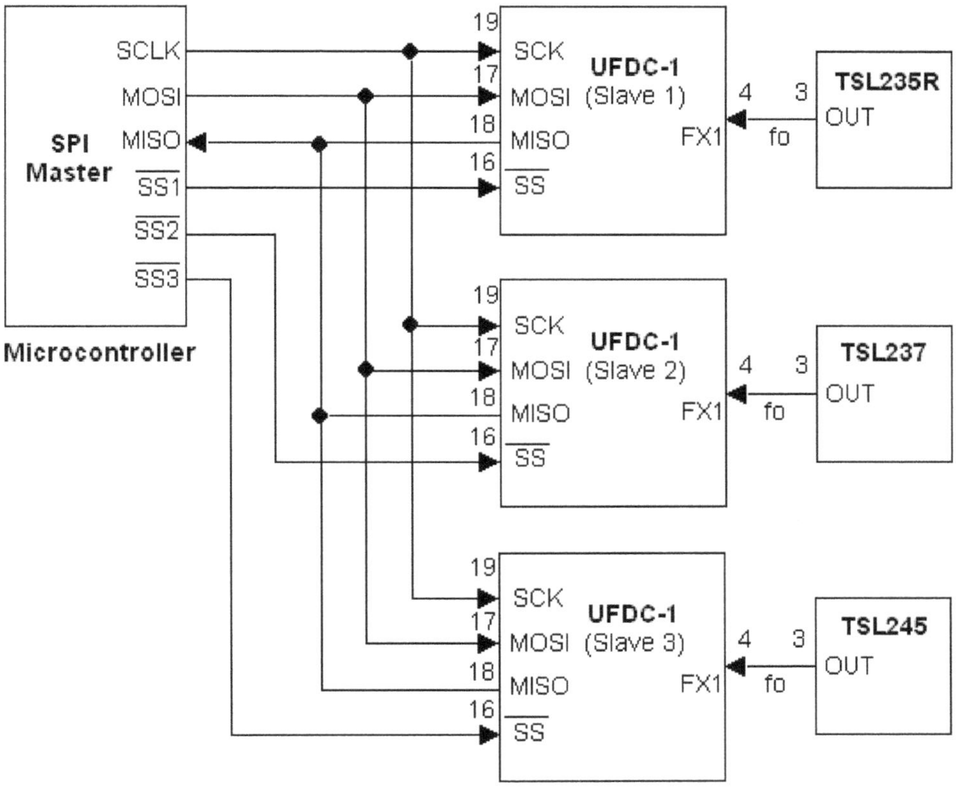

Fig. 5.11 Single master, multiple slave SPI implementations.

Appropriate commands for SPI communication to the UFDC-1 are the same as it was shown in Figure 5.10. The nominal clock frequency is 100 kHz and 400 kHz. The maximal possible clock frequency is 2 MHz, however, additional experimental investigation is necessary in this case.

SPI's full duplex capability and fast data rates make this interface very efficient and simple for single master - single slave applications. In practical applications, the requirement for dedicated slave select signals severely limits the number of slave devices that can be connected to a microcontroller.

The SPI is better suited than I^2C for applications that are naturally thought of as data streams (as opposed to reading and writing addressed locations in a slave device). An example of a "stream" application is data communication between microprocessors or digital signal processors. Another is data transfer from analog-to-digital converters. SPI can also achieve significantly higher data rates than I^2C. It really gains efficiency in applications that take advantage of its duplex capability, such as the communication between a "codec" (coder-decoder) and a digital signal processor, which consists of simultaneously sending samples in and out.

The I^2C bus has more complex protocol put it at a disadvantage in single master-single slave applications. Its weakness turns into strength if a larger number of slave devices needs to be connected or a multi-master system is needed.

The UFDC-1 works well with these both buses and due to this ability different frequency output optoelectronic sensors can be connected to these popular in sensors world buses.

The described design approach for digital optoelectronic sensors and sensor systems based on the UFDC-1 lets significantly simplify the design process, reduce time to market and production price and produce color and light sensors as well as multiparamters sensors with high metrological performances. The UFDC-1 can be used for both designs in the same time: for maximum resolution and accuracy as well as for maximum data-acquisition rate.

In comparison with the direct microcontroller interfacing the proposed approach also lets to eliminate many design problems connected with the use of advanced measurement methods for frequency measurements, microcontroller selection, its programming in Assembler, and additional error components due to so-called program-dependent or software-related effects.

5.4 Digital Optoelectronic Sensors and Sensor Systems Based on USTI IC

5.4.1 Light- and Color-to-Digital Converters

The CMOS level quasi-digital outputs of various optoelectronic sensors (light- and color-to-frequency converters, see Table 5.1) allow also a direct connection to the USTI [1, 21]. Let consider a design example based on the light-to-frequency converter S9705 [2, 7]. Both components are shown in Figure 5.12 and circuit diagram of optoelectronic sensor system example based on these components is shown in Figure 5.13.

Fig. 5.12 USTI in 32-lead TQFP package (1), and light-to-frequency converter S9705 (2).

Other optical sensors, for example, colour sensor TCS230 from TAOS (USA) [1] or reflective colour sensor OPB780 from OPTEK Technology [22] can also be interfaced by the same manner. Two frequency output sensors can be connected to the USTI at the same time.

A software example for the RS232 interfacing slave connection mode for two optical sensors (light sensor S9705 and colour sensor OPB780) is shown in Figure 5.14.

The command 'A02' sets the relative error for frequency-to-digital conversion [23] and should be use only once. Appropriate command 'M' sets the frequency measurement mode in the 1st and 2nd channels. The command 'S' starts measurement in appropriate channel. The command 'C' checks the measurements status and returns the value 'b' if the measurement in progress or the value 'r' if the measuring results is ready. The last command 'R' reads results. The use of 'C' command is very important at low frequencies measurements. In opposite side, there is a risk to get a previous result instead of the new one.

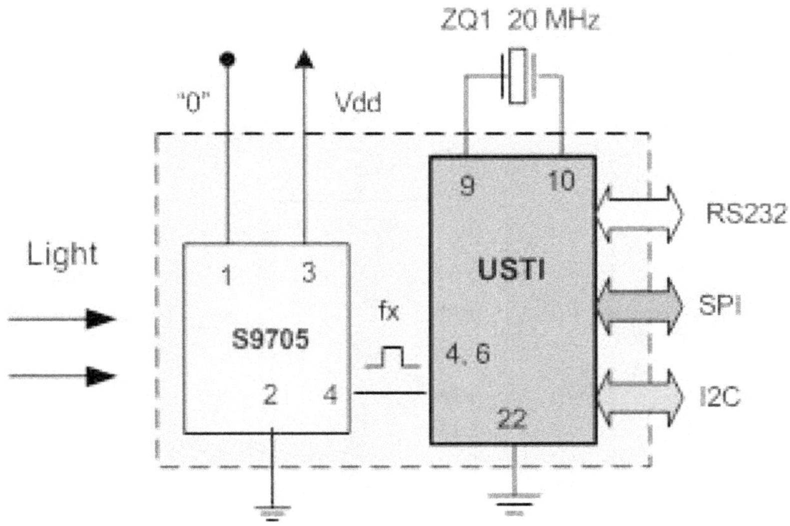

Fig. 5.13 Circuit diagram of optoelectronic sensor system.

>A02	;Set the relative error 0.25 %
>M00	;Set up a frequency measurement mode in the 1st channel
>S	;Start a frequency measurement (light sensor)
>C	;Check the measurement status ('r'-ready, 'b' -in progress
>R	;Read a result of frequency measurement in Hz
>462987.345	

>M0E	;Set up a frequency measurement mode in the 2nd channel
>S	;Start a frequency measurement (colour sensor)
>C	;Check the measurement status ('r' -ready, 'b'-in progress
>R	;Read a result of frequency measurement in Hz
>37005.0119	

Fig. 5.14 Commands for RS232 communication mode at light and colour measurements by the USTI.

The aim of experimental investigation described below was to determine main metrological performances of the designed optical sensor system based on the light-to-frequency converter S9705 (Hamamatsu, Japan) and USTI IC. The measuring set-up is shown in Figure 5.15.

Preliminarily, the USTI has been calibrated at laboratory temperature range (+25.3 ^{0}C to +26.4 ^{0}C) in order to eliminate additional systematic error due to quartz oscillator trimming inaccuracy (calibration tolerance) and a short term temperature instability [24].

The USTI has been connected to a PC, where terminal software Terminal v1.9b was running.

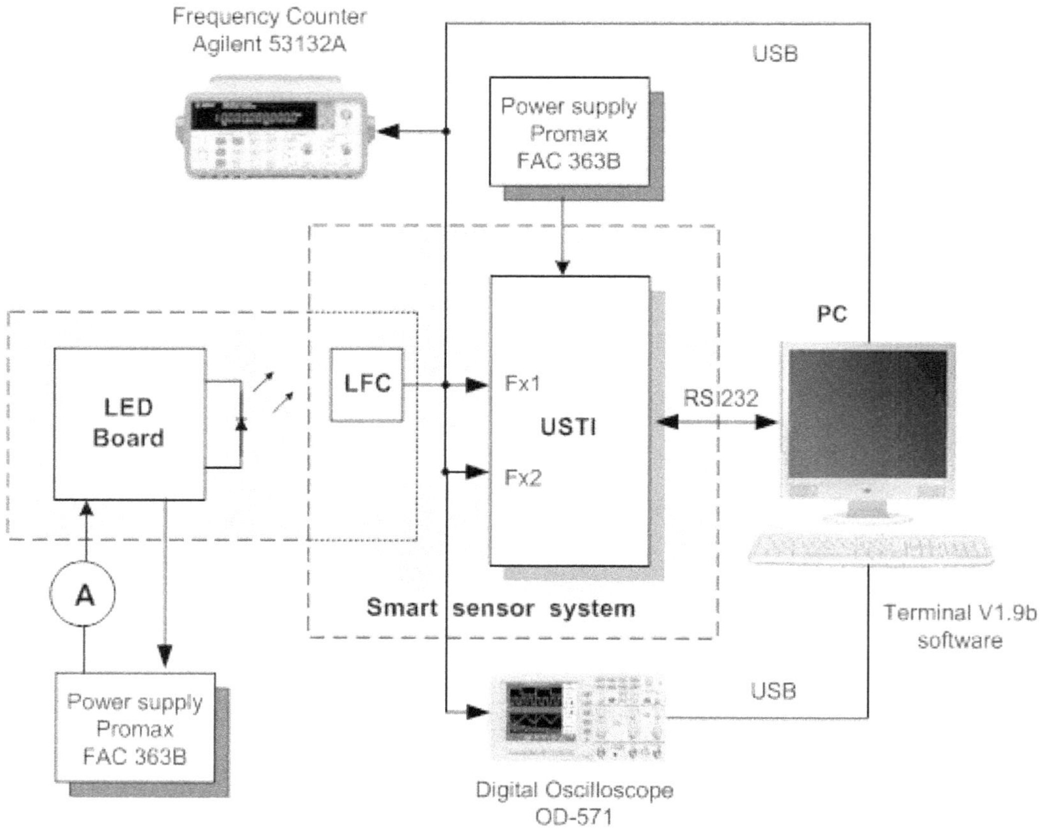

Fig. 5.15 Measurement set-up for optoelectronic sensor system investigation.

The light-to-frequency converter S9705 has mounted on a LED evaluation board together with a white light diode, the light intensity of which was set-up with the help of current source (Promax FAC 363B) and changing by a potentiometer with 25 μA steps. The current through this diode was measuring by an ampermeter (Figure 5.15).

The LED evaluation board photo and its circuit diagram are shown in Figures 5.16 and Figure 5.17 respectively.

The output of LFC was directly connected to the USTI. The frequency counter Agilent 53132A was used for frequency measurements in parallel with the USTI, and digital oscilloscope - for wave form visualization at LFC's output/USTI input. The frequency measurements have made for minimal and maximal possible frequencies of LED evaluation board: 5 Hz and 462 kHz respectively for both cases: without and with Schmidt trigger (74HC14D). Oscillograms of investigated sensor's output signals are shown in Figures 5.18 – 5.23.

The dependence of LFC's output frequency on current through the white light diode on the LED evaluation board is shown in Figure 5.24.

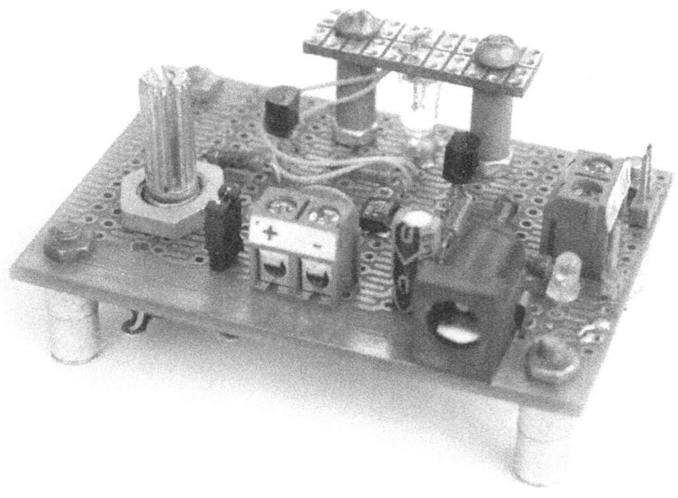

Fig. 5.16 LED evaluation board.

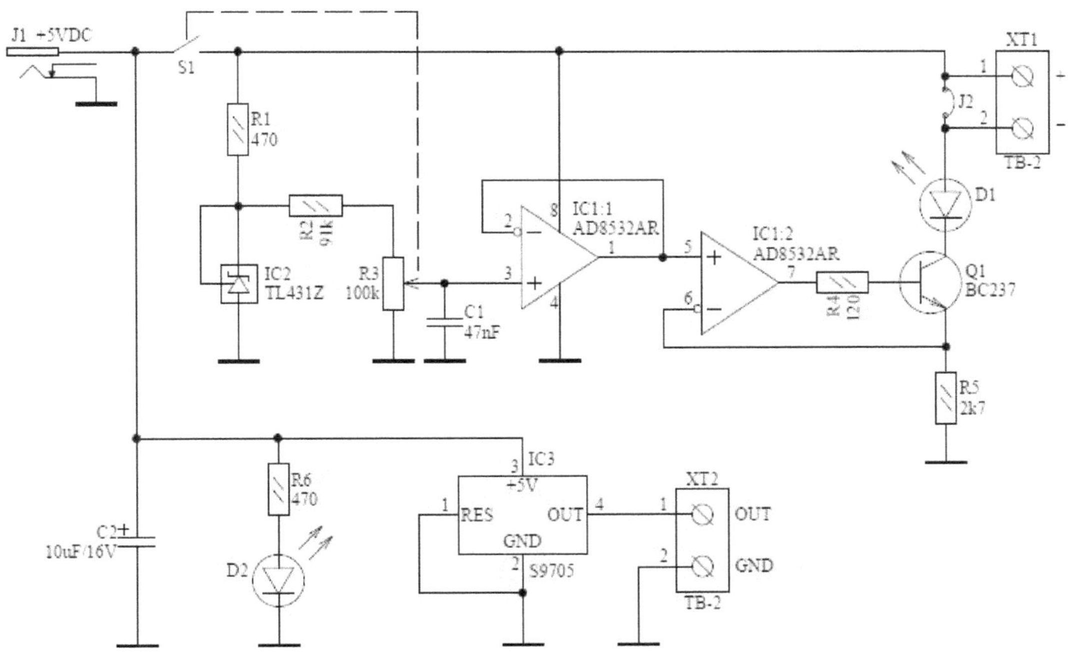

Fig. 5.17 LED evaluation board circuit diagram.

Each of investigated frequencies where measured 60 times and classical statistics was used for results processing. Measuring results for maximal and minimal frequencies for both: without and with Schmidt trigger are shown in Figure 5.25 and 5.26.

The χ^2 test for goodness of fit test was applied to investigate the significance of the differences between observed data in the histograms and the theoretical frequency distribution for data from the Gaussian distribution law.

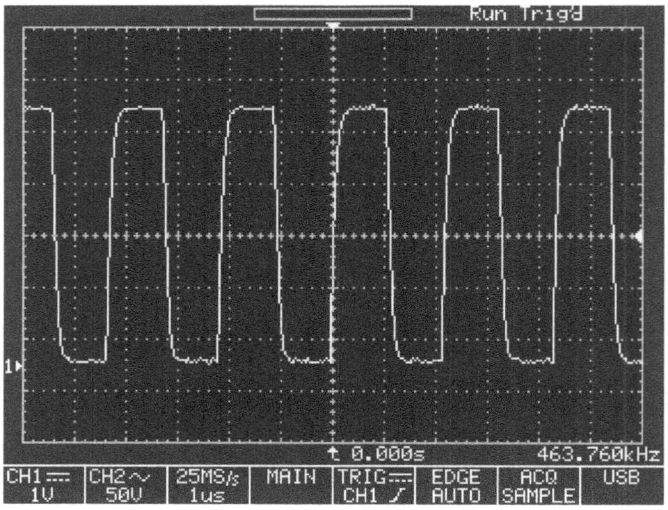

Fig. 5.18 Oscillograms of maximal frequency signal (~ 463 Hz).

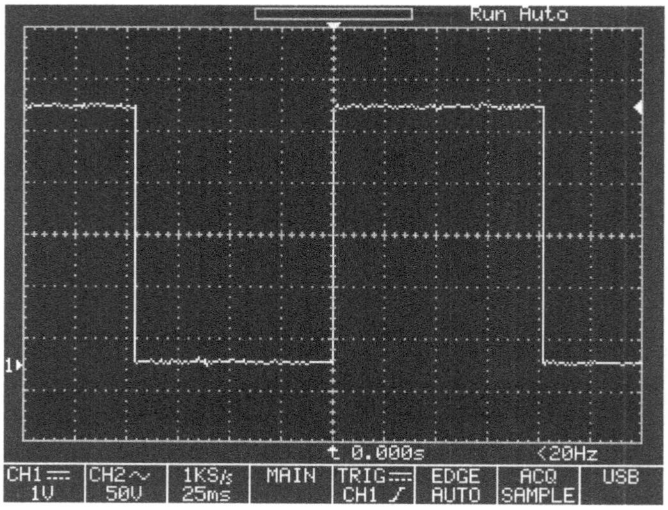

Fig. 5.19 Oscillograms of minimal frequency signal (~ 5 Hz).

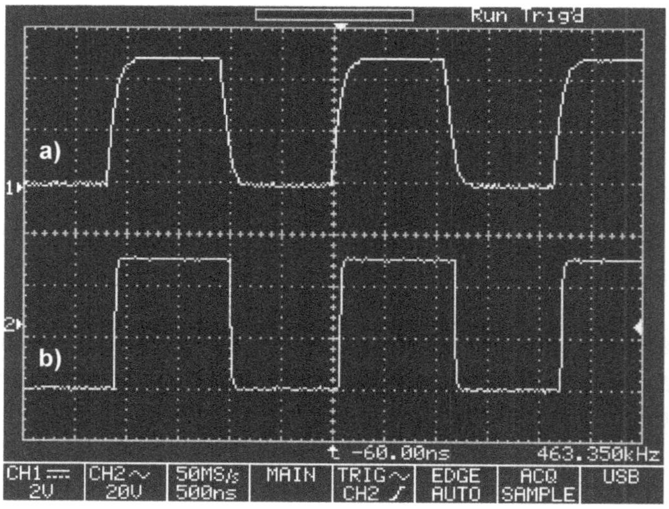

Fig. 5.20 Oscillograms at USTI input for frequency signal ~463 Hz: without Schmidt trigger (a), and with Schmidt trigger (b).

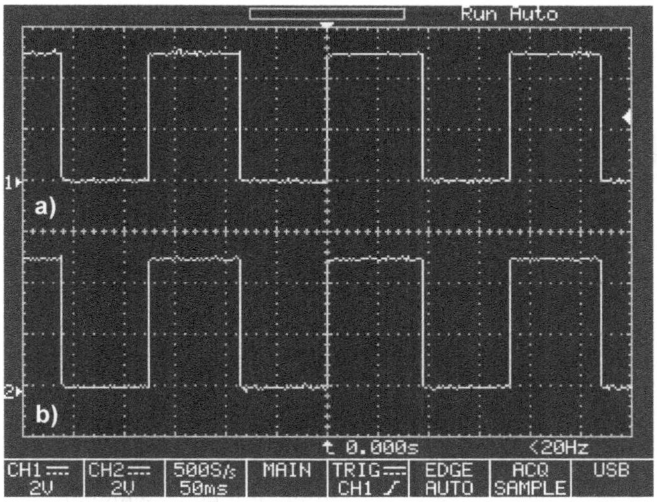

Fig. 5.21 Oscillograms at USTI input for frequency signal ~5 Hz: without Schmidt trigger (a), and with Schmidt trigger (b).

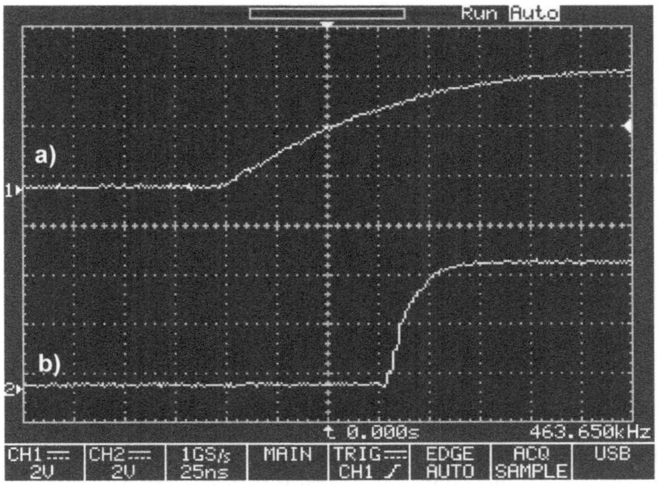

Fig. 5.22 Front of 463 Hz input pulse signal: without Schmidt trigger, rise time 110 ns (a), and with Schmidt trigger, rise time 24 ns (b).

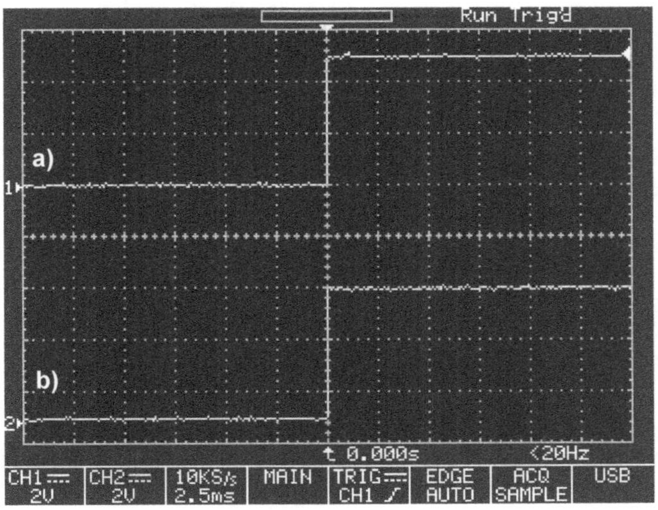

Fig. 5.23 Front of 5 Hz input pulse signal: without Schmidt trigger, rise time 80 μs (a), and with Schmidt trigger, rise time 70.7 μs (b).

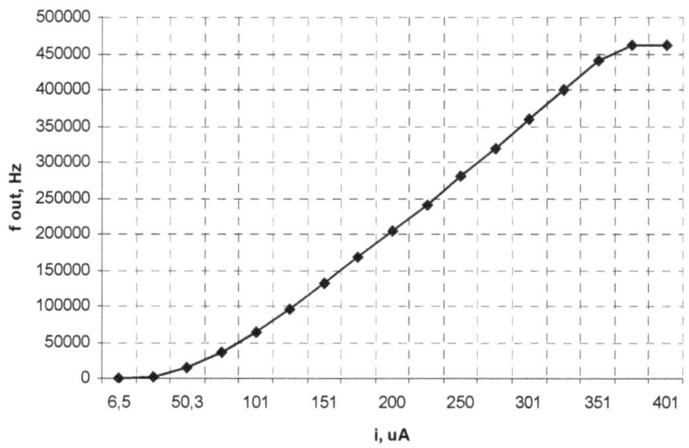

Fig. 5.24 Output frequency vs. current through a white light diode.

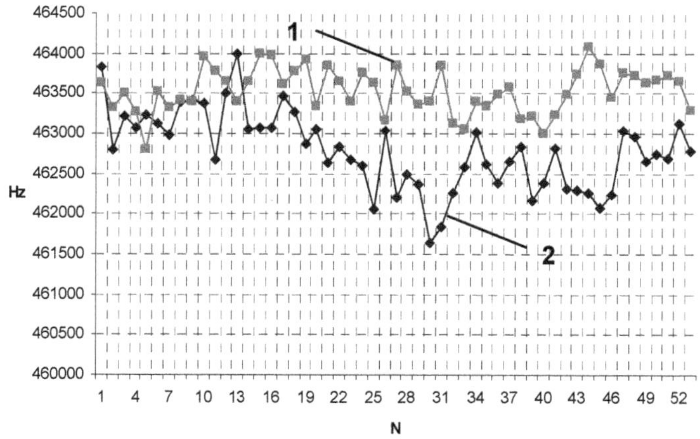

Fig. 5.25 Measuring results for maximal frequency ~ 463 Hz: with Schmidt trigger (1), and without Schmidt trigger (2).

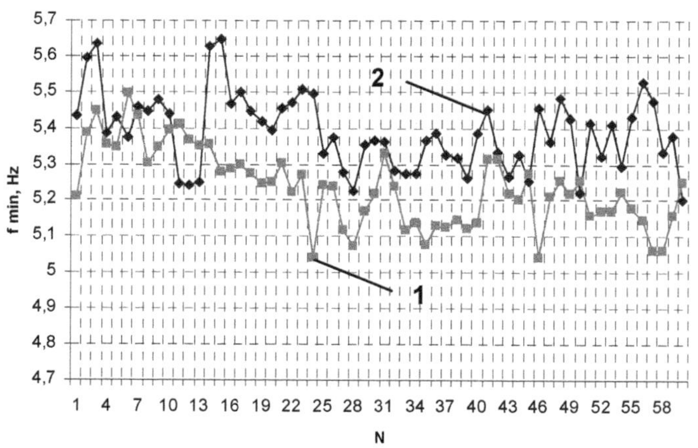

Fig. 5.26 Measuring results for minimal frequency ~ 5 Hz: with Schmidt trigger (1), and without Schmidt trigger (2).

The number of equidistant classes was calculated according to the following equation:

$$k = 1.9 \times N^{0.4}, \tag{5.2}$$

where N is the number of measurements [25].

At probability P = 97 %, and 6 equidistant classes (k=6), the hypothesis of Gaussian distribution law can be accepted for all sets of measurement data. The statistical characteristics are adduced in Table 5.3 and 5.4.

Table 5.3 Statistical characteristics at 463 Hz frequency measurements.

Parameter	463 kHz	
	Without Schmidt trigger	**With Schmidt trigger**
Number of measurements, N	53	60
Minimum f_x (min)	461653.265	464151.555
Maximum f_x (max)	463991.336	0.0062
Sampling Range, f_x (max) - f_x (min)	2338.0705	1354.2603
Arithmetic Mean	462788.685	463572.681
Variance	234229.738	6.6E-0009
Standard Deviation	483.9729	283.4972
Coefficient of Variation	956.2286	1635.1932
Confidence interval for arithmetic mean at P=97 %	$462644.42 < f_x < 462932.95$	$463493.257 < f_x < 463652.105$
Relative error, %	0.014	0.16
χ^2 – test (S) at: k=6; P = 97 % χ^2_{max} = 8.9	1.7272	2.5423
Hypothesis about Gaussian distribution	Accepted	Accepted

Table 5.4 Statistical characteristics at 5 Hz frequency measurements.

Parameter	5 Hz	
	Without Schmidt trigger	**With Schmidt trigger**
Number of measurements, N	60	60
Minimum f_x (min)	5.2014	5.041
Maximum f_x (max)	5.6466	5.4959
Sampling Range, f_x (max) - f_x (min)	0.4452	0.454
Arithmetic Mean	5.3899	5.236
Variance	0.0109	0.1071
Standard Deviation	0.1045	0.0001
Coefficient of Variation	51.577	48.907
Confidence interval for arithmetic mean at P=97 %	$5.3606 < f_x < 5.4192$	$5.206 < f_x < 5.266$
Relative error, %	0.54	0.57
χ^2 – test (S) at: k=6; P = 97 % χ^2_{max} = 8.9	6.6726	1.8498
Hypothesis about Gaussian distribution	Accepted	Accepted

As it is shown from the tables, the Schmidt trigger does not increased accuracy of frequency-to-digital conversion.

The proposed design approach for digital optoelectronic sensors and sensor systems based on the USTI IC gives a unique opportunity to create various OEM sensor systems with high metrological performances.

5.4.2 Sensor System for Automatic Paper Type and Thickness Detection

One more interesting application of quasi-digital optoelectronics sensors and USTI IC is a low-cost sensor system for automatic paper type and thickness detection [26, 27].

Modern photo printers have optical sensors that automatically detect a type and thickness of paper. These sensors are able to recognize plain paper, coated paper, glossy/photo paper and transparencies in the in-tray. Whether using plain paper or specialty papers such as photo papers or transparencies, the paper type will be detected and the printer will automatically make the appropriate driver settings to produce a high quality print. This detection is not always a perfect process - the sensor nearly always reads "plain paper" when it shouldn't. For instance, it will misread paper that is marked, scratched, or wrinkled; paper that has letterhead or markings at the top; and paper that is dark (or that contains metallic filaments). Photo printers will print in lower quality (or "draft mode") for plain paper, because they cannot handle the resolution and ink quantity that is needed for a real photo-quality image. In addition, reflective configuration type sensors give an opportunity to recognize only photo paper, plain paper or transparencies but not its real thickness; and the sensing process adds up to 5 s to the print process.

In order to eliminate the mentioned problems a new, low cost, automatic, transmissive configuration paper thickness sensing system with an increased reliability and short detection time should be designed.

The designed sensor system consists of visible light source (red light emitting diode (LED)), low-cost light-to-frequency converter (LFC) TSL237 (TAOS, Inc., USA) [28], and universal sensors and transducers interface (USTI) IC [21, 23].

The LED is an ideal monochromatic light source for such application because of its high energy efficiency, small size, low operating voltage and cost [29].

A traditional automatic paper-type reflective sensor configuration is shown in Figure 5.27 (a); but in order to be able detect a paper's thickness and count the number of paper sheets in in-tray it is expediently to use a transmissive type sensor with a configuration shown in Figure 5.27 (b). The automatic paper-type sensor works by the following way. An LED shines onto the surface of the paper. The transmissive through the paper light is captured and recognized. This gives the printer information about the paper thickness and characteristics for the particular paper. This information is compared to a reference table of paper types that the printer has stored internally. Then, based on the information, the printer determines color mapping, half-toning, and number of print passes to produce the best output for the paper being used [27].

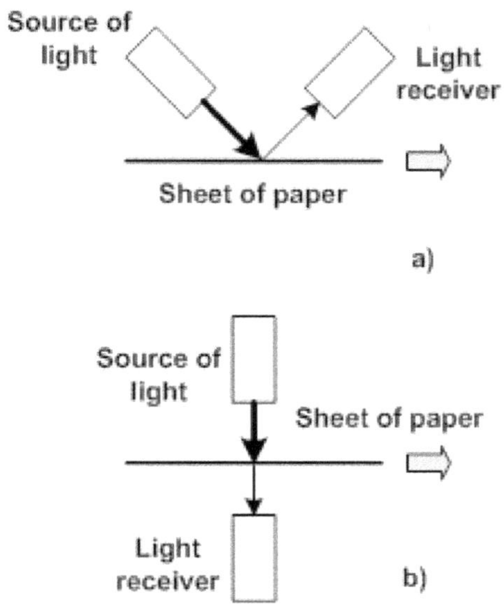

Fig. 5.27 Automatic paper-type sensor configurations: (a) tradition reflective type sensor; (b) proposed transmissive type sensor.

Nevertheless analog photodiodes or light-to-voltage converters can be used as a sensing element in such sensor systems, an output informative parameter as frequency on light-to-frequency converter's output has many advantages in comparison with analog output of sensing elements [30].

In the traditional solution, the current from a photodiode is very small (μA), making it susceptible to picking up noise, particularly if the transimpedance amplifier is separated from the photodiode by a considerable distance. In some applications, it is necessary to add shielding around the photodiode to keep electromagnetic interference and radio frequency interference from interfering with the signal. However, additional shielding is not needed with a light-to-frequency converter as long as adequate power supply is provided [31].

The TSL237 light-to-frequency converter combines a silicon photodiode and a current-to-frequency converter on a single monolithic CMOS integrated circuit. Output is a square wave (50 % duty cycle) with frequency directly proportional to light intensity (irradiance) on the photodiode. The frequency output allows direct interface to a microcontroller, UFDC, USTI or other logic circuitry. The device has been temperature compensated for the ultraviolet-to-visible light range of 320 nm to 700 nm and responds over the light range of 320 nm to 1050 nm [28]. The frequency at the output pin (OUT) is given by equation (5.1).

The dark frequency f_D is a constant error term in the output frequency calculation resulting from leakage currents, and is independent of light intensity. The TSL237 die is trimmed to minimize the magnitude of this dark frequency component so that it can be neglected in the transfer function calculation. In many applications, measurement of the actual dark frequency may be impractical due to measurement times ranging from several seconds to several minutes, and the fact that some devices may never transition (zero dark frequency).

The output of the device can be changed in a wide frequency range from 0.1 Hz to 2 Hz (dark frequency typical range) and from 0.5 to 1 MHz (maximum output frequency) and is designed to drive a CMOS logic input over short distances. Due to a wide dynamic range of ~120 dB this device is well suited for this paper thickness sensor system application.

The choice of interface and measurement technique depends on the desired resolution and data-acquisition rate. Maximum resolution, conversion speed and accuracy will be obtained by using the USTI especially designed for such kind of sensors applications. The designed sensor system is shown in Figure 5.28 and its main components - in Figure 5.29.

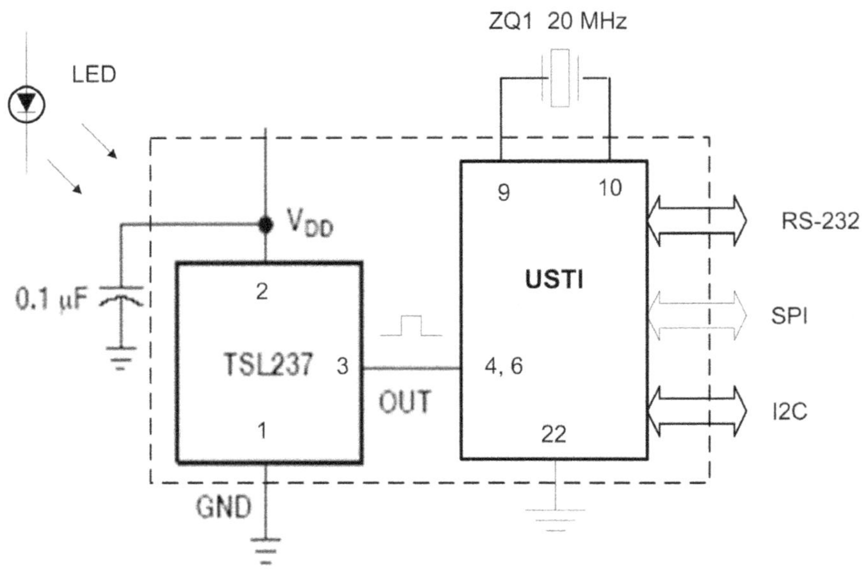

Fig. 5.28 Automatic paper thickness sensor system circuit diagram.

Fig. 5.29 Main sensor system components: light sensor TSL237 (1); LED (2); USTI (3).

Power-supply lines of TSL237 must be decoupled by a 0.01 μF to 0.1 μF capacitor with short leads placed close to the TSL237 (Fig. 5.28). A low-noise power supply is required to minimize jitter on output pulse.

Taking into account that the USTI has two identical channels, another light sensor TSL237 can be connected to the second measuring channel (for example, in the reflective type configuration working with the same LED) or color sensor with frequency output, for example, TCS230. Such advanced multisensors system configuration lets improve significantly reliability for paper-type recognition and make possible to control a paper quality at paper manufacturing processes and real-time measurement of paper sheet content at paper sorting and recycling.

The commands and appropriate comments for the USTI working with the RS232 interface (slave mode) in two-channel configuration for paper type and thickness determination are shown in Figure 5.30.

```
>A03            ; Set the relative error for the 1st channel δ= 0.1 %
>M00            ; Chose the frequency measuring mode in the 1st channel
>S              ; Start measurement
>C              ; Check result status
>R              ; Read result in Hz
100000.0254

>A03            ; Set the relative error for the 2nd channel δ= 0.1 %
>M0E            ; Chose the frequency measuring mode in the 2nd channel
>S              ; Start measurement
>C              ; Check result status
>R              ; Read result in Hz
70128.966
```

Fig. 5.30 Commands for USTI working with the RS232 interface (slave mode) in two-channel configuration.

The command "C" between "S" and "R" checks the measurement result status. It returns "r" if the result is ready and "b" if the measurement is in a progress. It is especially important at measurements in low frequency range.

The similar sensor systems can be designed also based on any frequency output light or infrared sensors such as TSL230, TSL235, TSL238, TSL245 (*TAOS*), S9705 (*Hamamatsu*) and MLX75304 (*Melexis*) [9]. These low-cost sensors have become attractive for such application, combining with the high-performance frequency-to-digital conversion based on the USTI to achieve the required accuracy and increased conversion time at a lower overall system cost.

The measurement set up that was used in experimental investigations is based on the light-to-frequency evaluation module from TAOS, Inc. [32]. A motherboard with an appropriate device-specific daughterboard, which were used in experiments is shown in Figure 5.31. The mother board was connected via USB port to a host PC running the Windows compatible host software application, which was used mainly for the ambient light subtraction in order to get rid of the effect of ambient light, and strobe LED control during all measurements. According to the algorithm, each time the sensor is measured, and extra measurement is made with the LED off. The value "Ambient" is then subtracted from the actual frequency measurement. The calculation used is a follows:

$$\text{Reading} = (\text{Signal} + \text{Ambient}) - \text{Ambient}. \quad (5.3)$$

Fig. 5.31 Evaluation module's motherboard (1) with TSL237 daughterboard (2), and USB connector (3).

The sensor's output was directly interfaced to the first USTI's channel working in a frequency measurement mode with programmable constant relative error 0.1 % in the whole frequency range. The USTI was connected via RS232 to the same host PC running the Terminal V1.9b software. The measurement set up for automatic paper thickness sensor system investigation is shown in Figure 5.32.

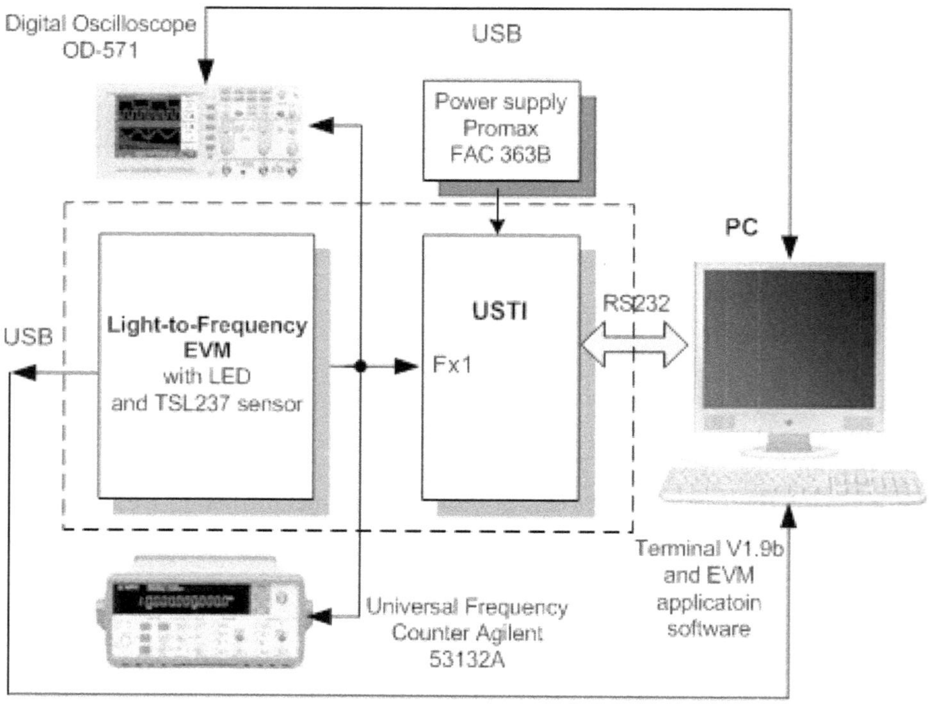

Fig. 5.32 Measurement set-up for automatic paper thickness sensor system investigation.

During experiments 8 paper patterns with different thickness from 0.086 to 0.217 mm including standard office types of paper 70 and 80 g/mm^2 were investigate (Figure 5.33). The oscillograms at sensors output correspond to paper sheets with maximal and minimal thickness are shown in Figure 5.34 (a, b). The dependence of sensor output on paper thickness is shown in Figure 5.35. A reference table of paper types can be stored in the USTI's or printer's memory.

Fig. 5.33 Paper patterns 0.086 mm (a), and 0.217 mm (b) at λ = 465 nm and P ≤ 3.258 mW white light source.

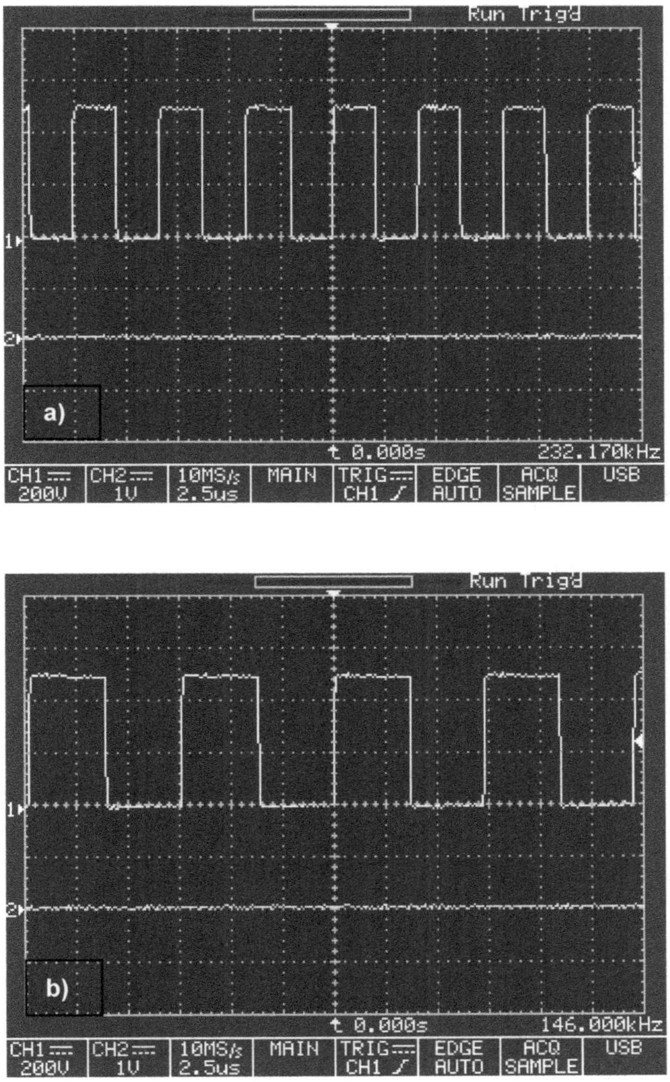

Fig. 5.34 Sensor output signals at 0.086 mm (a), and 0.217 mm (b) paper sheet thickness.

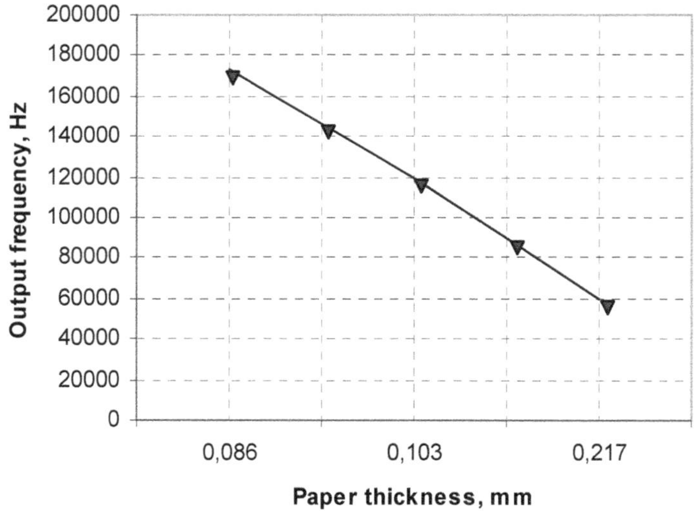

Fig. 5.35 Sensor output vs. paper thickness.

In addition to the paper thickness the developed sensor can detect the number of paper sheets of the same thickness, for example, office paper with 0.095 mm thickness (Figure 5.36). The maximum counted number of sheets is N=24 for 0.086 mm thickness and N=20 for 0.095 mm thickness at the distance between LED and LFC in 15 mm. This number can be increased by decreasing the distance between light source and sensing element. In spite of the fact that at $N > 8$ the dependence of frequency on number of sheets is not so well expressed, the USTI is capable to distinguish such changes due to a high resolution and low absolute error (± 2 Hz at f_x=2 000 Hz) at relative error ± 0.1 %.

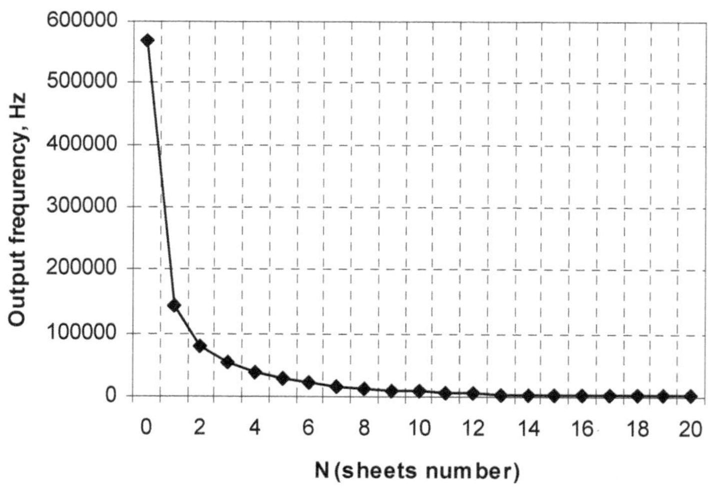

Data Table:

N	1	2	3	4	5	6	7	8	9	10
f_x, Hz	143000	81100	55500	39400	29500	22400	17350	12600	10350	8300

N	11	12	13	14	15	16	17	18	19	20
f_x, Hz	6900	5800	4600	3980	3400	2850	2510	2150	2000	1900

Fig. 5.36 Sensor output vs. sheets number (at 0.095 mm thickness).

The detection time in both experiments together with communication time does not exceed 15 ms at ± 0.1 % error for frequency-to-digital conversion.

The designed sensor can be also used for detection of paper sheets sticking in order to eliminate two leafs feeding at he same time.

In order to validate a high accuracy for frequency-to-digital conversion in a wide frequency range the frequency measurements for the maximum possible sensor output frequency $f_{x\ max} \approx 520$ kHz (paper absent status) and the minimum possible sensor output frequency $f_{x\ min} \approx 0.27$ Hz (low dark frequency f_O=f_D) for the light sensor were taken every second by the USTI and high precision calibrated counter (Agilent 53132A) until totaling 60 measurements. The measuring results are shown in Figure 5.37 (a, b). The measurements errors were evaluated from histograms and appropriate statistical characteristics. In both cases it does not exceed the programmable relative error $\delta \leq \pm 0.1$ %, therefore it can be neglected in comparison with the light sensor's error Statistical characteristics are adduced in Table 5.5.

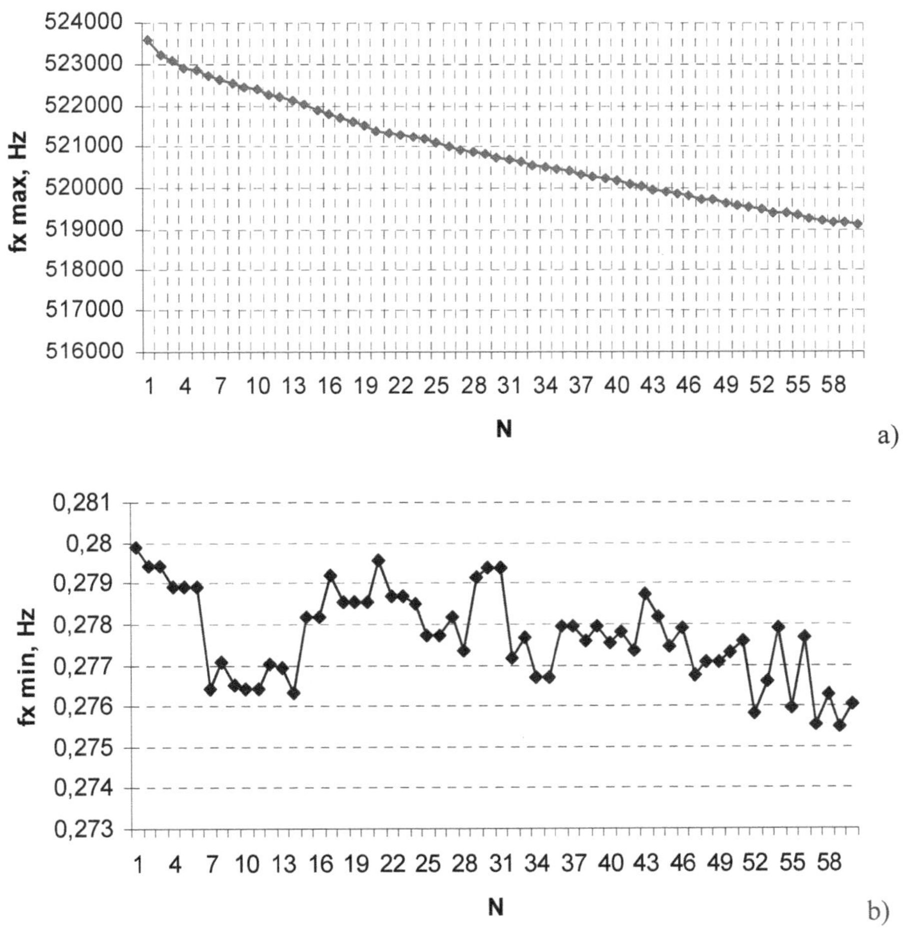

Fig. 5.37 Measurement results for $f_{x\ max} \approx 520$ kHz and $f_{x\ min} \approx 0.27$ Hz.

The χ^2-test for goodness of fit was applied to investigate the significance of differences between observed data in histograms and the theoretical frequency distribution for data from a normal population. At five equidistant classes (k=5) and a probability P = 97 %, according to the χ^2–test, $S < \chi^2_{max}$, where S = 6.1584 and 1.1586 is the sum of deviations between the data set and the assumed distribution for minimum and maximum frequency measurement accordingly; χ^2_{max} = 7.0 is the maximal possible argument of the χ^2 distribution. Hence, the hypothesis of Gaussian (normal) distribution can be accepted in both cases.

Table 5.5 Statistical characteristics.

Parameter	$f_{x\ max}$	$f_{x\ min}$
Number of measurements, N	60	60
Minimum f_x (min), Hz	519114.393	0.2755
Maximum f_x (max), Hz	523594.121	0.2799
Sampling Range, f_x (max) - f_x (min), Hz	4479.7277	0.0044
Median	0	0
Arithmetic Mean, Hz	520887.935	0.2777
Variance	1497556.04	1.2E-0006
Standard Deviation	1223.7467	0.0011
Coefficient of Variation	425.6501	251.5468
Confidence interval for arithmetic mean at P=97 %	$520545.093 \leq f_x \leq 521230.777$	$0.2774 \leq f_x \leq 0.278$
Relative error, %	± 0.07	± 0.1

The χ^2-test results are shown in Table 5.6, and distribution functions changes - in Figure 5.38 (a, b).

Table 5.6 χ^2-test results: for f_{xmax} (a) and f_{xmin} (b).

Classes	Observed incidence b[i]	Distribution function changes w[i]	Expected incidence c [i]	Deviation A[i]
1. 520010.33931	18	0.236644	14.20	1.018
2. 520906.28486	15	0.269337	16.16	0.083
3. 521802.2304	12	0.266525	15.99	0.996
4. 522698.17595	9	0.157958	9.48	0.024
5. 523594.12149	6	0.056031	3.36	2.070

a)

Classes	Observed incidence b[i]	Distribution function changes w[i]	Expected incidence c [i]	Deviation A[i]
1. 0.27638	7	0.116967	7.02	0
2. 0.27726	14	0.22875	13.72	0.006
3. 0.27813	17	0.308403	18.50	0.122
4. 0.27901	14	0.228827	13.73	0.005
5. 0.27989	8	0.093383	5.60	1.025

b)

Due to low cost and minimum hardware the designed paper thickness sensor system can be used in high quality modern photo- as well as in office budget laser or ink jet printers. A short detection time (~ 15 ms) do not introduce an additional significant time to the print process in comparison with traditional automotive paper thickness sensor system.

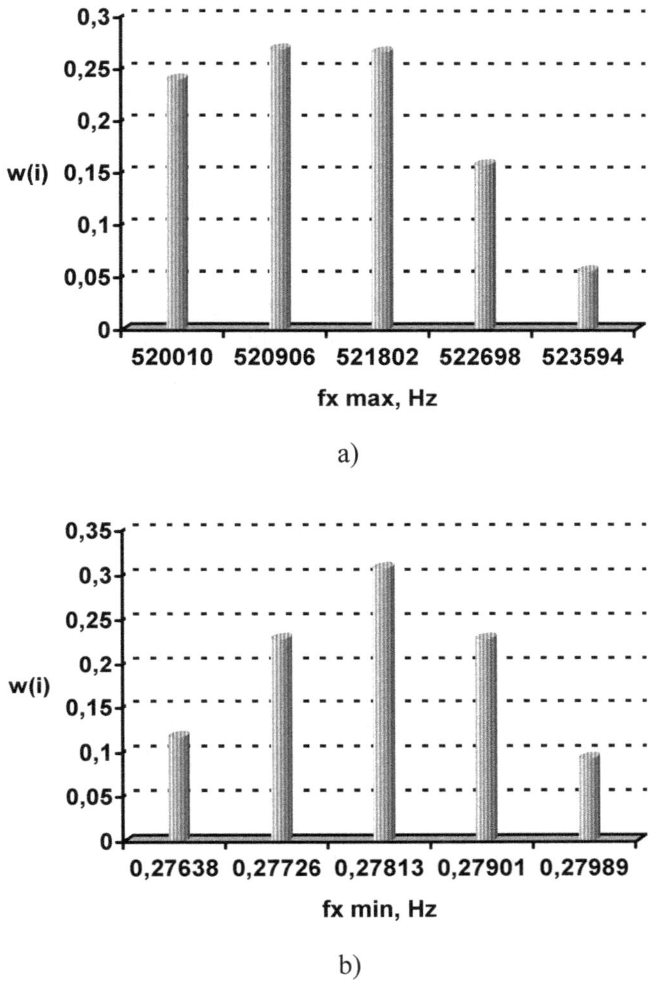

Fig. 5.38 Distribution functions w[i] changes.

The designed automotive paper thickness sensor system has a wide dynamic range of 5 000 000 : 1, immunity against high noises, high resolution and minimum component interface. Due to low price, minimum possible conditioning and interfacing hardware such sensor can be used not only in photo- but also in office laser and ink printers. In turn, low-cost light- and color-to-frequency converters have become attractive for different applications, combining with high-performance frequency-to-digital conversion based on the USTI to achieve the required accuracy, true digital output according to three popular serial sensors buses at a lower overall system cost.

Further research aims towards a fully integrated light-, color-to-frequency converters and USTI in CMOS technology as well as creation multisensor systems for different applications on its basis.

5.4.3 Non-Contact, Short Distance Measuring System

Let consider one more example of quasi-digital optoelectronic sensors application in short distance measuring systems [33]. Complex processes of industrial automation and robotic systems need information about placement of different objects. For this purpose noncontact distance measuring systems are used for placement and distance determination. As usually, existing measuring systems are based on laser ultrasonic or radar devices [34], and complete measuring systems cost some hundred dollars. The low measuring range for such sensor systems are limited by 15-20 mm [34]. Relatively low-cost measuring systems based on inductive displacement sensors have practically unlimited low measuring range (~2 mm) but the limited high measuring range up to 10-20 mm [34]. However, all of these sensor systems have high resolution, accuracy and short reaction time.

Low-cost short distance measuring systems with a price up to some tens dollars can be built based on the optical principle, infrared light sources as IRED and appropriate infrared light-to-frequency converters [2]. Such systems are using infrared optical sensors with output signal proportional to a distance between sensing element and object. Frequency output sensors have many advantages in comparison with analog (voltage or current) output sensors [10, 35]. For example, a low-cost short distance measuring system described in [35] uses frequency output optical sensor (infrared light source and infrared light-to-frequency converter), has 35-60 mm measuring range, and measuring time from 0.25 to 1 s.

Very often in industrial automation and robotic systems it is necessary to measure distance from some millimeters to some tens centimeters with a high speed by a non-contact reflected light way. Existing industrial sensors and measuring systems [34] do not cover the mentioned distance range (typically, their measuring range is beginning from 10 mm) or/and they are relatively expensive. In order to overcome these disadvantages, an inexpensive, noncontact distance measuring systems with extended measuring range and high measuring speed was built and tested.

The designed sensor system consists on an infrared light source (IRED), infrared light-to-frequency converter (IR LFC) TSL245R [36] and Universal Sensors and Transducers Interface (USTI) integrated circuit, which is well suitable for such applications. The infrared light-to-frequency converter combines a silicon photodiode with square area of 1.36 mm^2) and a current-to-frequency converter on a single monolithic CMOS integrated circuit. The output is a square wave (50 % duty cycle) with a frequency directly proportional to the light intensity. Because the output is TTL compatible, it allows direct interface to the USTI. The IR LFC is characterized for operation over the temperature range of −25 °C to +70 °C and is supplied in a 3-lead plastic side-looker package with an integral visible-light cutoff filter and lens [36]. The device responds over the infrared light range of 800 nm to 1100 nm and has a wide frequency range from ~0.4 Hz (dark frequency) to ~500 kHz. The output frequency can be calculated according to the equation (5.1).

The optical sensing element operates with reflections. The light emitted by the IRED is reflected by the target object and a fraction of it comes back and is detected by the IR LFC. More signal is reflected when an object is closer. The output of the sensor reveals the distance of an object. The light intensity is converted into a frequency by means of current-to-frequency converter and then the USTI converts the frequency to digital according to one of three popular serial interfaces: RS232, SPI and I^{2}C. The circuit diagram of sensor system is shown in Figure 5.39.

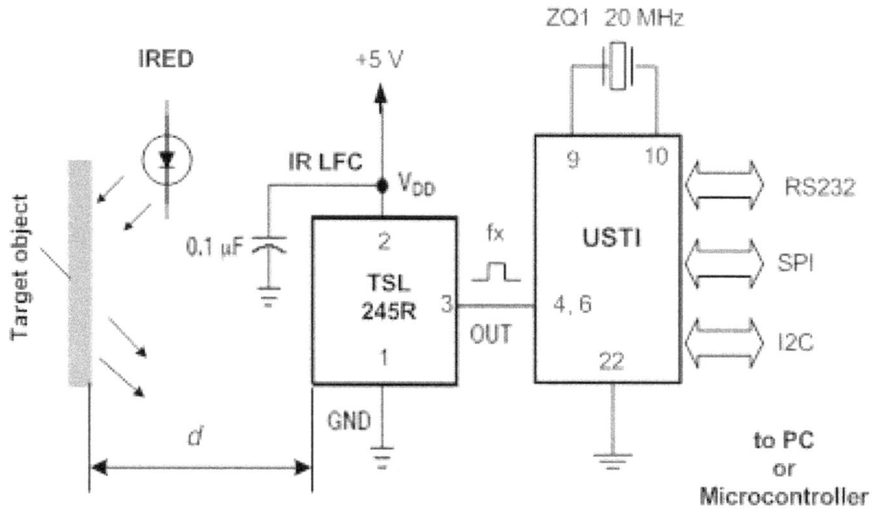

Fig. 5.39 Short distance measuring system: *d* - a distance to be measured; IRED - infrared light source; IR LFC - infrared light-to-frequency converter; USTI- Universal Sensors and Transducers Interface.

The appropriate commands for USTI working in the RS232 slave communication mode and measuring the frequency are shown in Figure 5.40.

```
> M00        ; Frequency measurement in the 1st channel
> A06        ; Relative error set up (δ= ± 0.01 %)
> S          ; Start measurement
> C          ; Check the USTI status (b-busy, r-ready)
> R          ; Read result fx:

191063.5233
```

Fig. 5.40 USTI commands for frequency measurement (RS232 slave communication mode).

Due to a wide dynamic range, high accuracy and non-redundant conversion time the USTI IC is well suited for low-cost, non-contact short distance measuring systems.

The measurement set up for experimental investigation of distance measuring system is shown in Figure 5.41. Both IRED and IR LFC are mounted on the TAOS LTF EVM TSL245R daughterboard installed in the sockets B of the LTF EVM motherboard [32], which was connected to a PC with the help of USB interface. For one's turn the motherboard with the daughterboard were mounted on the electronic digital caliper Z22855 Powerfix, which was used for distance set up and measurement with the absolute error ± 0.02 mm and resolution ± 0.01 mm (Figure 5.42).

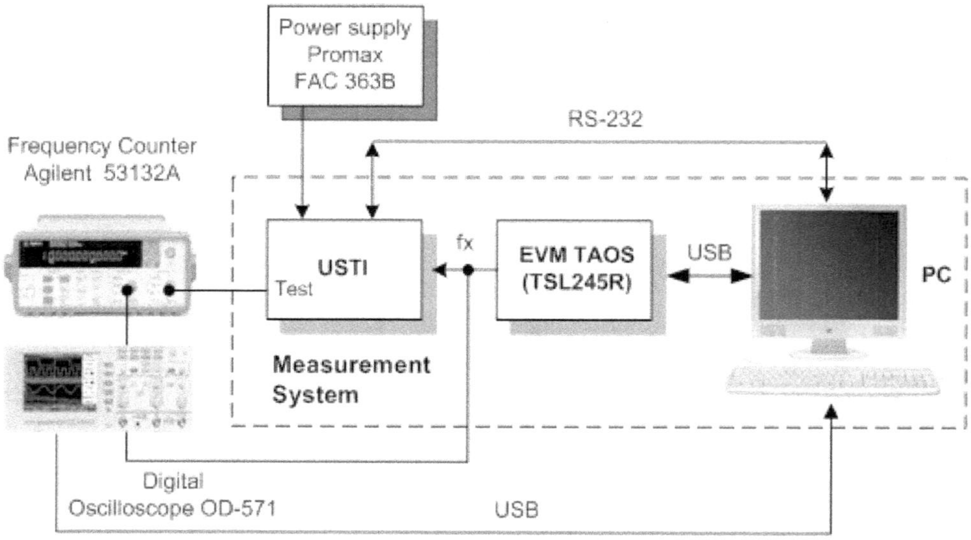

Fig. 5.41 Measurement set up for experimental investigation of distance measuring system.

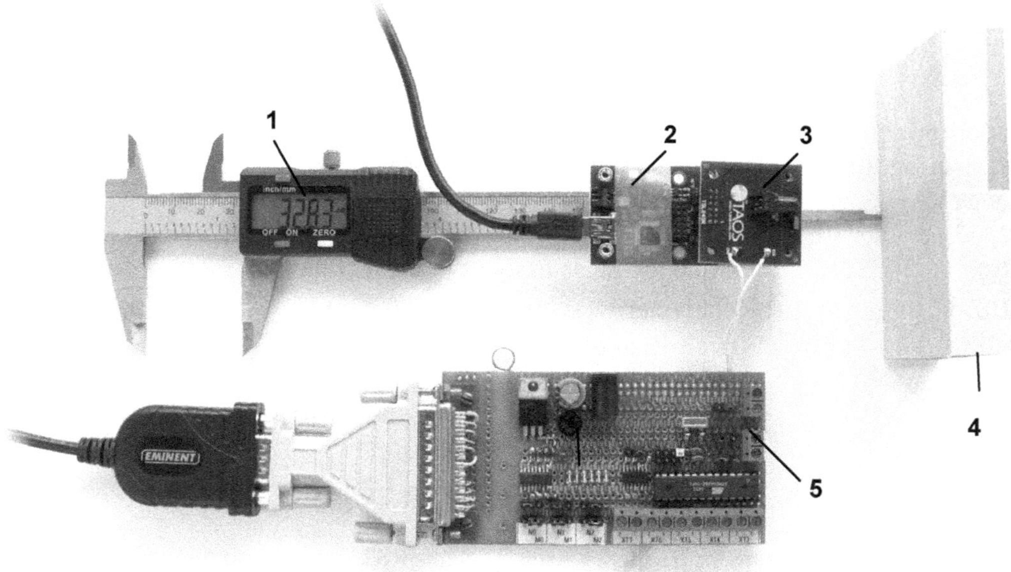

Fig. 5.42 A short distance sensor system: 1- electronic digital caliper Z22855 Powerfix; 2- LTF EVM motherboard; 3- TAOS LTF EVM TSL245R daughterboard; 4-target object; 5-USTI Evaluation Board.

The USTI was calibrated before in laboratory conditions at +22.7 ^{0}C (calibration constant is Δ_f =10002493.9580 Hz) in order to eliminate a systematic errors due to trimming inaccuracy (calibration tolerance) and aging for the low-cost, 20 MHz quartz crystal oscillator and short time temperature instability [24]. The sensor system was supplied at +5V dc by a Promax FAC-363B power supply. Digital oscilloscope OD-571 was used for signal waveforms monitoring. The measured by USTI frequency values were sent to a PC via an RS232 interface implemented with the ST202D integrated circuit mounted together with the USTI on evaluation board. The high precision calibrated universal counter Agilent 53132A with the ultra high stability oven with temperature stability <2.5×10^{-9} was used for comparative precision measurements of frequency at the TSL245R output in parallel to the USTI.

The Graphical User Interface (GUI) was implemented with the help of terminal software Terminal V1.9b for Windows and TAOS LTF EVM software (Figure 5.43). The Terminal V1.9b was used for connection and data exchange between the USTI and PC according to the RS232 interface, and the LTF EVM software was use for connection and data exchange between the motherboard, daughterboard and PC according to the USB interface. In addition, the LTF EVM software was used for rough measurements of frequency at the TSL245R output.

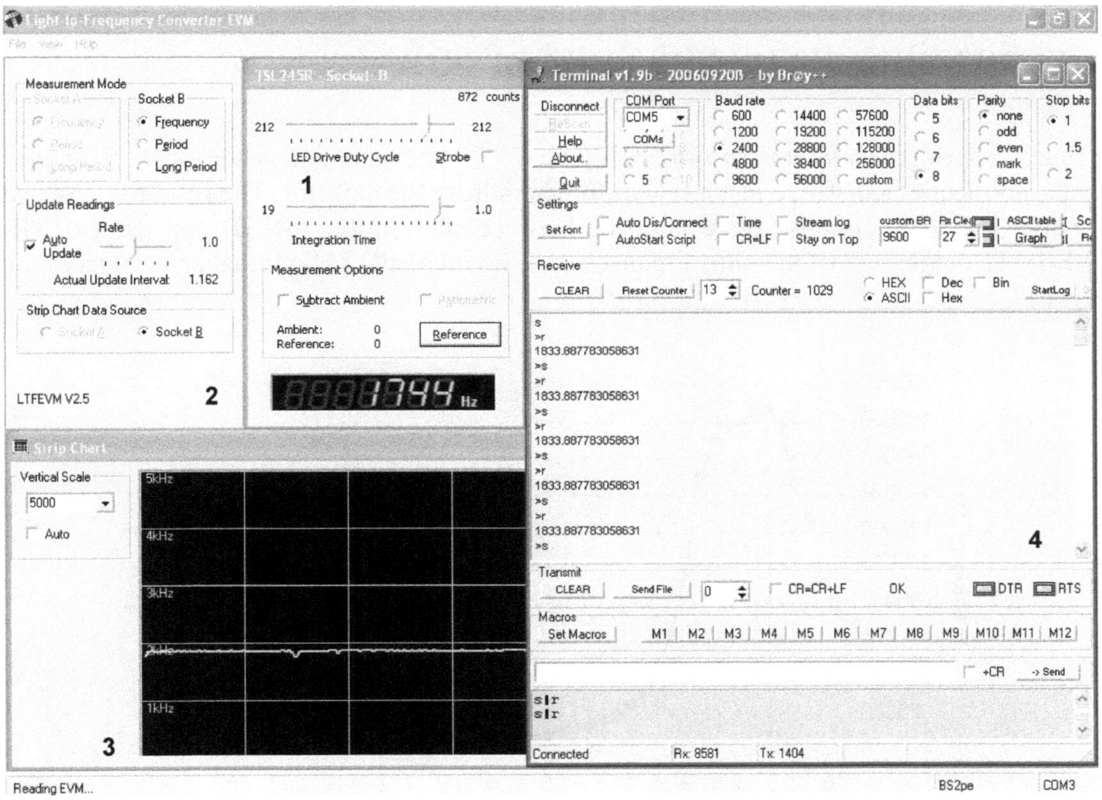

Fig. 5.43 GUI measuring software: LTF EVM software: 1 -daughterboard window; 2- window with a control section; 3 - strip chart recorder window; 4 - Terminal V1.9b for Windows.

Measurements were made from 1.5 to 215 mm distances range. The frequency on the sensor's output was changed proportional to the reflected light and measured in the ~1.8 kHz to 190 kHz frequency range for a target object with the standard calibration reflective surface (18 % reflective gray paper, for which most sensors are calibrated), Figure 5.44.

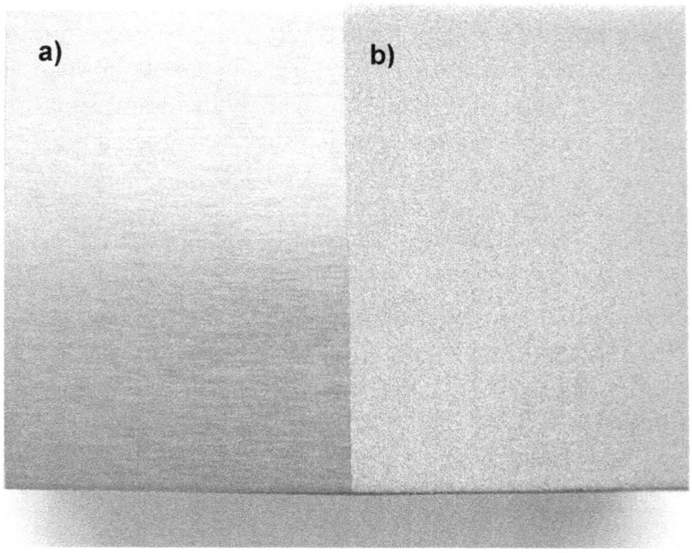

Fig. 5.44 Target object with the duraluminium surface (a), and standard calibration reflective surface (b).

The measuring results are shown in Figure 5.45. The graph shows a peak frequency measurement at a distance just a few millimeters from the detector. The response drops at shorter distances because there is space between the emitter (IRED) and detector (IR LFC). The closeness of the object reduces the amount of IRED light that can reflect to the IR LFC.

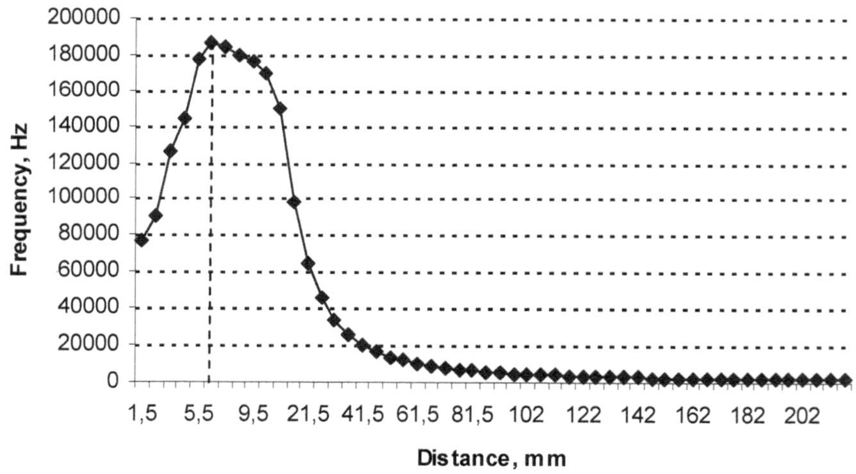

Fig. 5.45 Distance from target object to sensing element (IR LFC).

In the case if a target object is made from the duraluminium (Figure 5.44, a) more light is reflected from the object's higher reflectivity surface and output frequency will be increased in 3 times. In turn, the full scale of distances also will be increased a little.

The determined resolution for this measuring system is 0.01 mm, and response at 75 mm, for example, is 34 Hz/0.01 mm. The time of measurement calculated according to the equations (2.5), (2.6), where conversion time should be calculated according to the following equation:

$$\begin{cases} t_{conv} = \dfrac{1}{f_x} \quad if \quad \dfrac{N_\delta}{f_0} \prec T_x \\ t_{conv} = \dfrac{N_\delta}{f_0} + (0 \div T_x) \quad if \quad \dfrac{N_\delta}{f_0} \geq T_x \end{cases}, \tag{5.4}$$

where $N_\delta = 1/\delta$ is the number proportional to the required programmable relative error δ; $T_x = 1/f_x$ is the period of converted frequency; f_0= 625 kHz is the internal reference frequency of the USTI [37].

For the RS232 interface communication at 38400 baud rate and ±0.01 % programmable relative error the time of measurement is not more than $T_{meas\ max}$ = 26.35 ms, where t_{conv} = 16.5 ms, t_{comm} = 6.25 ms and t_{calc} = 3.6 ms. This time of measurement is less in 3.4 times in comparison with the conversion time of charge-balancing integrating type 16-bit ADCs (90 ms typical), that are used in various light-to-digital converters [38]; and less in 9.5 – 38 times in comparison with measuring systems described in [35].

Minimal and maximal frequencies at the sensor's output (Figure 5.46 a, b) also were measured by the USTI with the minimum possible constant relative error $\delta = \pm$ 0.0005 %.

Frequencies were measured by the USTI and universal counter Agilent 53132A in parallel until totaling 65 measurements. The measurement error has been evaluated from appropriate statistics characteristics. Statistical characteristics are adduced in Tables 5.7 and 5.8, and densities of distribution are shown in Figures 5.47 (a, b) respectively. The relative error does not exceed the programmable relative error ± 0.0005 % in both cases.

The χ^2-test for goodness of fit was applied to investigate the significance of the differences between observed data in the histograms and the theoretical frequency distribution for data from a uniform population. For five equidistant classes and a probability P = 97 %, according to the χ^2 –test, $S < \chi^2_{max}$, where S is the sum of deviations between the data set and the assumed distribution and χ^2_{max} is the maximal possible allowable argument of the χ^2 distribution. Hence, the hypothesis of uniform distribution can be accepted in the both cases.

Some additional design considerations should be used at distance measuring systems design. It is a choice of IR LED and its current stabilization; composition of the target object, etc. Other design considerations are connected with mechanical issues.

An additional infrared- or visible light-to-frequency converters [2] can be connected to the second channel of the USTI in order to extend it functionality and application areas.

The designed sensor system can be used for a variety of purposes, including absolute distance measurement, thickness calculation, slope and deformation control, linear distance measurement, position control, profile logging, centering control, and diameter/eccentricity measurement.

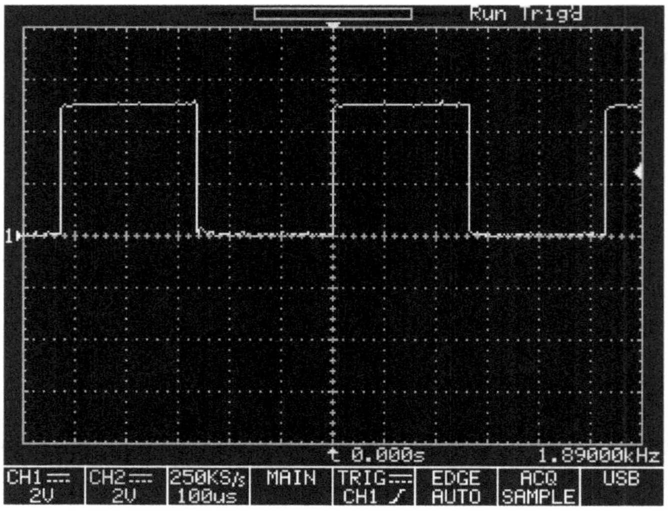

a)

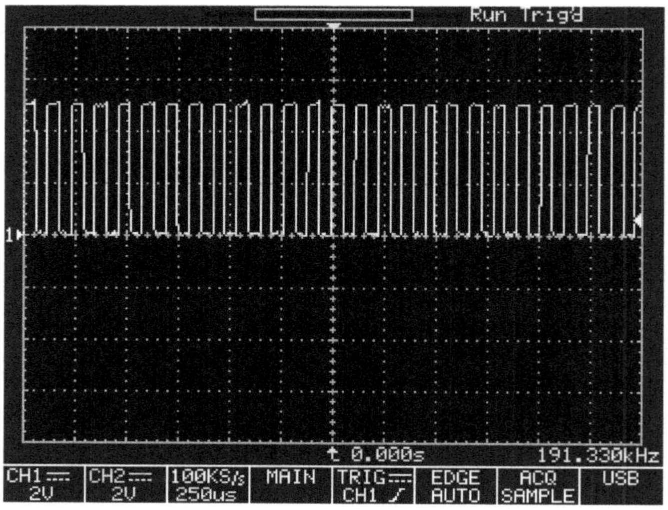

b)

Fig. 5.46 Oscillograms at sensor's output at $f_{x\,min}$ = ~ 1.83 kHz (a), and $f_{x\,max}$ = ~ 190 kHz (b).

Table 5.7 Statistical characteristics at $f_{x\,max}$ measurements.

Parameter	**Value**
Minimum $f_{x\,max}$ (min), Hz	190630.952
Maximum $f_{x\,max}$ (max), Hz	191115.705
Sampling Range, $f_{x\,max}$ (max)- $f_{x\,max}$ (min), Hz	484.7524
Median	0
Arithmetic Mean, Hz	190853.372
Variance, Hz	19236.0482
Standard Deviation, Hz	138.6941
Coefficient of Variation	1376.0744
Confidence Interval at probability P = 97 %	190816.04 < $f_{x\,max}$ < 190890.704
Relative error, %	0.00019 < 0.0005

Table 5.8 Statistical characteristics at $f_{x\ min}$ measurements.

Parameter	Value
Minimum $f_{x\ min}$ (min), Hz	1831.6609
Maximum $f_{x\ min}$ (max), Hz	1832.9903
Sampling Range, $f_{x\ min}$ (max)- $f_{x\ min}$ (min), Hz	1.3294
Median	0
Arithmetic Mean, Hz	1832.2021
Variance, Hz	0.1367
Standard Deviation, Hz	0.3697
Coefficient of Variation	4955.401
Confidence Interval at probability P = 97 %	1832.1026 < $f_{x\ min}$ < 1832.3017
Relative error, %	0.0005

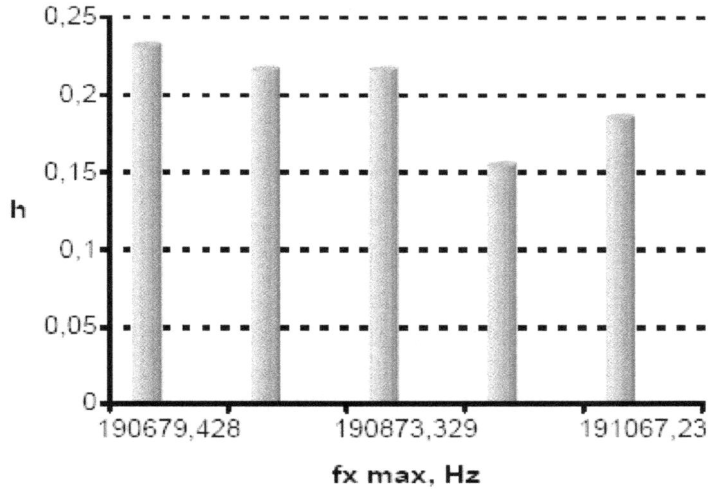

a)

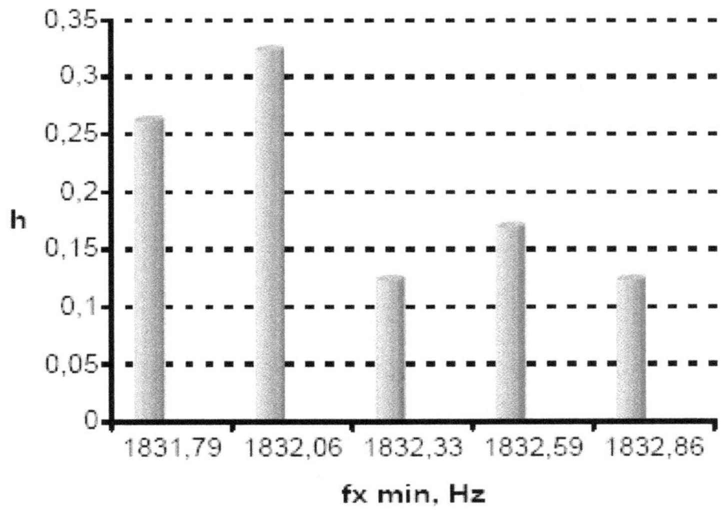

b)

Fig. 5.47 Density of distribution for $f_{x\ max}$ (a), and $f_{x\ min}$ (b) measurements.

The designed measuring system demonstrates high metrological performances at minimum possible hardware and relatively inexpensive elements costs. Performance improvements are achieved due to novel, precision modified method of the dependent count for frequency measurements with a non-redundant conversion time, which is used in the Universal Sensors and Transducers Interface IC.

Such measuring systems can be used also in many applications for measurement of geometrical quantities as tilt, deformation, diameter, etc. and proximity sensing applications.

5.5 References

[1]. Yurish S. Y., Smart and Intelligent Optoelectronic Sensor Systems: OEM Design Approach, in *Proceedings of the 2nd International Conference on Sensor Device Technologies and Applications (SENSORDEVICES' 2011)*, 21-27 August 2011, French Riviera, Nice/Saint Laurent du Var, France, pp. 38-43.

[2]. List of Optical Sensors Manufacturers at Sensors Web Portal: http://www.sensorsportal.com/HTML/SENSORS/OptoSens_Manufacturers.htm

[3]. Yurish S. Y., Intelligent Opto Sensors' Interfacing Based on Universal Frequency-to-Digital Converter, *Sensors & Transducers*, Vol.56, Issue 6, June 2005, pp.326-334.

[4]. Light-to-frequency: http://www.taosinc.com/Productfamily.aspx?id=2&SD=ltf

[5]. Color Sensors: http://www.taosinc.com/Productfamily.aspx?id=5&SD=cs

[6]. Intelligent Opto Sensor Data Book. *Texas Instruments*, SOYDE02B, 1996.

[7]. Light-to-frequency Converter Photo IC S9705, *Hamamatsu*, 2007.

[8]. MLX75304 Light-to-Frequency SensorEyeC, Preliminary Datasheet, *Melexis*, 2008.

[9]. TSL230RD, TSL230ARD, TSL230BRD Programmable Light-to-Frequency Converters, TAOS054P, *Texas Instruments*, October 2007.

[10].Kirianaki N. V., Yurish S. Y., Shpak N. O., Deynega V. P., Data Acquisition and Signal Processing for Smart Sensors, *John Wiley & Sons*, 2002.

[11].Yurish S. Y., Program-oriented Methods and Measuring Instruments for Frequency-Time Parameters of Electric Signals, *PhD Thesis,* State University Lviv Polytechnic Lviv, 1997 (In Ukrainian).

[12].Reverter F., Microcontroller-based Interfaces for Quasi-Digital Sensors, *PhD Thesis,* Technical University of Catalonia (UPC Barcelona, Spain), 2004.

[13].Integrated Digital Ambient Light Sensor and Proximity Sensor ISL29015, Datasheet, *Intersil*, 2008.

[14].Industry's Lowest-Power Ambient Light Sensor with ADC, MAX9635, *Maxim*, 2010.

[15].Low Power Ambient Light-to-Voltage Non-Linear Converter, ISL29102, Data Sheet, FN6483.1, *Intersil*, July 2010.

[16].List of Voltage-to-Frequency Converters Manufacturers at Sensors Web Portal: http://www.sensorsportal.com/HTML/SENSORS/VFC_Manufacturers.htm

[17].Jim Williams, Designs for High Performance Voltage-to-Frequency Converters, Application Note 14, *Linear Technology*, March 1986.

[18].Universal Frequency-to-Digital Converter (UFDC-1), Specification and Application Note, *Technology Assistance BCNA 2010, S. L.,* 2011.

[19].TCS3200, TCS3210 Programmable Color Light-to-Frequency Converters, TAOS099, *TAOS*, July 2009.

[20].TSL238 High-Sensitivity Light-to-Frequency Converter, TAOS073J, *TAOS*, September 2008.

[21].Universal Sensors and Transducers Interface (USTI), Datasheets, *Technology Assistance BCNA 2010, S. L.*, 2011.

[22].Reflective Colour Sensor Assembly OPB780Z, *Optec Technology, Inc*., USA, 2009.

[23].Universal Sensors and Transducers Interface (USTI). Specification and Application Note, *Technology Assistance BCNA 2010, S. L*., Spain, 2011.

[24].Yurish S. Y, Advanced Automated Calibration Technique for Universal Sensors and Transducers Interface IC, in *Proceedings of the IEEE International Instrumentation and Measurement Technology Conference (I²MTC' 2009)*, Singapore, 5-7 May 2009, pp.402-405.

[25].Novitskiy P. V., Zograf I. A., Errors Estimation of Measurement Results, *Energoatomizdat*, 1991 (in Russian).

[26].Yurish S. Y., Low-Cost Automatic Paper-Type Sensor Based on Universal Sensors and Transducers Interfacing ICs, in *Proceedings of 22nd International Conference EUROSENSORS XXII*, 7-10 September 2008, Dresden, Germany, pp. 425-428.

[27].Yurish S. Y., Sensor System for Automatic Paper Thickness Detection Based on Universal Sensors and Transducers Interface, *Sensor Electronics and Microsystems Technologies*,3/ 2009, pp.70-78.

[28].TSL237 High Sensitivity Light-to-frequency Converter, TAOS 052G, USA, 2007.

[29].Hai Yang, Xue Wei, Xianghui Liang, Mingwei Su, Xiaohua Lu, A SoC and LED based Reconfigurable Subminiature Spectrometer for Hand-held Measurement Applications, *Measurement*, 41, 2008, pp. 44-54.

[30].S. Y. Yurish, Sensors and Transducers: Frequency Output Versus Voltage Output, *Sensors & Transducers*, Vol. 49, 11, 2004, pp. 302-305.

[31].Ray King, LFT Technology Simplifies Light Sensing, *Control Engineering*, 6/1/2007.

[32].Light-to-frequency EVM. User's Guide, *TAOS, Inc*., USA.

[33].Yurish S. Y., Non-contact, Short Distance Measuring System for Wide Applications, in *Proceedings of IMEKO XIX World Congress*, Lisbon, Portugal, 6-11 September 2009, pp.643-647.

[34].B. Duval, Advances in Analog Distance Sensing, *Sensors Magazine*, September 2004.

[35].P. Arce, M. G. Fernandez, A. Ferrando, M. I. Josa, A. Molinero, J. C. Oller, E. Calvo, C. F. Figueroa, T. Rodrigo, I. Vila, and A. L. Virto, A Simple Method for Short Distance Measurements, *Nuclear Instruments and Methods in Physics Research*, A 517, pp.378-380, 2004.

[36].TSL245R Infrared Light-to-Frequency Converter, TAOS060D, September 2007.

[37].Yurish S. Y., Novel Universal Frequency-to-Digital Converters and Sensors Interfacing Integrated Circuits, *Sensor Electronics and Microsystems Technologies*, n.3, pp.80-90, 2008.

[38].Light Sensor Applications, Application Note AN1422.0, *Intersil*, 2008.

Chapter 6

Digital Temperature Sensors and Sensor Systems

6.1 Quasi-digital Temperature Sensors: State-of-the-Art

Temperature is one of the most widely measured variables in industrial, consumer, and computer applications. It is an analog quantity, but digital systems often use temperature to implement measurement, control, and protection functions. Integrated temperature sensors play an important role in different measuring systems, microsystems, and biomedical applications.

According to *MarketResearch.com's* report entitled '2011 Temperature Sensors' [1], the future for IC temperature sensors appears positive. There are a wide variety of applications for IC temperature sensors, including high volume and fast growing items such as computers, mobile phones, LCDs, and personal electronics such as cameras and MP3 players. Other applications include power supplies, white goods, and HVAC. Also IC temperature sensors are found on many printed circuit boards and in most devices that have a hard drive.

For now, it appears that the automotive market will yield more attractive opportunities for temperature sensor suppliers because of the advancements in technology, the emphasis upon improving fuel economy and lowering emissions, and the wide range of possible applications.

This report estimates that the current market for temperature sensing ICs is valued at $573 million, and it is growing at a compound annual rate of 5 %, which should increase market size to $739 million by year 2016 [1].

New analysis from *Frost & Sullivan,* World Temperature Sensors and Transmitters Market, finds that market earned revenues amounting to $3,635.2, million in 2009 and estimates this to reach $5,306.9 million in 2015 [2].

In most applications in which temperature is measured, the measured value must be converted from analog to digital form. By providing a digital output, digital sensors do not require external circuitry for linearization and data conversion [3, 4].

Quasi-digital temperature sensors make optimal solution for many sensing applications and sensor systems especially when the sensor needs to be placed in a harsh environment or when the measured signal needs to be transmitted over a long distance.

Today many manufactures produce frequency (period, duty-cycle) output temperature sensors. It is *Analog Devices* with its popular TMP03/04/05/06 sensors; *Maxim* with pulse-width modulation (PWM), period and frequency output sensors MAX6576/6577, MAX6666/67, MAX6672/73 and MAX6676/77 (PWM temperature sensors provide a square wave output whose duty cycle is proportional to the temperature); *Smartec* with its duty-cycle output temperature sensor SMT160-30; *Sea-Bird Electronics* with its high precision frequency output sensors SBE3plus, SBE 3F and SBE8; *Slope Indicator* with its VW precision frequency output sensors; *Melexis* with its infrared thermometers MLX90614 with a customizable PWM output for continuous reading; *Environdata, Hosberg* and *Spica Technology* with its frequency output temperature sensors TA40, Flex-T and T/FRQ respectively, and other manufacturers [5, 6]. Main performances of quasi-digital temperature sensors from different manufacturers are shown in Table 6.1.

Another big group of frequency output temperature transducers are so-called quartz thermometers based on the principle that the resonant frequency of a material such as quartz is a function of temperature, thus enabling temperature changes to be translated into frequency changes. The temperature-sensing element consists of a quartz crystal enclosed within a probe (sheath). The crystal is connected electrically so as to form the resonant element within an electronic oscillator. Measurement of the oscillator frequency therefore allows the measured temperature to be calculated. Quartz thermometers have a very linear output characteristic over the temperature range between - 40 °C and +230 °C. The measurement resolution of temperature is 0.1°C.

As it is visible form the Table 6.1 modern quasi-digital temperature sensors have high accuracy and wide dynamic frequency range. So, SBE3plus temperature has ± 0.003 % relative FS error. The frequency-to-digital converter should have ± 0.0003 % or at least ± 0.0006 % error in order to be neglected. Quasi-digital temperature sensors can be simply interfacing to microcontroller. But it means the use of program-oriented conversion methods. These introduce additional error components due to so-called program-dependent or software-related effects [5, 7-9], for example, the error due to delay of reaction to interruption and error of shift in time of the response for interruption. Both components take place at any waveform of signals, including rectangular. Very often these components exceed the quantization error and it is necessary to use non-trivial methods and optimization counter techniques for its minimization.

Other frequency output temperature sensor has a broad frequency range from 0.5 to 1.3 MHz corresponds to -50 °C ... +120 °C temperature range [10]. The temperature sensors THSENS-F (Table 6.1) also has a broad output frequency range from 400 kHz to 1.8 MHz corresponds to -50 °C ... +150 °C temperature range, and ±2 °C absolute error.

Table 6.1 Quasi-digital temperature sensors performances.

Sensor	Max. Temp. Error, °C,	Temp. Range, °C	Resolution, bits	Output	Output Range
Analog Devices					
TMP03	± 1.5	-40 to +100	16	PWM	*
TMP04	± 1.5	-40 to +100	16	PWM	*
TMP05	± 0.5	-40 to +150	12	PWM	*
TMP06	± 0.5	-40 to +150	12	PWM	*
Maxim Integrated Products					
MAX6576	± 3.0	-40 to +125	*	Period	0.0023 … 0.26 s
MAX6577	± 3.0	-40 to +125	*	Frequency	14.57...1592.6 Hz
MAX6666	± 1.0	-40 to +125	11	PWM	*
MAX6667	± 1.0	-40 to +125	11	PWM	*
MAX6672	± 3.0	-40 to +125	*	PWM	*
MAX6673	± 3.0	-40 to +125	*	PWM	*
MAX6676	± 1.5	-40 to +125	*	PWM	*
MAX6677	± 1.5	-40 to +125	*	PWM	*
MicReD					
THSENS-F	± 2.0	-50 to +150	*	Frequency	0.2...1…1.8 MHz
Sea-Bird Electronics					
SBE 3F	± 0.001	-5 to +35	*	Frequency	2...6 kHz
SBE 3plus	± 0.001	-5 to +35	*	Frequency	2...6 kHz
SBE 8	± 0.01	-3 to +30	16	Frequency	0,1 ... 200 Hz
Slope Indicator					
VW	± 0.3	-20 to +80	*	Frequency	*
Smartec					
SMT160-30	± 0.7	-45 to +130	*	Duty-cycle	1...4 kHz
Melexis					
MLX90614	± 0.5	-20 to +120	10	PWM	*
Honsberg					
Flex-T	± 1 % (FS error)	0 to +250	*	Frequency	2000 Hz (max)
Environdata					
TA40	± 0.2	-20 ... +60	0.025°C	Frequency	0 ... 40 Hz
Spica Technology					
T/FRQ	± 1.5	-40 ... +80	0.02°C	Frequency	58…90 Hz

* - information is not available

A temperature sensor based on a temperature-to-frequency relaxation oscillator circuit for extended temperature range from –190 °C to +210 °C has frequency output range from 15 kHz to 2 kHz. The circuit employed a high temperature silicon-on-insulator precision timer CHT-555 and other specially selected, temperature-stable commercial-off-the-shelf parts as thin-film platinum RTD, solid tantalum and ceramic capacitors [11]. Another temperature sensor based on a temperature-to-frequency relaxation oscillator circuit contains a monolithic quad operational amplifier HT1104 (*Honeywell*); high precision, thin film platinum RTD in a ceramic package; the resistors consisted of high-temperature precision-film elements; and the capacitor comprised of multi-layer stacked units made of NPO/COG ceramic for optimum thermal stability [12]. It has a wide –195 °C to +200 °C temperature range and 2.3 kHz to 9.2 kHz frequency output.

The Universal Frequency-to-Digital Converter (UFDC) [13] and Universal Sensors and Transducers Interface (USTI) [14] Series of ICs let eliminate these design problems, because of it uses the advanced conversion method to convert different frequency-time parameters of signal to digital. Such design approach lets significantly simplify the design process, reduce time to market and production price and produce data acquisition (DAQ) systems for quasi-digital temperature sensors with high metrological performances. Let consider quasi-digital temperature sensors and its interfacing to frequency-to-digital converters in details.

6.2 Digital Temperature Sensor Systems Based on Quasi-digital Sensors and FDC ICs

As it was marked in the review [15] 'the past few decades have witnessed an explosive growth in sensors and sensor-based applications, which have led to a greater demand for sensor interfacing integrated circuits (ICs)'. Additionally, the dominance of electronic control over mechanical control further influences the growth of sensor interfacing ICs.

Due to a big number of existing quasi-digital temperature sensors with frequency, period, duty-cycle and pulse width modulated outputs currently available in the market [6], the task of design of appropriate interfacing ICs suitable for creation low-cost, intelligent sensors systems was very important and well-timed. Today, the task has been solved due to introduction in marked UFDC, USTI and FDCP Series of ICs.

Let consider some temperature sensors systems based on PWM, frequency, period and duty-cycle output semiconductor sensors that are available now on a modern temperature sensors market, for example, TMP03/04/, TMP05/06, MAX6673, MAX6577, MAX6576, and SMT160-30 respectively.

The TMP03/TMP04 are monolithic temperature detectors that generate a modulated serial digital output that varies in direct proportion to the temperature of the device [16]. The digital output of the TMP04 is CMOS/TTL compatible, and is easily interfaced to the input of UFDC-1, UFDC-1M-16, USTI or FDCP. The open-collector output of the TMP03 is capable of sinking 5 mA. The TMP03 is best suited for systems requiring isolated circuits utilizing optocouplers or isolation transformers. The output of the TMP03 is a square wave with a nominal frequency of 35 Hz (± 20 %) at 25°C. The output signal is shown in Figure 6.1.

Fig. 6.1 TMP03 output format.

The temperature should be calculated according to the following equations:

$$T(°C) = 235 - \left(\frac{400 \times T1}{T2}\right) \tag{6.1}$$

$$T(°F) = 455 - \left(\frac{720 \times T1}{T2}\right) \tag{6.2}$$

The time interval T1 (pulse width) and T2 (space interval) can be easily read by the mentioned FDCs, with the above calculations performed in software by PC connected through a COM-port or master microcontroller in case of I^2C or SPI buses. T1 is nominally 10 ms, and compared to T2 is relatively insensitive to temperature changes. A useful worst-case assumption is that T1 will never exceed 12 ms over the specified temperature range. The output frequency vs. temperature is shown in Figure 6.2 (a) and T1 and T2 time intervals vs. temperature - in Figure 6.2 (b).

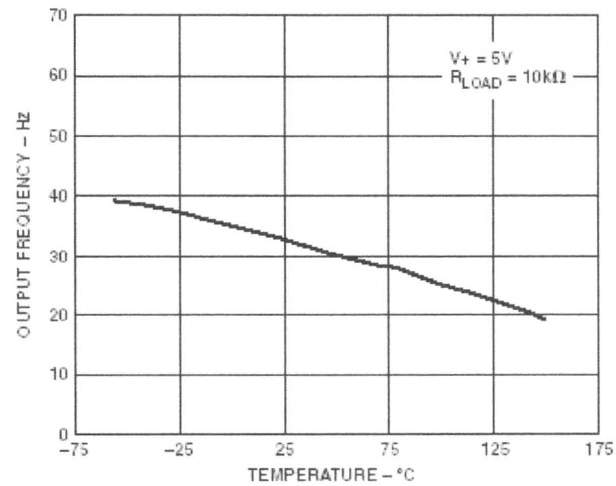

a)

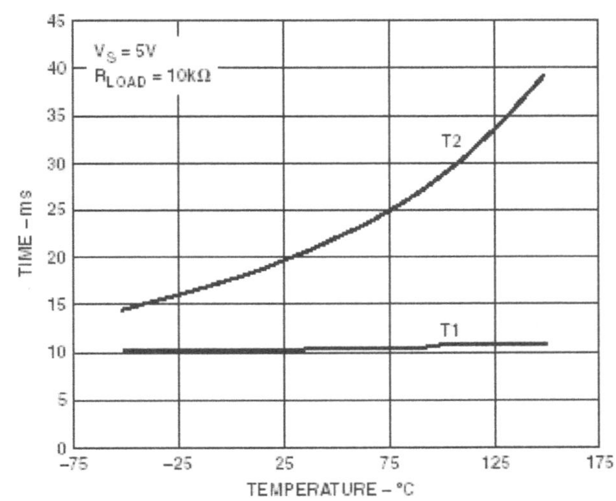

b)

Fig. 6.2 Output frequency vs. temperature (a), and T1 and T2 time intervals vs. temperature (Reproduced from [16]).

The FDCs simply measures the *T1* and *T2* time intervals then PC or microcontroller calculates the temperature using the equations (6.1) or (6.2).

The UFDC-1 can directly convert pulse width and pulse space into digital in the same channel. Three different approaches to perform PWM measurements can be identified [17]:

- Measure the pulse width t_p and pulse space t_s in the same, single period of the PWM signal;
- Measure the pulse width t_p in one period and the pulse space t_s in another period of the PWM signal;
- Measure the period duration $T_x = t_p + t_s$ in one period and pulse space t_s in another period of the PWM signal. The pulse width t_p then should be calculated as $t_p = T_x - t_p$.

The metrological analyze shows that the best results are obtained by using the first and the second methods. The second method was implemented in the UFDC-1 based temperature sensor systems based on various quasi-digital temperature sensors, for example, shown in Figure 6.3. The same experimental results were obtained also by G. Horak, *et al.* and described in [18]. Appropriate commands (RS232 interface) for the UFDC-1 are shown in Figure 6.4.

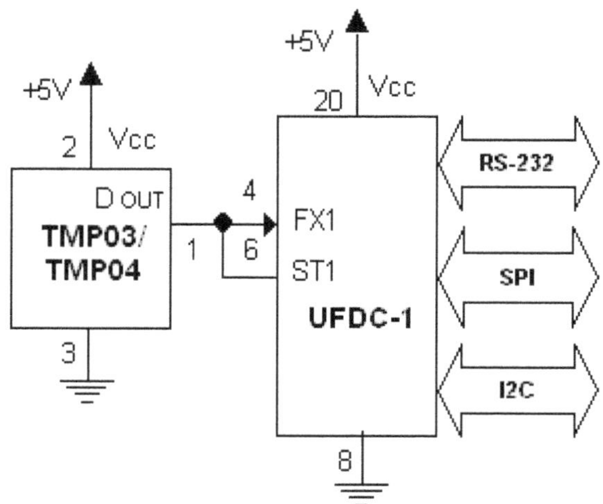

Fig. 6.3 TMP03/TMP04 interfacing to the UFDC-1.

Example:

>MB	Pulse interval T1 measurement mode
>S	Start a measurement
>C	Check result status
>R	Read a result for T1
>MC	Space interval T2 measurement mode
>S	Start a measurement
>C	Check result status
>R	Read a result for T2

Fig. 6.4 Appropriate UFDC-1 commands (RS232) for T1 and T2 time intervals measurement.

Taking into account the sensors' relative error ± 1.5 % FS the UFDC-1 conversion error should be chosen 0.1 % in order to be neglected.

The **TMP05/TMP06** are monolithic temperature sensors that generate a modulated serial digital output (PWM), which varies in direct proportion to the temperature of the devices [19]. The pulse width (T1) of the PWM remains static over all temperatures, while the space interval (T2) varies by the same way as for TMP03/TMP04 quasi-digital temperature sensors (Figures 6.1, 6.2).

The output of the TMP05/TMP06 is CMOS/TTL compatible, and is easily interfaced to the input of FDCs. Pulse-width modulated (PWM) output gives a square wave whose ratio of pulse width to space interval is proportional to temperature. The typical period is 116 ms at 25°C. This format is easily decoded by a PC (RS232 interfacing) or master microcontroller (I^2C or SPI bus communication modes) into either °C or °F values.

The temperature should be calculated according to the following equation:

$$T(°C) = 421 - \left(751 \cdot \frac{T1}{T2}\right) \tag{6.3}$$

The pulse width T1 and space interval T2 are values easily read by the UFDC-1. In standard interfacing mode to the UFDC-1, the TMP05/TMP06 should be used in continuously converting mode with appropriate conversion rate [19]. In order to use the shot or daisy-chain mode the custom UFDC design should be ordered.

Figure 6.5 shows how the TMP05/TMP06 quasi-digital sensors can be interfaced to the UFDC-1 IC. This corresponds to the continuously converting mode (floating FUNC pin) and the nominal conversion rate T1/T2=40/76 (floating CONV/IN pin). The appropriate UFDC-1 commands are the same as in the Figure 6.4.

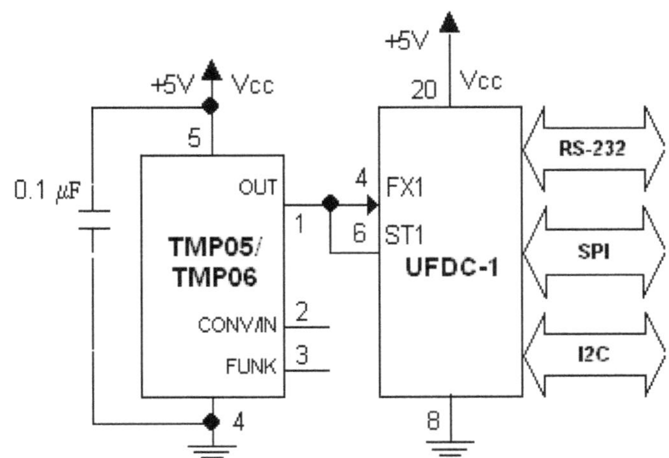

Fig. 6.5 TMP05/TMP06 interfacing to UFDC-1: T1 and T2 time intervals measurements.

The TMP03/TMP04 and TMP05/TMP06 quasi-digital temperature sensors can be also interfaced to the USTI IC by the same manner (hardware interfacing).

The **MAX6576/MAX6577** are low-cost, low-current quasi-digital temperature sensors [20]. The MAX6576 converts ambient temperature into a 50 % duty-cycle square wave with a period proportional to absolute temperature (°K):

$$T(°C) = \frac{Tx(\mu s)}{Ks} - 273.15, \tag{6.4}$$

where Ks is the scalar multiplier.

The MAX6577 converts the ambient temperature into a 50 % duty-cycle square wave with a frequency a square wave with a frequency proportional to absolute temperature:

$$T(°C) = \frac{fx(Hz)}{Ks} - 273.15 \tag{6.5}$$

The period/frequency range (the scalar multiplier Ks) of the output square wave can be selected by hard-wiring the two time-select pins (TS0, TS1) to either V_{DD} or GND.

The oscillograms on MAX6576 and MAX6577 sensors' outputs are shown in Figure 6.6 and Figure 6.7 respectively.

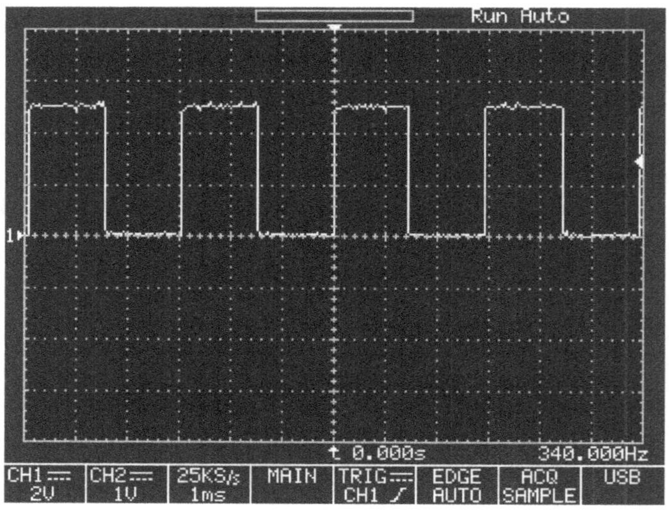

Fig. 6.6 Oscillograms on MAX6576 sensor's output (period ~ 3 ms).

Figure 6.8 shows the MAX6576 interfacing to the UFDC-1. In this example, TS1 and TS0 are tied to ground to select a scalar multiplier of 10 μs/°K. The MAX6576 converts the ambient temperature into a square wave with a period that is equal to the absolute temperature. The UFDC-1 reads the period of the square-wave sensor's output. The PC (RS232 communication mode) or master microcontroller (I^2C or SPI buses communication modes) calculates the temperature in °C according to equation (6.4).

Figure 6.9 shows the MAX6577 interfaced with the UFDC-1. In this example, TS1 is tied to ground and TS0 is tied to V_{DD} to select a scalar multiplier of 1Hz/°K. The MAX6577 converts the ambient temperature into a square wave with a frequency that is equal to the absolute temperature. The UFDC-1 reads the period of the square-wave sensor's output. The PC (RS232 communication mode) or master microcontroller (I2C or SPI buses communication modes) calculates the temperature in oC according to equation (6.5). The result of frequency measurement equals to temperature in oK. So, this mode is also very useful for digital temperature sensor design with oK unit. No any calculation for temperature is necessary in this case.

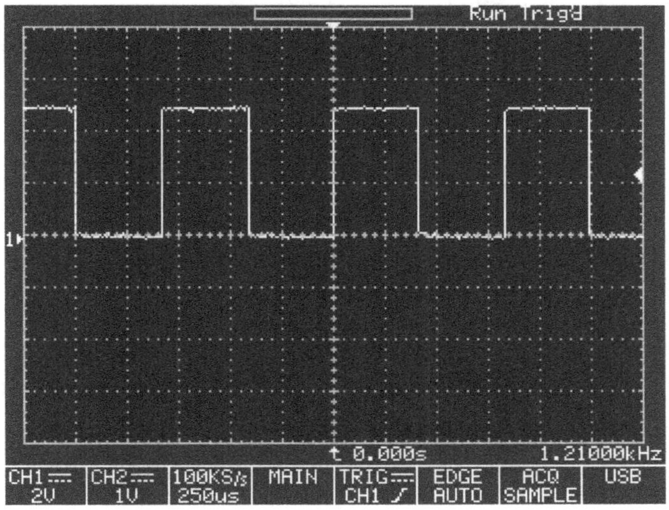

Fig. 6.7 Oscillograms on MAX6577 sensor's output (frequency ~ 1.21 kHz).

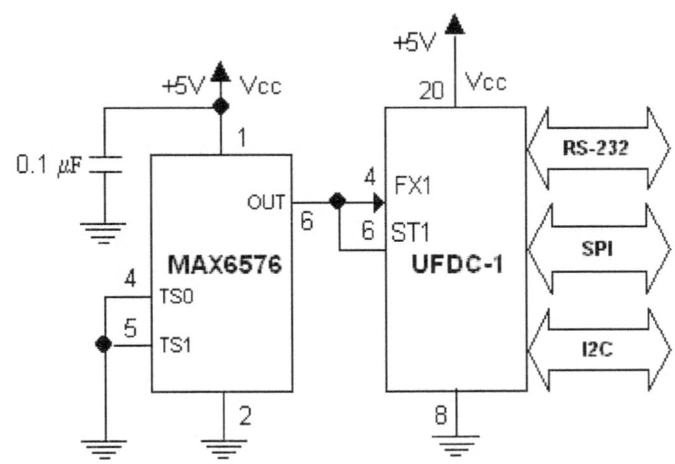

Fig. 6.8 Quasi-digital, period output temperature sensor MAX6576 interfacing to the UFDC-1 IC.

Taking into account the wide conversion frequency range of UFDC-1 it is not matter of fact which of scalar multiplier K_s is used.

These sensors can be also working with the USTI and FDCP ICs. Appropriate commands should be used to measure frequency and period. Similar to the UFDC-1 the USTI can serves two different quasi-digital temperature sensors at the same time.

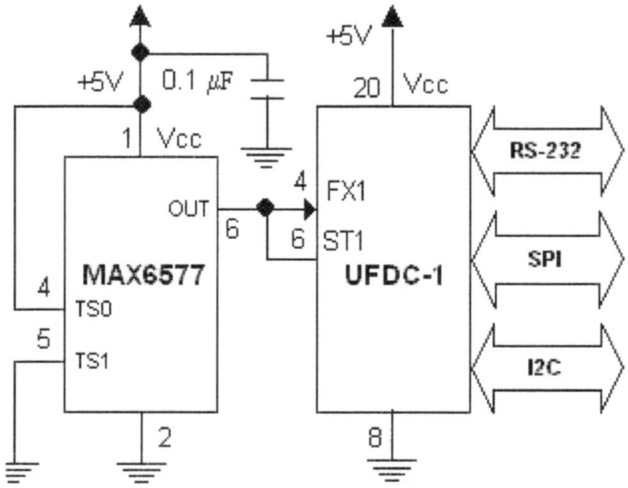

Fig. 6.9 Quasi-digital, frequency output temperature sensor MAX6577 interfacing to the UFDC-1 IC.

Interfacing boards for quasi-digital temperature sensors: MAX6576 (1), MAX6577 (2) and MAX6673 (3) experimental investigations in climatic chambers are shown in Figure 6.10.

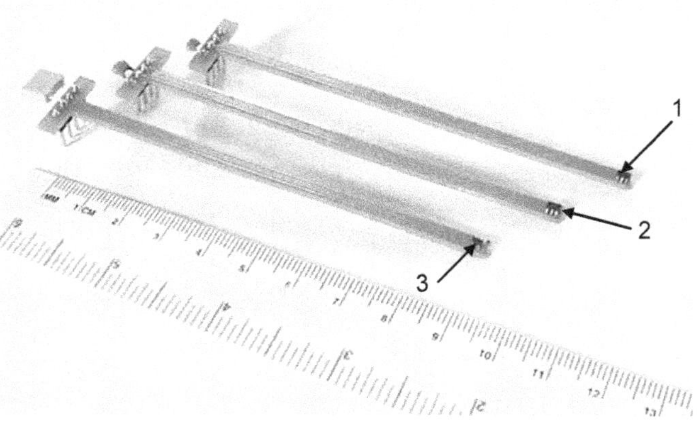

Fig. 6.10 Interfacing boards for quasi-digital temperature sensors: MAX6576 (1), MAX6577 (2) and MAX6673 (3).

The **MAX6666/MAX6667** are high-accuracy, low-cost, low-power quasi-digital temperature sensors with a single-wire output. The MAX6666/MAX6667 convert the ambient temperature into a ratiometric PWM output with temperature information contained in the pulse width modulated output square wave. The MAX6666 has a push-pull output and the MAX6667 has an open-drain output [21]. The output is a square wave with a nominal frequency of 35 Hz (± 20 %) at +25 °C. The output format is decoded as in equation (6.1), where T1 is fixed with a typical value of 10 ms and T2 is modulated by the temperature. These temperature sensors can be interfacing to the UFDC-1 by the same manner as the TMP03/TMP04 sensors from *Analog Devices*.

Other models of PWM output sensors are **MAX6672/MAX6673** low-current quasi-digital temperature sensors with a single-wire output. These temperature sensors convert the ambient temperature into a 1.4 kHz PWM output, which contains the temperature information. The MAX6672 has an open-drain output and the MAX6673 has a push-pull output [22].

The square wave output waveform time ratio contains the temperature information. The output is a square wave with a nominal frequency of 1.4 kHz at +25 °C. The output format is shown in Figure 6.11.

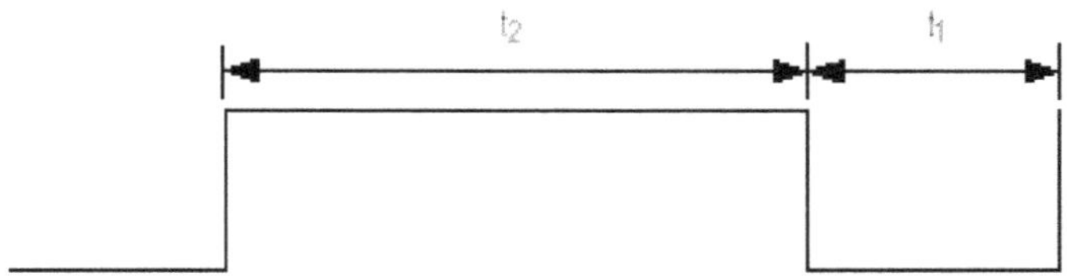

Fig. 6.11 MAX6672/MAX6673 temperature sensors output format.

The temperature is obtained with the following formula:

$$T(°C) = -200 \cdot \left(0.85 - \frac{t_1}{t_2}\right)^3 + \left(425 \cdot \frac{t_1}{t_2}\right) - 273, \qquad (6.6)$$

where t_1 is a fixed value and t_2 is modulated with the temperature.

For temperatures greater than +50 °C, the temperature error is primarily first order and the following equation can be used:

$$T(°C) = \left(425 \cdot \frac{t_1}{t_2}\right) - 273 \qquad (6.7)$$

These temperature sensors also can be interfacing to the UFDC-1 and USTI by the same manner as, for example, the TMP03/TMP04 sensors from *Analog Devices*.

A temperature sensors system based on quasi-digital sensor MAX6673 and universal frequency-to-digital converter UFDC-1 were investigated experimentally in the climatic chamber CTS T-40/50 in the working temperature ranges of sensors from - 40 ^{0}C to +130 ^{0}C and +125 ^{0}C respectively with the step 10 ^{0}C at relative humidity 48.2…54.5 %. The temperature inside of the climatic chamber was setup and maintained with the ±0.5 ^{0}C absolute error. For each of investigated temperature 60 measurements by means of UFDC-1 and precision universal frequency counter Agilent 53132A were made. Absolute and relative errors and statistical characteristics were determined. The climatic chamber and measuring equipments are shown in Figure 2.16.

Oscillograms at MAX6673 temperature sensor's output at - 40^0 C, +20 ^{0}C and +125 ^{0}C are shown in Figure 6.12 (a), (b) and (c) respectively.

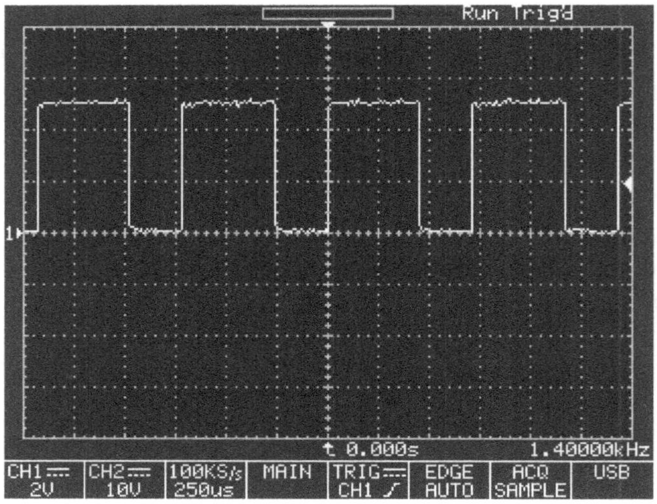

(a)

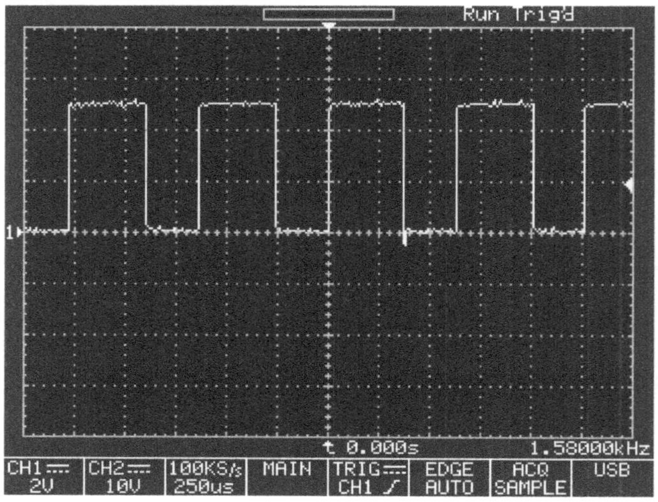

(b)

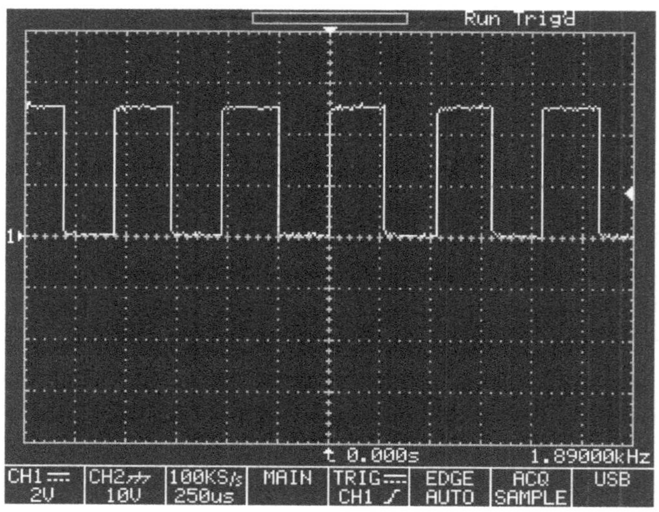

(c)

Fig. 6.12 Oscillograms at MAX6673 temperature sensor's output: at - 40 ^{0}C (a), + 20 ^{0}C (b) and +125 ^{0}C (c).

Main statistical characteristics at pulse width and pulse space measurements at +120 ^{0}C on MAX6673 temperature sensor output adduced in Table 6.2.

Table 6.2 Main statistical characteristics at pulse width and pulse space measurements.

Parameter	**t_p, μs**	**t_s, μs**
Number of measurements, *N*	60	60
Minimum *t.* (min)	439.0625	258.8125
Maximum *t.* (max)	449.375	262.875
Sampling Range, *t* (max)- *t* (min)	10.3125	4.0625
Median	0	0
Arithmetic Mean	444.1698	259.8865
Variance	6.1175	0.6898
Standard Deviation	2.4734	0.8306
Coefficient of Variation	179.5822	312.9082
Relative error, %	0.16	0.089
Confidence interval	$443.4769 \le t_p \le 444.8627$	$259.6538 \le t_s \le 260.1191$
S	12.0816	14.7858
χ^2	7.0	7.0

Taking into account that the sum of deviations between the data set and the assumed distribution $S > \chi^2$, the hypothesis about Gaussian distribution can not be accepted in both cases. The densities of distributions are shown in Figure 6.13 (a, b).

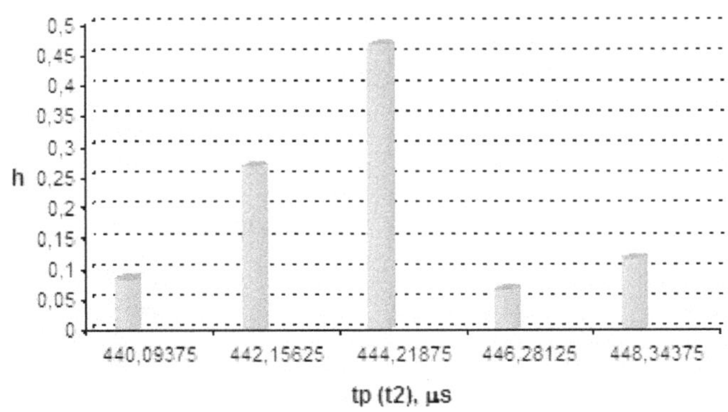

(a)

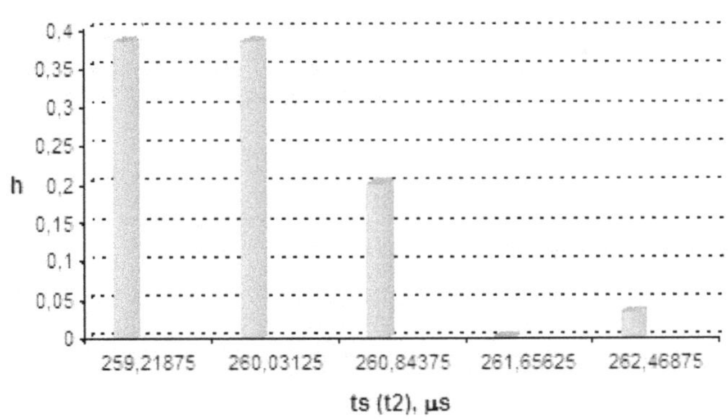

(b)

Fig. 6.13 Density of distribution at pulse width t_p (a) and pulse space t_s (b) measurements.

The **MAX6676/MAX6677** are high-accuracy, low-power quasi-digital temperature sensors with a single-wire output. The MAX6676/MAX6677 convert the ambient temperature into a ratiometric PWM output with temperature information contained in the output square wave. The MAX6676 has an open-drain output and the MAX6677 has a push-pull output [23]. The output is a square wave with a nominal frequency of 1.8 kHz (±20 %) at +25 °C. Its output signal is shown in Figure 6.14.

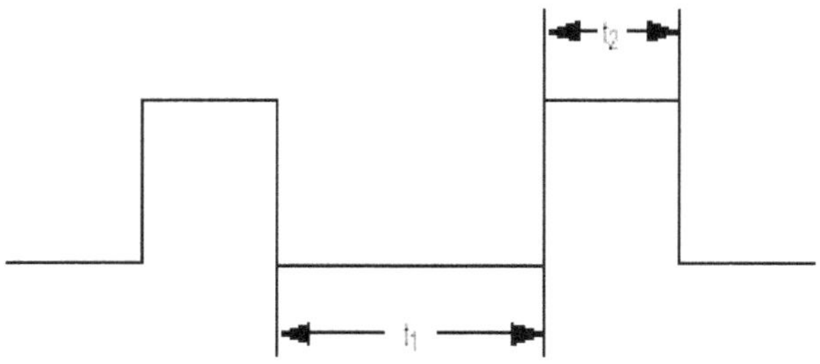

Fig. 6.14 MAX6676/MAX6677 sensor's output signal.

The output format is decoded as follows:

$$T(°C) = 398.15 \cdot \left(\frac{t_1}{t_2} \right) - 273.15, \tag{6.8}$$

where t_1 is fixed with a typical value of 0.24 ms and t_2 is modulated by the temperature.

The MAX6676/MAX6677 sensors interfacing to the UFDC-1 is shown in Figure 6.15 The resistor is required only for the MAX6676 sensor.

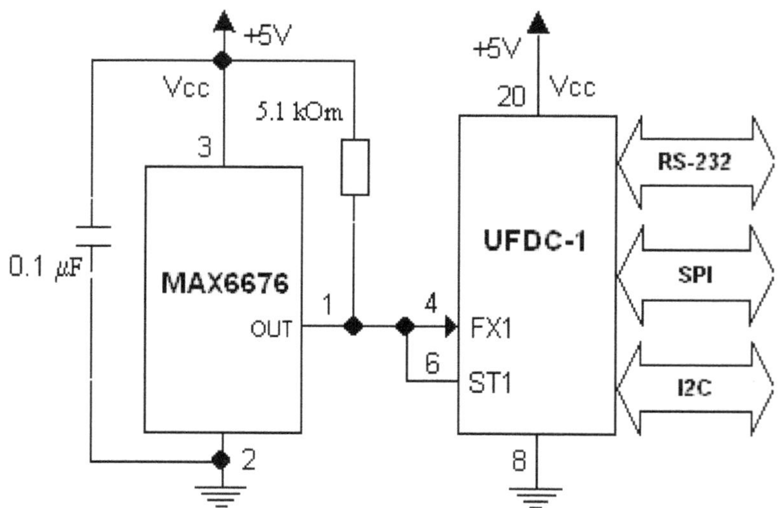

Fig. 6.15 MAX6676 sensor interfacing to UFDC-1.

The temperature sensor **SMT160-30** is a sophisticated full silicon temperature sensor with a quasi-digital (duty-cycle) output [24]. The output of the sensor can handle cable length up to 20 meters. This makes the SMT160-30 very useful in remote sensing and control applications and its output can be easy interpretable by both UFDC-1 and USTI ICs.

The duty-cycle of the output signal is linearly related to the temperature (Figure 6.16) and the temperature can be calculated according to the following equation:

$$T(°C) = \frac{D.C. - 0.320}{0.0047}, \qquad (6.9)$$

where $D.C.$ is the duty-cycle.

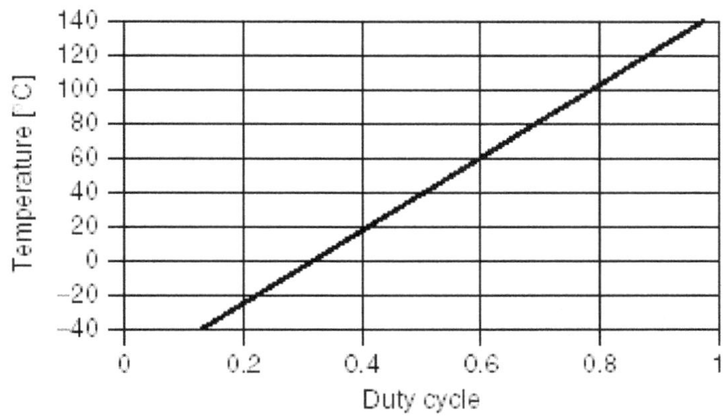

Fig. 6.16 Temperature vs. duty-cycle for SMT160-30 temperature sensor.

Appropriate commands (RS232 interface) for the UFDC-1 operation mode for duty-cycle – to digital conversion are shown in Figure 6.17

```
>M4       ; Duty-cycle measurement initialization
>S        ; Start a measurement
>C        ; Check result status
>R        ; Read a result for T1

0.4585
```

Fig. 6.17 Appropriate UFDC-1 commands (RS232) for duty-cycle measurement.

Oscillogram at SMT160-30 temperature sensor's output at +20 ^{0}C is shown in Figure 6.18 and the dependence of absolute error on measuring temperature is shown in Figure 6.19 Main statistical characteristics at +20 ^{0}C adduced in Table 6.3.

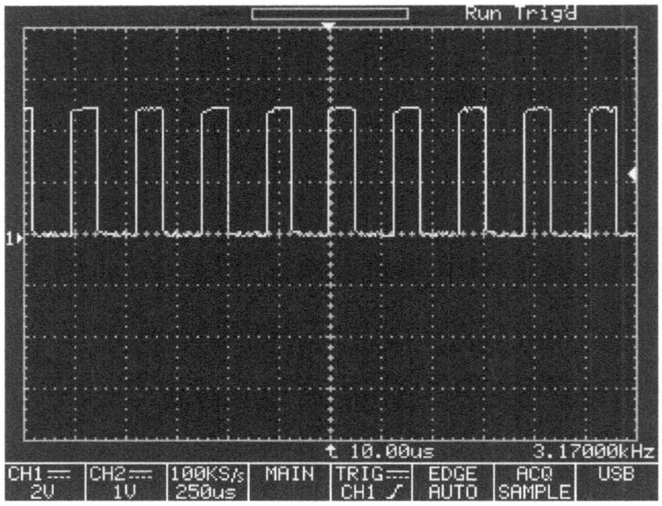

Fig. 6.18 Oscillogram at SMT160-30 temperature sensor's output: at +20 ^{0}C.

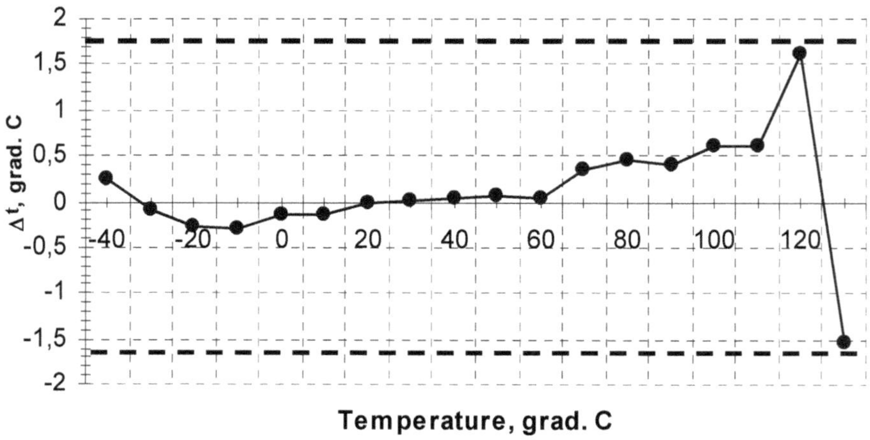

Fig. 6.19 Dependence of absolute error on measuring temperature.

Table 6.3 Main statistical characteristics at duty-cycle measurements.

Parameter	Duty-cycle, %
Number of measurements, N	60
Minimum $D.C.$ (min)	0.4585
Maximum $D.C.$ (max)	0.4588
Sampling Range, $D.C.$ (max)- $D.C.$ (min)	0.0003
Median	0
Arithmetic Mean	0.4586
Variance	6.6E-0009
Standard Deviation	0.0001
Coefficient of Variation	5666.7026
Relative error, %	0.044
Confidence interval	$0.4586 \leq D.C. \leq 0.4587$
χ^2 – test (S) at: k=5; P = 97 % χ^2_{max} = 7.0	5.1699

The χ^2-test for goodness of fit was applied to investigate the significance of the differences between observed data in the histograms and the theoretical frequency distribution for data from a normal population. Taking into account that the sum of deviations between the data set and the assumed distribution $S < \chi^2$, the hypothesis about Gaussian distribution can be accepted. The density of distribution is shown in Figure 6.20.

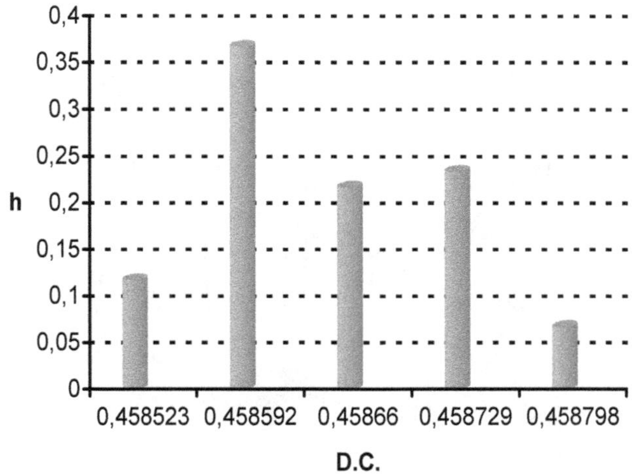

Fig. 6.20 Density of distribution at duty-cycle measurements.

Oscillograms at SMT160-30 temperature sensor's output at -40 ^{0}C and + 130 ^{0}C are shown in Figure 6.21 and 6.22 respectively.

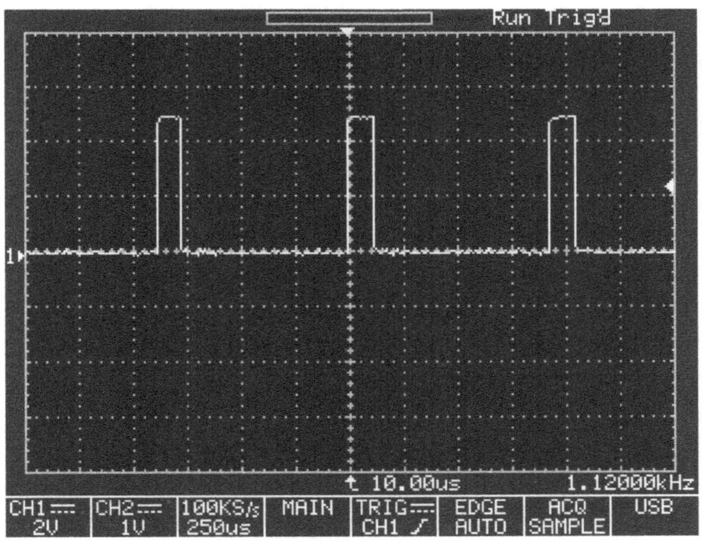

Fig. 6.21 Oscillograms at SMT160-30 temperature sensor's output: at - 40 ^{0}C.

The main distinction of frequency output temperature sensors from *Sea-Bird Electronics* [6] is its very high accuracy. The frequency-to-digital converter should have one order less or at least in five times less conversion error than the sensor's error in order to be neglected in comparison with the sensor error. Taking into account the high accuracy and working frequency range let see what we have if are using the standard counting method or indirect counting method (both methods are classical and using widely till now) for frequency-to-digital conversion. In order to maintain the accuracy for SBE 3F and SBE 3plus sensors, the standard counting technique for frequency-to-digital conversion needs 166.7 s conversion time. The UFDC-1 is based on the advanced conversion method with non-redundant conversion time and can guarantee the requested accuracy at 0.66 s conversion time.

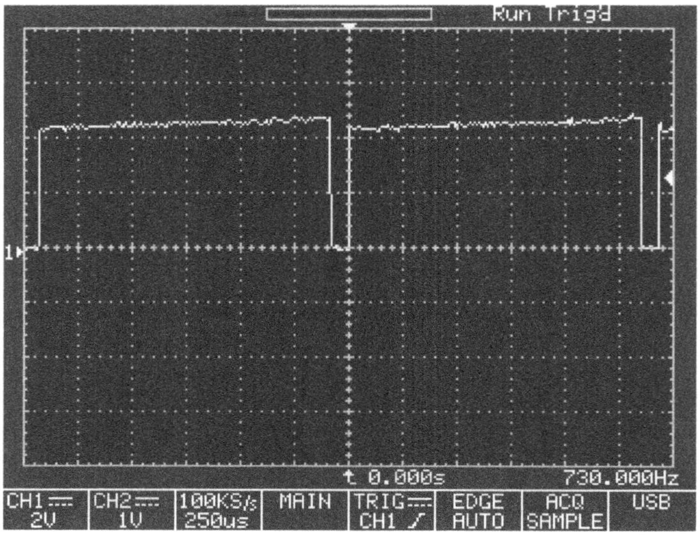

Fig. 6.22 Oscillograms at SMT160-30 temperature sensor's output: at +125 ^{0}C.

The UFDC-1 is working well with the SBE8 temperature sensor. In this case the programmable error for frequency-to-digital conversion of UFDC-1 should be chosen 0.0025 %. In case of SBE 3F and SBE 3plus sensors the USTI IC with 0.0005 % relative error should be used.

There are many ways to display the measurement results. Any terminal software for Windows can be used with the UFDC-1 and USTI ICs using the serial port of a PC, for example, the mentioned before Terminal V1.9b free software [25]. Other possibilities are Component Works 2.0 (ActiveX control) or LabView. Both software products are from the National Instruments. The virtual thermometer based on the SMT160-30 quasi-digital temperature sensor is shown in Figure 6.23.

The described design approach based on the UFDC-1 or USTI lets significantly simplify the design process, reduce time to market and production of digital temperature sensors and DAQ systems for quasi-digital temperature sensors with high metrological performances.

In comparison with the direct microcontroller interfacing the proposed approach also lets to eliminate many design problems connected with the use of advanced measurement methods for frequency-time measurements, microcontroller choice, its programming in Assembler and additional error components due to so-called program-dependent or software-related effects.

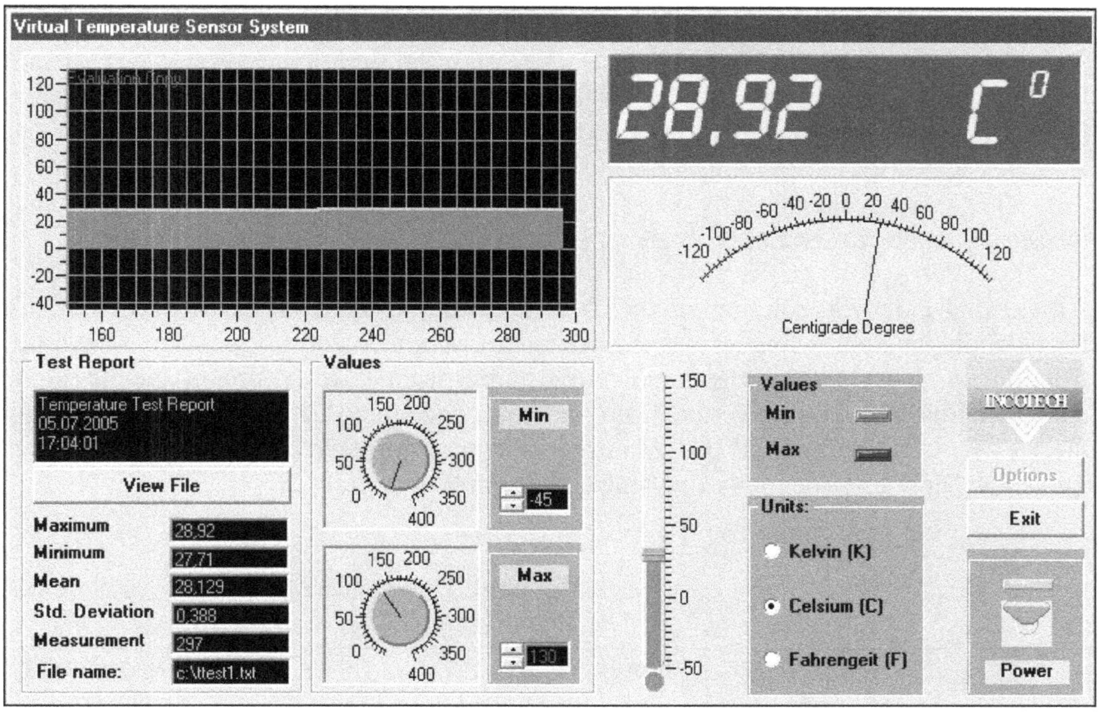

Fig. 6.23 Virtual thermometer based on SMT160-30 quasi-digital temperature sensor.

Each 2-channel UFDC or USTI IC can work with two different quasi-digital temperature sensors. In applications where multiple sensors are used, easy multiplexing can be obtained by low cost digital multiplexers.

Designed digital sensors and DAQ systems for quasi-digital temperature sensors based on the UFDC-1 and USTI can be used for different applications: temperature monitoring, remote temperature measurement, environmental control systems, industrial process control, thermal protection, etc.

6.3 Digital Temperature Sensor Systems Based on RTDs and Thermocouples

One more interesting application of the UFDC-1 IC in the smart temperature sensor is described in [26]. This paper reports the fabrication of nickel-based resistance temperature detectors (RTD), transducers and the implementation of the required electronics to build a smart temperature transducer to operate in oil pipeline. For the conditioning and processing circuitry design, one of approaches was to use an op-amp based oscillator and frequency-to-digital converter. The conditioning electronics is designed with off-the-shelf components, such as, for example, UFDC-1 IC. The RTD was used as a part of an oscillator circuit. This oscillator was a relaxation oscillator built with an operation amplifier, OPAMP. The transducer signal was transmitted as an FM signal, and therefore was not sensitive to additive noise. The FM signal was applied to a frequency-to-digital converter [26]. As the temperature changes, the resistance of the sensor changes, modifying the oscillating frequency. The frequency is converted to digital with a frequency-to-digital converter (UFDC-1). Such design approach assures better noise immunity in comparison with the traditional approach, when the current source, ADC and microcontroller have been used. The classical design approach's performance was not adequate for this application because of this technique needs much more hardware and suffers from noise and interference from other signals [26].

The RTD can be also directly interfaced to the USTI IC, working in the resistance measuring mode [27], see also Chapter 14.

All existing industrial solutions for RTD to frequency conversion based DAQ systems, for example, analog to frequency modules from *Omega*, analog-to-frequency converters from *Soclair Electronics,* programmable transmitter from *Nokeval*, programmable digital converter from *Loreme*, converters from *Megatron* and others [28], are working well with the UFDC, USTI and FDCP ICs. Main performances for such converters are shown in Table 6.4. More information is available at Sensors Web Portal [28].

Table 6.4 RTD-to-frequency converters performances.

Converter	Temp. Range, °C	Resolution, bits (°C)	Accuracy	Frequency Output Range, Hz
		Omega, Inc.		
DRP-8540	-100 ... +400	12 (0.1)	± 0.5 plus RTD conformity	100 ... 5 100
		Soclair Electronic, AG		
RTMF 70	*	*	± 0.2 %	0 ... 20 000
		Nokeval		
6746B	-60 ... +700	*	± 0.1 % of reading +0.1 °C	0 ... 2 500
		Loreme		
CNL48/F	-200 ... +600	*	± 0.3 °C	0 ... 10
		Megatron		
MIFN-PT100	*	*	± 0.5 % FS	0 ... 10 000

* - information is not available

Thermocouples are the most widely used temperature sensor in industrial manufacturing environments. Thermocouples consist of two wires made of different metals that are joined at one end, called the measurement junction. At the other end of the conductors, a reference junction is formed. When the measurement junction and the reference junction have different temperatures, a continuous current flows in the circuit. The resulting voltage is a function of the difference in temperature between the measurement and the reference junctions. The amount of voltage depends on the types of metals used. In comparison with a RTD, the thermocouple not so accurate and stable as RTD, but it can be used to measure extremely high temperatures.

Nevertheless thermocouples are typical analog output sensors, it is preferably to have more robust frequency output for such temperature sensors in many applications. To solve this design problem, preliminary voltage-to-frequency conversion should be used. After, the UFDC, USTI or FDCP ICs can be used to convert frequency-to-digital.

The frequency output can be obtained with the help of thermocouple signal conditioner AD594/AD595 and voltage-to-frequency converter AD654 [29]. The temperature-to-frequency converter is shown in Figure 6.24.

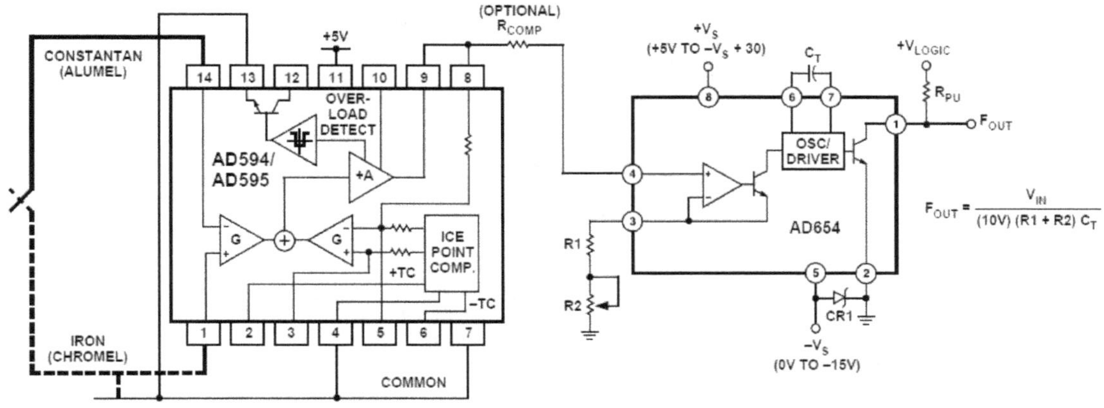

Fig. 6.24 Temperature-to-frequency converter [29].

This approach not only affords noise immunity over long transmission paths but also provides information which may be directly interfaced to a FDC. A low cost voltage-to-frequency converter, the AD654, converts the 10 mV/°C voltage output to a TTL compatible square wave output [29]. This sensor system provides temperature measurements from 0°C to +300 °C. Higher thermocouple temperatures will require a higher power supply voltage to maintain a maximum AD594/AD595 output swing of 2.5 V below the supply. A maximum usable frequency of 500 kHz resulting in 0.4 % nonlinearity. Other temperature ranges and output frequencies are achievable.

Other solutions for voltage output temperature sensors, based on discreet components and ICs are described in [30-35]. As it is visible from review, solutions for voltage-to-frequency conversions with high metrological performances were known for a long time. The bottle neck was the further frequency-to-digital conversion. Today, due to availability on the modern market Series of UFDC, USTI and FDCP ICs, such design problem can be solved by the most effective way.

The main metrological performances of various analog to frequency conversion modules from different manufacturers suitable for thermocouple DAQ systems industry design are shown in Table 6.5. Some of them are universal and can be used also for RTDs (see Table 6.4).

Nevertheless of a relatively broad frequency range, the UFDC, USTI and FDCP are well suitable for working with all mentioned converters.

Table 6.5 Thermocouple-to-frequency converters performances.

Converter	Temp. Range, °C	Accuracy	Frequency Output Range, Hz
Omega, Inc.			
DRP-8511Type J	0 … +500	±2 °C plus conformity error	100 … 5 100
DRP-8512 Type K	0 … +500	±2 °C plus conformity error	100 … 5 100
DRP-8611Type J	0 … +1000	±2 °C plus the error of J Thermocouple	10 … 1 100
DRP-8612 Type K	0 … +1250	±1.2 °C	5 … 2 520
Soclair Electronic			
TCMF 70 (all type)	All ranges	± 0.2 %	0 … 20 000
Nokeval			
6746B (all type)	All ranges	± 0.1 % of reading +0.1 °C	0 … 2 500
Loreme			
CNL48/F (all types)	All ranges	± (0.3 … 2) °C	0 … 10

* - information is not available

6.4 Digital Temperature Sensor Systems Based on Voltage-Output Sensors and VFC

Popular integrated temperature sensors, for example, L34 also can be interfaced with the UFDC and USTI ICs through an intermediate voltage-to-frequency converters [36, 37]. Such approach is very suitable for various remote sensing applications. For example, the LM34 can be used in conjunction with an LM131 voltage-to-frequency converter to perform the desired conversion from temperature to frequency. A relatively simple circuit which performs over a +3 °F to +300 °F temperature range is shown in Figure 6.25 [38].

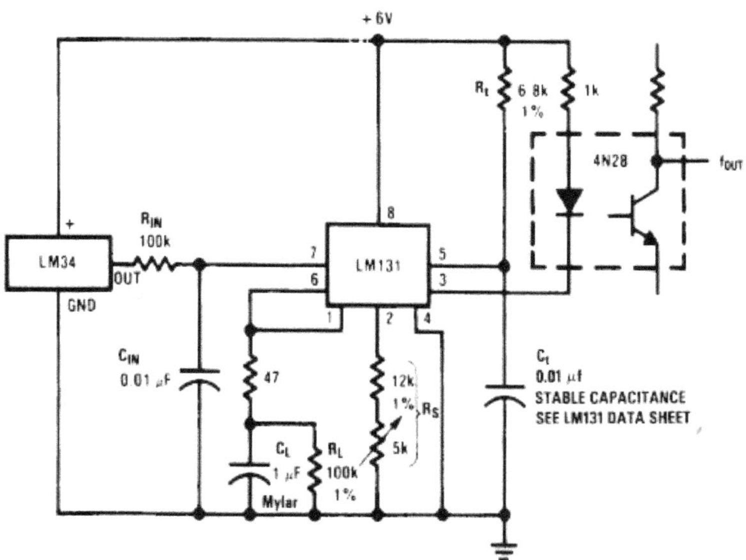

Fig. 6.25 LM34 with Voltage-to-Frequency Converter and isolated output (3 °F to +300 °F; 30 Hz to 3000 Hz) [38].

The output frequency of this circuit can be found from the equation:

$$f_{OUT} = \left(\frac{V_{IN}}{2.09V}\right) \cdot \left(\frac{R_S}{R_L}\right) \cdot \left(\frac{1}{R_t C_t}\right), \tag{6.10}$$

where resistor R_S is used to adjust the gain of the LM131. If R_S is set to approximately 14.2 kΩ, the output frequency will have a gain of 10 Hz/°F. Isolation from high common mode levels is provided by hanneling the frequency through a photoisolator. This circuit is also useful for sending temperature information across long transmission lines where it can be decoded at the receiving station [38].

6.5 References

[1]. 2011 Temperature Sensors, *MarketResearch.com*, July 2011.

[2]. Frost & Sullivan Evaluate World Temperature Sensor Market, *Sensors Magazine*, January 2010.

[3]. Yurish S. Y., Data Acquisition Systems for Quasi-Digital Temperature Sensors Based on Universal Frequency-to-Digital Converter, *Sensors & Transducers, Vol.57, Issue 7, July 2005, pp.341-351.*

[4]. Yurish S. Y., Smart Sensors for Electrical and Non-electrical, Physical and Chemical Variables: State-of-the-Art, in Smart Sensors and MEMS, ed. by Sergey Y. Yurish and Maria Teresa S. R. Gomes, *Kluwer*, 2004, pp. 1-50.

[5]. Kirianaki N. V., Yurish S. Y., Shpak N. O., Deynega V. P., Data Acquisition and Signal Processing for Smart Sensors, *John Wiley & Sons*, 2002.

[6]. List of Temperature Sensors Manufacturers at Sensors Web Portal: http://www.sensorsportal.com/HTML/SENSORS/TempSens_Manufacturers.htm

[7]. Yurish S. Y., Program-oriented Methods and Measuring Instruments for Frequency-Time Parameters of Electric Signals, *PhD Thesis,* State University Lviv Polytechnic Lviv, 1997 (In Ukrainian).

[8]. Reverter F., Microcontroller-based Interfaces for Quasi-Digital Sensors, *PhD Thesis,* Technical University of Catalonia (UPC Barcelona, Spain), 2004.

[9]. Yurish S., Reverter F., Pallàs-Areny R., Measurement error Analysis and Uncertainty Reduction for Period- and Time Interval-to-digital Converters based on Microcontrollers, *Measurement Science and Technology*, Vol.16, No.8, 2005, pp.1660-1666.

[10]. V. Székeley, M. Rencz, S. Török, Cs. Márta, L. Lipták-Fegó, CMOS Temperature Sensors and Built-in Test Circuitry for Thermal Testing of ICs, *Sensors and Actuators, A71*, pp. 10-18, 1998.

[11]. R. L. Patterson, A. Hammoud, M. Elbuluk, D. E. Culley, A Temperature Sensor Using a Silicon-on-Insulator (SOI) Timer for Very Wide Temperature Measurement, *Report NASA/TM*-2008-215159, May 2008.

[12]. R. L. Patterson, A. Hammoud, Use of COTS (Commercial-Off-The-Shelf) Parts for Wide-Range Temperature Sensing from Hot Jet Engine Distributed Control Systems to Cryogenic Space Missions, *NASA Report*, October 2007.

[13]. Universal Frequency-to-Digital Converter (UFDC-1), Datasheets, *Technology Assistance BCNA 2010, S. L.,* 2011.

[14]. Universal Sensors and Transducers Interface (USTI), Datasheets, *Technology Assistance BCNA 2010, S. L.,* 2011.

[15].Developments in sensor interfaces (Technical insights), *Frost & Sullivan*, 2007.
[16].Serial Digital Output Thermometers TMP03/TMP04, *Analog Devices*, USA.
[17].Yurish S. Y., Low-Cost, Smart Temperature Sensors Systems Based on Universal Frequency-to-Digital Converter, in *Proceedings of the IEEE Sensors Applications Symposium (SAS' 2009)*, 17-19 February 2009, New Orleans, LA, USA, pp.287-292
[18].G. Horak, V. Bilas, D. Vasić, Noise Analysis of the Pulse-width Modulator for Quasi-digital Sensors, *in Proceedings of the IEEE International Instrumentation & Measurement Technology Conference (I2MTC' 2008)*, 12-15 May 2008, Victoria, Vancouver Island, British Columbia, Canada, pp. 687-691.
[19].±0.5°C Accurate PWM Temperature Sensor in 5-Lead SC-70 TMP05/TMP06, *Analog Devices*, USA.
[20].SOT Temperature Sensors with Period/Frequency Output MAX6576/MAX6577, *Maxim*, 19-1484; Rev 0; 4/99, 1999, USA.
[21].High-Accuracy PWM Output Temperature Sensors MAX6666/MAX6667, *Maxim*, 9-2138; Rev 13, 8/09, 2009, USA.
[22].PWM Output Temperature Sensors in SC70 Packages MAX6672/MAX6673, *Maxim*, 19-2458; Rev 0; 10/02, 2002, USA.
[23].Low-Voltage, 1.8 kHz PWM Output Temperature Sensors MAX6676/MAX6677, *Maxim*, 19-2660; Rev 1; 1/03, 2003, USA.
[24].SMT160-30 Digital Temperature Sensor, Datasheet, smt16030n, *Smartec*, November 2010.
[25].Terminal V1.9b: http://www.sensorsportal.com/DOWNLOADS/Terminal.exe
[26].Edval J. P. Santos, Isabela B.Vasconcelos, RTD-based Smart Temperature Sensor: Process Development and Circuit Design, in *Proceedings of the 26th International Conference on Microelectronics (MIEL' 2008)*, Nis, Serbia, 11-14 May 2008, pp. 333 – 336.
[27].Universal Sensors and Transducers Interface (USTI), Specification and Application Note, *Technology Assistance BCNA 2010, S. L.*, 2011.
[28].List of DAQ Systems Manufacturers at Sensors Web Portal: http://www.sensorsportal.com/HTML/SENSORS/DAQ_Systems_Manufacturers.htm
[29].Joe Marcin, Thermocouple Signal Conditioning Using the AD594/AD595, Application Note, AN-369, *Analog Devices*, 1998, USA.
[30].Jim Williams, Designs for High Performance Voltage-to-Frequency Converters, Application Note 14, *Linear Technology*, March 1986.
[31].VTF, Voltage (and Temperature) to Frequency Converter, Ver. 990414-d1, *Tucson Amateur Packet Radio*, 1997.
[32].Jim Williams, Some Techniques for Direct Digitization of Transducer Outputs, Application Note 7, *Linear Technology,* February 1985.
[33].Jim Williams, Measurement and Control Circuit Collection, Application Note 45, *Linear Technology,* June 1991.
[34].Temperature to Frequency, Daqarta's Web Site, 2011: http://www.daqarta.com/dw_kkee.htm
[35].Thermocouple to Frequency, Daqarta's Web Site, 2011: http://www.daqarta.com/dw_kkjj.htm
[36].List of Voltage-to-Frequency Converters Manufacturers at Sensors Web Portal: http://www.sensorsportal.com/HTML/SENSORS/VFC_Manufacturers.htm
[37].Voltage (and Temparature) to Frequency Converter, *Tucson Amateur Packet Radio*, USA, 1997.
[38].LM34/LM35 Precision Monolithic Temperature Sensors, Application Note 460, AN009051, *National Semiconductor Corporation*, 2002.

Chapter 7

Digital Pressure Sensors and Transducers

7.1 Quasi-digital Pressure Sensors: State-of-the-Art

Pressure sensors and transducers are very widely spreaded in various industries. It can be found in numerous OEM applications, and they are used in process control, automotive industry, etc. A wide variety of pressure sensing technologies including MEMS is available today [1]. Pressure sensors are one of the main applications for MEMS technology in the automotive market.

According to the *Frost & Sullivan* market report [2] "MEMS-based pressure sensors are currently being used for measuring manifold air pressure and also for measuring the parameter for applications such as the fuel rail, fuel tank, and engine oil. Further, pressure sensors are also used for taking measurement in the side doors and for passenger detection. The pressure sensor market in the automotive industry is set for explosive growth, driven by the application of tire pressure monitoring systems. Tire pressure sensors are also expected to gain acceptance in Europe and Japan and also in luxury vehicles in the emerging economies" [2].

In 2009 MEMS pressure sensors dominate the market revenues with a whopping revenue share of 48.2 % [3]. The market revenues are expected to grow $ US 526.7 million in 2012 [2]. According to the *CIOL Bureau* smart pressure sensors are expected to emerge as the fastest growing product segment in the smart sensors market worldwide [4].

Since the mid 1980s the technology has progressed from uncompensated bulk micro machined piezoresistive pressure sensors to a Complementary Metal Oxide Semiconductor (CMOS) -integrated bulk micro machined pressure sensor with non volatile memory. This has enabled the addition of features such as three-wire interface and changeable rail voltages besides resulting in higher accuracy [2].

Pressure sensors and transducers convert the external pressure into the electrical output signal. In order to convert voltage signal to digital an embedded or external ADC should be used. As an alternative to the conventional ADC, it is possible to measure either a period (frequency) or a pulse width of an incoming square or rectangular wave signal [5]. In addition to eliminate the need for an ADC, a frequency output is conductive to applications requiring the sensor output to be transmitted over long distances, or where the presence of noise in the sensor environment is likely to corrupt an otherwise healthy signal… Frequency is not affected by resistive phenomena; in other words, the measured frequency will be the frequency transmitted at the output of a remote sensor [5].

The first monolithic integrated pressure sensor with the frequency output was designed and tested in 1971 in Case Western Reserve University [6]. Today many manufacturers produce frequency, pulse number and PWM output pressure sensors and transducers for different applications [7]. Performances of some modern quasi-digital pressure sensors and transducers are shown in Table 7.1. Much more quasi-digital sensors are at the Sensors Web Portal [7].

As it is visible from the Table 7.1 all existing quasi-digital sensors and transducers cove a broad pressure range suitable for various applications (industrial, medical, automotive, etc.). Several manufacturers produce frequency output transducers and pressure standards (for example, *Pressure Systems, Paroscientific, Inc.*, and others) with very low FS relative error ± 0.01 %. Besides frequency output pressure sensors and transducers, there are some interesting sensors with pulse number (*Omron*) and PWM (*ADZ Nagano*) outputs on the modern sensor market.

Many frequency output pressure sensors, transducers and interfacing circuits are described also in research articles and papers. So, a simple capacitance-to-frequency conversion circuit to directly read the capacitance changes due to applied pressure from the three types of sensors studied in this work: diaphragms of Kapton, stainless steel, and titanium is described in [8]. The frequency ranges for these tree type of pressure sensors were: from 12.42 to 12.544 kHz (for pressure from 0 to 34 kPa, Kapton pressure sensor); from 4074 to 4082 Hz (for pressure from 0 to 180 kPa for stainless steel pressure sensor), and from 6033 to 6037.5 Hz (for pressure 0 to 182 kPa for tatinum pressure sensor).

The pressure transducers with frequency output described in [9] has pressure range from 0 to 40 kPa and frequency range from 410 to 520 kHz approximately.

An interfacing circuit for piezoresistive pressure sensors with frequency output described in [10] is designed for the commercial solid state pressure sensor SPX200 from *Sensym*, and lets to obtain the pressure range from – 200 to +200 kPa and frequency output from 18.2 to 21.7 kHz.

A low-cost circuit with time interval output for pressure measurement based on the Maxim 7556 timer reported in [11] is based on resistance-to-time converter. At resistance changing from 0 to 1 kΩ, the output time interval is changing from 0 to 700 μs. Clear, that this circuit has no real direct digital output but only a quasi-digital output, and additional hardware must be used for the further time interval-to-digital conversion.

A monolithically integrated surface micromachined touch mode capacitive based pressure sensor with frequency output is described in [12]. The capacitance interface circuit consists of C/I and I/F converters, and has 300-900 Hz output at 0-60 psi pressure range.

Table 7.1 Quasi-digital pressure sensors and transducers performances.

Sensor	Pressure Range	Relative FS error, %	Output
Chezara (Ukraine)			
VT1201	0.5 … 180 MPa	± 0.25 (mean square error)	15 … 22 kHz
VT 1202	0.5 … 60 MPa	± 0.15 (mean square error)	15 … 22 kHz
Kulite			
EFT-1-1000	1.7; 3.5; 7; 17; 35; 70; 170; 350 Bar	± 2	5 … 20 kHz
Druck Incorporated			
RPT 410	600 ÷ 1100 mbar	± 0.05	600 … 1100 Hz
Omega			
PX106 Series	0 … 41.4 kPa; 0 … 1.38 MPa	± 1	1 … 6 kHz
Omron			
D8M-R1	0 … 196.13 Pa	*	80 … 300 kHz
D8M-D1/D2	0 … 5.88 kPa	*	Pulse count, 1 pulse/9.81 Pa (1/0.0014 psi)
D8M-D82	0 … 4.9 kPa	*	Pulse count, 1 pulse/9.81 Pa (1/0.0014 psi)
Paroscientific, Inc.			
8DP	10 … 700 m	± 0.01	37 … 42 kHz
8B	1400 … 7000 m	± 0.01	37 … 42 kHz
181KT	0 … 700 m	± 0.02	30 … 42 kHz
2000 Series	1 … 35 bar	± 0.01	30 … 42 kHz
3000 Series	69 Bar	± 0.01	30 … 42 kHz
4000 Series	138 … 2757 bar	± 0.01	30 … 42 kHz
5300 Series	0 … 20.7; 0 … 41; 0 … 124 kPa	± 0.01	30 … 42 kHz
Pressure Systems			
960 Series	103 … 3447 kPa	± 0.01	30 … 45 kHz
Seamap			
GDL	0 ÷ 40 m	± 1	6 … 10 kHz
ADZ Nagano			
SML-MO2	0 … 400 MPa	± 0.5	PWM (0.5 … 5 kHz); frequency (0.2 … 2.2 kHz)
Goodrich, Corp.			
2011T	1 38/50/80 mm.Hg	± 0.02	160 … 180 kHz
Apollo			
AP708	0 … 60 mBar, 0 … 1 bar	± 0.1	40 … 70 kHz
Environdata			
BP10	750 … 1050 mBar	±3 mbar (absolute error)	15.25 … 19.75 Hz
Seismic Stuff			
AGP2702F, AGM1704M	0 to 40 m; 0 to 3000 PSI	± 0.1 m; ± 10 PSI	2…6 kHz; 2…5 kHz
Quartzdyne			
QHB, SPB, QMB	0 … 35 000 psia (2 415 bar)	0.015 ... 0.03	10 ... 100 kHz

"*" - information is not available.

The frequency output sensitivity is between 5.0 - 25.0 Hz/psi in the linear pressure range of 8–60 psi.

Verba R. *et al* reports a pressure transducer with multiplexed frequency transmitter [13]. It is based on pressure capacitance sensor and hybrid IC with frequency output. The transducer designed to 0 … 10 kPa up to 0 … 6 MPa pressure ranges and 0.5 … 0.25 % FS error.

If sensors can provide a frequency modulated signal that is linearly proportional to the applied pressure being measured, then an accurate system is a good solution for many challenging sensing applications. In addition to the cost saving of such system, this design concept offers additional benefits to remote sensing applications and sensing in electrically noisy environments as well as other benefits of frequency as informative parameters like wide dynamic range, multiparametrisity, simplicity of communication and interfacing, etc.

Taking into account the high accuracy of modern frequency output pressure sensors and transducers (the FS relative error is 0.01 %) and broad frequency range (from some hundreds Hz up to some hundreds kHz), it is evidently that no any classical frequency-to-digital conversion methods can be used. In order to be neglected, the frequency-to-digital conversion relative error must be in order less (or at least in five times less) than the sensor's error. Therefore, the frequency must be measured with 0.001 % or at least, 0.002 % relative error. Let see what will happen at the use of standard (classical) counting technique. For 30 kHz measurement with quantization error 0.001 % it will be necessary 3.33 s. This conversion time is much more than sensor's response time and is redundant for all rest frequencies from output frequency range. Much better result gives any advanced frequency-to-digital conversion method but customer or manufacturer must buy a license for the use any of these methods. In addition, even the advanced method will be chosen for microcontroller based pressure digital sensor design, the developer must take into account many addition components of conversion error. It is the error because of the latency time between the occurrence of the interrupt request and the beginning of the interrupt machine cycle, the error due to the time delay between the beginning of interrupt machine cycle and the start or stop the built-in time/counter, etc [14]. Very often, such software related components of error are exceeded or equal to the quantization error [15, 16]. It needs very careful error components analyze and different methods for its reduction.

In order to simplify a design process of digital pressure sensors and transducers and reduce time-to-market and prime cost, it is expediently to use the Universal Frequency-to-Digital Converter (UFDC-1) [17] or Universal Sensors and Transducers Interface (USTI) IC [18].

7.2 Digital and Smart Pressure Sensors and Sensor Systems

7.2.1 Interfacing with Frequency Outputs Pressure Sensors and Transducers

The UFDC-1 and USTI ICs [17, 18] work well with all frequency ranges of modern quasi-digital pressure sensors and transducers. It has a programmable conversion error from 1 up to 0.001 % and 0.0005 % respectively, advanced communication possibilities (RS232 interface, SPI and I2C bus capabilities). The UFDC-1 and USTI based sensor systems are extremely cost-effective. The level of sensor intelligence that can be obtained for only a few percents of the total cost has made the UFDC-1 and USTI the elements of choice for MEMS based pressure sensors and transducers. Very often, such MEMS are based on sensors contain diffused piezoresistors, the output parameters of which are temperature depended. The temperature error can be compensated due to the use of built-in quasi-digital temperature sensors, for example, frequency output temperature sensor THSENS-F from *MicRed* [19]. Many modern frequency output pressure sensors and transducers have built-in temperature sensors. For example, it is in all series of pressure sensors from *Paroscientific, Inc.* Such kind of multiparameter sensors can be used in different applications like meteorological stations, medical devices, automotive industry, etc. Temperature signal is a frequency output with a 45 ppm/°C sensitivity within the band 168 kHz to 172 kHz [20-21].

These multiparameter sensors can be directly connected to the UFDC-1 as it is shown in Figure 7.1 for 8000 Series of frequency output depth sensors from *Paroscientific, Inc.* [22, 23]. The temperature frequency signal is connected to the first channel of the UFDC-1, the pressure frequency signal is connected to the second channel. Appropriate commands (RS232 interface) for the UFDC-1 operation mode for frequency–to–digital conversion in both channels with 0.001 % relative conversion error are shown in Figure 7.2. The different conversion error can be chosen for each of channel.

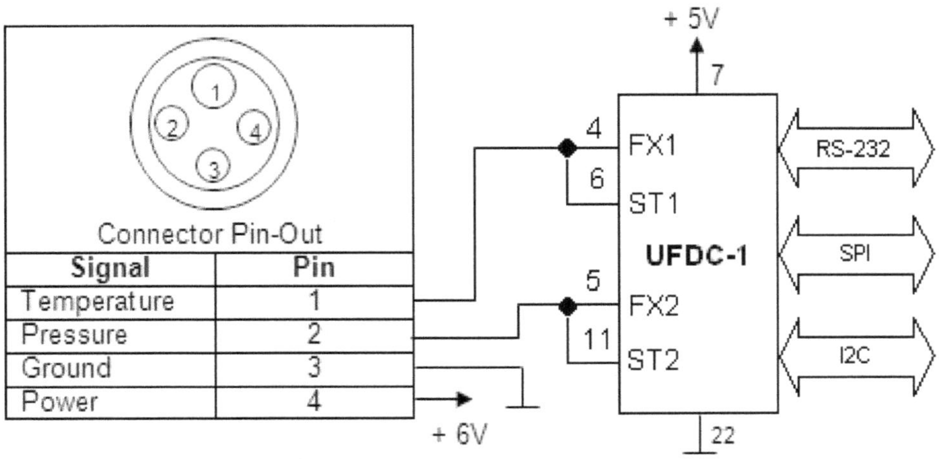

Fig. 7.1 Connection diagram for 8000 Series of frequency output depth sensors from *Paroscientific, Inc.*

```
>M0   ; Frequency measurement initialization in the first channel
>A0   ; Choose the conversion error 0.001 %
>S    ; Start a measurement
>C    ; Check result status: returns 'r' if ready or 'b' if busy
>R    ; Read a result proportional to temperature

>ME   ; Frequency measurement initialization in the second channel
>A0   ; Choose the conversion error 0.001 %
>S    ; Start a measurement
>C    ; Check result status: returns 'r' if ready or 'b' if busy
>R    ; Read a result proportional to pressure
```

Fig. 7.2 Appropriate UFDC-1 commands (RS232) for frequency-to-digital conversion in both cannels with 0.001 % relative error.

If no any embedded sensor is built in a pressure sensor, any temperature sensors from the list of quasi-digital temperature sensors manufacturers [24] can be used in the design of such multisensor systems.

Precision pressure sensors with embedded temperature sensors are also producing by the Quartzdyne, Inc. [7, 25]. All QUARTZDYNE® Pressure Transducers contain three quartz crystal sensor elements (Figure 7.3). The first of these is sensitive primarily to exposed pressure, the second responds to temperature, and the third has minimal sensitivity to either pressure or temperature.

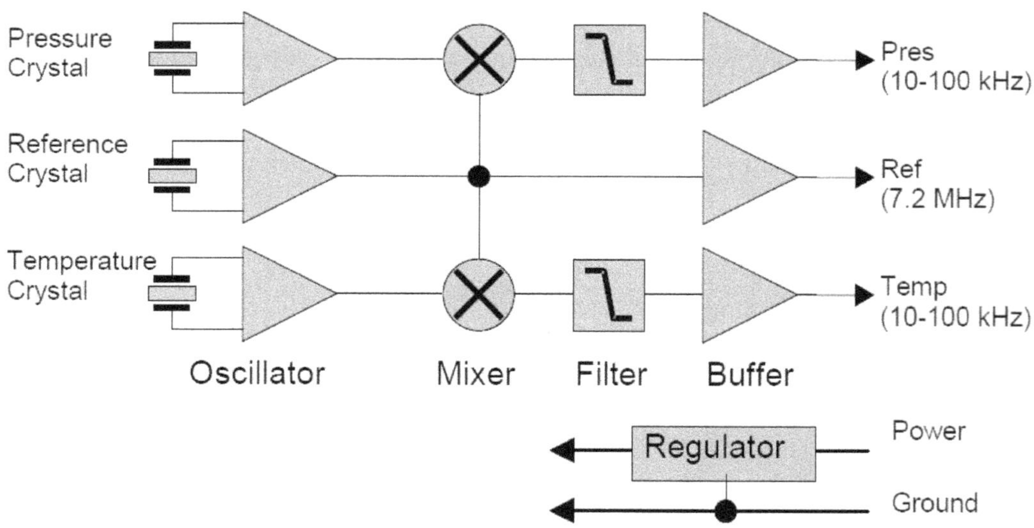

Fig. 7.3 Oscillator circuit block diagram [25].

The circuit consists of three oscillators, two mixers, buffering circuitry and a regulator. The outputs are two low frequency (10-100 kHz) sensor output signals and a high frequency (7.2 MHz) reference signal. The transducer can be connected to UFDC-1 or USTI IC and power supply. The high frequency reference signal may be used to count the two low frequency signals using the "reference-based" coefficients. Pressure is computed using the two low frequency signals, the unique calibration coefficients for the transducer, and the algorithm (see [25]). Temperature may also be calculated from the frequencies. Pressure frequency output is chanted from 22 kHz to 70 kHz in 0 … 20 000 psi pressure range, and temperature frequency output – from 62 kHz to 25 kHz in 0 … + 200 ^{0}C temperature range.

One more interesting feature of the UFDC-1 or USTI based systems is a self-adaptation. The quantization error can be changed during the conversion process depend on condition of measurement in order to have, for example, minimum possible conversion time at critical points of pressure range. More details about intelligent pressure sensor systems design is available in Chapter 17.

The same design approach can be used for other high accurate pressure sensors and transducers shown in Table 7.1 or on Sensors Web Portal [7]. For relative errors more than 0.001 % appropriate conversion error for the UFDC-1 should be selected.

Figure 7.4 shows an LCD pressure gauge that is made with a frequency output integrated pressure sensor and the UFDC-1.

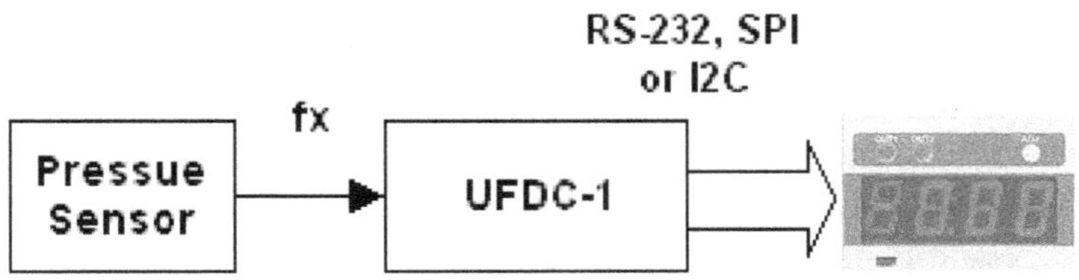

Fig. 7.4 LCD pressure gauge based on UFDC-1.

The interfacing between the UFDC-1 and LCD indicator can be made according to RS232 interface, SPI or I^{2}C serial buses depend on the LCD.

7.2.2 Interfacing with Pulse Number Outputs Pressure Sensors and Transducers

Another interesting sensors are solid-state pressure sensors D8M-D1, D8M-D2 and D8M-D82 from *Omron* [7]. They outputs generate number of pulses proportional to the pressure. For example, 0 kPa = 30 pulses; 0.59 kPa = 60±32 pulses; 1.96 kPa = 200±24 pulses, etc. Such sensors also can be connected directly to the UFDC-1 (the first channel) because of the converter has a pulse number counting mode. Appropriate commands (RS232 interface) for the UFDC-1 and USTI operation modes for pulse number counting are shown in Figure 7.5 and 7.6 respectively.

In comparison with the UFDC-1 commands, the USTI's commands have two bytes for mode number. In both cases no any relative error is necessary to set up for the pulse counting method.

```
>MD   ; Pulse count initialization
>S    ; Start a measurement
>C    ; Check result status: returns 'r' if ready or 'b' if busy
>R    ; Read a result proportional to pressure
```

Fig. 7.5 Appropriate UFDC-1 commands (RS232) for pulse counting mode.

```
>MOD  ; Pulse count initialization
>S    ; Start a measurement
>C    ; Check result status: returns 'r' if ready or 'b' if busy
>R    ; Read a result proportional to pressure
```

Fig. 7.6 Appropriate USTI commands (RS232) for pulse counting mode.

7.2.3 Digital Pressure Sensors and Transducers with Voltage-to-Frequency Conversion

Since many pressure sensors provide a voltage output, it should be converted to a frequency proportional to the sensor output voltage (Figure 7.7). Assuming the sensor's analog voltage output is proportional to the applied pressure, the resulting frequency will be linearly related to the pressure being measured. Although many timing circuits can perform voltage-to-frequency conversion (VFC), most of the simple typos provide neither the accuracy nor stability needed for reliability encoding a sensor signal quantity [5]. Fortunately, there are many commercially available VFC ICs [26] and discreet ICs based solutions [27] that can handle this function with high metrological performances.

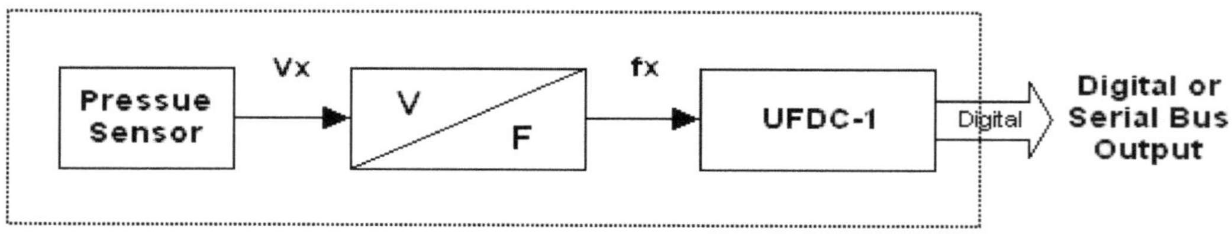

Fig. 7.7 Digital pressure sensor with voltage-to-frequency conversion.

Some interesting detailed examples, for example, using *Motorola's* pressure sensors series and *Analog Devices* VFC AD654 or any voltage output pressure sensors (0-10 V full-scale voltage range) and VFC32 from *Burr-Brown* are described in [5] and [28] accordingly. VFCs outputs can be directly connected to the UFDC-1 in order to produce digital output. Taking into account a very wide frequency range of the UFDC-1 and USTI any full scale frequency range for VFCs can be used.

A relatively straightforward circuit for converting pressure of MPX2000 series sensor to frequency is described in [29, 30]. It is based on the non-synchronous voltage-to-frequency converter AD654 with input voltage up 14 V, FS output frequency 500 кHz and accuracy 0.4 %/kHz.

7.2.4 Digital Pressure Sensors with PWM and Duty-Cycle-to-Digital Conversion

For some applications, a frequency modulated (FM) or pulse width modulated (PWM) output is more desirable than a frequency output. In comparison with a frequency output signal the duty-cycle is rather immune to interfering signals, such as spikes [31], and the ratio does not depend on the absolute value of any component [32].

Both FM and PWM outputs inherently have better noise immunity. Generally, FM outputs are more widely accepted than PWM outputs, because PWM outputs are restricted to a fixed frequency. With either an FM or PWM output, the UFDC-1 and USTI ICs can be used to convert such frequency-time parameters of electrical signal into a digital.

The stable PWM output could be obtained with simple, inexpensive circuitry, a PWM output would be a cost-effective solution for noisy environment/remote sensing applications while incorporating the advantages of frequency outputs [33]. The pulse width modulated output pressure sensor design is described in details in [33]. It was made for a *Freescale Semiconductor, Inc.* MPX5000 series pressure sensor (Figure 7.8). The pulse width output vs. pressure is shown in Figure 7.9. In order to produce digital output according to the RS232 interface or serial buses SPI or I2C, the UFDC-1 or USTI ICs should be used in the design. The appropriate commands and interfacing circuits for PWM and duty-cycle-to-digital conversions were described in Chapter 7 for applicable semiconductor quasi-digital temperature sensors.

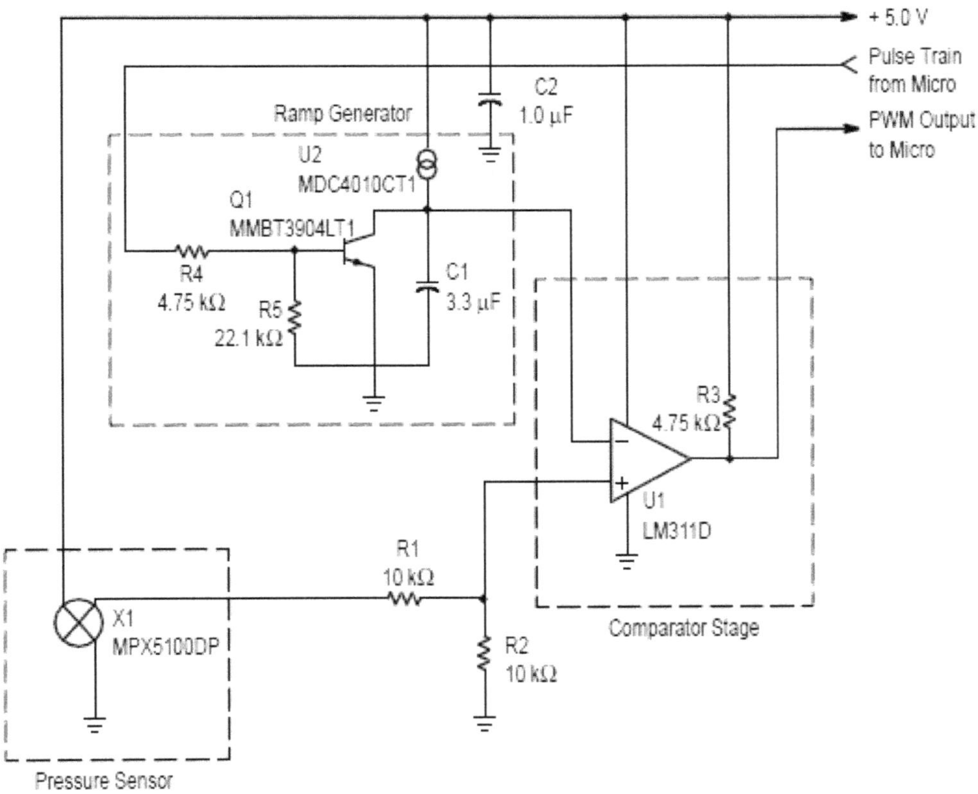

Fig. 7.8 Quasi-digital PWM output pressure sensors [33].

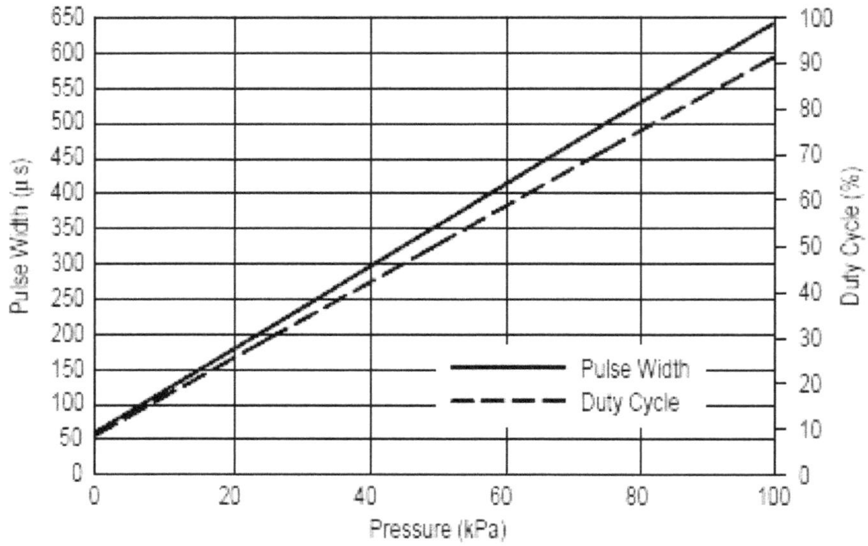

Fig. 7.9 PWM output pressure sensor linearity data [33].

Combining silicon micromachining designs and processes with the advanced CMOS UFDC-1 and USTI ICs overcomes many of the early limitations of single-chip sensors. This co-integrated unit can be produced using, for example, co-integration technology or hybrid technology. By eliminating the need for ADC, the frequency (duty-cycle or PWM)-to-digital conversion schemes reduce the systems complexity. The results are high-performance single-chip pressure sensors with truly digital output (RS232 interface) or bus output (SPI or I2C) at significant reduction of production costs, time-to-market and simplification of the design process.

In comparison with the direct microcontroller interfacing the described design approach also lets to eliminate many design problems connected with the microcontroller selection, its programming in low level Assembler language and additional error components due to so-called program-dependent or software-related quantization effects.

7.3 References

[1]. Smart Sensors and MEMS, ed. by Sergey Y. Yurish and Maria Teresa S.R. Gomes, *Kluwer Academic Publishers*, 2004.

[2]. MEMS Pressure Sensors in the Automotive Industry, *Frost & Sullivan*, February 2007.

[3]. MEMS Sensors Market Geared for Growth - Emerging Trends and Opportunities, *Frost & Sullivan*, February 2010.

[4]. Smart Sensors to Reach $7.8 bn by 2015, *CIOL Bureau*, October 2010.

[5]. Baum J., Frequency Output Conversion for MPX2000 Series Pressure Sensors, *Sensors*, Vol.14, No.4, April 1997, pp.34-39.

[6]. Blaser E. M., Ko W. H., Yon E. T., A Miniature Digital Pressure Transducer, in *Proceedings of the 24th Annual Conference on Engineering in Medicine and Biology*, November, 1971, Las Vegas, Nevada, USA.

[7]. List of Pressure Sensors and Transducers Manufacturers at Sensors Web Portal: http://www.sensorsportal.com/HTML/SENSORS/PresSens_Manuf.htm

[8]. Sung-Pil Chang, Mark G. Allen, Demonstration for integrating capacitive pressure sensors with read-out circuitry on stainless steel substrate, *Sensors and Actuators A*, Vol. 116, 2004, pp. 195–204.
[9]. Osadchuk V. S., Osadchuk A. V., Transducer of Pressure with a Frequency Output, *Electronics and Electrical Engineering*, Vol.67, No.3, 2006, pp.9-12.
[10].Vlassis V., Siskos S., An Interfacing Circuit for Piezoresistive Pressure Sensors with Frequency Output, *International Journal of Electronics*, Vol.87, No.1, 2000, pp. 119-127.
[11].Carmen Busoch Morlan, Brian Opady Buafull, Germ´an Morales Miranda, and Angel Regueiro-Gomez, A Low-Cost Circuit with Direct Digital Output for Pressure Measurement, *IEEE Transaction on Instrumentation and Measurement*, Vol.48, No.4, 1999, pp.817-819.
[12].Shuwen Guo, Jun Guo, Wen H. Ko, *Sensors and Actuators A*, Vol. 80, 2000, pp. 224 - 232.
[13].Vrba R., Sveda M., Marecek K., Pressure Transducer with Multiplexed Frequency Transmitter, in *Proceedings of the Sensors for Industry Conference (SIcon /04)*, New Orleans, USA, 27-29 January 2004, pp.7-10.
[14].Yurish S. Y., Reverter F., Pallàs-Areny R., Measurement error analysis and uncertainty reduction for period- and time interval-to-digital converters based on microcontrollers, *Measurement Science and Technology*, Vol.16, No.8, 2005, pp.1660-1666.
[15].Yurish S.Y., Program-oriented Methods and Measuring Instruments for Frequency-Time Parameters of Electric Signals, *PhD Thesis,* State University Lviv Polytechnic Lviv, 1997 (in Ukrainian).
[16].Reverter F., Microcontroller-based Interfaces for Quasi-Digital Sensors, *PhD Thesis,* Technical University of Catalonia (UPC Barcelona, Spain), 2004.
[17].Universal Frequency-to-Digital Converter (UFDC-1), Specification and Application Note, *Technology Assistance BCNA 2010, S. L,.* 2011.
[18].Universal Sensors and Transducers Interface (USTI), Specification and Application Note, *Technology Assistance BCNA 2010, S. L,.* 2011.
[19].http://www.micred.com
[20].http://www.paroscientific.com/overview.htm#transducers
[21].http://www.paroscientific.com/oceanography.htm
[22].Yurish S. Y., Intelligent Digital Pressure Sensors and Transducers Based on Universal Frequency-to-Digital Converter (UFDC-1), *Sensors & Transducers*, Vol. 60, Issue10, October 2005, pp.432-438.
[23].http://www.paroscientific.com/pdf/Series8000.pdf
[24].List of Temperature Sensors Manufacturers at Sensors Web Portal: http://www.sensorsportal.com/HTML/SENSORS/TempSens_Manufacturers.htm
[25].Operating Manual for QUARTZDYNE® Frequency Output Pressure Transducers, *Quartzdyne, Inc.*, November 2009.
[26].List of Voltage-to-Frequency Converters Manufacturers at Sensors Web Portal: http://www.sensorsportal.com/HTML/SENSORS/VFC_Manufacturers.htm
[27].Jim Williams, Designs for High Performance Voltage-to-Frequency Converters, Application Note 14, *Linear Technology*, March 1986.
[28].Kirianaki N.V., Yurish S.Y., Shpak N.O., Deynega V.P., Data Acquisition and Signal Processing for Smart Sensors, *John Wiley & Sons*, 2002.
[29].Warren Schultz, Interfacing Semiconductor Pressure Sensor to Microcomputer, Motorola Semiconductor Application Note, AN1318D, *Motorola, Inc.*, 1997.
[30].Jeff Baum, Frequency Output Conversion for MPX2000 Series Pressure Sensors, Motorola Semiconductor Application Note, AN1316D, *Motorola, Inc.*, 2000.

[31].Meijer G. C. M. , Concepts and Focus Point for Intelligent Sensor Systems, *Sensors and Actuators A,* 41-42, 1994, pp.183-191.

[32].Middelhoek S., French P. J., Huijsing J. H., Lian W. J., Sensors with Digital or Frequency Output, *Sensors and Actuators,* Vol. 15, 1988, pp.119-133.

[33].Jacobsen E., Baum J., Using a Pulse Width Modulated Output with Semiconductor Pressure Sensors, *Freescale Semiconductor Application Note*, AN1518, Rev 2, 05/2005.

Chapter 8

Digital Humidity, Dew Point and Moisture Sensors

8.1 Quasi-digital Humidity, Dew Point and Moisture Sensors: State-of-the-Art

A relative humidity is one of main parameter in environmental and meteorological measurements, air control in different climatic systems, building or home monitoring, etc. [1]. In addition, humidity measurements are used in different industrial manufacturing processes for paper, food, textile, semiconductors, etc. [2].

Relative humidity (RH) sensors have become ubiquitous over the past 20 years and almost universally used in OEM applications where RH sensing is required [3], for example, in industrial, commercial, and weather telemetry applications.

Many designed and described in the literature humidity sensors and sensor systems are based on humidity sensing elements and appropriate readout electronics that generate a voltage output. Then, an analog-to-digital converter is used in order to produce a digital output [4-6]. However, the rapid growth of the microelectronics field and availability of inexpensive but high performance UFDC-1 and USTI integrated circuits [7, 8] and humidity, moisture, dew point sensors and sensing modules with frequency output [9-11] have led to an increased interest in the implementation of low cost self-adaptive smart humidity sensors and sensor systems, which use a frequency as an informative parameter of humidity sensor's output.

There are four main types of humidity sensors: capacitive, resistive, thermal conductivity humidity and quartz crystal microbalance (QCM) based humidity sensors [12, 13, 9, 10]. All are suitable for digital humidity sensors and systems design based on UFDC-1 and USTI ICs. Capacitance or resistance sensing element must be included into a relaxation circuit in order to produce quasi-digital (frequency) output for UFDC-1 IC. The humidity sensors based on the QCM can be directly interfacing to the mentioned both ICs (see

Chapter 12). In addition, capacitance and resistance sensing elements can be directly interfacing to the USTI, since the last one has special modes for capacitance and resistance measurements (see Chapters 13 and 14 respectively).

The thermal conductivity humidity sensors (or absolute humidity sensors) measure the absolute humidity by quantifying the difference between the thermal conductivity of dry air and that of air containing water vapor [12]. Such sensors consist of two matched negative temperature coefficient (NTC) thermistor elements in a bridge circuit. Taking into account that the USTI can be directly interfacing to passive resistive bridges (see Chapter 15), such approach should be promised as new one for read-out circuitry and interface design for thermal conductivity digital humidity sensors.

Today, many manufacturers produce different humidity, dew point and moisture sensor modules and transducers [11]. Although an RH sensor might seem a simple device, its proper operation is highly dependent on careful humidity calibration. Performances of some modern humidity sensors and transducers with frequency and PWM outputs are shown in Table 8.1. More quasi-digital humidity sensor modules and transducers are available at Sensors Web Portal [11].

As it is visible from the table and literature review, the output frequency range of modern quasi-digital humidity, dew point and moisture sensors is from some part of kHz to some hundred kHz. This frequency range can be converter to digital by discrete logical digital circuits or by microcontroller. The first design solution is not optimal for today because of needs more hardware expenses. The second design solution needs minimal possible hardware but a designer must take into account a lot of different addition components of conversion error due to so-called program-dependent or software-related effects [14-16]. As usually, the designer is not an expert in frequency measurements and an optimal frequency-to-digital converter design is not a trivial task. In order to simplify a design process of digital humidity sensors and reduce time-to-market and prime cost, the UFDC-1 and USTI ICs should be used [7, 8].

In order to use these ICs for a high frequency output humidity sensors, for example, a ~ 50 MHz frequency output capacitive humidity sensor integrated with micro heater and ring oscillator circuit and fabricated by a CMOS–MEMS technique [17], an external prescaller (devider by 16) should be used. In this case it will be possible to measure frequencies up to 120 … 144 MHz respectively [7, 8].

8.2 Digital Humidity Sensing Modules Design

Digital output humidity sensors can be created based on a humidity-to-parameter-to-frequency-to-digital conversion. Taking into account that the error of sensing elements is from 2 to 5 %, the analog low-cost integrated circuits, for example, timer 555 or 556 can be used as a parameter-to-frequency converter in order to produce a quasi-digital sensor output [18-24]. A quasi-digital (frequency-output) humidity sensor based on the HS11XX capacitance humidity sensing element from *Humirel* and TLC555 timer is shown in Figure 8.1.Then, the frequency should be converted into a digital output (RS232, SPI, I^2C, etc.) with the help of a frequency-to-digital converter, for example, UFDC-1 or USTI ICs [7, 8].

Table 8.1 Main performances of quasi-digital humidity, dew point and moisture sensor modules.

Sensor Module	Performances		
	Humidity Range, % RH	**Relative error, %**	**Frequency Output, kHz**
	Blue Earth, LLC.		
MiniCap2 + oscillator circuit	10 … 90	*	10 … 200
	E+E Elektronik, GmbH		
EE05 Series, HC200	10 … 90	± 3 at 20 °C	61.1 … 48.6
EE25 Series, HC1000	0 … 100	± 3	*
	Galltec+Mela, GmbH		
Humidity Frequency Converter FOM 6/2	10 … 90	± 3	57.9 ... 48.4
	Humirel		
HTF3100	*	± 3 at 55 %RH	*
HTF3130	10 ... 95	± 3 at 55 %RH	7.155 … 6.210
HTF3223	10 ... 95	± 5 at 55 %RH	9.560 … 8.030
HTF3225	*	± 5 at 55 %RH	*
HTF3226	10 ... 95	± 5 at 55 %RH	9.44 … 8.070
HTF3226LF	10 ... 95	± 5 at 55 %RH	9.49 … 8.225
HTF3227	*	± 3 at 55 %RH	*
	Measurement Specialties		
HTF3000LF	0 … 100	± 3 at 55 %RH	6.56 ... 6.64
	Kurabe		
KN-1050	0 … 95	± 5	4.95 … 5
	Bitron		
ATS	5 … 95	± 5	*
	Michell Instruments		
I7000 & 7400 (Hygrosmart)	0 … 100	±2 at 20 °C	PWM, 5 … 30 Hz
	Gefran		
S7000 HYGROSMART	0 … 100	± 2	PWM, 5 … 30 Hz
	Hygrosens Instruments		
HYTEMOD-KFRQ-OG	20 … 90	± 3	30 … 50
HYTEMOD-EFRQ-OG	20 … 90	± 3	30 … 50
HYTEMOD-EFRQ-HEL	20 … 90	± 3	30 … 50
	U.P.S.I.		
G-UCN/TUNC.32, G-UCN/TUNC.34	2 … 98	± 2	7.176 … 5.826
	Environmental Measuring Systems		
CS 616 - soil moisture sensor	*	± 2.5	Up to 70
SMX - soil moisture sensor	*	*	0.724 … 4.59
	Vishay		
DPS 0603/1206 (Dew point sensor + 555 timer)	10 … 98	*	0.01 … 85

"*" – information not available.

The UFDC or USTI ICs carry out the frequency-to-digital conversion of the physical parameters of humidity and temperature in two channels, which are made available via an I^2C, SPI or RS232 interfaces as digital values.

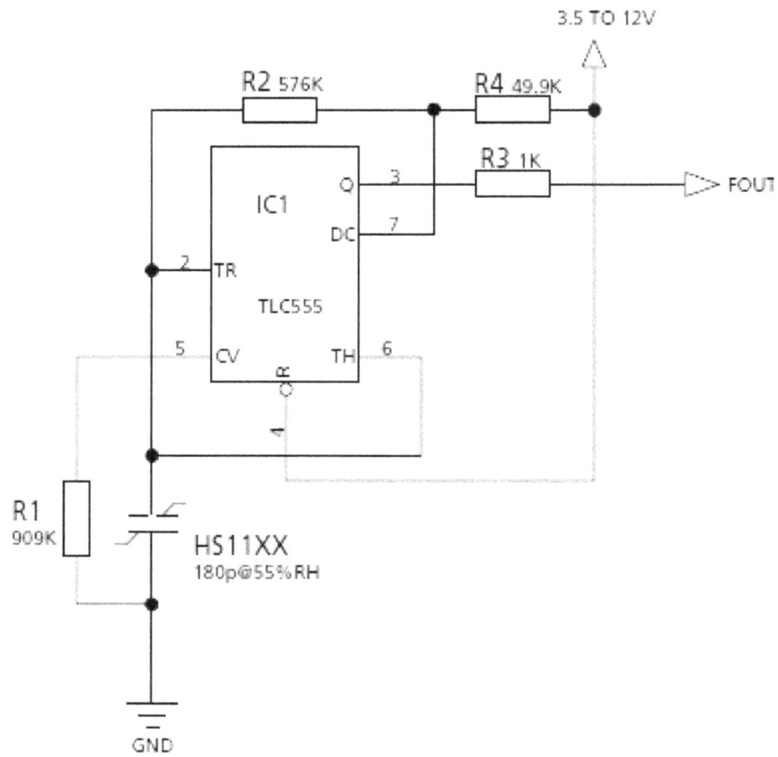

Fig. 8.1 Quasi-digital (frequency-output) humidity sensor based on HS11XX capacitance humidity sensing element [19].

The digital interfaces corresponds to the RS232 standard (from 300 to 38400 data transfer rate for the slave mode); I^2C standard (up to 100 kHz clock rate), SPI standard (from 100 kHz to 2 MHz) and can be used together with other I^2C or SPI components connected at the same bus. Simultaneous operation of up to 8 × 2 = 16 sensors is possible on the same I^2C bus [7, 8].

Let consider some design examples of digital humidity sensor modules based on these universal integrated circuits [25].

The MiniCap2 is a low cost, user calibrated, capacitive type relative humidity sensor. It features fast wet-up and dry-down response, good long-term stability, and sensitivity. This sensor can be used in a simple oscillator circuit to provide a frequency output that is inversely proportional to relative humidity [26]. In order to produce digital output according to RS232 serial interface or obtain bus capabilities according to the SPI and I2C serial buses the UFDC-1 or USTI should be used in such designs. Both ICs work well in all frequency range of modern humidity, dew point and moisture sensor modules. Its quantization error does not depend on the frequency and can be chosen from the appropriate range [7, 8]. Therefore, in order to be neglected, this relative conversion error must be in order low (or at least in five times less) in comparison with the sensor's error. Therefore, in other words ± (0.25 … 0.5) % quantization error should be enough.

Appropriate commands (RS232 interface) for the UFDC-1 IC working in the frequency measuring mode with ± 0.25 % relative conversion error are shown in Figure 8.2.

```
>M4   ; Frequency measurement initialization in the 2nd channel
>A2   ; Choose the conversion error 0.25 %
>S    ; Start a measurement
>C    ; Check result status: returns 'r' if ready or 'b' if busy
>R    ; Read a result

8050.097845
```

Fig. 8.2 Appropriate UFDC-1 commands (RS232 interface) for frequency-to-digital conversion with 0.25 % relative error.

The EE05 small size humidity transmitter is designed for low cost applications such as climate control, humidifiers and dehumidifiers, storage monitoring, processing of granulated materials. The frequency output with TTL level makes EE05 ideal for data acquisition, control systems and data loggers as there is no need for A/D conversion. The well proved *E+E* humidity sensing element HC200 guarantee excellent long-term stability and resistance against chemical pollutants [27]. The connection example for data logging or data acquisition is shown in Figure 8.3.

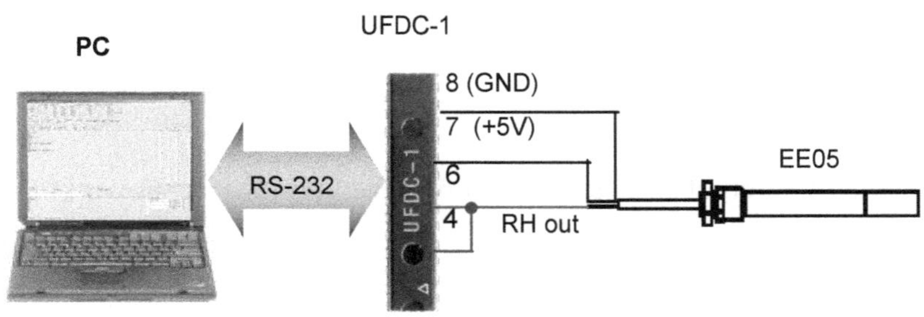

Fig. 8.3 The connection diagram for data logging or data acquisition with humidity transducer EE05 and UFDC-1.

The *Galltek +MELA* humidity frequency converter FOM 6/2 is an OEM subassembly, which converts the humidity signal into a calibrated frequency signal [28]. The frequency at the output as a function of relative humidity is shown in Figure 8.4.

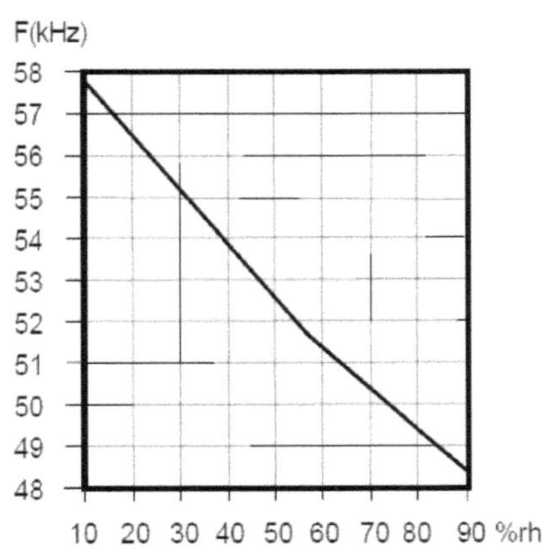

Fig. 8.4 Frequency at the output as a function of relative humidity.

The digital output humidity sensors (RS232, SPI or I^{2}C), data acquisition system or data logger can be easy built by using the MELA-capacitive humidity sensing elements, the humidity frequency converter, UFDC-1 or USTI IC.

Based on the rugged HTS2030SMD humidity sensor, HTF3130 is a dedicated humidity and temperature transducer from *Humirel (Measurement Specialties, Inc.,)* designed for OEM applications where a reliable and accurate measurement is needed. It features a very small size for easy, cost-effective mechanical mounting [29]. Direct interface with the UFDC-1 and USTI ICs is made possible with the module's linear frequency output.

The output frequency is determined according to the following equation:

$$F_{out} = 7314 - 16.79 \cdot RH + 0.0886 \cdot RH^2 - 0.000358 \cdot RH^3, \quad (8.1)$$

with F_{out} in Hz and RH in %.

The HF 3223/HTF 3223 is a dedicated humidity and temperature transducer designed for OEM applications based on the rugged HS1101 humidity sensor [30]. Its signal output can be calculated according to the following formula:

$$F_{out} = 9740 - 18 \cdot RH, \quad (8.2)$$

with F_{out} in Hz and RH in %.

The frequency at the output as a function of relative humidity is shown in Figure 8.5.

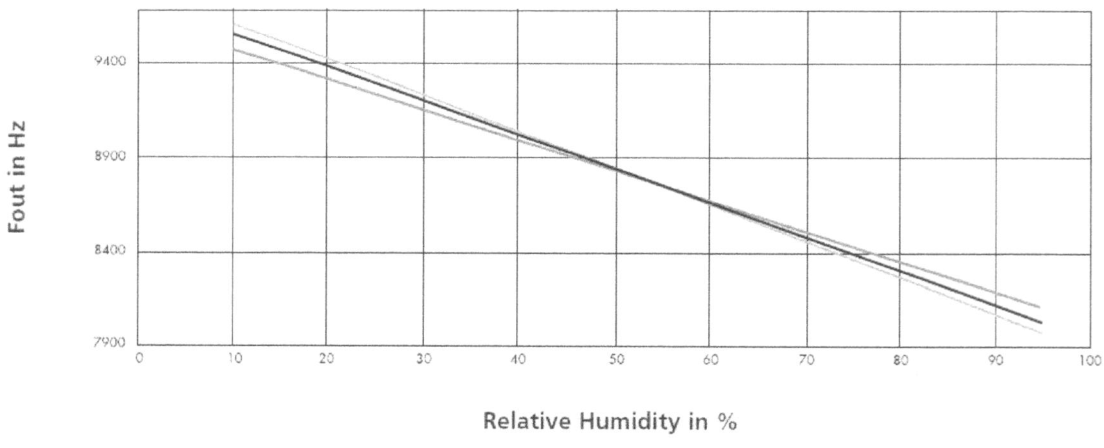

Fig. 8.5 Frequency output vs. relative humidity for HF 3223.

A dedicated humidity and temperature transducer HTF 3226 is also based on the rugged HS1101 humidity sensor from *Humirel*. It has the same frequency output vs. RH dependence as it is shown in Figure 8.5. The signal output can be determined according to the following equations:

$$F_{out} = 9600 - 15.8 \cdot RH \text{ - for the linear reference curve} \quad (8.3)$$

and

$$F_{out} = 9570 - 14.28 \cdot RH - 0.015 \cdot RH^2 \text{ - for the second order curve} \quad (8.4)$$

Another humidity and temperature transducer from *Humerel* - HTF3226LF - is based on the rugged HS1101LF humidity sensor [31]. This transducer also has the same output characteristic as it is shown in Figure 8.5. The signal output can be determined according to the following equations:

$$F_{out} = 9595 - 14.8 \cdot RH \text{ - for the linear reference curve} \quad (8.5)$$

and

$$F_{out} = 9659 - 18.79 \cdot RH - 0.0404 \cdot RH^2 \text{ - for the second order curve} \quad (8.6)$$

The output characteristic of humidity-to-frequency converter KN-1050 from *Kurabe* is shown in Figure 8.6 It is equipped with a high performance relative humidity sensor of variable capacitance type [32]. The connection diagram to the UFDC-1 is shown in Figure.8.7.

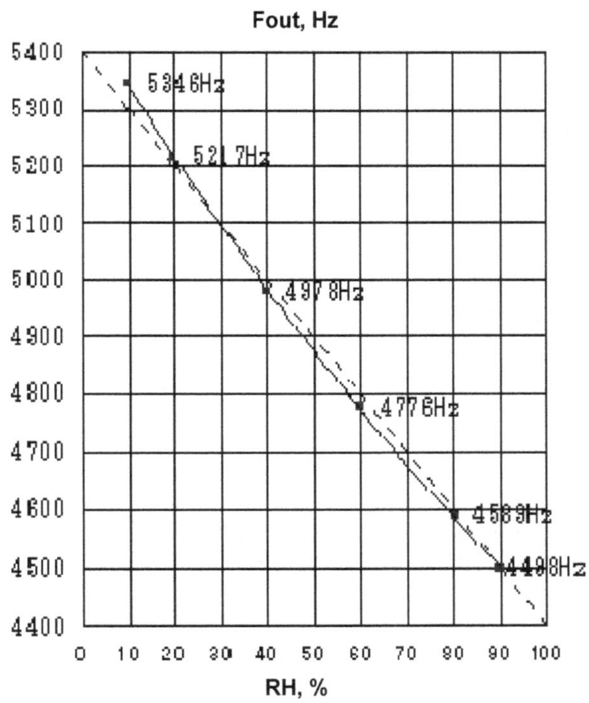

Fig. 8.6 The output characteristic of humidity-to-frequency converter KN-1050 at 25 °C.

Very often modern humidity sensor modules are combined with temperature sensors. As usually, Pt1000 or NTC thermistor are used as a temperature sensing element working in range from – 40 °C up to +85 °C [21, 27, 29-31]. In this case the temperature measurement error is only ± 3 %. In order to expand temperature range and increase the accuracy of temperature measurements, the UFDC-1 or USTI should be used with any semiconductor temperature sensors [33]. Taking into account that the UFDC-1 and USTI have two measuring channels, no any additional hardware is necessary. Any quasi-digital temperature sensor can be connected directly to its first channel to convert frequency, period, duty-cycle or PWM signal to digital. A humidity transducer should be connected to its second channel to convert the frequency to digital. A possible connection diagram for multisensors systems is shown in Figure 8.8.

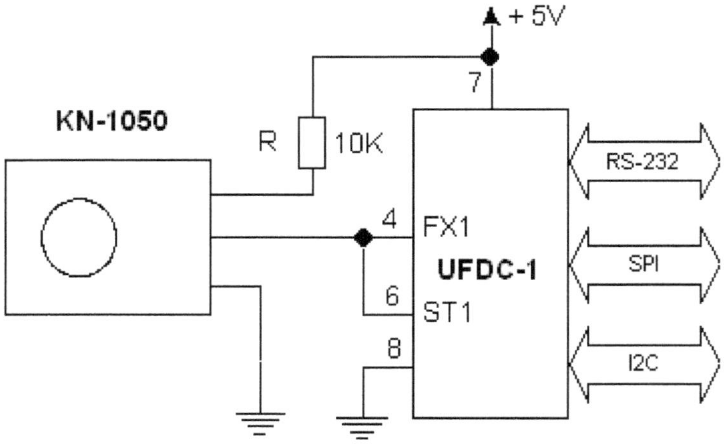

Fig. 8.7 The KN-1050 connection diagram to the UFDC-1.

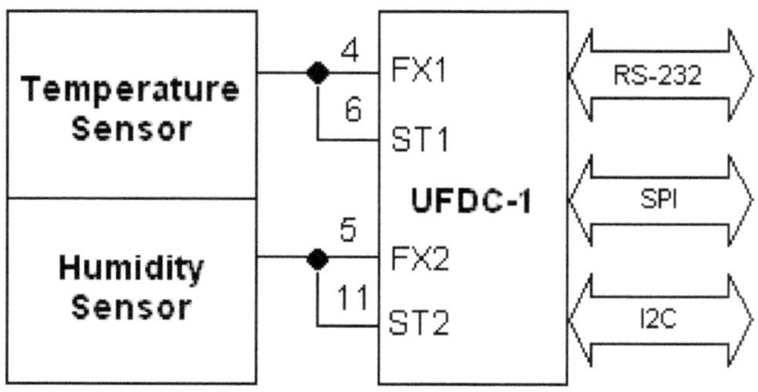

Fig. 8.8 Temperature and humidity multisensors system.

Appropriate commands (RS232 interface) for the UFDC-1 working in frequency measuring mode (humidity measurement, HTF3130 sensor, the second channel) and period conversion (temperature measurement, sensor MAX6576, the first channel) is shown in Figure 8.9.

The relative error and measuring mode should be set up only once in each of channel. After, it will be enough to repeat cyclically the following commands consecution: *S, C* and *R*.

As usually, the relative humidity error of sensing element depends on temperature. The temperature error component can be corrected in such multisensors systems.

Taking into account SPI and I^2C bus capabilities of the UFDC-1 and USTI such multisensors systems can be used widely in different systems for climatic control and monitoring, for example, in storage rooms, museums, etc.

```
>M1   ; Period measurement in the 1st channel (temperature conversion with MAX6576)
>A2   ; Choose the conversion error 0.25 % for the 1st channel
>S    ; Start a measurement
>C    ; Check result status: returns 'r' if ready or 'b' if busy
>R    ; Read a result (period proportional to the temperature)

>ME   ; Frequency measurement in the 2nd channel (humidity conversion with HTF3130)
>A2   ; Choose the conversion error 0.25 % for the for the 2nd channel
>S    ; Start a measurement
>C    ; Check result status: returns 'r' if ready or 'b' if busy
>R    ; Read a result (frequency proportional to the humidity)
```

Fig. 8.9 UFDC-1 commands (RS232) for period- and frequency-to-digital conversion.

The USTI IC can also directly convert resistance and capacitance to digital [8]. It means, that embedded temperature sensing elements from humidity module such as NTC and Pt temperature sensors can be directly read-out by the USTI without any additional hardware.

Besides frequency output, some humidity sensors have PWM output, for example, S7000 (*Gefran*), I7000 and 7400 (*Michell Instruments*) [34], see Table 8.1 and Figure 8.10. The pulses on its outputs are shown in Figure 8.11.

Fig. 8.10 I7000 & 7400 (Hygrosmart) interchangeable sensor module for relative humidity [34].

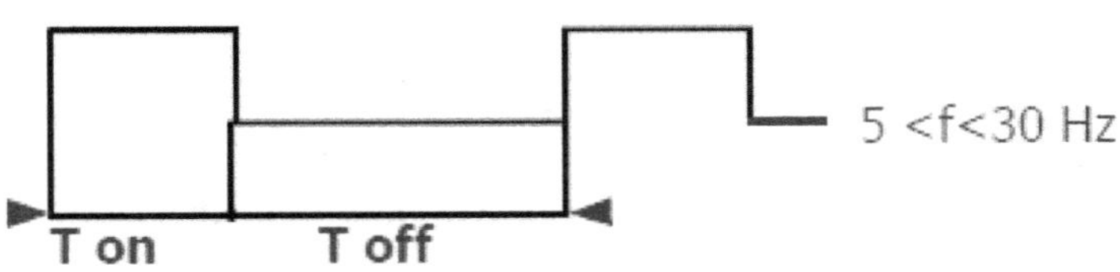

Fig. 8.11 PWM signal on sensor's output.

The relative humidity should be calculated according to the following equation:

$$\%RH = \lfloor (T_{on}/T_{off}) \times 2049 - 1 \rfloor / 16.2 \tag{8.7}$$

Appropriate commands (RS232 interface) for the USTI working in pulse width and space measuring modes are shown in Figure 8.12.

```
>MOB   ; Pulse width measurement in the 1st channel
>S     ; Start a measurement
>C     ; Check result status: returns 'r' if ready or 'b' if busy
>R     ; Read result (Ton)

>MOC   ; Pulse space measurement in the same 1st channel
>S     ; Start a measurement
>C     ; Check result status: returns 'r' if ready or 'b' if busy
>R     ; Read result (Toff)
```

Fig. 8.12 USTI commands (RS232 interface) for pulse width and pulse space measurements.

For pulse width and pulse space measurements it is not necessary to set up the relative error. Such measurements are made with maximal possible accuracy. The relative error for pulse width and pulse space measurements can be calculated according to (3.1).

8.3 Experimental Results

The aim of this experimental investigation was to demonstrate high metrological performance of designed digital humidity sensor system. Readers can easy to reproduce such experiments in their laboratories.

The designed sensor system consists of dedicated humidity and temperature transducer HTF3130 [29] and USTI that especially designed for precision frequency–to–digital conversion and signal conditioning in such applications (Figure 8.13). The transducer consists of rugged HTS2030SMD humidity sensor and NTC temperature sensor. The frequency on the fist transducer's output proportional to the relative humidity (10 %RH ... 95 %RH), and resistive on the second output – to the temperature (- 40 ^{0}C...+85 ^{0}C). Oscillogram at humidity sensor's output at 55 RH % is shown in Figure 8.14. The sensor system can be interfaced to a PC or work autonomously in a master communication mode (RS232).

The circuit diagram of humidity and temperature sensors system is shown in Figure 8.15 and commands for RS232 communication mode at adaptive measurement of humidity and temperature by the USTI – in Figure 8.16. Taking into account that maximal resistance of NTC temperature sensor R_x = 262960 Ohm [29], the reference calibration resistor R_c should be chosen form the range $0 < R_x < R_c$ and is 263 Ohm.

Fig. 8.13 Main components of humidity smart sensor systems: (1) – humidity sensing module with frequency and resistive outputs; (2) – USTI.

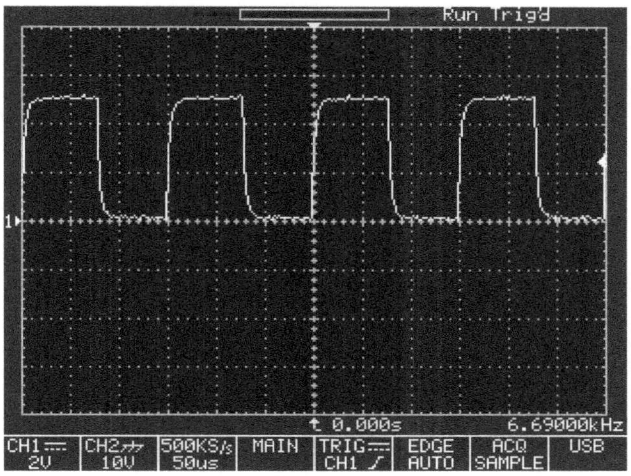

Fig. 8.14 Oscillograms at humidity sensor's output at 55 % RH.

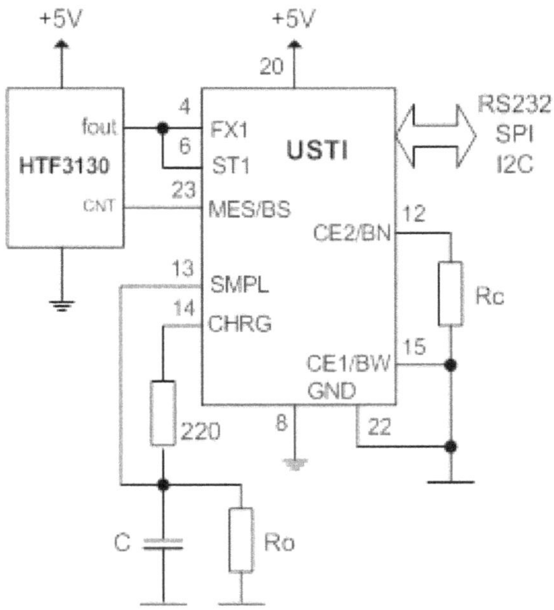

Fig. 8.15 Circuit diagram of humidity and temperature sensors system.

```
>M00         ; Set up a frequency measurement mode in the 1st channel
>A02         ; Set up the relative error 0.25 %
>S           ; Start a frequency measurement (humidity)
>C           ; Check result status: returns 'r' if ready or 'b' if busy
>R           ; Read a result of frequency measurement (humidity)
>M10         ; Set a resistance Rx measurement mode
>E263000.0   ; Set the reference value of Rc=263 kΩ
>W1B         ; Set the charging time 100 ms
>S           ; Start a resistive measurement (temperature)
>C           ; Check result status: returns 'r' if ready or 'b' if busy
>R           ; Read a result of resistive measurement (temperature)
```

Fig. 8.16 Commands for RS232 communication mode at adaptive measurement of humidity and temperature by the USTI.

The capacitance $C \geq 0.002/R_c$ (F) and the time of capacitance charge $T = 2200 \times C$ (sec). The values R_0 and T must be set up with the appropriate commands E and W respectively (Figure 8.16).

Humidity sensor system was investigated experimentally in the climatic chamber CTS T-40/60 at +25 ^{0}C (Figure 8.17). Experimental results are shown in Figure 8.18 Statistical characteristics at frequency measurements (55 RH %) and density of distribution are shown in Table 8.2 and Figure 8.19 The relative error of frequency measurement by the USTI was not more than ± 0.0077 % in all frequency range from 6200 to 7155 Hz.

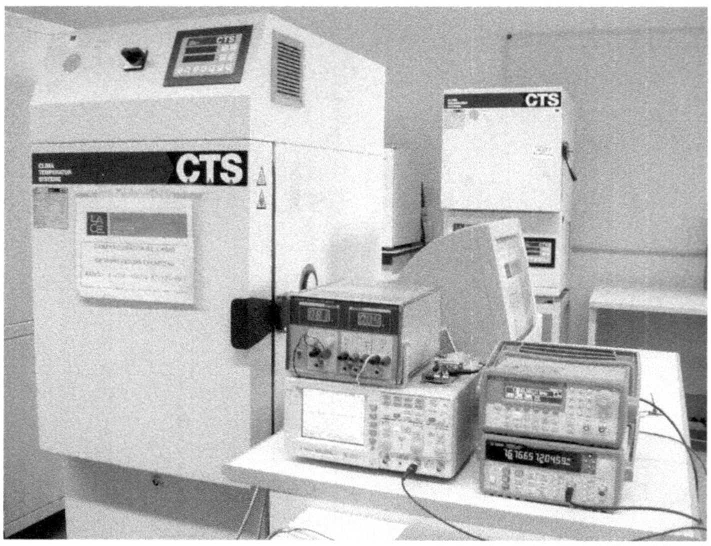

Fig. 8.17 Climatic chamber CTS 40/60 and measuring equipment.

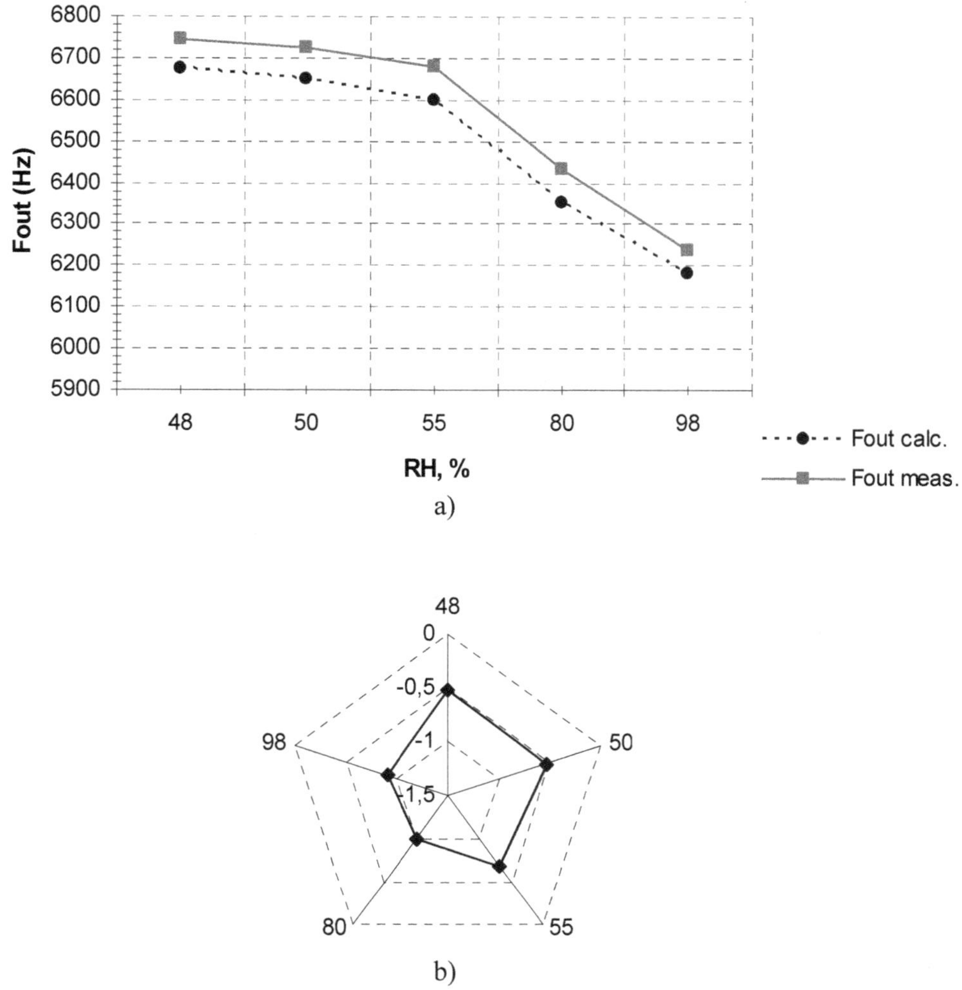

Fig. 8.18 Dependences of output frequency (a) and absolute error (b) on relative humidity RH, %.

Table 8.2 Statistical characteristics at f_x measurements (55 RH %).

Parameter	Value
Number of measurements, N	60
Minimum $f_{x\ min}$, Hz	6678.0817
Maximum $f_{x\ max}$, Hz	6683.5299
Sampling Range, $f_{x\ max} - f_{x\ min}$, Hz	5.4481
Median	0
Arithmetic Mean, Hz	6681.2234
Variance, Hz	3.3891
Standard Deviation, Hz	1.8409
Coefficient of Variation	3629.2505
Confidence Interval at probability P = 97 %	$6680.7076 \le f_x \le 6681.7391$
Relative error, %	$0.0077 < 0.01$
χ^2 – test (S) at: k=5; P = 97 % χ^2_{max} = 10.9	10.6667
Hypothesis about uniform distribution	$S < \chi^2_{\max}$ (accepted)

k is the number of equidistant classes;
S is the sum of deviations between the data set and the assumed distribution;
$\chi^2_{\max}$ is the maximal possible allowable argument of the χ^2 distribution.

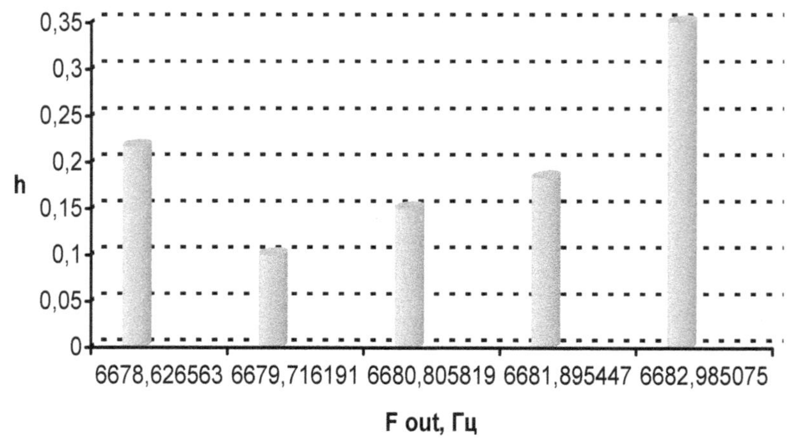

Fig. 8.19 Density of distribution for frequency output measuring results at 55 % RH.

The difference between calculated humidity and measured valuse is due to a difference in the temperature operating range (Figure 8.18 a). The typical response look-up table data were obtained according to the polynomial reference curve for modeled signal output (F_{out} = 7314 - 16.79×RH + 0.0886×RH^2 - 0.000358×RH^3) at +25 ^{0}C [29]. The real experiments were carry out at +25 ^{0}C ± 0.5 ^{0}C.

The χ^2-test for goodness of fit was applied to investigate the significance of the differences between observed data in the histograms and the theoretical frequency distribution for data from a uniform population. Taking into account that the sum of deviations between the data set and the assumed distribution $S < \chi^2$, the hypothesis about uniform distribution can be accepted.

Taking into account that the USTI has two independent channels, it is possible to use any quasi-digital temperature sensor with frequency, priod, duty-cycle or PWM output [33] instead of resistive NTC or RTD temperature sensors. The temperature sensor can be used in such sensor system for signal drift compensation.

A capasitance-type humidity sensing element aslo can be connected directly to the USTI because of IC can convert a capacitance-to-digital. In this case one external resistor and reference capasitor should be used [8].

Due to high metrological performance of the UFDC-1 and USTIIC ICs they are also suitable for interfacing of any humidity sensors based on quartz crystal microbalance (QCM) gravimetric devices, for example, described in [9, 10, 13].

A humidity, dew point and moisture sensor devices created in this way offers the advantage of an optimum performance for the user over an entire application range, even at the lower range limit at around 0 percent relative humidity as well as at extreme high humidity and dew formation. This approach also results in lower manufacturing costs, as the individual components and UFDC or USTI ICs, as well as the sensing element and carrier substrate, can be tested before assembly. This translates into lower costs for end-users such as system designers.

In comparison with the direct microcontroller interfacing the proposed design approach also lets to eliminate many design problems connected with the microcontroller selection, its programming in Assembled language and additional error components due to so-called program-dependent or software-related effects.

8.4 References

[1]. Matos, R., Cabral, S., Bravo, N., Cordeiro, A., Palma, J., Digital Temperature and Humidity Monitoring Unit for Remote Applications, in *Proceedings of the IEEE International Symposium on Consumer Electronic (ISCE' 2008)*, 14-16 April 2008, pp.1- 4.

[2]. Hideo Shibata, Masahiro Ito, Masahiro Asakursa, Kenzo Watanabe, A Digital Hygrometer Using a Polyimide Film Relative Humidity Sensor, *IEEE Transactions on Instrumentation and Measurement*, Vol. 45, No. 2, April 1996, pp.564-569.

[3]. Robert M. Brown, Ensuring RH Sensor Repeatability with Capacitance Testing, *Sensors*, July 1997.

[4]. Sung P. Lee, Ji G. Lee, Joong W. Chang, Joo N. Kim , Sang H. Lee, Shaestagir Chowdhury, Integrated Sensor System for Humidity Sensing of Robots Using Analog Mixed CMOS Technology, in *Proceedings of the IEEE International Workshop on Robotic and Sensors Environments (ROSE' 2007)*, Ottawa, Canada, 12-13 October 2007, pp.1-4.

[5]. Giovanni Bucci, Marco Faccio, Carmine Landi, The Implementation of a Smart Sensor Based on Piece-Linear A/D Converter, in *Proceedings of the IEEE Instrumentation and Measurement Technology Conference*, Ottawa, Canada, May 19-21, 1997, pp. 1173-1177.

[6]. Xiaojing Shi, Hiroki Matsumoto, Kenji Murao, A High-Accuracy Digital Readout Technique for Humidity Sensor, *IEEE Transactions on Instrumentation and Measurement*, Vol.50, No.5, October 2001,pp. 1277-1282.

[7]. Universal Frequency-to-Digital Converter (UFDC-1), Specification and Application Note, *Technology Assistance BCNA 2010, S. L.*, 2011.

[8]. Universal Sensors and Transducers Interface (USTI), Specification and Application Note, *Technology Assistance BCNA 2010, S. L.*, 2011.

[9]. Galatsis, K., Wenmin Qu, Wlodarski W., Quartz Crystal Microbalance Humidity Sensor with Porous Electrodes, in *Proceedings of Conference on Optoelectronic and Microelectronic Materials Devices*, 1999, pp. 373 – 375.

[10].H. Ito, S. Kakuma, R. Ohba, K. Nod, Development of a Humidity Sensor using Quartz Crystal Microbalance, in *Proceedings of Annual Conference SICE*, Fukui, Japan, 4-6 August 2003, pp.1175-1178.

[11].List of Humidity and Moisture Sensors Manufacturers at Sensors Web Portal: http://www.sensorsportal.com/HTML/SENSORS/HumiditySens_Manufacturers.htm

[12].Denes K. Roveti, Choosing a Humidity Sensor: A Review of Three Technologies, *Sensors*, Vol.18, No.7, July 2001.

[13].Jaruwongrungsee K., Wisitsoraat A., Tuantranont A., Lomas T., Humidity Sensor Utilizing Multiwalled Carbon Nanotubes Coated Quartz Crystal Microbalance, in *Proceedings of the 2nd IEEE International Nanoelectronics Conference (INEC' 2008)*, pp.961-964.

[14].Yurish S. Y., Program-oriented Methods and Measuring Instruments for Frequency-Time Parameters of Electric Signals, PhD Thesis, *State University Lviv Polytechnic*, Lviv, 1997 (in Ukrainian).

[15].Reverter F., Microcontroller-based Interfaces for Quasi-Digital Sensors, PhD Thesis, *Technical University of Catalonia* (UPC Barcelona, Spain), 2004.

[16].Yurish S. Y., Reverter F., Pallàs-Areny R., Measurement error analysis and uncertainty reduction for period- and time interval-to-digital converters based on microcontrollers, *Measurement Science and Technology*, Vol.16, No.8, 2005, pp.1660-1666.

[17].Ching-Liang Dai, A capacitive humidity sensor integrated with micro heater and ring, oscillator circuit fabricated by CMOS–MEMS technique, *Sensors and Actuators B*, Vol. 122, 2007, pp. 375–380.

[18].Smart Sensors and MEMS, ed. by Sergey Y. Yurish and Maria Teresa S.R. Gomes, *Kluwer Academic Publishers*, 2004.

[19].Relative Humidity Sensor HS 1100/ HS 1101, Technical Data, HPC001, Rev. 7, *Humirel*, June 2002.

[20].Application Note for Dew Sensors (DPS 0603/1206), Application Note 28772, *Vishay Beyschlag*, August 2007.

[21].Temperature and Relative Humidity Sensor HTS2030SMD, Technical Data, HPC055 Rev0, *Humirel*, June 2004.

[22].Karim Ben Dhia, Calibration of Frequency output module Using CMOS 555 timer & HUMIREL Humidity Sensor: HS1100/HS1101/HTS2010SMD, Application Note, HPC031revC, *Humirel*, May 2005.

[23].HS1101LF - Relative Humidity Sensor, HS1101LF HPC052_G, *Measurement Specialties, Inc*., April 2008.

[24].HCH-1000 Series Humidity Sensor Calibration and Output Circuit Data, *Honeywell*, 2008.

[25].Yurish S. Y., Digital Humidity Sensors and Data Logger Design Based on Modern Frequency-to-Digital Converter, *Sensors & Transducers*, Vol. 59, Issue 9, September 2005, pp.419-425.

[26].MiniCap2 - Humidity Sensor (http://www.blueearthresearch.com/msensors.htm).

[27].EE05 Series of Humidity and Temperature Transmitter with Frequency Output, *E+E Elekronik*.

[28].Product Info Sheet No. B 1.6 Humidity Sensing Elements, Modules Humidity Frequency Converter, *Galltec Mess- und Regeltechnik GmbH MELA Sensortechnik GmbH*, March 2004.

[29].HTF3130LF – Temperature and Humidity Module, HTF3130LF HPC101 Rev B, *Measurement Specialties, Inc.*, April 2008.

[30].Temperature and Humidity Module HF 3223 / HTF 3223, Technical Data, HPC014 Rev G, *Humirel*, July 2001.

[31].HTF3226LF – Temperature and Humidity Module, HTF3226LF HPC053 Rev E, *Measurement Specialties, Inc.,* April 2008.

[32].Humidity-to-Frequency Converter KN-1050, Technical Note, CAT.No.KN-1050.

[33].List of Temperature Sensors Manufacturers at Sensors Web Portal: http://www.sensorsportal.com/HTML/SENSORS/TempSens_Manufacturers.htm

[34].Humidity and Dew-Point Instruments, I7000 & 7400 (Hygrosmart) Interchangeable Sensor Module for Relative Humidity, *Michell Instruments*, p. 98.

Chapter 9

Digital Accelerometers, Inclinometers and Gyroscopes

9.1 Quasi-digital Accelerometers and Inclinometers: State-of-the-Art

The measurement of acceleration or its derivatives such as vibration, shock, tilt, inclination, etc. has become very wide used in different applications [1]. Accelerometers with frequency output are known since the beginning of 60th [2]. The past decade MEMS accelerometers began to become commonplace in the market.

According to market analyst firm *Yole Developpement*, overall MEMS market expected to grow from \$8 billion+ in 2010 to \$18 billion in 2015 [3]. According to *Frost & Sullivan*, MEMS accelerometers are one of the most key MEMS products along with pressure sensors and gyroscopes. Among the main MEMS-based sensors, accelerometers accounted for 25.6 percent of the total market revenues for MEMS sensors [4].

There are many manufacturers producing quasi-digital single chip accelerometers with frequency or duty-cycle outputs [5]. Some quasi-digital accelerometers also have been described in the literature [6-8]. The surface micromachined products provide the sensing element and the signal conditioning circuitry on the chip, and require only a few external components. Some manufactures have taken this approach one step further by converting the analog output into a quasi-digital format such as duty-cycle, frequency or pulse-modulated signal (PWM). This does not only lift the burden of designing of the fairly complex analog circuitry for the sensor, but also reduces the cost and the board area [9]. Main features of quasi-digital accelerometers from different manufacturers are shown in Table 9.1. More information about quasi-digital accelerometers and inclinometers is available at Sensors Web Portal [10, 11].

Very often such accelerometers are called "digital" accelerometers but really they should be called "quasi-digital" because of they have frequency, duty-cycle or PWM outputs.

Table 9.1 Quasi-digital accelerometers' performances.

Device	Number of Axis	Range, g	Sensitivity Accuracy, %	Output
		Analog Devices		
ADXL202	2	± 2	± 16	Duty-cycle
ADXL210	2	± 10	± 20	Duty-cycle
ADXL212	2	± 2	*	Duty-cycle
ADXL213	2	± 1.2	± 10	Duty-cycle
		Honeywell		
RBA500	*	± 70	*	Frequency
SA500	*	± 80	*	Frequency
		Kionix		
KXG20	2	± 2	*	Duty-cycle
		MEMSIC, Inc.		
MXD2020E	2	± 1	*	Duty-cycle
MXD2125	2	± 2	*	Duty-cycle
MXD6025	2	± 2	*	Duty-cycle
MXD6125	2	± 0.5 ÷ 2	*	Duty-cycle
		Silicon Designs, Inc.		
1010	1	± (2 ÷ 200)	± 2	Pulse density
2010	1	± (2 ÷ 200)	± 2	Pulse density
2420	3	± (2 ÷ 200)	± 2	Pulse density
		DIGI SENS AG		
BB / ED21		± (0.3 ÷ 7.7)	< 0.03 % (max. error)	Frequency: 8 … 19 kHz

'*' – information not available.

Such devices can be directly interfacing to modern microcontrollers but frequency, duty-cycle and pulse width measurements with high resolution are not trivial tasks. A designer must take into account many factors or so-called program-dependent or software-related effects [12-14] that influence on the conversion error, for example, the error due to delay of reaction to interruption and error of shift in time of the response for interruption. Very often, these components exceed the quantization error. In order to simplify a design process of digital accelerometers and inclinometer and reduce time-to-market and primary cost, the UFDC-1 or USTI ICs should be used [15-17].

9.2 Digital Accelerometers Design

9.2.1 Accelerometers based on quasi-digital sensors

Analog Devices manufacturers three models of dual axis MEMS accelerometers ADXL202, ADXL210, ADXL212 and ADXL213 with duty-cycle modulated outputs.

The ADXL202E is a low-cost, low-power, complete 2-axis accelerometer with a measurement range of ±2 *g* [18]. The ADXL202 can measure both dynamic acceleration (e.g., vibration) and static acceleration (e.g., gravity). The outputs are Duty-Cycle Modulated (DCM) signals whose duty-cycles (ratio of pulse width to period) are proportional to the acceleration in each of the two sensitive axes (Figure 9.1). These outputs may be measured directly with both UFDC and USTI ICs, requiring no A/D converter or glue logic.

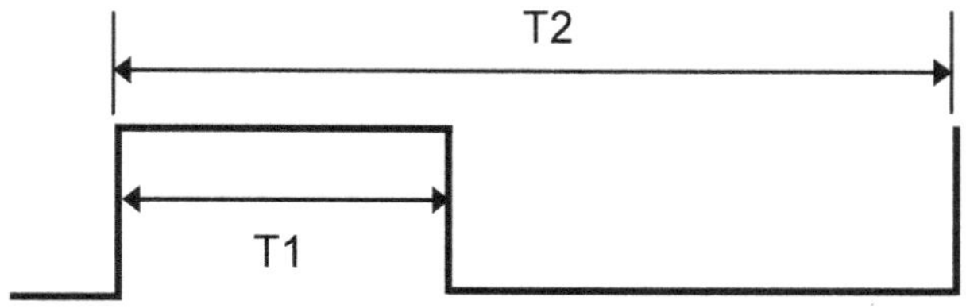

Fig. 9.1 ADXL202, ADXL210, ADXL212 and ADXL213 output signal.

The DCM period *T2* is adjustable from 0.5 ms to 10 ms via a single resistor (R_{SET}) by choosing a value between 100 kΩ and 2 MΩ. As outlined in the data sheet, the nominal duty-cycle output of the ADXL202 is 50 % at 0 g and 12.5 % duty-cycle change per *g*. Therefore to calculate acceleration from the duty- cycle:

$$A\ (g)=\frac{(T1/T2)-0.5}{12.5\%} \tag{9.1}$$

If the 0 *g* duty-cycle output of the ADXL202 is other than 50 %, and/or the duty-cycle changes more or less than 12.5 % per *g*, the acceleration calculation will be inaccurate. In practice, the 0 *g* output and the sensitivity of the ADXL202 vary somewhat from device to device (see the data sheet for details). So, this formula can only be used for low accuracy measurements. For higher accuracy measurements, the actual offset and scale values must be substituted. Two simple methods of calibration to find the actual offset and scale values are described in [19].

Note that while *T2* is nominally constant, it does change over temperature and contains some jitter. For systems that do not require resolutions of better than 100 mg, *T2* may be measured only once. For more accurate measurements, several *T2* measurements should be made and averaged. The average should be updated periodically to account for *T2* drift over temperature [19].

The bandwidth of the ADXL202 may be set from 0.01 Hz to 6 kHz via capacitors C_X and C_Y. The functional block diagram is shown in Figure 9.2.

Please note, that the UFDC-1 can measure duty-cycle for only one axis, for example, X in the 1st UFDC-1's channel, while the USTI can measure the duty-cycle in both channels, and therefore, to serve two axis: X and Y.

In order to choose right values of external components for ADXL202, the Interactive Design Tools -*Accelerometers: ADXL202 Calculator* is recommended for the use, taking into account high metrological UFDC's and USTI's performances [15, 16] especially related to the duty-cycle output and acquisition rate. The screen of Interactive Design Tools is shown in Figure 9.3.

Many cases exist where it's important to know the temperature of the ADXL202. Sometimes temperature information is necessary for control or data logging requirements or for high precision tilt sensor applications (for temperature compensation of the zero g drift of the accelerometer). A simple method of embedding the temperature information from a TMP36 voltage output temperature sensor in the period acceleration output of the ADXL202 is described in [20]. No A/D converter is required, and no additional I/O is necessary.

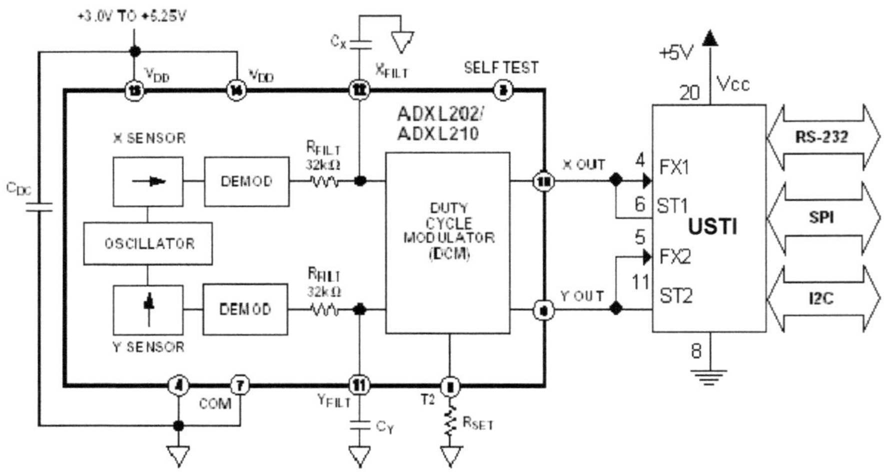

Fig. 9.2 ADXL202 to UFDC interfacing functional diagram.

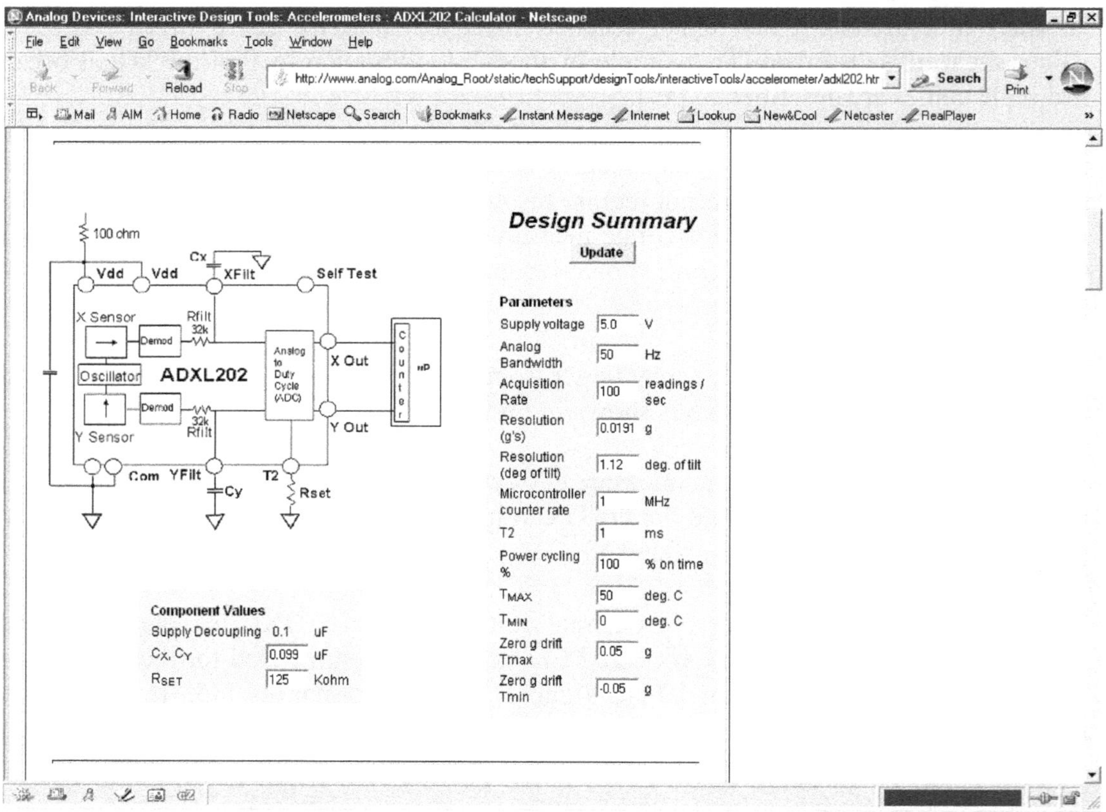

Fig. 9.3 Interactive design tools -accelerometers: ADXL202 calculator.

An interesting solution to improve MEMS accelerometers performance using the similar technique is described in [21]. The block diagram of sensor systems is shown in Figure 9.4. The sensor system contains accelerometer ADXL202E, temperature sensors with analog output TMP36, UFDC-1 converter and microcontroller PIC16F84A. The main features of this sensor system include a dual measurement of static acceleration and temperature and the ability to automatically adjusted, in real time, the data acquisition rate based on the noise amplitude evaluated during the auto-calibration phase.

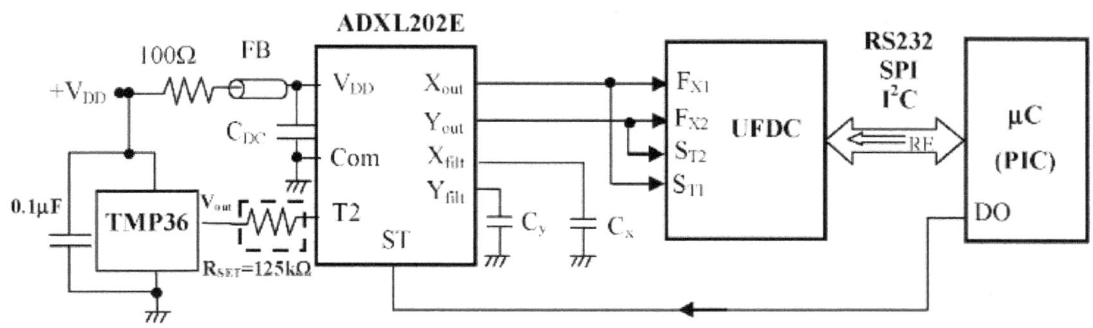

Fig. 9.4 Block diagram of sensor systems [21].

The ADXL210 is a low cost, ±10 g, dual-axis accelerometer with a digital output, all on a single monolithic IC [18]. The ADXL210 will measure accelerations with a full-scale range of ±10 g, making it suitable for tilt measurement. It is an improved, smaller version of the ADXL210. The ADXL210 to USTI interfacing functional diagram is shown in Figure 9.5. The ADXL210's output signal is shown in Figure 9.1.

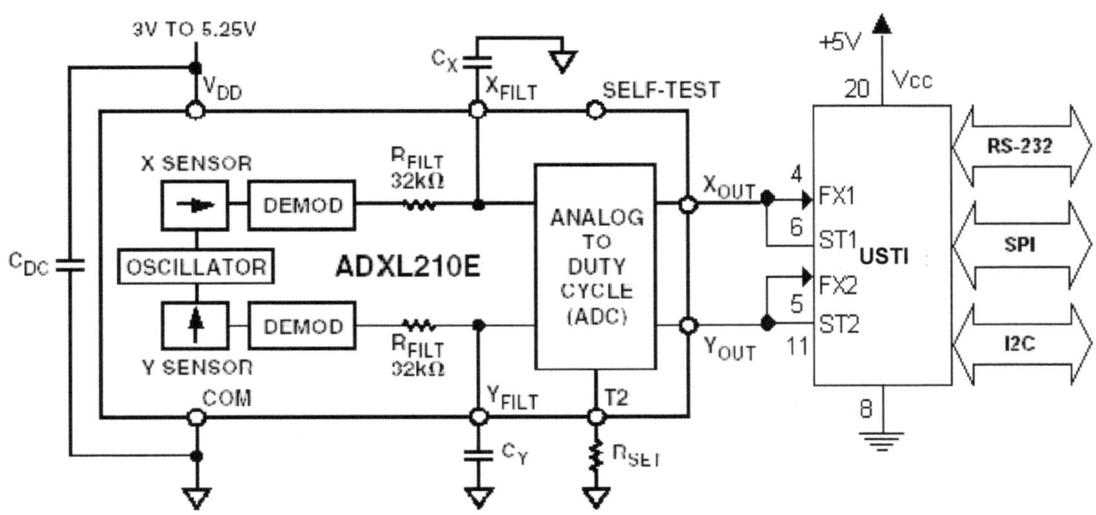

Fig. 9.5 ADXL210 to USTI interfacing functional diagram.

In order to calculate acceleration from the duty-cycle the following equation should be used:

$$Acceleration\ (g) = \frac{(T1/T2 - 0.5)}{4\%} \tag{9.2}$$

The ADXL213 is a precision, low power, complete dual axis accelerometer with signal conditioned, duty-cycle modulated outputs, all on a single monolithic IC [22]. The ADXL213 measures acceleration with a full-scale range of ±1.2 g (typical). The ADXL213 can measure both dynamic acceleration (e.g. vibration) and static acceleration (e.g. gravity).

The outputs are quasi-digital signals whose duty-cycles (ratio of pulse width to period) are proportional to acceleration (30%/g). The duty-cycle outputs can be directly measured by the UFDC or USTI without an A/D converter or glue logic.

The ADXL213 greatly improves upon the ADXL202 in two critical specifications: zero-g stability over temperature, and sensitivity accuracy. Bandwidths of 0.5 Hz to 250 Hz may be selected to suit the application.

The acceleration can be calculated according to the following formula:

$$Acceleration\ (g) = \frac{(T1/T2 - 0.5)}{30\%} \tag{9.3}$$

The calculation should be done by a master microcontroller or PC. The ADXL213 to USTI interfacing functional diagram is shown in Figure 9.6. The output signal is the same as shown in Figure 9.1.

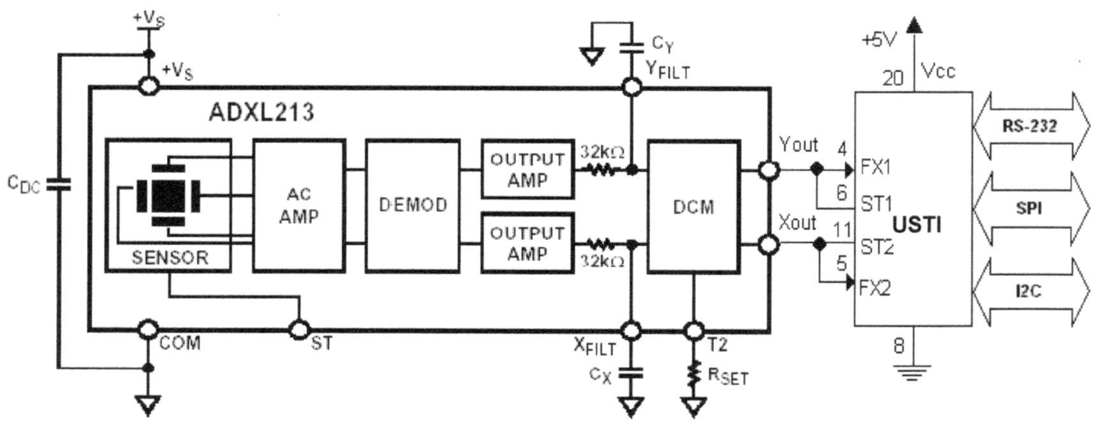

Fig. 9.6 ADXL213 to USTI interfacing functional diagram.

The low cost ± 2 g dual axis accelerometer ADXL212 is the newest quasi-digital accelerometer from Analog Devices, Inc. [23]. It is a high precision, low power, complete single- and dual-axis accelerometers with duty cycle modulated outputs, all on a single, monolithic IC. The ADXL212 measures both dynamic acceleration (such as vibration) and static acceleration (such as gravity). The outputs are digital signals whose duty cycles (ratio of pulse width to period) are proportional to acceleration (12.5%/g) in each of the two sensitive axes (Figure 9.1). The duty cycle outputs can be directly measured by a

USTI or UFDC-1 ICs without an analog-to-digital converter or glue logic. The output period is adjustable from 0.5 ms to 10 ms via a single resistor (RSET). The acceleration should be calculated according to equation (9.1). The ADXL212 accelerometer should be interfaced to the USTI by the same manner as it is shown in Figure 9.6. The user selects the bandwidth of the accelerometer using Capacitors C_X and C_Y at the XFILT and YFILT pins. Bandwidths of 0.5 Hz to 500 Hz can be selected to suit the application. The T2 should be set up as T2 (s) = R_{SET}/125 MΩ.

The KXG20 silicon micromachined linear accelerometers from *Kionix* consist of a sensor element and an ASIC [24]. The sensor element functions on the principle of differential capacitance. Acceleration causes displacement of a silicon structure resulting in a change in capacitance. An ASIC, using a standard CMOS manufacturing process, detects and transforms changes in capacitance into a duty-cycle signal, which is proportional to acceleration. Output is in the form of a pulse width modulated duty-cycle signal offering broad application and design flexibility. The acceleration can be calculated according to the following equation:

$$Acceleration\ (g) = \frac{(T1/T2 - 0.5)}{20\%} \tag{9.4}$$

The 0 *g* condition corresponds to 50% of duty-cycle (T1/T2 = 0.5). The application schematic is shown in Figure 9.7 Outputs *X* (pin 8) and *Y* (pin 9) should be connected to the USTI or UFDC-1 by the same manner as it is shown in Figures 9.2, 9.5 and 9.6.

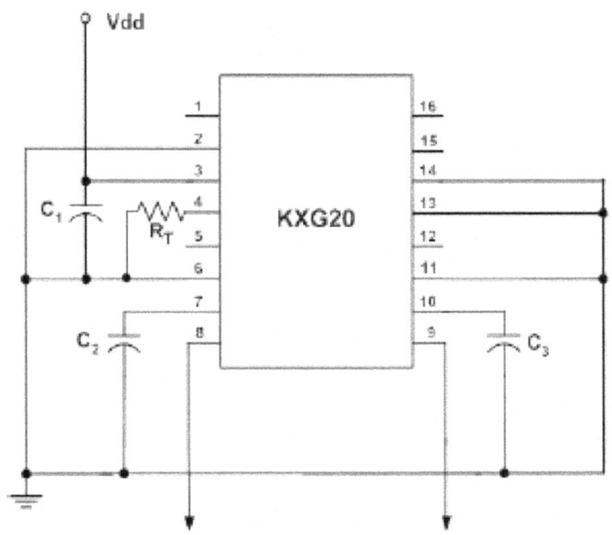

Fig. 9.7 Application schematic of KXG-20 accelerometer.

External capacitors *C2* and *C3* are used to set the -3dB filter point for each sensor output. In a typical application, the desired bandwidth will be determined by the fastest signal needing to be measured. It is recommend to use 0.1 μF for decoupling capacitor *C1*. In order to calculate the *C2* and *C3* for the sensor the following equation should be used:

$$C2 = C3 = \frac{1}{2\pi \cdot 32000 f_{BW}}, \qquad (9.5)$$

where f_{BW} is the sensor bandwidth frequency needed in application (typically from 10 Hz to 1500 Hz).

To set the digital output, data rate R_T can be calculated with this formula:

$$R_T = \frac{437 \cdot 10^6}{f_{PWM} \cdot V_{dd}}, \qquad (9.6)$$

where f_{PWM} is the frequency of digital output data rate (from 100 to 2000 Hz) $f_{PWM} = 1/T2$.

The custom design consideration for this model of accelerometer is the same as considered above for quasi-digital accelerometers from *Analog Devices, Inc.*

Other quasi-digital accelerometers, first of all RBA500 and SA500 frequency output accelerometers from Honeywell and some series of accelerometers (inclinometers) from *Rieker Electronics, Inc.* [5, 11] also can be directly interfacing to the UFDC and USTI ICs. The acceleration in Honeywell's devices is measured as a function of the frequency difference between two vibrating quartz beams [25]. The RBA500 accelerometer is primarily used to supplement GPS navigation systems. It is a good choice where frequency output, high-g, small size, low power, and light weight are necessary. The output of the accelerometer is also thermally compensated through the use of an internal temperature sensor and Honeywell-supplied coefficients; when integrated over time, delta Velocity is directly provided to the user's system.

Accelerometers from *Rieker Electronics, Inc.* are available with pulse-width or frequency- modulated outputs.

Dual axis CMOS accelerometers from *MEMSIC, Inc.*, for example, MXD2125GL/HL/ML/NL provide two digital outputs that are set to 50% duty-cycle at zero *g* acceleration. The outputs are digital with duty-cycles (ratio of pulse width to period) that are proportional to acceleration [26]. The duty-cycle outputs can be directly interfaced to the UFDC or USTI. The acceleration can be calculated according to the equation (9.4). The MXD2125GL/HL/ML/NL has two duty-cycle outputs (x, y). The acceleration is proportional to the ratio T1/T2. The zero *g* output is set to 50 % duty-cycle and the sensitivity scale factor is set to 20 % duty-cycle change per *g*. These nominal values are affected by the initial tolerance of the device including zero *g* offset error and sensitivity error. This device is offered from the factory programmed to either a 10 ms period (100 Hz) or a 2.5 ms period (400 Hz). Other models of quasi-digital accelerometers MXD2020E [27], MXD6025 and MXD6125 from the *MEMSIC, Inc.* are differ by the ranges of accelerations and equations for the acceleration calculation. So, the acceleration for the MXD6125 must be calculated according to the equation (9.1).

The noise level is one determinant of accelerometer resolution. The second relates to the measurement resolution of the UFDC and USTI ICs when decoding the duty-cycle output. The actual resolution of the acceleration signal is limited by the time resolution of the ICs used to decode the duty-cycle. Taking into account the high resolution of the

UFDC and USTI converters, the accelerometer's noise floor may set the lower limit on the resolution.

The appropriate USTI's commands for duty-cycle measurement in two channels (RS232 interface) are shown in Figure 9.8.

```
>M04        ; Duty-cycle measurement initialization in the 1st channel
>S          ; Start a measurement
>C          ; Check result status: returns 'r' if ready or 'b' if busy
>R          ; Read a result for the X axis

0.7568

>M14        ; Duty-cycle measurement initialization in the 2nd channel
>S          ; Start a measurement
>C          ; Check result status: returns 'r' if ready or 'b' if busy
>R          ; Read a result for the Y axis

0.2432
```

Fig. 9.8 Appropriate USTI commands (RS232) for duty-cycle measurement.

It is not necessary to set up the relative error, because of in duty-cycle measuring mode the maximum possible accuracy is used. The relative error can be calculated according to (3.1). The measuring mode should be selected only once for each of channel. Then, the following sequence of commands should be repeated cyclically for each of channel: S, C and R.

The Model 1010 from *Silicon Designs, Inc.* is a low-cost, integrated accelerometer for use in zero to medium frequency instrumentation applications. It produces a digital pulse train in which the density of pulses (number of pulses per second) is proportional to applied acceleration. It requires a single +5V power supply and a TTL/CMOS level clock of 100 kHz - 1 MHz [28]. The output is ratiometric to the clock frequency and independent of the power supply voltage.

The *Silicon Designs, Inc.* also produces the single- and triaxial digital accelerometer modules 2010 and 2420 respectively [29, 30] with pulse density outputs.

The *Digital Sensor Technology* (*Digi Sens AG*, Switzerland) manufactures accelerometer Type BB / ED 21 with frequency output [31]. The sensing element is essentially a silicon differential capacitor, in which the deflection of the middle plate (seismic mass) is measured. The signal is then converted electronically, so as to give an acceleration output in the form of a 5V TTL frequency signal. The temperature is also measured in the sensor, allowing a computed temperature correction of both zero and gain. The BB accelerometer was developed specifically for dynamic weighing.

Due to its standard CMOS technology the UFDC and USTI ICs (IP) can be easy embedded by manufacturers into its accelerometers in order to produce really digital single IC chip device with high metrological performances, wide functionality and communication capabilities.

9.2.2 Acceleration to Frequency and PWM Circuits

Low cost monolithic accelerometers with voltage output may be paired with a circuit whose output changes with frequency (V/F) to provide a TTL level frequency output. The UFDC and USTI ICs can be easily convert frequency to digital and sent down a long transmission to compute the applied acceleration by master microcontroller or PC.

A high performance acceleration-to-frequency circuits can be built based on analog accelerometers ADXL05, ADXL50 and voltage-to-frequency converter AD654 from *Analog Devices, Inc.* [32]. Accelerometers directly convert any applied acceleration into an analog output voltage which then controls the output frequency of an AD654 low cost voltage-to-frequency converter IC.

One of the most popular applications of accelerometers is in tilt/inclination measurement. An accelerometer uses the force of gravity as an input to determine the inclination angle of an object. ADXL05 accelerometer can be also connected to a low cost CMOS 555 timer to provide a frequency output. Such solution can be used for frequency output accelerometer tilt sensor design [32].

A pulse width modulated (PWM) outputs very suitable for remote sensing and noisy environments can be obtained from analog signal with the help of analog or discrete circuit solutions [33, 34].

A voltage output of popular analog accelerometers can be also easy converted to noise resistance frequency or duty-cycle output signal and then to be interfacing directly to the UFDC or USTI ICs.

9.3 Digital Inclinometers Design

At design and development of new sensors and sensor systems, a low price, minimum possible hardware, high metrological reliability, ‘intelligent’ features and short time to market are main requirements especially for automotive industry that has a great demand for miniaturized low cost inclination (tilt) sensors [35]. Several publications have discussed the design and implementation of inclinometers for different applications [36-42], however traditional inclinometers have difficult to meet all above-mentioned demands and up to now no exists commercial solution for a full range, dual-axis inclinometer-sensor, which fulfils these requirements. In addition, existing inclinometers do not support any ‘intelligent’ features. Only a few inclinometers on modern sensors market are compatible with the IEEE 1451 standard that determines the TEDS formats. So, the task of design and development of low-cost, full range, smart inclinometers that can be an alternative to the expensive existing products is well-timed and issue of current importance.

The designed and investigated sensor system consist of 2-axis acceleration sensor on a single IC chip ADXL210E [18] from *Analog Devices, Inc.* and 2-channel Universal Sensors and Transducers Interface (USTI) integrated circuit [16, 43]. The ADXL210E is a low-cost, low-power accelerometer with quasi-digital output whose duty cycles (ratio of pulse width to period) are proportional to acceleration. Nevertheless today many manufactures produce accelerometers with analog (voltage) output, quasi-digital signals like frequency (period), PWM and duty-cycle have many well-known advantages as accelerometers informative parameter in comparison with analog signals. It is high noise immunity, high power signal, high reference accuracy, wide dynamic range, multiparametricity, simple interfacing, integration and coding [44]. In comparison with a sensor's frequency output the duty-cycle output has also additional advantages. It is rather immune to interfering signals, such as spikes [45], and the ratio does not depend on the absolute value of any component [46].

The ADXL210E's duty cycle converter has a resolution of approximately 14 bits; better resolution than the accelerometer itself [18]. The actual resolution of the acceleration signal is, however, limited by the time resolution of the counting devices used to decode the duty cycle. The faster the counter clock, the higher the resolution of the duty cycle and the shorter period can be for a given resolution.

The duty cycle output of ADXL210E is directly interfaced to the USTI that especially designed for precision (up to 0.005 %) duty-cycle – to – digital conversion and signal conditioning in such applications. The USTI is based on the novel method for duty-cycle measurements [47], has a programmable quantization error and non-redundant conversion time. According to this method the maximal relative error at duty cycle measurement can be calculated according to the following equation:

$$\delta_{D.C.} = \frac{N_\delta \cdot T_0 + \tau_x}{\tau_x \cdot N_\delta}, \tag{9.7}$$

where $N_\delta = 1/\delta$ is the number indirectly proportional to the programmable relative error of period T_x measurement (from 1 to 0.0005 %); T_0 is the period of USTI's reference frequency $f_0 = 1/T_0$=625 kHz; τ_x is the pulse width.

The USTI also adds 'intelligent' features in the inclinometer and can store a TEDS (according to the IEEE 1451 standard) with accelerometer's calibration constants and USTI's quantization error, which can be programmed.

The duty cycle period of ADXL210E is adjustable and was chosen in this design via a single resistor R_{SET} = 620 kΩ equal to ~5 ms [48]. In order to improve noise performance a narrow bandwidth of 100 Hz was chosen with the help of two external capacitors $C_X = C_Y$=0.047 μF. The functional block diagram of inclinometer is shown in Figure 9.5 and main inclinometer's components in Figure 9.9.

The appropriate commands for USTI working in RS232 communication mode and measuring the duty cycle are shown in Figure 9.8.

The ADXL210E will require calibration for application such as inclinometer. The simple calibration set up is shown in Figure 9.10. It includes the inclinometer mounted on the laser level HG 909A and 0-360^0 calibration scale-disc.

Fig. 9.9 Main inclinometer components: miniature accelerometer with external components mounted on PCB (1), and universal sensors and transducers interface (USTI) integrated circuit (2).

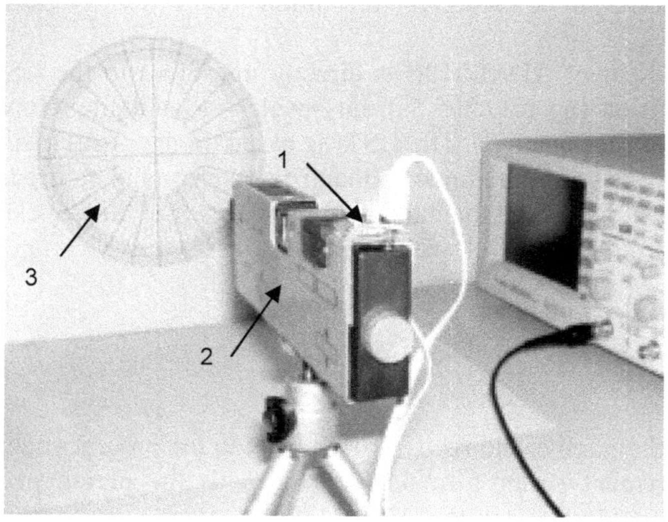

Fig. 9.10 Simple calibration set up: inclinometer (1); laser level HG 909A (2); 0-360^0 calibration scale-disc (3).

For some inclinometer applications one axis measurement is enough. In this case instead of USTI the Universal Frequency-to-Digital Converters (UFDC-1 and UFDC-1M-16) can be used [15, 49]. These ICs can measure the duty cycle only in one (the 1st) channel.

By the same manner described above, inclinometers can be also based on any accelerometers with quasi-digital output, for example, duty cycle, frequency and pulse density output accelerometers considered above. True digital output inclinometers with SPI, I2C and RS232 interfaces can be designed based on the T6 incremental inclinometer from the *US Digital*, the PWM output NG – type inclinometers from the *Nordic Transducer*, the PWM output SCA830 inclinometer from the *VTI Technologies*, etc. [11].

The aim of experimental investigation was to determinate some metrological performances including an error for duty cycle – to – digital conversion. The measurement set up is shown in Figure 9.11. The USTI was calibrated before at +23.8 ^{0}C. The sensor system was supplied at +5V dc by a Promax FAC-363B power supply. Digital oscilloscope OD-571 was used for signal waveforms monitoring. The measured duty-cycle values were sent to a PC via an RS232 interface implemented with the help of the ST202D integrated circuit. The user interface was implemented with the help of terminal software Terminal V1.9b for Windows.

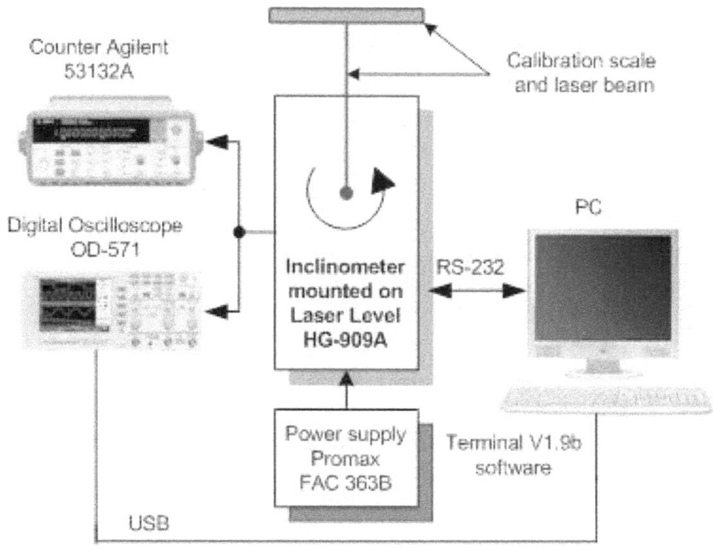

Fig. 9.11 Experimental measuring set up.

Duty-cycle measurements were taken every second by USTI and high precision calibrated counter (Agilent 53132A) until totaling 60 measurements. The measurement error was evaluated from histograms and appropriate statistical characteristics. The measuring results for $D.C._X = 17$ % and $D.C._Y = 49$ % (see oscillograms in Figure 9.12) are shown in Figure 9.13 and appropriate statistical characteristics in Table 9.2.

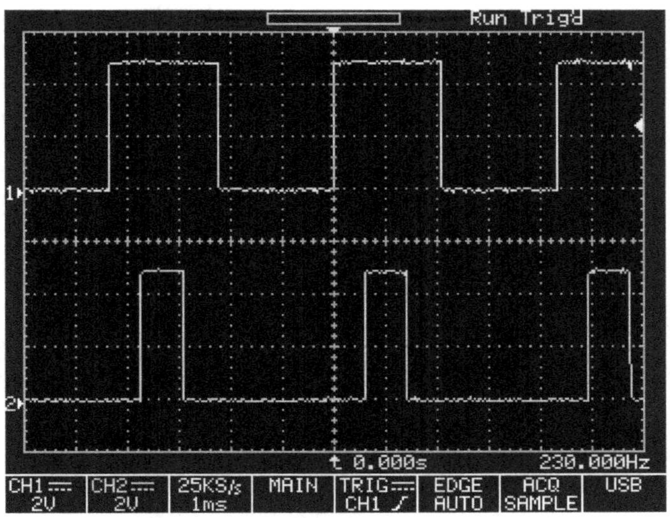

Fig. 9.12 Oscillograms of output signals: X-axis (1) and Y-axis (2).

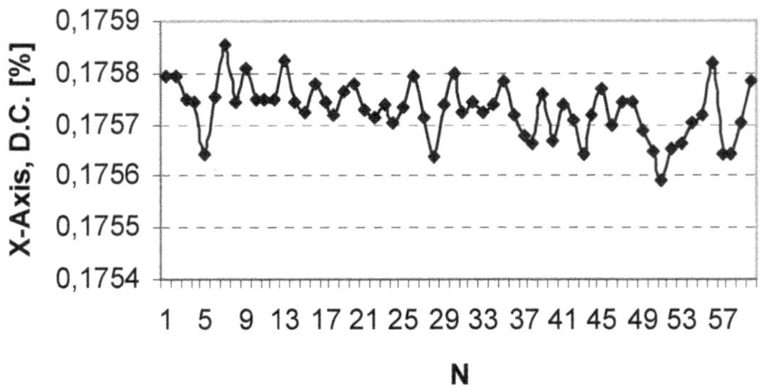

(a)

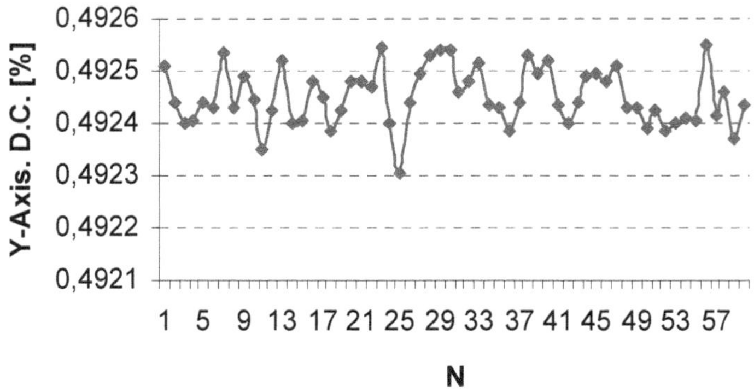

(b)

Fig. 9.13 Duty-cycle results of measurement for X- (a) and Y (b)-axis at inclinometer's calibration position.

Table 9.2 Statistical Characteristics of Measurement Results.

Parameter	**Value**	
	X-Axis	*Y-Axis*
Number of measurements, N	60	60
Minimum f_x (min)	0.1756	0.4923
Maximum f_x (max)	0.1759	0.4926
Sampling Range, f_x (max)- f_x (min)	0.0003	0.0002
Median	0	0
Arithmetic Mean	0.1757	0.4925
Variance	2.8E-9	2.9E-9
Standard Deviation	0.0001	0.0001
Coefficient of Variation	3296.8533	9204.0289
Relative error, %	0.11	0.041
χ^2 – test (S) at: k=5; P = 97 % χ^2_{max} = 7.0	6.1584	4.084
Hypothesis about Gaussian distribution	S < χ^2_{max} (accepted)	S < χ^2_{max} (accepted)

k is the number of equidistant classes;
S is the sum of deviations between the data set and the assumed distribution;
χ^2 max is the maximal possible allowable argument of the χ^2 distribution.

The χ^2-test for goodness of fit was applied to investigate the significance of the differences between observed data in the histograms and the theoretical frequency distribution for data from a normal population. Taking into account that the sum of deviations between the data set and the assumed distribution $S < \chi^2$, the hypothesis about Gaussian distribution can be accepted. The frequency distribution is shown in Fig. 9.14.

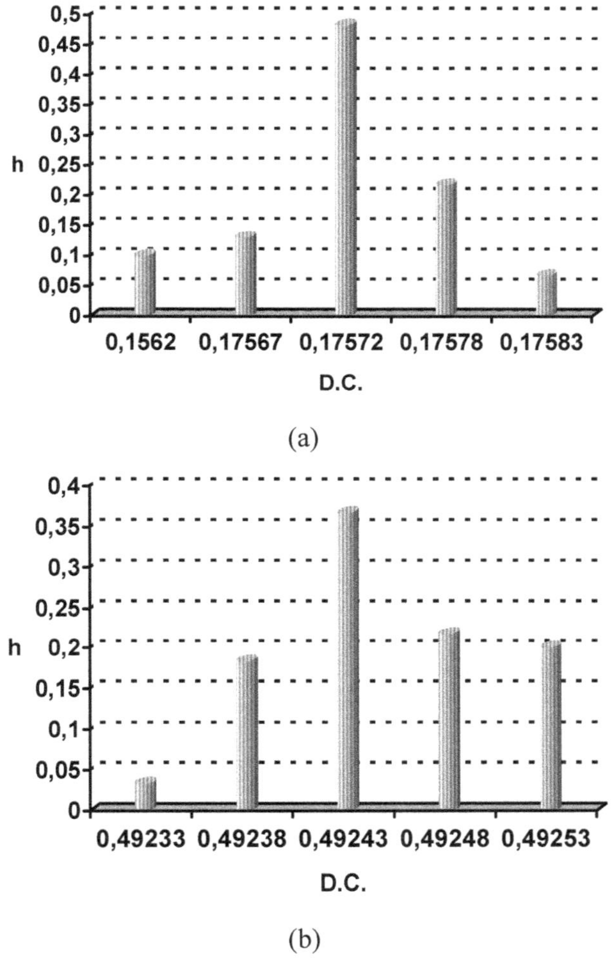

Fig. 9.14 Frequency distribution: X-Axis (a); Y-Axis (b).

The dependence of sensor's signal over angle of the test system corresponding to several automotive applications is shown in Fig. 9.15

The design methodology, described above lets create low-cost inclinometers with several intelligent functions. The device based on a novel measuring and signal conditioning concept. It contains a standard two-axis accelerometer with duty-cycle output and Universal Sensors and Transducers Interface integrated circuit with programmable features [16].

The USTI can store the TEDS according to the IEEE 1451 standard with accelerometer's calibration constants and USTI's programmable quantization error in its memory. Experimental results have confirmed high metrological performance of designed sensor system. The developed inclinometer is suitable for different applications including automotive to determine the position of automobiles with intelligent hand brake systems; load-dependent regulation of automatic gears, chassis and over-roll detection.

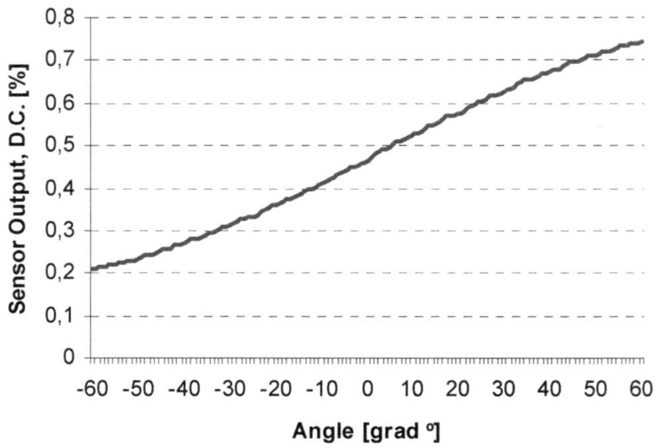

Fig. 9.15 Sensor signal over angle of the test system.

9.4 Digital Gyroscopes Design

There is a wide variety of gyroscope types based on different principles: from mechanical to microelectronic. Vibratory gyroscopes are interesting from fabrication point of view. They can be implemented in MEMS technology. Such gyroscopes are using vibrations of structures together with the Coriolis effect to measure angular motion in one form or another [50]. The vast majority of conventional vibratory gyroscopes provide an output signal that is proportional to the amplitude of the sense mode vibration, i. e. the measured vibration amplitude is proportional to the applied rate. The amplification of the small Coriolis vibration signal in MEMS gyroscopes make noise a considerable issue. Vibrating gyroscope designers focused on minimizing dimensions or changing the resonators' geometrical form or material, but always with the same classic detection method (amplitude detection), and only few of them were focused on frequency detection [51].

Non-amplitude gyroscopes use changes in natural frequency or induced phase differences of the proof mass vibrations to measure rotation rate. The result is that output signals are frequency or phase modulated and makes the gyroscope less affected by noise [52]. Frequency is generally easier to measure compared to amplitudes, which typically require high-performance ADC.

A direct frequency output vibrating gyroscope has not yet been seen on the market. However, some patents and articles have been granted and published [52, 53]. As it was written in [53] Three main reasons make frequency output preferable than amplitude output gyroscopes. First, it is well known that noise affects the amplitude of the transmitted signal by increasing or reducing it, more than the frequency. In rough environments (such as aeronautics and space), noise sources are numerous, and if it is necessary to get accurate gyroscopic measurements, a gyroscope with a direct frequency output seems to be vital [53]. Second, in classic vibrating rate gyroscopes (amplitude detection), the sensitivity depends on the amplitude of the driving vibrating mode, the difference between natural frequencies of the two modes, and on the quality factor when this difference is small. Also, the sensitivity depends on the geometrical form of the resonator. In the direct frequency output vibrating gyroscope, the sensitivity depends only on the difference between the natural frequencies of the two vibrating modes. In classic vibrating gyroscopes, the problem of superharmonic resonances is solved by reducing the excitation (or driving) amplitude, but since the Coriolis acceleration is proportional to this amplitude, this means that the sensitivity of the gyro decreases. In frequency output gyroscopes, the sensitivity is independent of the excitation amplitude [53]. Finally, the integration of frequency output rate gyroscope with quasi-digital accelerometers and further frequency-time parameters to digital conversion with the help of UFDC and USTI ICs [15, 16], will bring a lot of benefits for many applications including inertial navigation systems, as it simplifying the signal processing electronics and reduce noise.

The block diagram of direct frequency output vibrating gyroscope [53] is shown in Figure 9.16.

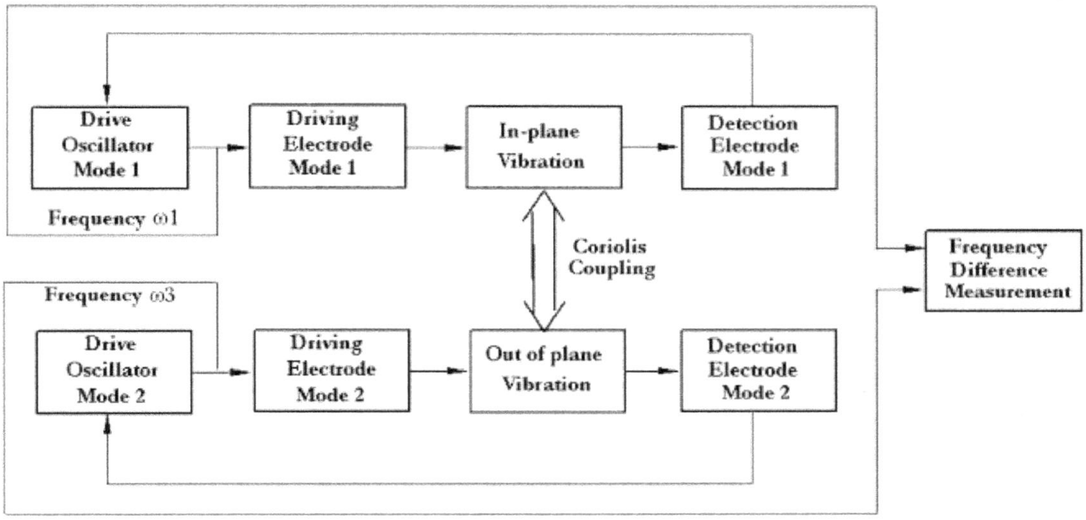

Fig. 9.16 Block diagram of direct frequency output vibrating gyroscope [53].

The frequency difference can be directly measured by both: UFDC and USTI ICs. An example of commands for frequency difference measurement by the USTI is shown in Figure 9.17.

```
>M06  ; Frequency differences measurement initialization (mode 6)
>A0A  ; Set up conversion error ± 0.0005 %
>S    ; Start measurement
>C    ; Check result status: returns 'b' if busy, 'r' if ready
>R    ; Read result (fx1-fx2)

2.8745532
```

Fig. 9.17 Appropriate set of the UFDC-1 commands (RS232 interface) for frequency differences measuring mode.

The frequency signal f_{x1} is coming on FX1, ST1 inputs and signal f_{x2} - on FX2, ST2 inputs.

9.5 References

[1]. Rob O'Reilly, Harvey Weinberg, The Five Motion Senses: MEMS Inertial Sensing to Transform Applications, *Sensors*, January 1, 2010

[2]. Talon W. J., Digital Integrating Accelerometer Provides Velocity Data, *Space Aeronautics*, No.4, 1963.

[3]. MEMS Growth in the Post-Recession Economy, *Sensors*, April 2010.

[4]. MEMS Sensors Market Geared for Growth - Emerging Trends and Opportunities, *Frost & Sullivan*, February 2010.

[5]. List of Accelerometers Manufacturers at Sensors Web Portal: http://www.sensorsportal.com/HTML/SENSORS/Accelerometers_Manuf.htm

[6]. Bin Xie, Chen-Yang Xue, Wen-Dong Zhang, Ji-Jun Xiong, Bin-Zhen Zhang, Jie Hu, A GaAs Micromachined Accelerometer with Frequency Output Based on Resonant Tunneling Diodes, in *Proceedings of the 8th International Conference on Solid-State and Integrated Circuit Technology (ICSICT '06)*, 23-26 October 2006, Shanghai, China, pp. 545 - 547.

[7]. Pei Xu, Jun Liu, Wendong Zhang, A Measuring Circuit for MEMS Resonant Accelerometer (www.paper.edu.cn/index.php/default/scholar/downpaper/liujun-9).

[8]. Deyong Chen , Zhengwei Wu, Lei Liu, Xiaojing Shi and Junbo Wang, An Electromagnetically Excited Silicon Nitride Beam Resonant Accelerometer, *Sensors*, Vol.9, 2009, pp. 1330-1338.

[9]. Bowling S., Richey R., Two Approaches to Measuring Acceleration, *Sensors*, 17, No.2, 2000.

[10].Accelerometers web section at Sensors Web Portal (http://www.sensorsportal.com): http://www.sensorsportal.com/HTML/SENSORS/Accelerometers.htm

[11].Inclinometers web section at Sensors Web Portal (http://www.sensorsportal.com): http://www.sensorsportal.com/HTML/SENSORS/Tilt_Sens_Manuf.htm

[12].Yurish S. Y., Program-oriented Methods and Measuring Instruments for Frequency-Time Parameters of Electric Signals, *PhD Thesis,* State University Lviv Polytechnic Lviv, 1997 (In Ukrainian).

[13].Reverter F., Microcontroller-based Interfaces for Quasi-Digital Sensors, *PhD Thesis,* Technical University of Catalonia (UPC Barcelona, Spain), 2004.

[14].Yurish S.Y., Reverter F., Pallàs-Areny R., Measurement error analysis and uncertainty reduction for period- and time interval-to-digital converters based on microcontrollers, *Measurement Science and Technology*, Vol.16, No.8, 2005, pp.1660-1666.

[15].Universal Frequency-to-Digital Converter (UFDC-1), Specification and Application Note, *Technology Assistance BCNA 2010, S. L,*. 2011.

[16].Universal Sensors and Transducers Interface (USTI), Specification and Application Note, *Technology Assistance BCNA 2010, S. L,*. 2011.

[17].Yurish S. Y., Practical Circuits and Interface Techniques for MEMS Accelerometers with Quasi-Digital Output, *Sensors & Transducers*, Vol.58, Issue 8, August 2005, pp.352-359.

[18].Low Cost 62 g/610 g Dual Axis iMEMS® Accelerometers with Digital Output, ADXL202/ADXL210, Rev. B, *Analog Devices, Inc.*, 1999.

[19].Weinberg H., Using the ADXL202 Duty Cycle Output, Application Note, AN-604 *Analog Devices*, USA.

[20].Harvey Weinberg, Embedding Temperature Information in the ADXL202's PWM Outputs, Application Note, AN-600, Rev.0, *Analog Devices, Inc.*, USA, 2002.

[21].J. M. Dias Pereira, Carlos Banha, Octavian Postolache, P. Silva Girao, Improving Accelerometers Performance Using Smart Sensing Techniques, in *Proceedings of the 5th IEEE Conference on SENSORS 2006,* Daegu, Korea, 22-25 October, 2006, pp.654-657.

[22].Low Cost ±1.2 g Dual Axis Accelerometer ADXL213, Rev. A, *Analog Devices, Inc.*, 2010.

[23].Precision ±2 g Dual Axis, PWM Output Accelerometer ADXL212, *Analog Devices, Inc.,* 2011.

[24].KGX-20 Preliminary Specification, *Kionix*, 2003.

[25].Accelerex® RBA500 Accelerometer, EXP030, *Honeywell International, Inc.*, USA, August 2005.

[26].Ultra Low Noise, Low offset Drift ±1 g Dual Axis Accelerometer with Digital Outputs, MXD2125GL/HL, MXD2125ML/NL, *MEMSIC, Inc.*, USA, 2004.

[27].Ultra Low Noise, Low offset Drift ±1 g Dual Axis Accelerometer with Digital Outputs MXD2020E/FL, MEMSIC, Inc., USA, 2004.

[28].Model 1010, Digital Accelerometer, *Silicon Designs, Inc.*, USA, September 2007.

[29].Model 2010 Digital Accelerometer Module, *Silicon Designs, Inc.,* USA, September 2007.

[30].Model 2420 Triaxial Digital Accelerometer Module, *Silicon Designs, Inc.,* USA, September 2007.

[31].Accelerometer Type BB / ED 21, *Digital Sensor Technology*, S004E, August 2008.

[32].Kitchin C., Quinn D., Sherman S., Acceleration to Frequency Circuits, AN-411, Application Note, *Analog Devices, Inc.,* USA.

[33].Ferran Bayes, PWM Circuit Uses One Op Amp, *EDN*, 6 July 2000 (http://www.edn.com/article/CA47000.html?1=1).

[34].Jacobsen E., Creating a PWM Output Sensor Using a Field-Programmable Analog Array, *Sensors*, May 1998.

[35].D. Benz, T. Botzelmann, H. Kück and D. Warkentin, On Low Cost Inclination Sensors Made from Selectively Metalized Polymer, *Sensors and Actuators A*, Vol. 123-124, pp. 18-22, 2005.

[36].L. Hua-Wei, M. Tao and Q. H. Meng, Design and Implementation of a Digital Inclinometer, in *Proceedings of the IEEE International Conference on Mechatronics & Automation*, Niagara Falls, Canada, July 2005, pp. 489-492.

[37].G. Qinglei, L. Huawei, M. Shifu and H. Jian, Design of a Plane Inclinometer based on MEMS Accelerometer, in *Proceedings of the International Conference on Information Acquisition*, 9-11July 2007, Jeju City, Korea, pp. 320-323.

[38].D. Lapadatu, S. Habibi, B. Reppen, G. Salomonsen, T. Kvisterøy, Dual-axes Capacitive Inclinometer/Low-g Accelerometer for Automotive Applications, in *Proceedings of the 14th IEEE International Conference on Micro Electro Mechanical Systems (MEMS' 2001)*, Interlaken, Switzerland, 2011, pp. 34-37.

[39].U. Mescheder and S. Majer, Micromechanical Inclinometer", *Sensors and Actuators A*, Vol. 60, 1997, pp. 134-138.

[40].J. Leavitt, High Bandwidth Tilt Measurement Using Low-Cost Sensors, *IEEE Transactions on Mechatronics*, Vol.11, No.3, June 2006, pp. 320-327.

[41].H. H. Djambazian, C. Nerguizian, V. Nerguizian, and M. Saad, 3D Inclinometer and MEMS Acceleration Sensors, in *Proceedings of the IEEE International Symposium*

on Industrial Electronics, Montreal, Quebec, Canada, 9-12 July 2006, Vol. 4, pp. 3338-3342.

[42].D. H. Parker, F. R. Schwab, J. W. Shelton and T. L. Weadon, Calibration and Modeling of a Dual-axis Inclinometer, *Precision Engineering*, Vol.29, 2005, pp.381-385.

[43].Yurish S. Y., Low-Cost, Dual-Axis Smart Inclinometer, in *Proceedings of 7th IEEE Conference on Sensors IEEE SENSORS 2008*, 26-29 October 2008, Lecce, Italy, pp. 1488-1491.

[44].Yurish S. Y., Sensors and Transducers: Frequency Output Versus Voltage Output, *Sensors & Transducers*, Vol.49, Issue 11, November 2004, pp.302-305.

[45].G. C. M. Meijer, Concepts and Focus Point for Intelligent Sensor Systems, *Sensors and Actuators A*, Vol. 41-42, 1994, pp. 183-191.

[46].S. Middelhoek, P. J. French, J. H. Huijsing and W. J. Lian, Sensors with Digital or Frequency Output, *Sensors and Actuators*, Vol. 15, 1988, pp. 119-133.

[47].Yurish S. Y., Method for Duty-cycle Measurement, Patent No. 79854 (UA), 2007.

[48].Dual-axis Accelerometer Evaluation Board ADXL210EB, *Analog Devices, Inc.*, USA, 2003.

[49].High-Speed Universal Frequency-to-Digital Converter UFDC-1M-16, Specification and Application Note, *Technology Assistance BCNA 2010, S. L.*, 2011.

[50].A. C. Dreyer, Modeling of MEMS Vibratory Gyroscopes Utilizing Phase Detection, M.Sc. Thesis, *University of Stellenbosch*, South Africa, March 2008.

[51].H. F. Tiersten, D. S. Stevens, and P. K. Das et al., Acoustic surface wave accelerometer and rotation rate sensor, in *Proceedings of the IEEE Ultrasonic Symposium*, 1980, pp. 692–695.

[52].Vibratory Gyroscope Utilizing a Frequency-based Measurement and Providing a Frequency Output, Patent WO 2010/051560 A1, US, Priority Data: 3 November 2008.

[53].Hassan Moussa and Roger Bourquin, Theory of Direct Frequency Output Vibrating Gyroscopes, *IEEE Sensors Journal*, Vol. 6, No. 2, April 2006, pp. 310-315.

Chapter 10

Digital Magnetic Sensors

10.1 Quasi-digital Magnetic Sensors: State-of-the-Art

The global market for magnetic sensors has been growing at a steady pace, both in terms of technology and applications. The sensors include a range of products that find applications in industrial and science frontiers as well as personal and commercial use. Position sensors, speed sensors and record heads in hard disk drives in computers are the most commonly used magnetic sensor types currently in vogue. According to *Reportlinker.com* report, the world magnetic sensors markets to reach US $2.73 billion in 2016 at a CAGR of 9.6 %, and according to a new Report by *Global Industry Analysts, Inc.*, the global market for magnetic sensors is forecast to reach over US $3.2 billion by the year 2017, driven by the rapidly burgeoning demand from the automotive and cell phone end-use segments [1].

The silicon magnetic sensor market will show strong growth as demand surges from the automotive, industrial and consumer markets, according to *iSuppli Corp*. The market research firm predicts that global silicon magnetic sensor revenue will reach US $1.4 billion in 2013 [2]. Growing forward, the market will reach US $1.9 billion in 2015 a CAGR of 12 % [3]. In the automotive body segment alone, the number of magnetic sensors and switches will grow to 9.4 % in 2013 [2]. Revenue derived from the use of magnetic sensors in automotive motors will reach $ 160.3 million in 2012, up a solid 38.2 percent from $ 116.0 million this year, extending the steady rise of the market over the last three years. And although revenue growth after 2012 will moderate to the single digits, the five-year compound annual growth rate from 2010 to 2015 still will equate to a robust 16 percent. By 2015, magnetic sensor revenue in automotive motors will amount to $ 193.6 million.

From the six sensor signal domains the magnetic one takes an important place. Today many manufacturers produce different magnetic sensors and devices on its basis: Hall effect sensors, magnetoresistors, magnetodiodes, GMR, SQUID, magnetometers, navigation compasses, etc. for different applications [4-11].

Magnetic sensors differ from most other detectors in that they do not directly measure the physical property of interest. They detect changes or disturbances in magnetic fields that have been created or modified, and from them derive information on properties such as direction, presence, rotation, angle, or electrical currents. The output signal of these sensors requires some signal processing for translation into the desired parameter [12].

Smart magnetic sensors with sensing elements and associated electronics such as amplification and signal conditioning on the same die are the latest trend. Such digital sensors can be built based on so-called quasi-digital magnetic sensors with frequency, duty-cycle or PWM output. This solution lets to achieve many advantages due to useful properties of frequency-time domain signal as informative parameters of sensor's output. Many manufacturers produce such quasi-digital sensors [13] and according to International Frequency Sensor Association (IFSA) studies its part is continuously growing during the last years.

Onboard sensor electronics can increase signal levels to significant voltages with the least pickup of interference. It is always best to amplify low-level signals close to where they are generated. Converting analog signals to digital (switched) outputs within the sensor is another way to minimize electronic noise. The use of comparators and digital outputs makes the nonlinearity in the output of sandwich GMR materials of less concern [13].

The interface-electronic circuit required to connect quasi-digital sensors to data acquisition systems. But as it was shown in the review of integrated frequency-to-digital converters [14] no one can be used for all frequency ranges and accuracies of modern quasi-digital sensors because of existing converters are based on classical frequency-to-digital conversion methods, its quantization error is dependent on the frequency range and/or they have redundant conversion time. In order to eliminate these disadvantages and meet modern requirements to interface-electronic circuits, the Universal Frequency-to-Digital Converter (UFDC-1) or Universal Sensors and Transducers Interface (USTI) ICs should be used to produce an effective design [15, 16].

Several quasi-digital magnetic sensors are known from the literature [17-20]. An advanced integrated magnetic-field sensors with frequency output is described in [17]. Arrays of 20 different types of sensor chips have been fabricated on two types of silicon wafers and have been tested. An average output frequency is from 60 to 150 kHz. The sensor's non-linearity is less than 0.1 % in the range of magnetic field up to 400 mT. The frequency output of these sensors provides noise-immune data transmission and a simple interface.

The frequency output magnetic sensor [18] working in 1.35 … 2.5 T flux density range and has from 212 to 330 kHz frequency signal on its output, directly proportional to the measurand.

A novel resonant quasi-digital magnetic field sensor exploiting the Lorentz force and providing a frequency output is described in [19]. The mechanical resonator, a cantilever structure, is embedded as the frequency-determining element in an electrical oscillator. The characterized devices show a sensitivity of 60 kHz/Tesla at their resonance frequency f_0 = 175 kHz and a short-term frequency stability of 0.025 Hz, which corresponds to a resolution below 1 μT. The quasi-digital nature of the frequency output makes it less sensitive to the quality of the electronics (e.g., linearity or noise). Moreover, a frequency can easily be converted to a digital code and can be measured with very high precision [19].

A CMOS magnetic field to frequency converter has two voltage-controlled ring oscillators whose output frequency differences linearly vary with the magnetic field perpendicular to the chip surface (Figure 10.1) [20].

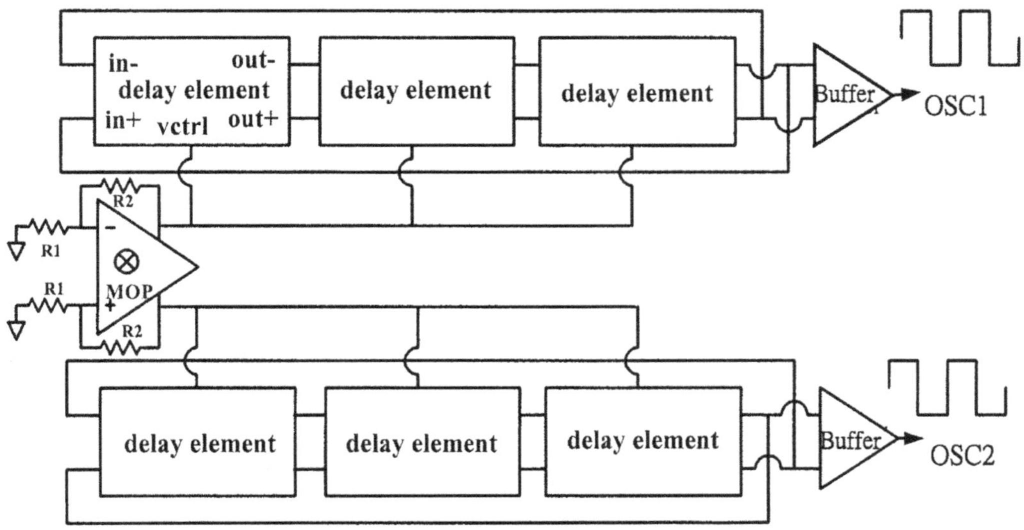

Fig. 10.1 Magnetic sensor with frequency outputs [20].

The frequency output (frequency difference) can be further processed by the UFDC-1 or USTI ICs circuits to achieve the high resolutions (see below).

Taking into account the both ICs UFDC and USTI have fabricated using standard CMOS process as well as magnetic sensors with frequency output(s) [17, 19, 20] these electronic components can be fabricated together to produce a single chip digital magnetic sensor.

10.2 Digital Magnetic Sensor Systems Design

Let consider some existing industrial magnetic sensors and its interfacing to the UFDC-1 and/or USTI IC in order to produce digital output magnetic sensors with serial output according to the RS232 interface or SPI and I^2C serial buses.

Micronas produces HAL810, HAL85X and HAL710/730 universal programmable linear magnetic field Hall effect sensors [21]. The HAL810 is a member of the Micronas family of programmable linear Hall sensors with the linear duty-cycle output signal [22].

It can be used for angle or distance measurements if combined with a rotating or moving magnet. The major characteristics, such as magnetic field range, sensitivity, output quiescent signal (output duty- cycle at B = 0 mT), and output duty-cycle range are programmable in a non-volatile memory.

The linear output is provided as the duty cycle output signal with typically 125 Hz refresh rate proportional to the magnetic flux through the Hall plate and up to 11 bit resolution. The duty-cycle is defined as the ratio between the pulse width 's' and the period 'd' of the signal (Figure 10.2).

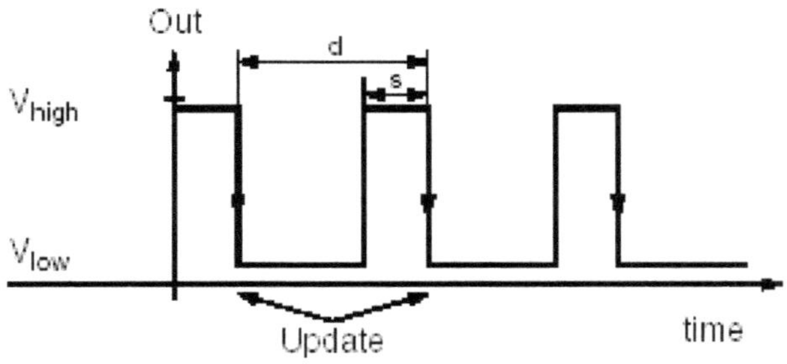

Fig. 10.2 HAL810 duty-cycle output.

Example for HAL810 sensor's output characteristics are shown in Figure 10.3 and interfacing circuit is shown in Figure 10.4.

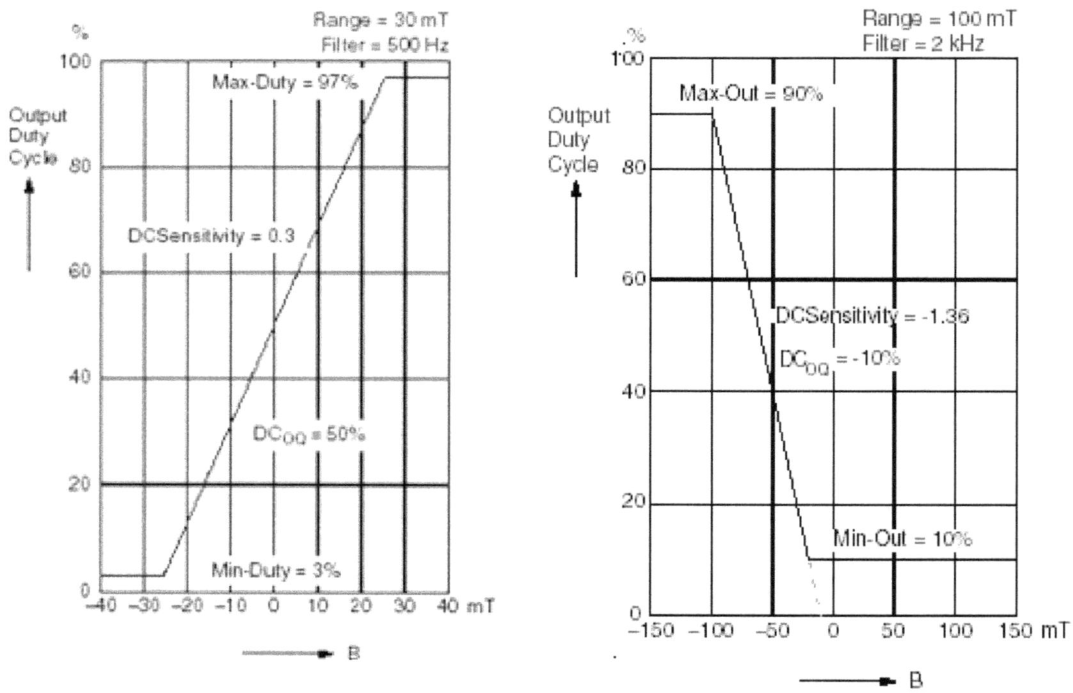

Fig. 10.3 Examples of HAL810 characteristics.

The appropriate set of the UFDC-1 commands (RS232 interface) for duty-cycle measuring mode is shown in Figure 6.17. The UFDC-1 converts the duty-cycle-to-digital with maximum possible accuracy. The quantization error can be calculated according to the equation (2.1).

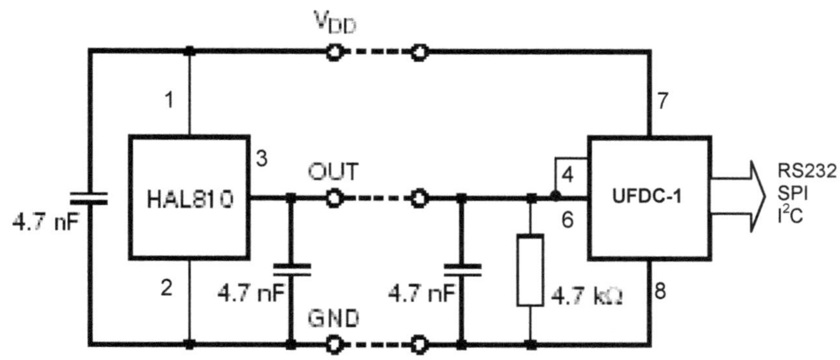

Fig. 10.4 HAL810 to UFDC-1 interfacing circuit.

Due to the sensor's versatile programming characteristics, the HAL 810 is the optimal system solution for applications such as: contactless potentiometers, rotary sensors, distance measurements, magnetic field and current measurement.

The HAL® 85x (HAL® 855 and 856) complement the existing Hall- effect sensor family HAL 8xx. Both universal magnetic field sensors provide an arbitrary output signal. The sensors are produced in submicron CMOS technology [23]. The sensors provide either a duty-cycle output or a serial Biphase-M output.

The HAL 710/730 Hall-Effect sensors with direction detection are also produced in CMOS technology. The sensors include two independent temperature-compensated Hall plates each equipped with active offset compensation and a comparator. The sensors provide two open-drain outputs – one for "count" and one for "direction" detection [24]. The HAL 710/730 is the optimal system solution for applications, such as Endposition detection, RPM measurement of motors in various applications and in flow meters.

There are also two other sensor from Micronas with duty-cycle output: Linear Hall-Effect Sensor HAL® 2850 and HAL®2859 [25, 26]. The HAL 2850 is a new member of the - Micronas varioHAL (HAL 28xy) family of programmable linear Hall-effect sensors [25]. It features a digital duty-cycle output with slew-rate control that enables a fast and robust data transfer in harsh environments. The duty-cycle signal can be directly decoded by any unit measuring a duty cycle of a rectangular signal, for example, UFDC and USTI ICs. The highest available frequency of output signal is 2 kHz with 12-bit resolution. It is customer-programmable in a range between 30 Hz and 2 kHz, with a certain resolution.

The HAL2859 is a factory-set linear Hall-effect sensor with duty-cycle output. The sensor features a temperature-compensated Hall plate with spinning current offset compensation, an A/D converter, digital signal processing, an EEPROM memory with redundancy to store Micronas calibration data, and protection devices at all pins. The internal digital signal processing is of great benefit because analog offsets, temperature shifts, and mechanical stress do not degrade digital signals. The HAL2859 is calibrated to a typical sensitivity of 0.67 % / mT, an offset of 50 % duty cycle and 2 kHz frequency output.

Generic applications include but not limited rotary position, leveling and angular measurements.

The *Allegro MicroSystems, Inc.* manufactures several Hall effect sensors with duty-cycle output, for example, A1351, A1354 and A1356. A high precision, programmable 2-wire linear Hall effect sensor IC A1354 with duty-cycle output proportional to the applied magnetic field [27]. Such output provides increased noise immunity compared to analog (voltage or current) output. This device is targeted at both the automotive and industrial sectors.

The *Speake & Co Llanfapley* produces FGM-series Magnetic Field Sensors: FGM – 1, FGM – 2 and FGM – 3 [28]. The FGM-X series of devices are very high sensitivity magnetic field sensors operating in the ±50 microtesla range (±0.5 oersted). They are simple, essentially three terminal devices, operating from a single +5 V supply, the connections being ground, +5V and output. The output is a robust 5 V rectangular pulse whose period is directly proportional to the field strength, (giving a frequency which varies inversely with the field), making it very easy to interface to UFDC-1 and USTI ICs. The typical period swing for the full range of an FGM-3 is from 8.5 ms to 25 ms (~120 kHz to ~50 kHz), a clear indication of its high sensitivity. The FGM-1 is essentially a miniature version of the FGM-3 with a slightly lower sensitivity. The FGM-2 is a single package containing two FGM-1 sensors at right angles to one another. Unlike Hall Effect field sensors, which are virtually unusable in this range because of their large temperature sensitivity, the FGM-series has a very low temperature coefficient. The typical FGM-3 response is shown in Figure 10.5 and typical response for the miniature types of sensor, the FGM-1 and FGM-2 is given in Figure 10.6.

Since the range covers the earth's field magnitude, multiple sensors can easily be arranged to provide compass orientation or full three-dimensional orientation systems, using the local earth's field as a reference, (-gimballed compass or virtual reality helmet devices.). Other applications include conventional magnetometry, earth field magnetometry, ferrous metal detectors, internal vehicle re-orientation alarm sensors, external vehicle or ship passage sensors, wreckfinders, non-contact current sensing or measurement, conveyor belt sensors or counters and in conjunction with small permanent magnets, movement and proximity sensors and ferrous impurity detectors for non-magnetic alloys.

Conventional magnetometry would include magnetic material measurement, B-H loop measurement, Preisach displays and archaeological artifact assessment [28].

The large pulse output gives considerable noise immunity permitting the use of transducers sited at long distances from the main system. The simplest method of making field measurements is to use the UFDC-1, UFDC-1M-16 or USTI IC directly on the output of the sensor, set to period measuring mode. Several useful applications are described in [28].

Many magnetic sensors form different manufacturers are intended for use as sensors for gear tooth wheels or magnetic encoders with a digital output signal (for example, AKL Sensors Series from *Rhopoint Component Ltd.*) The pulse output from the sensor corresponds with the gear teeth passing in front of it. When a gear tooth or magnetic pole is in front of the sensor, the sensor's output goes high; when the gear tooth or magnetic pole moves away, the output returns to low. This repeats at every tooth/pole, resulting in a pulse train output that provides speed information from the gear or encoder. The interfacing of rotation speed sensors to the UFDC-1 is described in Chapter 11.

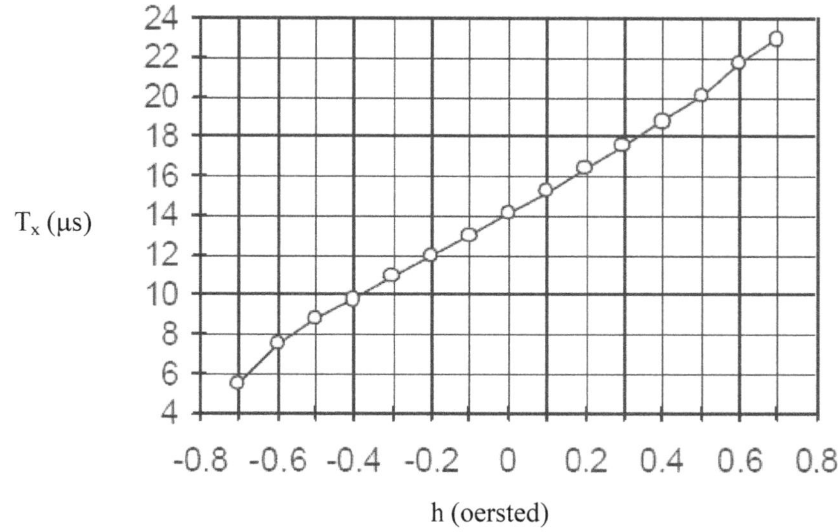

Fig. 10.5 Typical response for FGM-3 magnetic sensor: period vs. magnetic field [28].

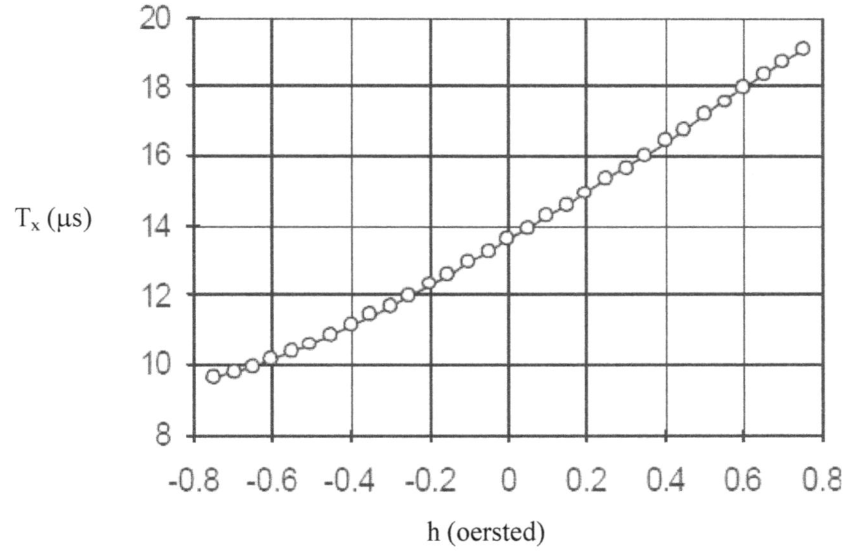

Fig. 10.6 Typical response for the miniature types of sensor FGM-1 and FGM-2: period vs. magnetic field [28].

The Earth Magnetic Field Sensor EMF-03 is produced by the *Xensor Integration* [29]. The EMF-03 is a revolutionary magnetic field transducer with accuracy better than one μT[1] and offers a one component compass solution as an alternative to magneto-resistive and inductive sensors/sensor bridges. The sensor output formats are RS232, SPI/microwire or duty-cycle. The sensor output can be connected to the UFDC or USTI ICs without any external components in order to supply another popular sensor communication standard - the I^2C bus. The sensor's block diagram is shown in Figure 10.7.

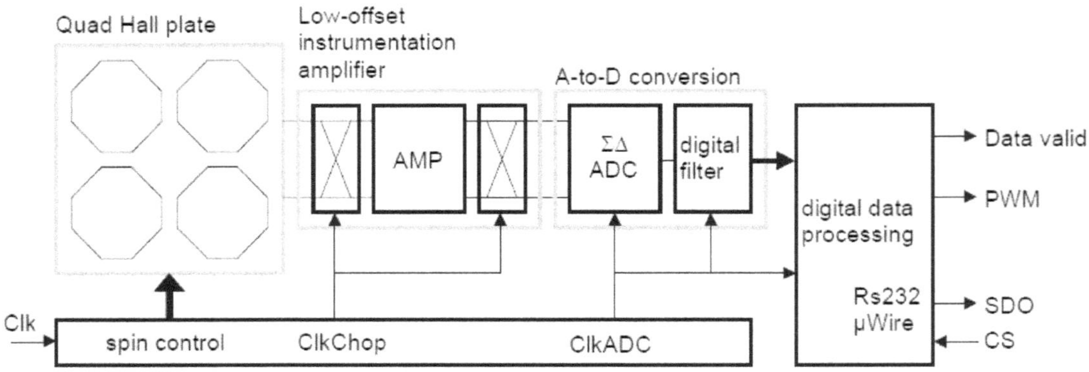

Fig. 10.7 The block diagram of EMF-03 Earth magnetic field sensor.

The duty-cycle at the sensor's output is proportional to the magnetic field:

$$D.C. = 50\% + \frac{Signal(T)}{338\mu T} \tag{10.1}$$

Two the same EMF-03 magnetic field sensors should be used to build a digital compass [29]. Two duty-cycle outputs of both EMF-03 sensors should be connected to the first and second channel of the USTI IC. The signals from the EMF03 sensors represent the raw X and Y vectors of the earth magnetic field in µT under a full circle of the compass in the X-Y plane.

The *BEI IDEACOD SAS* manufactures the TCW4 duty-cycle output magnetic absolute multi-turn modular sensor [30]. It has ± 0.3 % relative error, 16-bit total resolution, 10 to 90% duty-cycle output signal with 1 kHz frequency.

Two programmable CMOS Linear Hall Sensor IC with duty-cycle output MLX90291 and MLX90292 are also manufactured by the *Melexis* [31, 32]. Both sensors have 12-bit duty-cycle output signal proportional to the magnetic flux density.

Taking into account the continuously growing number of magnetic sensors with quasi-digital output during the last years, many research and development projects also focused on such kind of sensors.

The prototype of high resolution CMOS magnetic field to frequency converter circuit has been fabricated in a 0.5- m CMOS process and operated at a 5 V supply voltage [20]. The measured sensitivity is 24 kHz/mT and the power consumption is 5.1 mW. The small equivalent resolution of at least 20 T can be achieved. The frequency offset is 42 kHz when no magnetic field applied. Its nonlinearity within 120 mT is smaller than 0.56 %. However, the output frequency is a square term rather than a linear term related to B⊥. To compensate the nonlinearity, two ring oscillators are connected with the magneto-operational amplifier in a fully differential manner. One can obtain the difference of the output frequencies by frequency counters to obtain a linear term. This structure has two advantages. One is that the sensitivity is doubled. The other is that the even-order nonlinear terms can be cancelled if these two ring oscillators are well matched.

The frequency output (frequency difference) should be further processed by the UFDC-1 or USTI. There are two ways to measure the deference between two frequencies. The first way is to use a special frequency differences measurement mode. It lets to get result quickly, but not very precisely. The appropriate set of commands (RS232 interface) for the UFDC-1 working in the frequency difference measuring mode is shown in Figure 10.8.

The frequency signal f_{x1} is coming on FX1, ST1 inputs and signal f_{x2} - on FX2, ST2 inputs. The sign "minus" before the result of measurement testifies that $f_{x1} < f_{x2}$.

```
>M6   ; Frequency differences measurement initialization (mode 6)
>A4   ; Set up conversion error 0.05 %
>S    ; Start measurement
>C    ; Check result status: returns 'b' if busy, 'r' if ready
>R    ; Read result (fx1-fx2)
```

Fig. 10.8 Appropriate set of the UFDC-1 commands (RS232 interface) for frequency differences measuring mode.

The second way means two separate frequency measurements in each of channel and calculation the difference $\Delta f = f_{x1}$-f_{x2} with the help of an external master microcontroller or PC. This method lets to get very precision results with ≥ ±0.001 % relative error, but needs in twice more conversion time. Appropriate commands for such measuring mode are shown in Figure 10.9.

```
>M0   ; Frequency measurement initialization in the 1st channel
>A4   ; Set up conversion error 0.05 %
>S    ; Start measurement
>C    ; Check result status: returns 'b' if busy, 'r' if ready
>R    ; Read result (fx1)

>ME   ; Frequency measurement initialization in the 2nd channel
>A4   ; Set up conversion error 0.05 %
>S    ; Start measurement
>C    ; Check result status: returns 'b' if busy, 'r' if ready
>R    ; Read result (fx2)
```

Fig. 10.9 Appropriate set of the UFDC-1 commands (RS232 interface) for frequency measurements in two channels.

The part of quasi-digital magnetic sensors with frequency or duty-cycle output among all magnetic sensors is continuously growing. Today's market demonstrates very good examples with high metrological performances.

The use of quasi-digital integrated magnetic sensors and universal frequency-to-digital converter (UFDC-1) or universal sensors and transducers interface (USTI) ICs are making many new application areas technically and economically feasible. The special feature is not only eliminating the need for ADC and a very low noise but the potential to be integrated together in a frame of system-on-chip (SoC) to produce really digital output in accordance with the RS232 interface or SPI and I^2C serial buses. Therefore, these ICs can be parts of complex systems that contain a variety of additional sensors or chip-size magnetic sensor arrays. Such SoCs let significantly reduce production costs, time-to-market and simplify the design process.

10.3 References

[1]. Global Magnetic Sensors Market to Reach $3.2 Billion by 2017, According to a New Report by Global Industry Analysts, Inc., Press Release, *PRWeb*, 24 August 2011, (www.prweb.com/releases/magnetic_sensors/hall_sensor_ICs/prweb8571034.htm).

[2]. Nic Mokhoff, Demand to Surge for Silicon Magnetic Sensors in 2010, *EE Times*, 29 December 2009.

[3].Dixon R., Compasses Pick Up of Magnetic Sensor Market, MEMS and Sensors Special Report, *IHS iSuppli*, USA, 2011.

[4]. Yurish S. Y., Digital Magnetic Sensors Based on Universal Frequency-to-Digital Converter (UFDC-1), *Sensors & Transducers*, Vol.61 , Issue 11, November 2005, pp.446-450.

[5]. Ramsden E., Ten Easy Things to Do with Magnetic Sensors, *Sensors*, March 1997.

[6]. Law L., Magnetic Sensors and Timing Applications, *Sensors*, February 1998.

[7]. Smith C. H., Schneider R. W., Magnetic Field Sensing with GMR Sensors, Part 1: The Theory of Solid-State Magnetic Sensing, *Sensors*, September 1999.

[8]. Caruso M. J., Bratland T., Anisotropic Magnetoresistive Sensors Theory and Applications, *Sensors*, Vol. 16, No.3, March 1999.

[9]. Yi-Qun Li, An Innovative Passive Solid-State Magnetic Sensor, *Sensors*, Vol.17, No.10, October 2000.

[10].White T., A New Sensor ASIC for Changing Magnetic Fields, *Sensors*, Vol.17, No.3, March 2000.

[11].Rako P., Magnetic Measurement Tools Attract Attention, *EDN*, 24 July 2008.

[12].Caruso M. J., Bratland T., Smith C. H., R.Schneider, A New Perspective on Magnetic Field Sensing, *Sensors*, No.12, 1998.

[13].List of Magnetic Sensors Manufacturers at Sensors Web Portal: http://www.sensorsportal.com/HTML/SENSORS/MagSens_Manufacturers.htm

[14].Yurish S. Y., Kirianaki N. V., Pallas-Areny R., Low-Cost Precision Universal Programmable Frequency-to-Digital Converter, in *Proceedings of the IFAC Workshop on Programmable Devices and Systems (PDS' 2004)*, Poland, Cracow, 18-19 November 2004, pp.70-75.

[15].Universal Frequency-to-Digital Converter (UFDC-1), Specification and Application Note, Technology Assistance BCNA 2010, S. L,. 2011.

[16].Universal Sensors and Transducers Interface (USTI), Specification and Application Note, Technology Assistance BCNA 2010, S. L,. 2011.

[17].Gumenjuk S., Podlepetsky B., Advanced Integrated Magnetic-field Sensors with Frequency Output, *Sensors and Actuators A 62*, 1997, pp. 529-533.
[18].Gaman V. I., Drobot P. N., The Magnetic Sensor with the Frequency Output, in *Proceedings of the 5th International Conference APEIE-2000*, pp. 158-159.
[19].Robert Sunier, Tobias Vancura, Yue Li, Kay-Uwe Kirstein, Henry Baltes, Resonant Magnetic Field Sensor with Frequency Output, *Journal of Microelectromechanical Systems*, Vol. 15, No. 5, October 2006, pp.1098-1107.
[20].Shr-Lung Chen, Chien-Hung Kuo, Shen-Iuan Liu, CMOS Magnetic Field to Frequency Converter, *IEEE Sensors Journal*, Vol. 3, No. 2, April 2003, pp.241-245.
[21].Micronas web site: http://www.micronas.com/
[22].HAL® 810 Programmable Linear Hall-Effect Sensor, Data Sheet, DSH000034_003EN, *Micronas*, Edition 6 February 2009.
[23].HAL 85x Programmable Hall-Effect Sensors with Arbitrary Output, Product Information, *Micronas*, October 2002.
[24].HAL® 710, HAL® 730, Hall-Effect Sensors with Direction Detection, Data Sheet, DSH000031_002EN, Micronas, 13 October 2009.
[25].HAL® 2850 Linear Hall-Effect Sensor with PWM Output, Data Sheet, DSH000160_001EN, Micronas, 9 August 2011.
[26].Explore Intelligent Technologies, Sensors and Embedded Microcontrollers for Automotive and Industrial 2011, *Micronas*, June 2011.
[27].A1354 High Precision 2-wire Linear Hall Effect Sensor IC with Pulse Width Modulated Output, Allegro MicroSystems, Inc., 1 August 2011.
[28].FGM-Series Magnetic Field Sensors, SCL: www.speakesensors.com/PDF/detail.pdf
[29].Earth Magnetic Field Sensor EMF-03, *Xensor Integration*, 28 April 2005.
[30].Magnetic PWM Absolute Multi Turn Modular Sensors, TCW4 Range, *BEI IDEACOD SAS.*
[31].MLX90291 SMD Programmable Linear Hall Sensor IC Featuring PWM Output, Data Sheet, 3901090291, Rev 001, *Melexis*, April 2011.
[32].MLX90292 SMD Programmable Linear Hall Sensor IC PWM / PAS4 / PSI5-A, Rev. 1.0, *Melexis*, 3 June 2011.

Chapter 11

Rotational Speed Sensors

11.1 Rotational Speed Sensors: State-of-the-Art

There are many known rotational speed sensing principles and many commercially available sensors. The overwhelming majority of such sensors are based on the Hall-effect and magnetoresistors sensing principles or inductive sensing principle [1, 2]. There are also rotational sensors based on optical, fluidic and even on chemical principles [3].

According to the nature, rotational speed sensors are quasi-digital sensors. Output signal is a sine or square-wave output frequency of which is linearly dependent on the rotating speed and carries the information about it (Figure 11.1). Its frequency is proportional to the measured parameter. The inductive rotation speed sensor is shown in Figure 11.2.

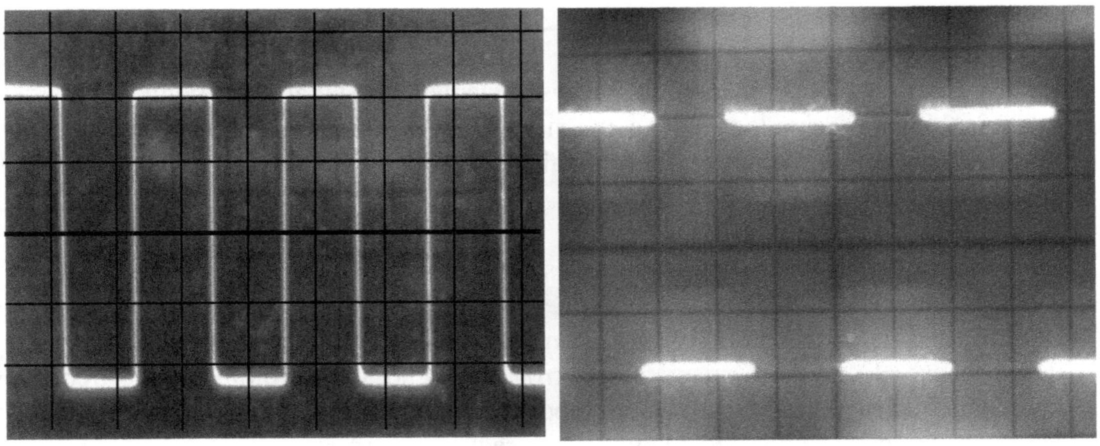

a) b)

Fig. 11.1 Waveform of rotation speed sensor's output: (a) - ASRS (*Autoelectronic*, Russia); (b) – *BALLUFF* (Germany).

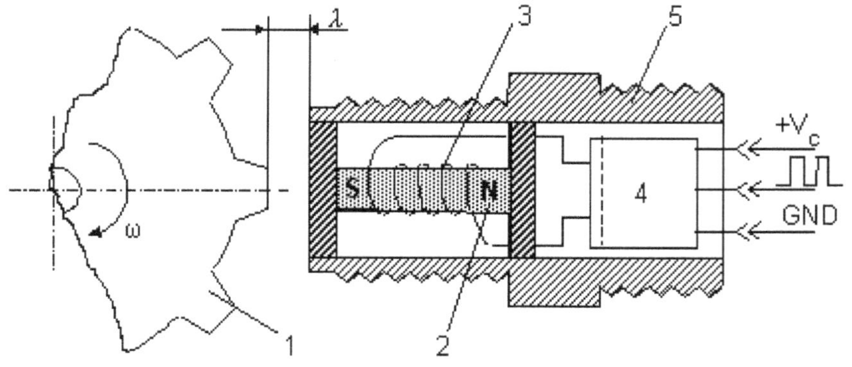

1 - gear tooth
2 - constant magnet
3 - inductance coil
4 - integrated circuit
5 - package
λ - air gap
ω – rotation speed (rpm)

Fig. 11.2 Inductive rotational speed sensor.

The measuring frequency of the rotation speed is given by

$$n_x = f_x \cdot \frac{60}{Z}, \tag{11.1}$$

where Z is the number of teeth; f_x is the frequency of rectangular pulses.

As usually, the rotation speed range is wide enough: from zero speed up to some thousands rpm. For example, the integrated rotational speed sensor KMI16/1 and KM18-4 from the Philips Semiconductors can measure rotational speed in 0 ... 25 kHz range with the maximum sensing distance 2 ... 2.7 mm [4, 5], which can be increased up to 4.5 mm in the rotational speed sensor for extended air gap applications KMI20/1. Sensor consists of a magnetoresistive sensor element, a signal conditioning integrated circuit in bipolar technology and a magnetized ferrite magnet. The frequency of the digital current output signal is proportional to the rotational speed of a gear wheel (see equation (11.1)).

The rotational speed measuring accuracy is determined by the frequency measurement's accuracy. The classical methods for frequency measurement (standard (direct) counting technique and indirect method of measurement) can not be used in such wide frequency range because of very high quantization error at the end and beginning of the frequency range. All disadvantages of classical methods are analyzed in details in our previous book [2]. There are many advanced methods for frequency and rotation speed measurement [6-13], but a customer must buy a license for the use it in his designs. It is true also for the combined conversion method [14]. In order to eliminate such kind of design problems it is expediently to use the UFDC-1 or USTI ICs [15, 16] based on advanced methods for frequency and rotational speed.

11.2 Digital Rotational Speed Sensors Design

The UFDC-1 and USTI ICs are ideally suited for rotational speed measurements [17]. It can directly measure a rotation speed and present the results in *rpm* units. The number of gear teeth *Z* (1…255) must be set up in advance. By default *Z*=1. Any rotation speed sensor with frequency output can be connected directly to these ICs. In the UFDC-1 IC, the first measuring channel should be used for this purpose, and any channel can be used in the USTI IC. The appropriate relative error of measurement should be chosen form the range ± (0.001 … 1) % for the UFDC-1 [15], and ± (0.0005 … 1) % for the USTI IC [16]. The appropriate set of commands for the UFDC-1 (UFDC-1M-16) ICs is shown in Figure 11.3.

```
>MA   ; Rotation speed measurement initialization
>Z0C  ; Set up Z=12(10)=C(16)
>A9   ; Choose the conversion error 0.001 %
>S    ; Start a measurement
>C    ; Check result status: returns 'b' if busy, 'r' if ready
>R    ; Read a result of measurement in rpm

1.6795
```

Fig. 11.3 Appropriate UFDC-1 commands (RS232 interface) for precision wide range rotation speed measurements with ± 0.001 % relative error.

Taking into account that IC can be used for a near zero speed measurement (~ 0 rpm) it is recommended to determine the minimal possible rotational speed according to the following equation:

$$n_{x\min} = \frac{3}{Z} \tag{11.2}$$

If after $1/_{fx\ min}$ = 20 s the command C still returns 'b' value it is necessary to make a decision, that rotational speed equals to 0 rpm or communication line is damaged.

Through RS232, SPI or I^{2}C buses the UFDC-1 and USTI can be connected to the LCD or LED indicators. It lets build precise wide speed range tachometers and tachometric systems with minimum possible hardware. The UFDC-1 and USTI ICs also can be connected to any master microcontroller or PC.

In sensor systems, the UFDC-1 and USTI can also measure a rotational acceleration. An external master microcontroller or PC calculates acceleration of rotation based on two measurements n_1 and n_2 of rotational speed and time interval for the second measurement t_2:

$$\varepsilon_x = \frac{n_1 - n_2}{t_2},\ rpm^2 \tag{11.3}$$

The sign "-" indicates deceleration of rotation, and "+" indicates acceleration of rotation.

Good examples of rotational speed sensors are dual- and single channel differential inductive position sensors PO2210, PO2211 and inductive position sensor with quasi-digital output IS1101 from *Posic* [18-20], Figure 11.4

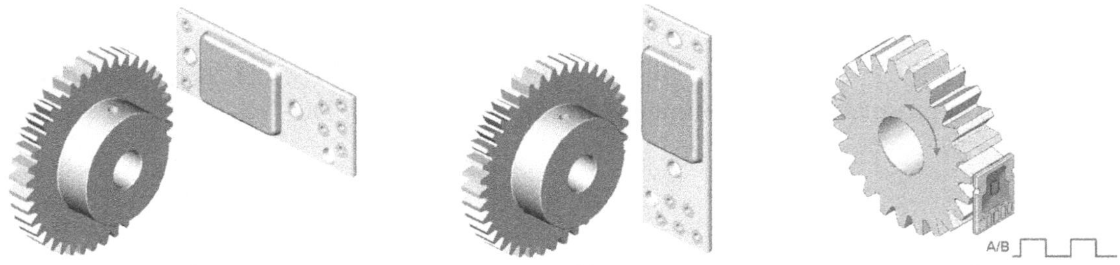

a) b) c)

Fig. 11.4 Inductive position sensors with quasi-digital outputs: PO2210 (a), PO2211 (b) and IS1101 (c) [18-20].

The inductive micro-sensor is an integrated version of the LVDT and resolver type of sensors [18]. The core is a sensor chip with one generator coil and two sets of detection coils (Figure 11.5). The detection coils are connected in a differential arrangement, to reject the common mode signal. The sensor also includes an electronic interface, which is composed of a high frequency excitation for the generator coil and two read-out channels for the two sets of detection coils (channels A and B). The read-out electronics extract the amplitude variation of the high frequency signal due to the presence of a metallic target. The output stage consists of a low-pass filter and a comparator. For a nominal target period of 1.3 – 1.5 mm (module 0.4 – 0.5), the outputs are two channels in quadrature (A quad B). Output signals A and B have a periodicity identical to the target gear. The ratio between the pulse-width (P) and the cycle-width (C) yields the duty cycle (D) for any of the two channels. The phase difference (f) between the channels A and B is defined by the ratio of the width between two rising edges over the cycle-width. The device has 0 to 10 kHz frequency range and 0 to 1 mm distance between sensor and target [18].

Differential inductive microcoil sensor PO2211 also has the same frequency range and air gap [19]. The generator micro-coil creates an AC magnetic field with a frequency around 1 MHz. The magnetic field couples into the two detection coils that are separated by a distance of 1 mm and that are connected in a differential configuration. When a ferromagnetic (steel) or electrically conducting (aluminum, brass, copper etc) object in front of the sensor deforms the AC magnetic field, a signal is measured at the outputs of the differential detection coils. The detection coil signal is amplified, demodulated, low-pass filtered and passed through a comparator. The comparator includes hysteresis in order to provide a stable switching point [19].

The IS1101 is a single-channel differential inductive position sensor suitable for counting on periodic targets like gears, cogwheels, racks, slotted discs, PCB-targets and for edge detection of metallic objects with larger dimensions. The target material can be either ferromagnetic (steel) or electrically conducting (aluminum, brass, copper). The sensor has 0 – 40 kHz frequency output and air gap up to 1.5 mm [20].

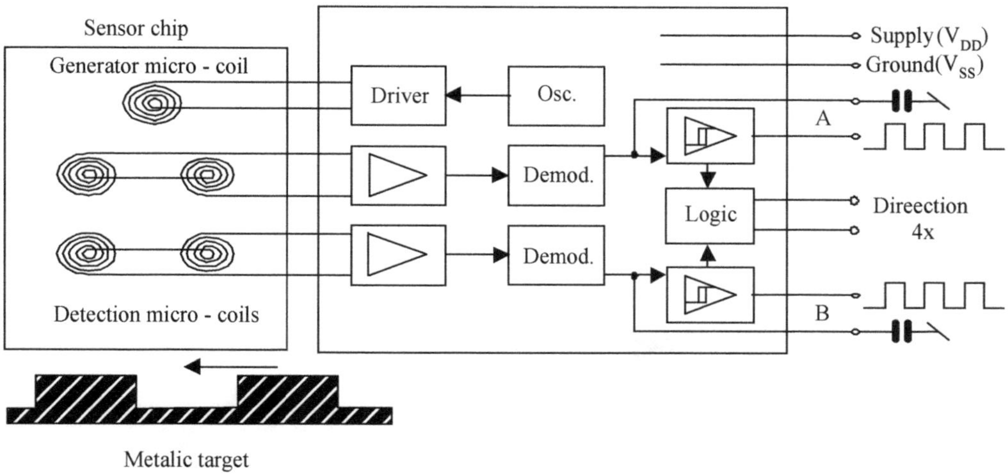

Fig. 11.5 PO2210 functional block diagram.

All these sensors can be easily connected to the UFDC-1 or USTI IC to produce a true digital output according to the RS232 interface, SPI or I²C serial buses. Due to minimum possible circuitry necessary for sensor output processing, the UFDC-1 and USTI ICs can be embedded into MEMS or SoC (Figure 11.6).

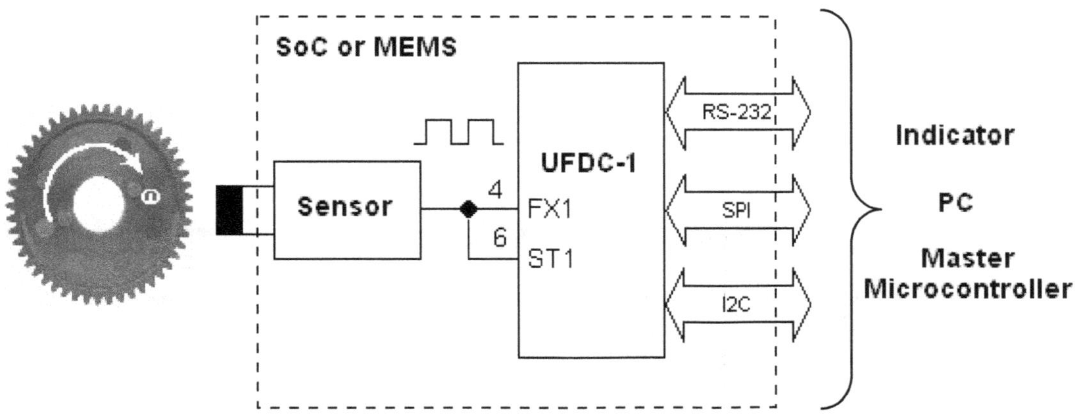

Fig. 11.6 SoC for rotation speed measurement.

One more interesting application based on the UFDC-1 or USTI ICs is self-adaptive intelligent rotational speed sensors. In such sensors, some parameters can be changed according to measurement conditions and algorithms. Therefore, due to programmable relative error and non-redundant conversion time of these ICs, rotational speed sensors on its basis can change accuracy for speed and opposite. This intelligent feature is very useful for Antilock Braking System (ABS). It lets select the minimum possible conversion time at critical rotational speed of wheels and maximum possible accuracy for the best engine optimization at the nominal rotational wheels speed. More details about such intelligent sensor systems are in Chapter 17.

There are many ways to display the rotational speed measurement results. Any terminal software for Windows can be used with the UFDC-1 and USTI ICs using the serial port of a PC. One of the possible choices is the Terminal V1.9b free software [21]. Other possibilities are the Component Works 2.0 (ActiveX control) or LabView Both software products are from the *National Instruments*.

An example of virtual specialized tachometer's GUI is shown in Figure 11.7 (a, b). Its software was written in Visual Basic 6.0 with the Component Works 2.0 (ActiveX control).

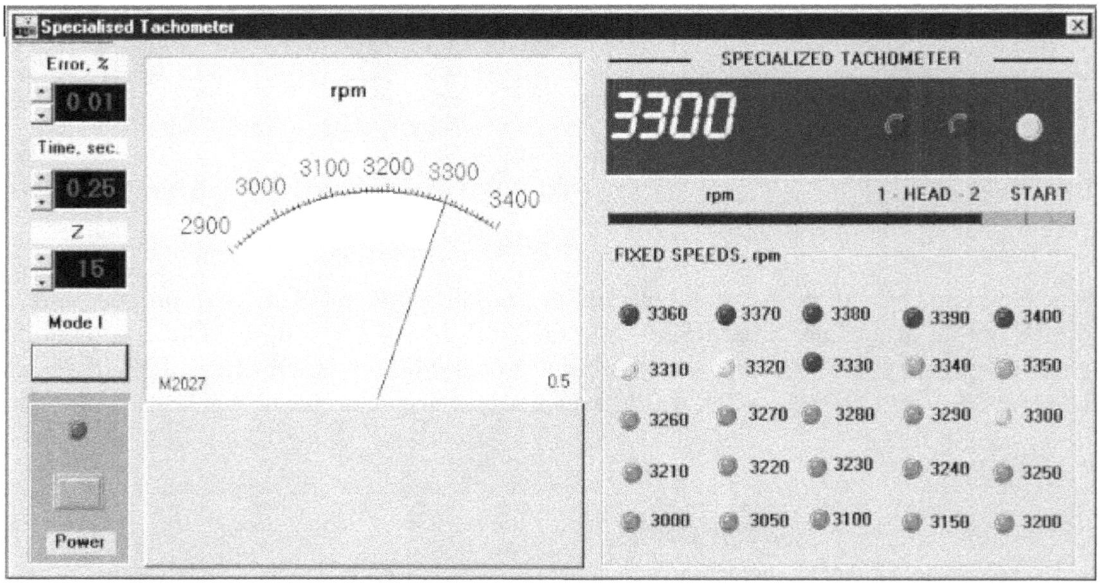

(a)

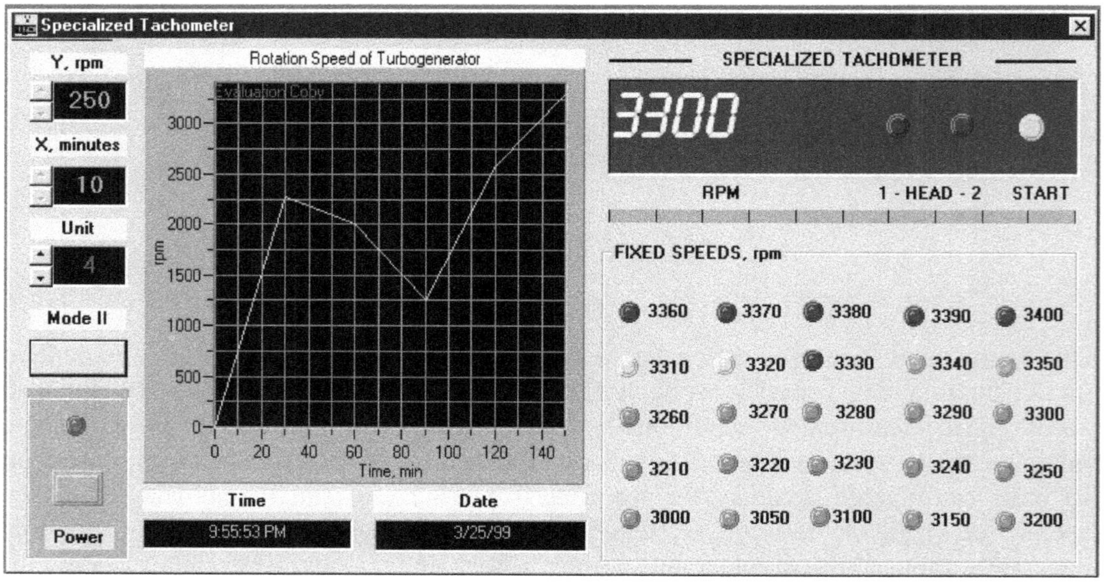

(b)

Fig. 11.7 Virtual tachometer's GUI example in the different working modes: circuit-breakers triggering tuning (a), and turbogenerator start.

The described design approach for a high precision, wide speed rotational sensing is based on the UFDC-1 and USTI ICs. It lets significantly simplify the design process, reduce time-to-market and production price and produce tachometers and tachometric systems with high metrological performances and digital self-adaptive intelligent rotational speed sensors.

In comparison with the direct microcontroller interfacing the proposed approach also lets to eliminate many design problems connected with the use of advanced methods for rotational speed measurement and SoC design.

11.3 References

[1]. List of Rotation Speed and Angular Position Sensors Manufacturers at Sensors Web Portal: http://www.sensorsportal.com/HTML/SENSORS/RotationSens_Manuf.htm

[2]. Kirianaki N. V., Yurish S. Y., Deynega V. P., Spak N. O., Data Acquisition and Signal Processing for Smart Sensors, *John Wiley & Sons*, Chichester, UK, 2002.

[3]. Akimitsu Ishihara, Shukuji Asakura, A Possible Motion Sensor with Frequency Output, in *Proceedings of the 18th Chemical Sensor Symposium*, 3-5 April, 1994, Tohoku University, Japan, Vol.10, Supplement A, pp.141-144.

[4]. KMI16/1 Integrated Rotational Speed Sensor, Data Sheets, *Philips Semiconductors* 15 May 1998.

[5]. KMI16/1 Integrated Rotational Speed Sensor, Data Sheets, *Philips Semiconductors*, 5 September 2000.

[6]. Ohmae T., Matsuda T., Kimiyama K., Tachikawa M. A microprocessor-controlled High-accuracy Wide-range Speed Regulator for Motor Drives *IEEE Transactions on Industrial Electronics*, Vol. IE-29, No.3, 1982, pp.207-211.

[7]. Bonet R. Design of High Performance Digital Tachometer with a Microcontroller, *IEEE Transactions on Instrumentation and Measurement*, Vol. 38, No. 6, 1989, pp.1104-1108.

[8]. Prokin M. Double Buffered Wide-Range Frequency Measurement Method for Digital Tachometers, *IEEE Transactions on Instrumentation and Measurement*, Vol. 40, No. 3, 1991, pp.606-610.

[9]. Prokin M. Extremely Wide-range Speed Measurement Using a Double-buffered Method, *IEEE Transactions on Industrial Electronics*, Vol. 41, No. 5, 1994, pp.550-559.

[10].Prokin M. Extremely Wide Speed Range Measurement System, *Review of Scientific Instruments*, Vol. 73, No.4, 2002, pp.1962-1964.

[11]. Prokin M. DMA Transfer Method for Wide-Range Speed and Frequency Measurement, *IEEE Transactions on Instrumentation and Measurement*, Vol. 42, No.4, 1993, pp. 842-846.

[12].Kirianaki N. V., Yurish S. Y., Shpak N. O., Advanced High-accuracy Wide-range Measurement Method of Dependent Count for Digital Tachometers and Tachometric Systems, in *Proceedings of the Joint IMEKO TC-1 & XXXIV MKM Conference*, Vol. III, Wroclaw, Poland, 2002, pp. 67-74.

[13].Yurish S. Y., Kirianaki N. V. Design of High Performances Digital Tachometers and Tachometric Systems Based on the Method of the Depended Count, in *Proceedings of the VII-th International Conference CADSM' 2003*, Lviv-Slavsko, Ukraine, 2003, pp. 378 – 381.

[14].Kirianaki N. V., Gayduchok R. M., Digital Measurements of Frequency-Time Parameters of Signals, *Vyshcha Shkola*, Lviv, 1978 (in Russian).
[15].Universal Frequency-to-Digital Converter (UFDC-1), Specification and Application Note, Technology *Assistance BCNA 2010, S. L.,* 2011.
[16].Universal Sensors and Transducers Interface (USTI), Specification and Application Note, *Technology Assistance BCNA 2010, S. L.,* 2011.
[17].Kirianaki N. V., Yurish S. Y., High Precision, Wide Speed Range Rotation Sensing with UFDC-1, *Sensors & Transducers*, Vol.59, Issue 9, September 2005, pp.426-431.
[18].Dual Channel Differential Inductive Position Sensor PO2210, Product Data, PO2210-PB-V1A, *Posic.*
[19].Differential Inductive Microcoil Sensor PO2211, Product Data, PO2211-PB-V1A, *Posic.*
[20].Inductive Position Sensor with Digital Output IS1101, Product Data, IS1101-PB-V1C, *Posic.*
[21].Terminal V1.9b: http://www.sensorsportal.com/DOWNLOADS/Terminal.exe

Chapter 12

Chemical Sensors and Biosensors

12.1 Quasi-Digital Chemical Sensors and Biosensors: State-of-the-Art

12.1.1 Chemical Sensors Review

Chemical sensors are becoming more and more important in any area where the measurement of concentrations of volatile compounds is relevant for both control and analytical purposes. They have also found many applications in sensor systems called electronic noses and tongues - arrays of sensors simulating the human olfactory and gustatory responses [1, 2]. Modern process control systems require accurate and precise chemical sensors, especially for assessing the quality of food, beverages, and pharmaceuticals. Specific detectors have been designed to filter out interfering substances and achieve a yes or no—detection of specific chemicals [3]. Tomorrow's chemical sensing needs will require more than the stand-alone sensors and signal conditioning circuits currently offered by many of today's sensor manufacturers. Customers are asking for turnkey chemical sensing systems and sensors that can be easily designed into their consumer gas detectors or OEM gas sensing products [4].

Micro-electro-mechanical systems (MEMS) and thin film technologies enable the integration of microelectronics circuits and multifunctional sensor arrays fabricated on silicon substrates for detecting chemical, mechanical, and physical parameters. Such integration offers new possibilities to decrease the cross-links between adjacent sensors [5].

Microfabricated chemical sensors offer greatly reduced power consumption, low operating voltage that is compatible with low-power CMOS devices, and rapid response time- all in a very small package size [4].

In the last ten-years, there has been much progress also in the area of chemical sensors, particularly in the sensitivity or detection limit. For example, detection limits down to single molecules have been demonstrated using different technologies [6]. However, in order to use sensors in the real-world environment outside of laboratories, many challenges remain. Among them, sensor saturation that limits the dynamic range, slow response time due to mass transport, and overexposure leading to incapacity of the sensor are a few common limitations [6].

According to new report by *Global Industry Analysts, Inc.* the world market for chemical sensors is projected to reach $17.28 billion by the year 2015 [7]. Demand for chemical sensors will be fuelled by newer applications for established sensor products; product innovation through the use of microfabrication and other high tech manufacturing techniques; and falling cost of high-performance sensor types [7].

Demand for chemical sensors in the US is forecast to increase 8.6 percent annually to more than $6 billion in 2014. Growth will be supported by continued strong demand for biosensors. Technological advances are resulting in unit price reductions, and have resulted in miniaturization and increased precision that allow new and wider applications. In the industrial market, demand for oxygen sensors for new vehicles is expected to grow, reflecting recovery in new automobile production. Continued regulation for environmental monitoring will also sustain growth in demand for chemical sensors through 2014 [8].

In the last decade, many quasi-digital chemical sensors were designed and described in the literature. Some interesting solutions are described briefly below.

An academic/industrial UK LINK project is currently being developing the handheld electronic nose (H^2EN) - an array of sensors simulating the human olfactory response. The design approach described in [9] places the resistive sensor in the feedback loop of a 'digital' RC oscillator the output of which is a square wave with a frequency inversely proportional to the sensor resistance.

Quartz crystal microbalance (QCM), surface- and bulk-acoustic waves based chemical sensors and biosensors as usually have frequency outputs. One of such solid-state SAW sensors is described in [2]. This sensor is an uncoated, high-Q piezoelectric quartz crystal with the natural resonating frequency of 500 MHz. Its surface temperature is controlled by a small thermoelectric element that cools the surface to promote the vapour condensation and then heats it for cleaning between analyses. The added mass of the analyte condensing on the crystal's surface lowers the vibration frequency in direct proportion to the amount of the condensate. This frequency is mixed with a reference frequency, and the intermediate frequency (typically 100 kHz) is counted by a microcontroller or universal ICs such as the UFDC or USTI [10, 11].

The acoustic gas sensor is described in [12]. In this sensor, the sound velocity is continuously measured with the high resolution in a gas-filled cell, by controlling the frequency of an oscillator via the transit time of the sound between an ultrasound transmitter and a receiver element. The microcontroller or the same universal ICs [10, 11] can handle frequency input signals directly, and also provides analog output signals by means of the pulse width modulation.

Chemical signal domain sensors can be used not only for measuring different chemical quantities and compositions of mixed gases, but also for measuring such a "non-chemical" quantity like the rotation speed [13], for which measuring the electrical and magnetic signal domain sensors are usually used. Electrochemical oscillations have often been observed in iron immersed in the solution containing phosphoric acid and hydrogen peroxide. This oscillation has been interpreted by the cross linkage between electrochemical reactions and mass transport processes in the vicinity of an electrode. Therefore, the oscillation frequency is expected to reflect the flow rate around the electrode. A rod of carbon steel S50C was mounted in the rotating disc apparatus. The intersection of the rod rotated concentrically. In the mixture of 0.5M phosphoric acid and 1.5M hydrogen peroxide the highly stable electrochemical oscillation appeared. The amplitude of the potential change was about 600 mV_{p-p}. The oscillating frequency was a sensitive function of the rotation rate in the range of 100 to 2 000 rpm [13].

More references and web links are available in the Chemical Sensors web section at Sensors Web Portal (http://www.sensorsportal.com) [14].

12.1.2 Biosensors Review

A biosensor is an analytical device which converts a biological response into an electrical signal. The term 'biosensor' is often used to cover sensor devices used in order to determine the concentration of substances and other parameters of biological interest.

Biosensors have many advantages, such as simple and low-cost instrumentation, fast response times, minimum sample pretreatment, and high sample throughput. Although biosensors are beginning to move toward field testing and commercialization in the United States, Europe, and Japan, relatively few have been commercialized. Increased research in this area demands the development of novel materials, new and better analytical techniques, and new and improved biosensors [15]. Some potential applications of biosensors are agricultural, horticultural and veterinary analysis; water and microbial contamination analysis; clinical diagnosis and biomedical applications; fermentation analysis and control.

According to a new report by *Global Industry Analysts, Inc.* global biosensors market to reach US $ 12 Billion by 2015 [16]. Expanding medical applications, increase in R&D activities, and emergence of new technologies such as microfluidics, and non-invasive biosensing alternatives are the other market propellers. In addition, key factors driving growth for glucose biosensors include growing diabetic population, user-friendly designs, and increased point-of-care applications [16].

The Analytical Review of World Biosensors Market form the *Frost & Sullivan* also reports the biosensors market growing to $ 14.42 billion in 2016 [17].

Research conducted by *Technavio* reveals that the Global Medical Biosensors Market is expected to grow at a CAGR of 12.6 percent during 2010-2014. The modern market is currently driven by growing integration of various technologies such as nanotechnology, microbiology etc., among other factors [18]. "With integration being the new buzz word in medical technology, medical biosensors are finding more and more applications. These medical devices have been innovating various technologies such lab-on-a-chip and micro electro mechanical systems", reports Technavio analyst.

As it was reported by the *BCC Research* the largest segment of the Global Microsensors Market, biochips, will reach $1.8 billion by the end of 2013, for a CAGR of 25.6 % and biochips market share is expected to increase from 17.2 % in 2007 to 21.6 % in 2013 [19].

Speaking about the USA biosensors market the demand for biosensors is projected to increase to $4.4 billion by 2014. The security and biodefense markets will make increasing use of biosensors, again supported by governmental regulation and the need for public safety [8].

During the past five years, biosensors have penetrated diverse markets. Biosensor companies have adopted microchip technology (small semiconductor-based crystal chips embedded with an IC capable of carrying out electronic functions) to reduce the size of their products [20]. A number of companies are trying to develop a single platform that can be used for multiple applications.

The technological trends in biosensors lead an appearance of so called Micro-Nano Bio Systems (MNBS) - smart systems combining microsensing and micoactuation, microelectronics, nano-materials, molecular biology, biochemistry, measurement, information and communication technologies [21], which will be able to:

- Increased intelligence (computation/decision power, sensing capabilities);
- Enhanced miniaturization and integration of devices and systems;
- Increased integration of bioactive components (molecular and cellular components, bio/nanochemistry) as well as processes.

The novel generation of MNBS must be:

- Smaller;
- Perform better;
- Faster;
- Cheaper;
- Delivering highly reproducible results;
- Increased sensitivity;
- Extremely and proven reliable.

Biosensor systems (lab-on-chip systems) awareness and use are limited by a number of factors among these are: miniaturization in biosensors poses technical challenges; some biosensors are still too expensive for commercial production. Fortunately, to eliminate these limitations and design a novel generation of MNBS is possible due to creation an integration platform based on quasi-digital (frequency-time) sensor signal domain, which is coming to change traditional analog (voltage and current) sensor signal domain.

In the last twenty years, many quasi-digital biosensors and biosensor systems were designed and described in the literature. For example, the data acquisition system for electrical conductivity images of biological tissues via contactless measurements is developed and described in [22]. Two different measurement systems are designed based upon measurement of impedance change of an eddy current type sensor. The first design (based upon excitation of a single coil probe) gives 100 Hz output frequency change for a conductivity change of 1.9 mS/cm and is tested at high frequency (1.4 MHz). Second design (based on one excitation and one receiving coil) operates at low frequency (15 kHz) and gives 18 mV change in output voltage for similar conductivity variation [22].

The modulator circuit based on voltage-to-frequency converter presented in [23] was designed to provide a simple and inexpensive portable means for recording bio-electric (typically electrocardiogram) signals. Such low-level (1 mV) bio-electric signals are characterized by low-frequency bandwidths (approximately 1 Hz to 100 Hz). The modulator circuit produces a frequency deviation of 200 Hz p-p.

The piezoelectric immunosensor for antigens detection in the picogram range is described in [15]. The change in frequency (ΔF) for 10 MHz AT-cut quartz crystal was from 2.5 to 22.5 kHz.

The MEMS-based smart pulse-coded biosensor for detection of determined levels of fluids or gases has been developed in University of Maryland, USA [24]. The sensor activation feeds a bridge which when in balance can be calibrated to designate levels of concentration of the fluid or gas components of interest. The bridge output can in turn be taken into a neural network which will form the smart portion of the system. The envisaged neural network is a bioinspired pulse- coded one based upon the neural-type cells and neural-type junctions. Inclusion of the neural network will allow the sensor to be used for real time monitoring and application to situations where feedback may be used to adjust for administering desired doses, for example of glucose [24].

More references and web links are available in the Biosensors web section at Sensors Web Portal (http://www.sensorsportal.com) [25].

12.2 Digital Sensor Systems Design Based on Quartz Crystal Microbalance and Other Chemical Principles

12.2.1 QCM-based Sensor Systems

Biosensors are chemical sensors that take advantage of the high selectivity and sensitivity of a biologically active material. It is well known that the resonant frequency of an oscillating piezoelectric crystal can be affected by a change in mass at the crystal surface. Piezoelectric immunosensors are able to measure a small change in mass [15].

Electrical signals from chemical and bio- transducers are often low and superimposed upon a relatively high and noisy (i.e. containing a high frequency signal component of an apparently random nature, due to electrical interference or generated within the electronic components of the transducer) baseline. The signal processing normally involves subtracting a 'reference' baseline signal, derived from a similar transducer without any biocatalytic membrane, from the sample signal, amplifying the resultant signal difference and electronically filtering (smoothing) out the unwanted signal noise. The relatively slow nature of the biosensor response considerably eases the problem of electrical noise filtration. The analogue signal produced at this stage may be output directly but is usually converted to a digital signal and passed to a microcontroller where the data is processed, converted to appropriate units and output to a display device or data store. However, it is more preferably to have a quasi-digital output instead of analog output in order to eliminate many readout and signal conditioning design problems.

The method automatically adjusts sampling duration according to the sensors' response to the analytes and converts the amplitude of the sensor output to a frequency output, giving us another opportunity to reduce noise in the signal [6]. The new system also provides us with a way to convert amplitude output of a sensor into frequency output, potentially can be exploited in noise reduction, and provides us with a new way to extract concentration and possibly other kinetic information24 of the interaction between the sensor element and the interested analyte [6]. In addition, the frequency-coded detection method automatically performs multiple samplings, enabling averaging and thus reducing noise/false alarms. It also reduces accidental damage to the sensor due to operator's error [6].

In the case of a sensor whose output is a quasi-digital, the representations of the output responses may be as follows: f (frequency); f/f_0 (relative frequency); $(f - f_0)$ (frequency deviation); $(f - f_0)/f_0$ (relative frequency deviation) [1].

Many modern "electronic noses", "tongues" and various precision applications for analytical chemistry are based on different chemical, mass variation sensors: quartz crystal microbalance (QCM), superficial SAW and bulk (BAW) acoustic wave sensors [9, 26-28]. All of these are built-up on a quartz piezoelectric substrate. Most gravimetric biosensors and immunobiosensors also use piezoelectric quartz crystals either as resonating crystals, or as bulk/surface acoustic wave devices [29-32].

The basic equations describing the relationship between the resonant frequency of an oscillating piezoelectric crystal and the mass deposited on the crystal surface have been derived by Sauerbrey, Stockridge, and Lostis [15]. Each followed a different path, but their final equations are similar, the Sauerbrey equation being the most widely accepted.

For quartz piezoelectric crystal (AT-cut) the crystal resonance frequency changes by Δf when a mass change Δm occurs on the crystal according to the Sauerbrey equation [30]:

$$\Delta f = -2 \cdot f_0^{\,2} \frac{\Delta m}{A \cdot \left(\mu_q \cdot \rho_q\right)^{1/2}} = -k \cdot \Delta m, \tag{12.1}$$

where f_0 is the fundamental frequency of the quartz crystal, Δm is the mass change, A is the piezoelectrically active area, ρ_q is the density of quartz (2.648 $g \cdot cm^{-3}$), μ_q is the shear modulus (2.947×10^{11} $dyn \cdot cm^{-2}$ for an AT-cut quartz) and k is the constant, which is determined by the quartz crystal type. The change in resonant frequency (Δf) in Hz is directly proportional to the mass adsorbed on the crystal surface. As usually, a crystal has a characteristic resonance frequency (f_0) in the MHz range, whereas a deposition causes a shift (Δf) of the resonance frequency up to a few hundred of ppm [26].

There are several methods and systems for the sensor resonance frequency shift measurement, reported in literature. E.g. the methods of impedance analyze, mentioned in [26] based on calculation of the response frequency variation on the basis of the impedance variation by mean of a vectorial impedance meter; the method of impedance analyze at which the amplitude-frequency characteristic is determined that is shifted when the resonance frequency is shifted [33]. An advance of such methods is a possibility to measure also a quartz crystal quality factor together with the resonance frequency. The main disadvantage is complex interface electronics. There is also a method of a frequency variation measurement via analog phase measurements [26] with the same demerit. Some other methods are based on digital signal processing (DSP) [33]. For example:

1) Methods based on spectral function analysis with the help of fast Fourier transform (FFT). The advantage of this method is a simple realization based on a DSP microprocessor, a disadvantage – low accuracy;
2) Most likelihood methods based on a signal approximation by the sine curve set or by the use of rejection filters. These methods have a high accuracy of frequency estimation but they need complex calculations;
3) Methods based on analyze of auto-correlation matrix of signal. Such method has a high accuracy, but there are strong requirements to the signal duration;
4) Interpolated FFT based methods, which also have a demerit – low conversion speed due to the necessity of some hundred periods processing.

Most suitable methods for the sensor resonance frequency shift measurement are so-called active (or self-excited oscillator) methods. A QCM sensor is used in a high frequency generator circuit as a stabilization element. The frequency of generator is changing according to the sensor resonance frequency [33]. Main difficulties of data acquisition systems design for such QCM sensors are high frequency signal digitizing, which require expensive ADCs. It is necessary, at first, to shift a high frequency signal to low (some kHz) frequency range. An example of such system is described in [9] and shown in Figure 12.1 This circuit solution is relatively complex and requires an ADC, DSP microprocessor, for example, ADS21XX from Analog Devices, and specialized custom made software for digital filtering, interpolation FFT, etc. [33].

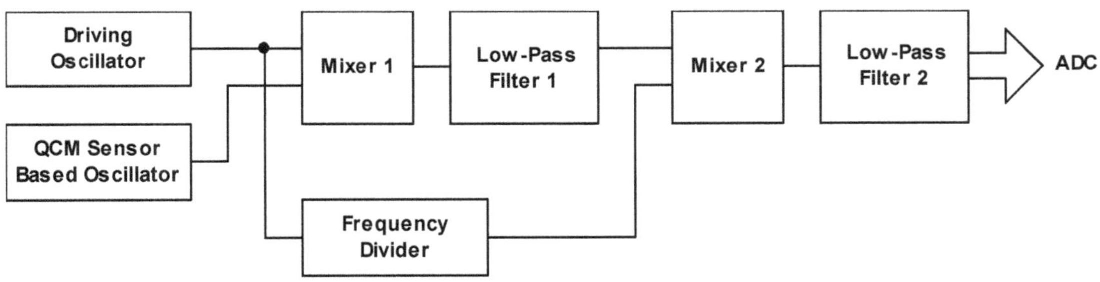

Fig. 12.1 Differential self-excited QCM sensor for liquids [33].

In addition, there are some limitations for the input signal and its frequency range. These mean that the frequency range is more or less determined, and measurand frequency is inside this frequency range. In other words, the frequency range, which contains a precise frequency value, should be known before [33]. As it was reported in [33] the measuring system based on the described method and circuit solution lets receive the absolute error ± 0.1 Hz at measurement of 1754 Hz. This corresponds to ± 0.0057 % relative error. The time of measurement is 0.1 s. However, as it was shown in [26], the frequency shift estimation in modern measurement systems for QCM sensors and other resonator-based bio- and chemical sensors must be performed in a reduced measurement time (less than 1/10 s).

The fully digital measuring system for study of the dynamic response of QCM sensors is described in [26]. The frequency shift is measured by a dedicated fully digital front-end hardware implementing a technique proposed in [34]. However, the system has a low accuracy with an uncertainty of 5 Hz in a range of 30 kHz that corresponds to ± 0.017 % relative error. The measurement time for this system is 20 ms × 10 = 200 ms because of frequency shifts are measured each 20 ms, 10 successive measurements are averaged to obtain a frequency sample. The uncertainty can be reduced only by increasing the measurement time via software programming [26] that is in contravention with the requirements for modern measuring and data acquisition systems for QCM sensors. The used method critical issue is also to have an almost uniform distribution of the initial phases of the streams. Some sufficient conditions to ensure the uniform distribution of the initial phases are provided in [34], and one of these is a specific relationship between the sampling frequency and measured frequency. But this relationship can be achieved only for the nominal frequency. As the frequency shift, even of about a few Hz from the nominal frequency, this condition is no more fulfilled, and the measurement system performance gets worse. In this case the error committed by the system is increased.

Another frequency measurement technique for quartz microbalance systems is described in [35]. It uses a period counting based on a D flip-flop mixer, directly applicable to quartz microbalances and quartz resonator-based sensors. The architecture of integrated period counting system is based on the indirect (period) measurement of a low frequency difference signal obtained by subtracting a high stability reference frequency from the unknown frequency to be measured. A digital mixer (D flip-flop based) circuit is used to obtain frequency differencing, eliminating the need for an analog mixer and low-pass filters. Using a method of reducing the risk of misclocking in the digital mixer the expected error of such system can be 0.06 ppm. But this measure leads to significant input frequency range narrowing (up to 7.006 ... 7.030 MHz) because of it needs a proper selection of the minimum difference of input frequencies that can reduce the probability of misclocking events to insignificance.

Very often the standard desktop frequency counters are used in QCM sensors-based systems in order to measure a resonance frequency [27]. As usually frequency counter-based systems has absolute error ± 1 Hz at 1 s measuring time. Such solution is expensive, needs complex automatization for experiment and is not suitable for industrial systems of analytic chemistry.

Based on the state-of-the-art review of frequency measurement and data acquisition systems for QCM and other resonator-based bio- and chemical sensors it is possible to formulate main requirements to various modern gravimetric analyze-based systems. Firstly, all such systems must have a low relative error of measurement (less than ± 0.001 %) constant in all frequency range, and high resolution. Secondly, the systems must have a possibility to measure both: resonance frequency (sometimes more than the reference frequency) and quartz crystal quality factor Q; to determine a time constant for chemical transient and different dynamic behaviors; and to exploit a kinetic for adsorption processes.

All these require the reduced time of measurement (less than 0.1 s). Such systems must have a wide frequency range without any restrictions for the input signal and let directly convert a frequency shift to digital without a preliminary conversion the unknown frequency to low frequency range. It should be also inexpensive and have simple interface electronics. At first sight some of requirements seem to be in contradiction but such problem was solved by the use of novel frequency measurement technique and integrated circuit on its basis.

The quartz microbalance system's efficiency is determined by the method for frequency measurements and device for its realization. For a high performance frequency measurement system designed for different QCM and other resonator-based bio- and chemical sensors it is expediently to use UFDC and USTI ICs based on the advanced digital method for frequency shift measurements [10, 11]. These novel ICs lets to measure a frequency shift Δf_x and a frequency deviation (absolute and relative). It means simple addition calculations according to the common equation [36]:

$$R = \frac{a \cdot f_x \pm b}{c}, \tag{12.2}$$

where *a, b* and *c* are constants, which should be set up in the same unit as a measurand. So, at frequency deviation measurement, a =0, c=0 and

$$\Delta f_x = f_x - b = \frac{n}{N} \cdot f_0 - b, \tag{12.3}$$

where the constant *b* should be the frequency of reference quartz resonator in the case of QCM systems.

A data acquisition system for differential self-excited QCM sensors is shown in Figure 12.2 [37, 38].

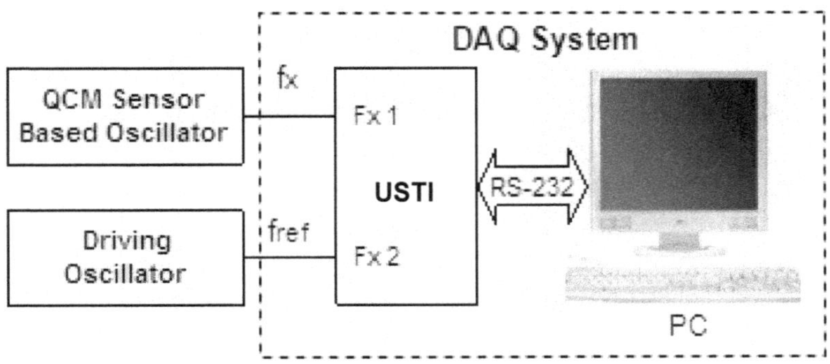

Fig. 12.2 DAQ system for differential self-excited QCM sensors.

It is based on the designed Universal Sensors and Transducers Interface (USTI) IC [11]. In comparison with the UFDC-1 [10] and UFDC-1M-16 [39], which is produced since 2004, the new integrated circuit has a special measuring mode for frequency deviation (absolute and relative) and extended frequency range up to 9 MHz (or 144 MHz with prescaling). It is based on the modified method of the dependent count and can measure frequency shift and deviation with the programmable relative error from ±1 to ±0.0005 % and non-redundant measurement time. The quantization error is not dependent on the measurand frequency. The relative error is constant in the whole frequency range and less than programmable relative error $\delta_x<\delta$. The USTI has a quartz-accurate automated calibration mode; RS232 (422), SPI and I^{2}C interfaces; master and slave communication modes. The internal reference frequency is f_0 = 625 kHz for a basic IC version and 16 MHz for a high speed IC modification, for example, USTI-EXT. Intelligent USTI's features include a self-adaptation possibility (a flexibility to change accuracy for speed and opposite during each of measurement). Custom designed versions of USTI can also store an IEEE 1451 Transducer Electronic Data Sheets (TEDS) in its flash memory with the aim to simplify a QCM sensor configuration in a different systems, and self-identification purpose.

In data acquisition systems based on the USTI any differential QCM or resonator-based sensor can be connected to the first IC's channel, and the driving oscillator – to the second channel. For such connections, the USTI should be used in a frequency difference measurement mode (f_{x1}-f_{x2}) with a sign determination. In this case f_x=f_{x1} and f_{ref}=f_{x2}. The sign "minus" before the result of measurement testifies that $f_{x1} < f_{x2}$. The mode for frequency difference measurement with prescaling (external divider by 16) is also possible for frequencies exceed 9 MHz.

Data exchange between the measuring system and PC computer can be established by the serial interface RS232 and any terminal window software, for example, freeware software Terminal v.1.9b for Windows [39]. Naturally, the LabVIEW software from National Instruments (USA) or similar can be also used in such systems to establish a communication between the USTI and PC computer as well as design different virtual instruments for QCM or any resonator-based sensors. Appropriate commands (at RS232 interface communication), which should be typed in the terminal window for the USTI working in the frequency shift measuring mode are shown in Figure 12.3.

```
>M06          ; Frequency difference measurement initialization
>A0A          ; 0.0005 % conversion relative error set up
>S            ; Start a measurement
>C            ; Check result status: returns 'r' if ready or 'b' if busy
>R            ; Read a result

7054.07537    ; Measurement result indication
```

Fig. 12.3 USTI commands for working in the frequency difference measuring mode (f_{x1}-f_{x2}).

In order to get more precision measuring results it is expediently to use a single frequency measuring mode (M00) for the f_x and f_{ref} and then calculate the frequency difference by an external microcontroller or PC.

Based on the received frequency shift measurement result a computer should calculate the appropriate mass and volume of chemical agent. The DAQ board prototype based on the USTI IC is shown in Figure 3.18.

In order to measure a frequency shift with the programmable relative error batter than ±0.01 %, the USTI should be calibrated before, or 1 ppm quartz crystal should be used. The IC provides so-called in-circuit quartz-accurate automated calibration [40]. The calibration is made by means of the communication interface RS232. The calibration sequence for the USTI is the following. The test command "T" toggles the IC to the calibration mode. A sequence of rectangular pulses with frequency $f_{osc}/2$ is generated at the TEST output. The frequency of this pulse sequence should be measured with the help of an external frequency counter with accuracy not worse than ±0.00005 % or, at least, ±0.0001%. The command "Fi.f" will put the clock constant value into the USTI. The calibration should be performed only once. The correction factor Δ is stored in the nonvolatile memory of the IC, which holds this value after power-off.

In some measuring systems instead of the driving oscillator, the numerical reference value of its frequency f_{ref} can be used. It this case, the USTI should be working in the absolute frequency deviation measuring mode to measure the frequency shift $\Delta f = f_{x1} - f_{ref}$. An example of such system is shown in Figure 12.4 [37. 38].

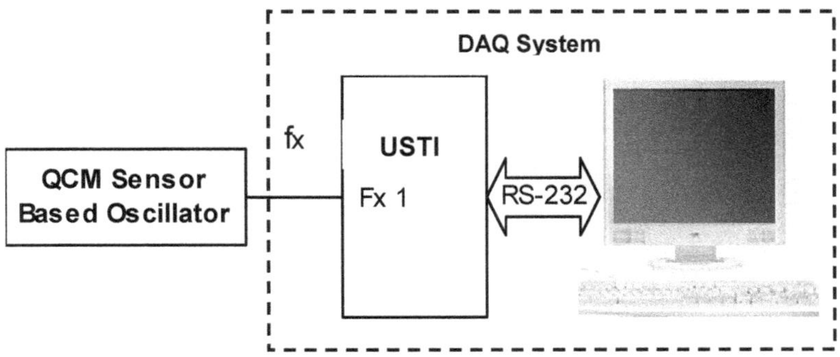

Fig. 12.4 Data acquisition system for QCM sensor [37, 38].

Appropriate commands for the USTI working in the absolute frequency deviation measuring mode are shown in Figure 12.5.

```
>M13             ; Frequency deviation measurement initialization in the 1st channel
>A00             ; Absolute deviation measurement with 5×10^-4 % relative error
>E7000000.34     ; Set up the reference frequency f_ref (Hz)
>S               ; Start measurement
>C               ; Check result status: returns 'r' if ready or 'b' if busy
>R               ; Read result

6000.7824        ; Measurement result indication
```

Fig. 12.5 USTI commands for working in the absolute frequency deviation measuring mode $(f_{x1}-f_{ref})$.

Similarly described above, in order to get more precision measuring results it is expediently to use a single frequency measuring mode (M00) for the f_x and then calculate the frequency difference (f_{x1}-f_{ref}) by an external microcontroller or PC.

12.2.2 Modeling and Experimental Results

The modeling results for time of measurement at relative error of measurement $\delta \in [0.01 \div 0.000005]$, reference frequency f_0 = 625 kHz and f_0 = 16 MHz, in measuring frequency range $f_x \in [0.8 \div 9]$ MHz are shown in Figure 12.6 (a, b) accordingly. As it is visible from the plot, time of measurement does not depend on the measurand frequency f_x and frequency difference in the high frequency range. It is non-redundant and mainly determined by the programmable relative error of measurement δ.

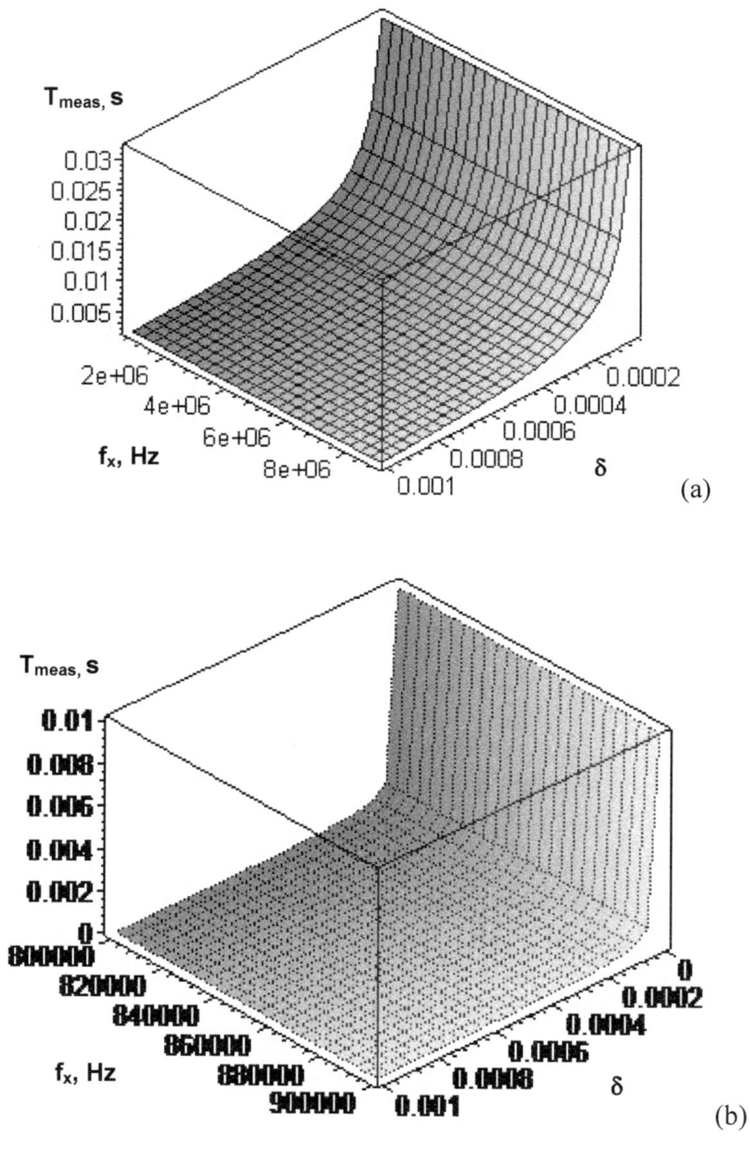

Fig. 12.6 The modeling results for time of measurement $T_{meas} = f(f_x, \delta)$ at f_0=625 kHz (a) and f_0=20 MHz (b).

The objective of experimental investigations was to determinate metrological performances for the USTI at precise frequency measurements in two channels correspond to measurements for differential self-excited QCM sensors. The measurement set up was the same as in is shown in Figure 3.24.

A square waveform pulse signal with frequency f_{x1}=9 000 000. 211 Hz corresponds to a driving oscillator's fundamental frequency was generated by the first functional generator Agilent 33220A and measured in the 2nd cannel FX_2 of USTI; and square waveform pulse signal with frequency f_{x2}= 8 999 800. 078 Hz corresponds to a QCM sensor based oscillator change in resonant frequency ($\Delta f = f_{x1} - f_{x2}$) in Hz is directly proportional to the mass adsorbed on the crystal surface was generated by the second functional generator Agilent 33220A and measured in the 1st cannel FX_1 of USTI. The IC was programmed to measure frequency with the minimal possible relative error ±0.0005 %. The supply voltage of the evaluation board was +5V dc, provided by the Promax FAC-363B power supply. Both frequencies were controlled by a frequency counter Agilent 53132A with ultra high oven stability internal time base. The two-channel digital oscilloscope Promax OD-571 monitored the signals waveforms. Before measurements, the USTI was calibrated in the working temperature range: +22.7 ... + 23.9 °C.

The measured values were sent to a PC via the RS232 (USB) interface implemented with the ST202D IC. The user interface was realized with the help of terminal software Terminal V1.9b for Windows. Each of measurements was taken until totaling 32 values. The measurement errors were evaluated from appropriate statistics. Results of frequency measurements for driving and QCM sensor based oscillator are shown in Figure 12.7. The statistical characteristics are adduced in Table 12.1 and appropriate densities of distributions are shown in Figure 12.8 (a, b). As it is visible from the table, the maximal relative error does not exceed the programmable $\delta_x < \pm$ 0.0005 %.

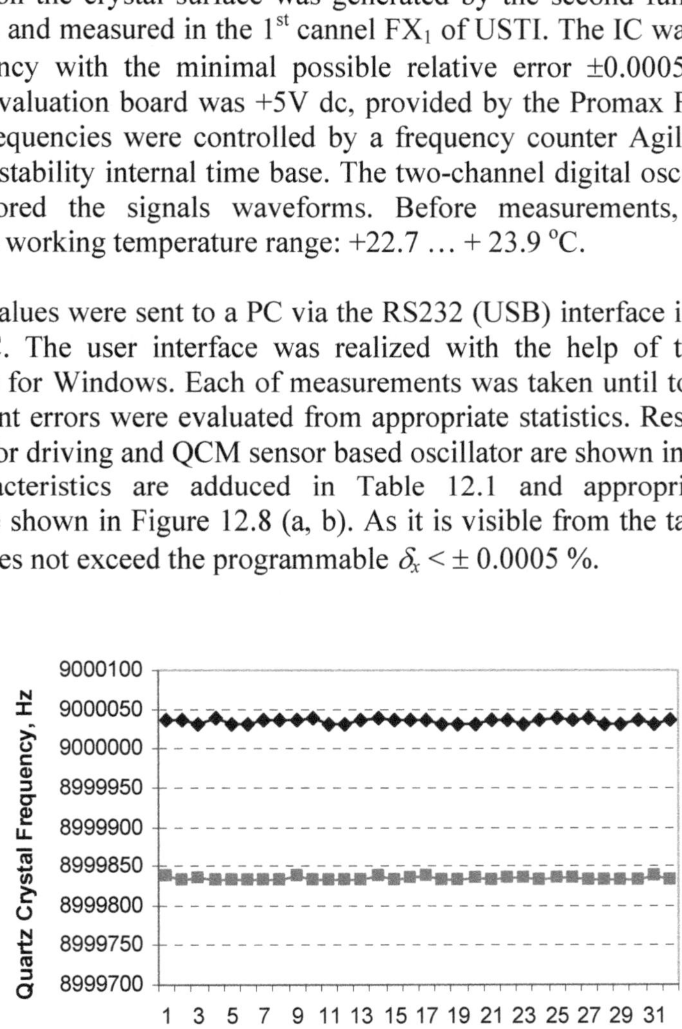

♦ - Driving oscillator; ■ - QCM sensor based oscillator

Fig. 12.7 Measurements results for driving and QCM sensor based oscillator.

In comparison with the frequency measurement system described in [33], the designed system at the same relative error of measurement $\delta = \pm 0.005$ % and frequency range lets receive the time of measurement 32 ms in comparison with 100 ms reported in [33]. In comparison with the frequency measurement system based on the all-digital frequency detector described in [26], the designed system at the same relative error of measurement $\delta = \pm 0.01$ % and frequency range lets receive the time of measurement 16 ms in comparison with 20×10=200 ms reported in [26], where frequency shifts are measured each 20 ms and 10 successive measurements are averaged to obtain a frequency sample.

Table 12.1 Statistical Characteristics.

Parameter	Value	
	QCM Sensor	Driving Oscillator
Number of measurements, N	32	32
Minimum f_x (min), Hz	9000032.2	8999832.15
Maximum f_x (max), Hz	9000038.45	8999838.4
Sampling range, f_x (max)- f_x (min), Hz	6.2515	6.2515
Median	0	0
Arithmetic mean, Hz	9000034.64	8999833.81
Variance	4.8753	5.6633
Standard deviation	2.208	2.3798
Coefficient of variation	4076067.12	3781815.29
Confidence interval at probability $P = 97$ %	$f_x \in$ [9000033.79÷9000035.49]	$f_x \in$ [8999832.9÷ 8999834.72]
Relative error, %	0.00036 < 0.00050	0.00043 < 0.00050

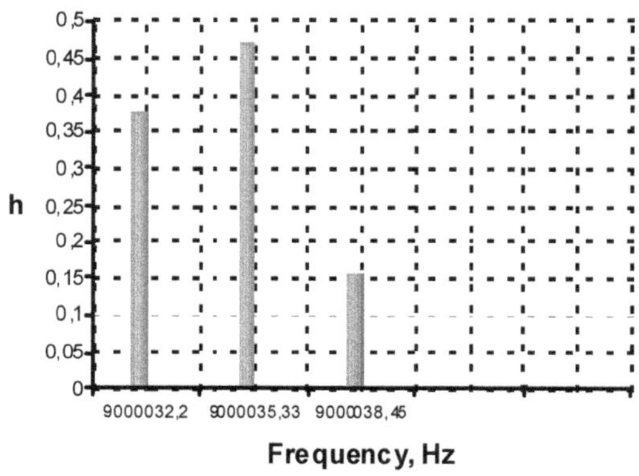

a)

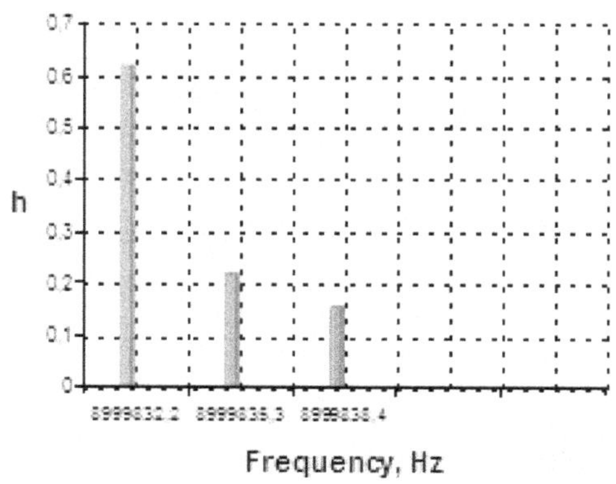

b)

Fig. 12.8 Density of distribution for driving (a) and QCM sensor based (b) oscillators.

The same time of measurement has also the integrated frequency measurement system based on the FPGA XC3142A from Xilinx [41]. It uses the classical indirect method for period measurement. This system was set up to take 250 periods for the measurement, which is close to the maximum allowed at the oscillator frequency of 1.23 kHz this equates to a measurement time of around 200 ms. The theoretical error in this system is significantly lower than the experimental results due to the noise in the system, which needs to be improved. Such property belongs to any system based on the period measurements.

In comparison with the frequency measurement technique described in [35], the strict requirements to the input signal, which limited its frequency range, are absent in the proposed measuring system. The frequency range can be from 0.05 Hz to 9 MHz (without prescaling) at constant quantization error in all frequency range, which does not exceed the programmable relative error of measurement and can be chosen, for example, ± 0.0005 % that corresponds to 5 ppm.

All these made the proposed frequency measurement system very suitable for QCM or resonator-based sensor interfacing in liquids and gases. Due to its non-redundant and reduced time of measurement the developed system can be used for a chemical transient time constant and different dynamic behaviors determination in order to exploit a kinetic for adsorption processes. Such kind of low-cost frequency measuring systems are based on the USTI can be used in full system-on-chip electronic noses and tongues, lab-on-chip and MNBS.

This approach has eliminated much of the hardware associated with most measuring systems for QCM sensors; there is no need for a microcontroller in the system and complex sensor interfacing electronics as it is directly compatible with an IBM PC compatible computers and use only one IC.

The data acquisition system for QCM and other (superficial SAW and bulk acoustic wave) resonator-based chemical and biosensors described in this article is based on the novel frequency measurement technique using the modified method of the dependent count and allows obtaining high metrological performances: programmable, constant in all frequency range, relative error of measurement up to ±0.0005 % that corresponds to 5 ppm; non-redundant, reduced conversion time of measurement and wide frequency range from 0.05 Hz to 9 MHz (144 MHz with prescaling). Due to all these the developed system can be used for a chemical transient time constant and different dynamic behaviors determination in order to exploit a kinetic for adsorption processes.

The system satisfies to all modern requirements and can measure both resonance frequency (sometimes more than the reference frequency) and quartz crystal quality factor Q for QCM sensors working, for example, in liquids and gases.

This design approach has eliminated much of the hardware associated with most measuring systems for QCM sensors; there is no need for a microcontroller in the system and complex sensor interfacing electronics as it is directly compatible with a PC.

12.2.3 Other Applications

The Universal Frequency-to-Digital Converter (UFDC-1) IC also can be used in different chemical sensors and biosensors based on various sensing principle. Here is two interesting examples for application in food industry and live farming.

A hand-held electronic tongue was developed for determining taste levels of astringency and umami in tea infusions in the National Taiwan University [42]. The sensing principles are based on quenching the fluorescence of 3-aminophthalate by tannin, and the fluorogenic reaction of o-phthalaldehyde (OPA) with amino acids to determine astringency and umami levels, respectively. Both reactions were measured by a single fluorescence sensing system with same excitation and emission wavelengths (340/425 nm). Functional block diagram of the hand-held fluorometer is shown in Figure 12.9.

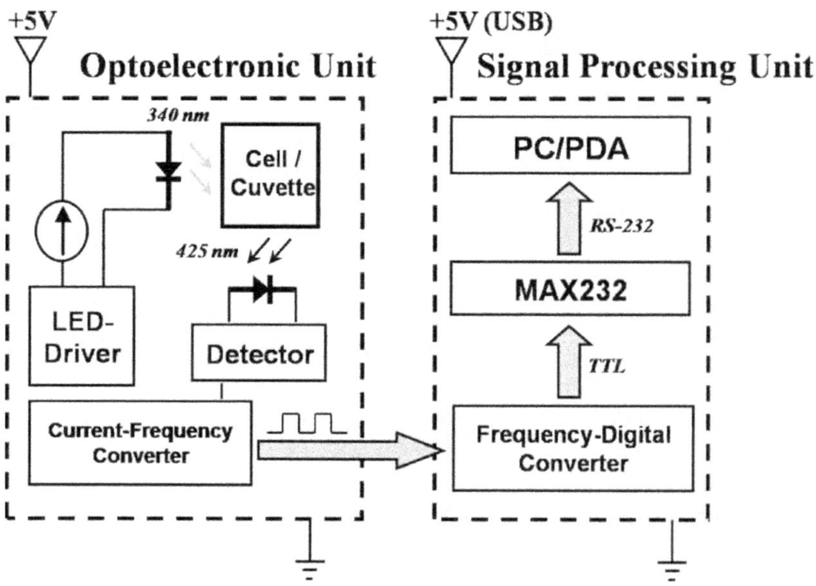

Fig. 12.9 Functional block diagram of the hand-held fluorometer [42].

A programmable color-to-frequency converter TCS230 from the *Texas Advanced Opto-electronic Solutions, Inc.* (USA) was used in the designed system to detect fluorescence. By combining a silicon photodiode array and a colour light-to-frequency converter, the TCS230 sensor outputs a frequency-modulated signal with 50 % duty cycle [43]. Such sensor with a frequency-modulated quasi-digital output is cheap and easily implemented in tiny electronic products without an A/D converter. Its frequency-modulated output signal can be directly processed by the Universal Frequency-to-Digital Converter (UFDC-1) IC. The frequency output range from 900 to 7000 Hz corresponds to the thiamine detection, and frequency range from 510 to 680 Hz corresponds to the tannic acid detection.

As mentioned above, the combination of a frequency output sensor and the UFDC-1 could simplify converting process to digital signal In this implementation, the UFDC-1 was used as an interfacing circuit between an optical sensor and a digital data processor. The communication protocol of the UFDC-1 was set up according to the Universal Asynchronous Receiver Transmitter (UART) convention by RS232 protocol in the BCD ASCII format, and the data transfer rate (baudrate) was set up to 2400 bps. The UFDC-1 output digital signals in TTL level (0/5V) to communicate with a PC via the MAX232 IC (HIN232CPZ, *Intersil Americas Inc.*). Additionally, the I^2C protocol can be implemented in the IC that is to be integrated in tiny portable electronic products, such as Personal Digital Assistants (PDA).

A program in the LabView software (LabView 7.1, *National Instruments, USA*) was developed for acquiring data at a sampling rate of 30 Hz via a RS232 interface in a USB hardware port of a PC.

Along with the signal conditioning simplifying, the UFDC-1 enhances the measuring performance, in terms of the detection range and the signal-to-noise ratio (SNR), beyond that of analog-to-digital (A/D) conversion. The dynamic range of UFDC-1 is not limited by the supply voltage of the IC and noises, and the signal is immune to noise because of the signal is in the frequency signal domain [43].

The hand-held taste assessment device provides a convenient, low-cost, fast measurement, and portable means to determine astringency and umami levels in partially fermented tea beverages.

A low-cost sensor system based on the Quartz Crystal Microbalance (QCM) and Universal Frequency to Digital Converter (UFDC-1) for screening semen samples in live farming was developed in the Nottingham Trent University (UK) [44]. Schematic diagram of the oscillator and frequency counter electronics is shown in Figure 12.10.

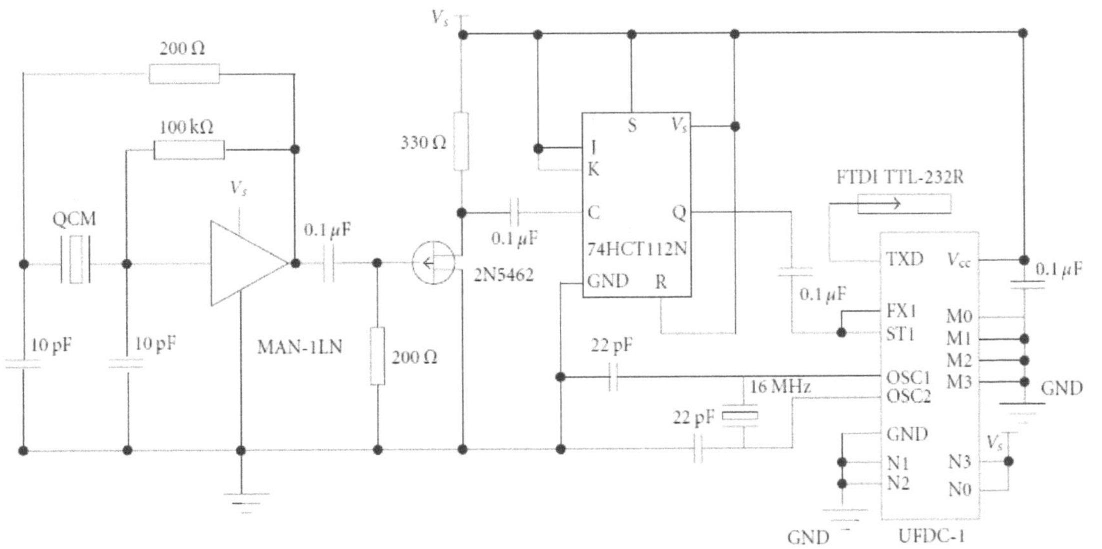

Fig. 12.10 Schematic diagram of the oscillator and frequency counter electronics [44].

The IC of Universal Frequency-to-digital Converter (UFDC-1) was used as a low-cost alternative for expensive Agilent's universal frequency counter, which is traditionally used in analytical chemistry laboratory measurements: 98.00 EUR (for development board EVAL UFDC-1/UFDC-1M-16) vs. 4 505.00 EUR (for Agilent 53132A 012 225 MHz Universal Counter with ultra high stability).

The UFDC's serial output is ideal for use with a USB to TTL serial cable (FTDI TTL-232R-3V3). A program was written in the free programming language Just Basic [45] to read the data from the COM port, extract the string for a single value, convert to a number, and store it to a spreadsheet file. Given the maximum frequency of the UFDC-1 without prescaling is 7.5 MHz, operation of the QCM in the "swim cell" at 10 MHz was above this frequency, and requires a divide by two prescalar. The thickness of the crystal reduces with increasing frequency and whilst the 10 MHz are sufficiently robust [45].

In addition to the oscillator, the circuit diagram in Figure 12.10 shows the integrated frequency counter and buffer circuit. The sinusoidal output of the oscillator has peak-to-peak amplitude around 7 V which is fed into an FET operated in switch mode between 0 V and +5 V from the USB. The digital signal is then fed into a divide by two circuit consisting of a single flip flop on a 74HCT112N and then into the F_{x1} input of the UFDC-1 IC operating in mode M0 – a single channel frequency measurement; for improved stability, the analogue part of the circuit (oscillator) was operated from a dc battery supply and LM3171 voltage regulator and the digital part of the circuit supplied from the +5 V of the USB.

The oscillator circuit presented is simple, inexpensive, and sufficiently stable for such an application. The UFDC-1 and RS232-USB cable, forms a complete low-cost and light-weight system and may provide a suitable platform for other QCM-based applications with the similar metrological performances as at the use of commercial frequency counters.

12.3 References

[1]. Andrea Orsini, Arnaldo D'Amico, Chemical Sensors and Chemical Sensor Systems: Fundamentals Limitations and New Trends, in Advances in Sensing with Security Applications, NATO Security through Science Series, 2006, Vol. 2, *Springer*, 2006, pp.69-94.

[2]. Edward J. Staples, A New Electronic Nose, *Sensors Magazine*, Vol.16, No.5, May 1999.

[3]. Edward J. Staples, Virtual Chemical Sensor Arrays and High-Resolution Olfactory Images: The zNose, *Sensors Magazine*, Vol.16, No.6, June 2001.

[4]. Stephan Trautweiler, New Silicon-Based Metal-Oxide Chemical Sensors, *Sensors Magazine*, Vol.16, No.9, September 1999.

[5]. Dentcho V. Ivanov, Advanced Sensors for Multifunctional Applications, *JOM-e*, Vol.52, No.10, October 2000.

[6]. Francis Tsow, Erica S. Forzani, and N. J. Tao, Frequency-Coded Chemical Sensors, *Analytical Chemistry*, Vol.80, 2008, pp.606-611.

[7]. Chemical Sensors: A Global Strategic Business Report, MCP-1718, *Global Industry Analysts, Inc.*, February 2010.

[8]. Chemical Sensors US Industry Forecasts for 2014 & 2019 Market Research Report, FED00047, *Freedonia Group*, January 2011.

[9]. Hatfield J. V., Daniels A. R., Snowden D., Persaud K. C., Payne P. A., Development of a Hand Held Electronic Nose (H2EN), in *Proceedings of the EUROSENSORS XIII*, 12-15 September 1999, The Hague, The Netherlands, pp. 215-218.
[10]. Universal Frequency-to-Digital Converter (UFDC-1), Specification and Application Note, *Technology Assistance BCNA 2010, S. L.*, 2011.
[11]. Universal Sensors and Transducers Interface (USTI), Specification and Application Note, *Technology Assistance BCNA 2010, S. L.*, 2011.
[12]. Hök B., Blückert A., Löfving J., Acoustic Gas Sensor with ppm Resolution, in *Proceedings of the EUROSENSORS XIII*, 12-15 September 1999, The Hague, The Netherlands, pp. 631-634.
[13]. Akimitsu Ishihara, Shukuji Asakura, A Possible Motion Sensor with Frequency Output, in *Proceedings of the 18th Chemical Sensor Symposium*, 3-5 April 1994, Tohoku University, Japan, Vol.10, Supplement A.
[14]. Chemical Sensors web section at Sensors Web Portal (www.sensorsportal.com): http://www.sensorsportal.com/HTML/SENSORS/Chemical.htm
[15]. Ashok Kumar, Biosensors Based on Piezoelectric Crystal Detectors: Theory and Application, *JOM-e*, Vol. 52, No.10, October 2000.
[16]. Biosensors in Medical Diagnostics: A Global Strategic Business Report, MCP-3310, *Global Industry Analysts, Inc.*, January 2011.
[17]. Analytical Review of World Biosensors Market, *Frost & Sullivan*, 8 June 2010.
[18]. Global Medical Biosensors Market 2010-2014, *Infiniti Research Limited*, February 2011.
[19]. Microsensors: MEMS, Biosensors and Nanosensors (IAS027A), *BCC Research*, April 2008.
[20]. Rajender Thusu, Strong Growth Predicted for Biosensors Market, *Sensors Magazine*, 1 October 2010.
[21]. FP7 ICT Work Programme 2011-2012, Objective ICT-2011.3.2, Smart Components and Smart Systems Integration.
[22]. Ahmad, T., Gencer, N. G., Development of a Data Acquisition System for Electrical Conductivity Images of Biological Tissues via Contactless Measurements, in *Proceedings of the IEEE 23rd Annual International Conference on Engineering in Medicine and Biology Society*, 2001, Vol. 4., pp.3380 – 3383.
[23]. W. Stephen Woodward, Bio-Potential-to-Frequency Converter/Modulator, *Electronic Design*, August 1999.
[24]. VLSI-MEMS Smart Pulse-Coded BioSensors, University of Maryland, MD, USA, http://www.ece.umd.edu/~newcomb/memsensr.html
[25]. Biosensors web section at Sensors Web Portal (http://www.sensorsportal.com): http://www.sensorsportal.com/HTML/SENSORS/Biosensors.htm
[26]. C. Di Nucci, A. Fort, S. Rocchi, L. Tondi, N. Uliviery, V. Vignoli, F. Di Francesco, M. B. Serrano-Santos, Study of the Dynamic Response of QCM sensors my means of a fast and accurate all-digital frequency detector, in Proceedings of the *IEEE Instrumentation and Measurement Technology Conference*, Anchorage, USA, 2002, (CD-ROM).
[27]. M. T. S. R. Gomes, Bulk acoustic wave sensors in chemical analysis, in Smart Sensors and MEMS, ed. by S. Y. Yurish, M. T. S. R. Gomes, Ed. , *Springer Verlag*, 2005, pp.421-445.
[28]. R. Lucklum, R. Borngraeber, P. Hauptmann, Acoustically Amplified Chemical QCM-sensor with Frequency and Attenuation Output, in *Proceedings of the Transducers '99*, Sendai, Japan, 1999, pp.1250-1253.
[29]. Inacio, J. N. Marat-Mendes, C. J. Dias, Development of a Biosensor based on a Piezoelectric Film, in *Proceedings of the European Conference on Solid-State*

Transducers (EUROSENSORS XVI), Prague, Czech Republic, 2002, Vol.3, pp.763-764.

[30].E. Katz, Electronic Biosensors based on Biomaterial-nanoparicle Hybrid Systems, in Smart Sensors and MEMS, ed. by S. Y. Yurish, M. T. S. R. Gomes, Ed., *Springer Verlag*, 2005, pp.447-474.

[31].S. R. Suri, Quartz Crystal based Microgravimetric Immunobiosensors, *Sensors & Transducers*, Vol. 66, Issue 4, April 2006, pp. 544-552.

[32].S. Kurosawa, H. Aizawa, M. Tozuka, M. Nakamura, J.-W. Park, Immunosensors using a Quartz Crystal Microbalance, *Measuring Science and Technology*, Vol. 14, 2003, pp. 1882-1887.

[33].L. A. Kuznetsov, M. V. Milonov, D. V. Nalivkin, A Method for High-precision Measurements of Vibration Frequency of Piezoelectric Sensors, *Sensors & Systems*, February 2004, pp.2-6 (in Russian).

[34].M. T. Hill, A. Cantoni, Precise All-digital Frequency Detector for High Frequency Signals, *IEEE Transactions on Communications*, Vol. 48, 2000, pp.1937-1944.

[35].Shankar, S. A. Morris, C. G., Hutchens, A Novel Measurement Technique for Quartz Microbalance Systems and Other Resonator-based Sensor Systems, in *Proceedings of the Sensors for Industry Conference (Sicon/02)*, Houston, Texas, USA, 2002, pp.134-138.

[36].N. V. Kirianaki, S. Y. Yurish, N. O. Shpak, V. P. Deynega. Data Acquisition and Signal Processing for Smart Sensors, Chichester, UK, *John Wiley & Sons*, 2001.

[37].Yurish S. Y., Intelligent Data Acquisition System for Quartz Crystal Microbalance and Other Resonator-Based Sensors, in *Proceedings of the IEEE International Workshop on Intelligent Data Acquisition and Advanced Computing Systems: Technology and Applications (IDAACS' 2007)*, 6-8 September 2007, Dortmund, Germany, pp. 2-7.

[38].Yurish S. Y., Low-Cost, Intelligent Data Acquisition System for QCM and Other Resonator-Based Bio- and Chemical Sensors, *International journal of Computing*, Vol.7, Issue 2, 2008, pp. 9-17.

[39].Terminal v.1.9b: http://www.sensorsportal.com/DOWNLOADS/Terminal.exe

[40].Yurish S. Y., Advanced Automated Calibration Technique for Universal Sensors and Transducers Interface IC, in *Proceedings of IEEE International Instrumentation and Measurement Technology Conference (I²MTC 2009)*, Singapore, May 5-7, 2009, pp.402-405.

[41].R. Stone, P. A. Payner, Measurement of the Response of Resistive Sensors using Logic Oscillators and Integrated Frequency Measurement, *Sensors and Actuators A: Physical*, Vol.76, Issue 1-3, 1999, pp.431-436.

[42].Kuang-Hua Chang, Richie L.C. Chen, Bo-Chuan Hsieh, Po-Chung Chen, Hsien-Yi Hsiao, Chi-Hua Nieh, Tzong-Jih Cheng, A Hand-held Electronic Tongue based on Fluorometry for Taste Assessment of Tea, *Bioelectronics*, Vol. 26, Issue 4, pp.1507-1513.

[43].TCS230 Programmable Color Light-to-Frequency Converter, Data Sheet, TAOS046, *TAOS, Inc.*, January 2003.

[44].Michael I. Newton, Shaun Atherton, Robert H. Morris, Simon M. Stanley, Carl R. Evans, David C. Hughes, and Glen McHale, Low Cost QCM Sensor System for Screening Semen Samples, *Journal of Sensors*, Vol. 2010, 2010, pp.1-5.

[45].Just Basic (http://www.justbasic.com/).

Chapter 13

Capacitive Sensors Interfacing

13.1 Capacitance-to-digital Converters: State-of-the-Art Review

Capacitive sensing elements are widely used in humidity, level, pressure, acceleration, position, proximity, biosensors, chemical sensors and transducers. In such sensor systems, physical, chemical or mechanical quantities are converter into capacitance values, which further should be converter into digital by an electronic circuit. As it was written in [1], successful commercial exploitation of highly miniaturized capacitive sensors is often inhibited by the presence of parasitic effects such as environmental noise and parasitic capacitance. Therefore, to alleviate the problems, it is important to integrate readout, interfacing and converting circuitry as directly as possible.

Commercially available capacitance humidity sensor Minicamp 2 (*GE Sensing*) has a capacitance ranges from 165.7 to 261.87 pF; humidity sensor KFS33-LC (*Hydrogen's Instruments, GmbH*) - from 305 to 355 pF; humidity sensor DSHS07 (*Smartec*) - from 280 to 380 pF and humidity sensor SMTHS10 from the same company has the capacitance range from 40 to 240 pF; capacitance humidity sensor with carbon nitride detecting element – from 350 to 900 pF [1].

Humidity sensors for technical gases as nitrogen, argon, helium, oxygen, hydrogen, etc. have typical capacitance range from 1 to 2.5 nF. However, existing, available on the market integrated circuits for capacitance-to-digital conversion have limited capacitance ranges and /or low accuracy. For example, integrated capacitance-to-digital converters AD7745/46 [2], AD7747 [3] and AD7150/51/52/53 [4] from *Analog Devices* have capacitance ranges that do not exceed one-two tens pF, only one I^2C or SPI interface and can work only in the slave communication mode. The capacitance converters of AD715X Series have only 12-bit resolution.

A capacitance meter based on the PIC16C622 microcontroller has also narrow capacitance range from 1 to 999 nF [5]. It does not cover a pF capacitive range and has a low accuracy (± 3.39 % average relative error). In the RLC meter, based on virtual counting technique, a sufficient accuracy is achieved only if capacitances are not to large or too small [6].

The ZSSC3123 cLite™ from *ZMDI* is a CMOS integrated circuit for capacitance-to-digital conversion (± 0.25 % FSO) and sensor-specific correction of capacitive sensor signals [7]. However, this signal conditioner is intended for capacitive sensors with capacitances only up to 260 pF. The capacitance-to-digital converters of ZSSC3122 and ZSSC31210 series also have a limited capacitance ranges (< 10 pF and < 260 pF respectively) [8, 9]. All ICs have two popular serial buses I^2C and SPI, but can work only in the slave communication mode, controlled by an external microcontroller.

The PCapØ1AK-0101xx is a CMOS capacitance-to-digital converter which is especially dedicated to capacitive humidity sensors [10]. The circuit has an integrated DSP to do the linearization of the humidity and temperature results by means of polynomials. Sensor and reference must be the same order of magnitude. Via a selectable serial I^2C or SPI interfaces, the chip can communicate with an external master [10]. The circuit diagram with the SPI interface is shown in Figure 13.1.

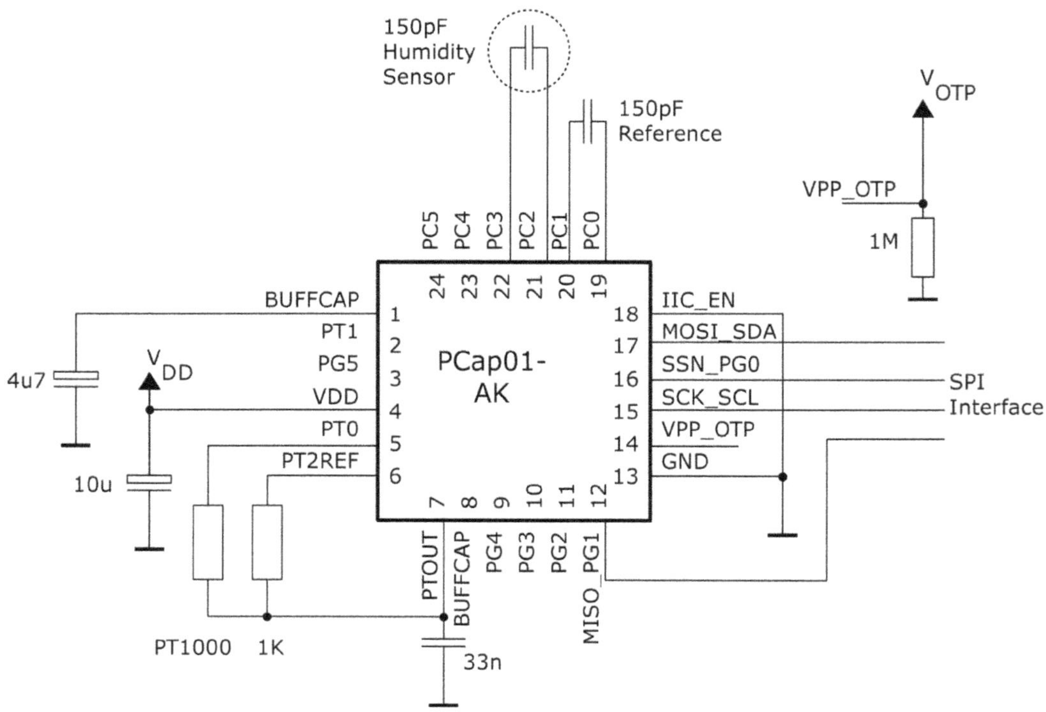

Fig. 13.1 Example schematics of PCap0 with SPI interface.

The Universal Transducer Interface (UTI) is a complete analog front end for low frequency measurement applications, based on a period-modulated oscillator. Sensing elements (capacitors) can be directly connected to the UTI without the need for extra electronics. Only a single reference element, of the same kind as the sensor, and external resistor for 300 pF capacitance range are required. The UTI outputs a microcontroller-compatible period-modulated signal [11].

High metrological performances for capacitance-to-digital conversion are reported in [12-14] but in the limited ranges of humidity sensors (from 10 to 330 pF, from 100 to 150 pF and from 149 to 206 pF respectively). Hence, a design of simple, universal, single chip capacitive sensor electronic interfacing circuit with high metrological performance and wide measuring range is an important and timely task.

The comparative performances of available commercial integrated capacitance-to-digital converters are shown in Table 13.1. The USTI IC is advantageously differ due to its extended capacitance range, reasonable average relative error, variety of serial interfaces and comfortable communication master mode for RS232 interface, which lets to use the IC without any microcontroller or microprocessor [15]. Owing to its wide range of measuring modes for various frequency-time parameters of electrical signals, the USTI can be used as an interfacing IC between the UTI and PC or UTI and microcontroller or DSP.

Table 13.1 Integrated Capacitance-to-Digital Converters.

IC	Capacitance Range	Relative Error, FS %	Interfaces	Communication Mode
		ZMDI		
ZSSC31210	2 pF to 260 pF	± 0.25	I^2C, SPI	Slave
ZSSC3122	< 10 pF	± 0.25	I^2C, SPI	Slave
ZSSC3123	< 260 pF	± 0.25	I^2C, SPI	Slave
		Smartec		
UTI 03	2 pF to 300 pF	n/a	Period modulated	Slave
		acam messelectronic, GmbH		
PCap01	1 pF to 1000 pF	n/a	I^2C, SPI	Slave
		Technology Assistance BCNA 2010, S. L.		
USTI	50 pF to 100 μF	*±0.036 %	I^2C, SPI, RS232	Slave and master (RS232)

* - average relative error in all measuring range; n/a – information is not available.

The direct capacitance sensors interfacing approach described below is based on the USTI IC application.

13.2 Direct Capacitive Sensors Interfacing

The universal sensors and transducers interfacing (USTI) integrated circuit (Figure 13.2) can work in a capacitance-to-digital conversion mode. It is based on a three-signal measurement technique, (initially published in [16]) at which the offset, reference and measurand values are converted into three time intervals by internal capacitance-to-time converter base on the internal comparator. The unknown capacitance should be calculated according to the following equation:

$$C_x = \frac{N_x - N_{off}}{N_{ref} - N_{off}} \cdot C_{ref}, \tag{13.1}$$

where N_x, N_{off} and N_{ref} are the numbers of reference frequency pulses counted during the measurand, offset cancelation and reference measurement stages respectively; C_{ref} is the precision reference capacitor.

Fig. 13.2 USTI IC in 32-pad, 5 × 5 mm MLF package.

The direct capacitive interface needs only a few external components as a reference capacitor C_{ref} and resistor R (Figure 3.16).

The converter has three popular serial interfaces RS232, SPI and I^2C for communication to PC or master microcontroller. It can work in a slave communication mode with all mentioned serial interfaces and a master communication mode for RS232 communication. In the last case no any master microcontroller neither external PC should be used for IC controlling. The USTI continuously generates converted results at its output.

The following design consideration should be taken into account for the external components. The resistance (in Ohm) should be calculated according to the equation (3.4), and the charging time (in seconds) should be calculated according to the following equation:

$$T = 2200 \times C_{ref}, \tag{13.2}$$

where C_{ref} is the reference capacitance in F.

The charging time should be set up with the help of USTI's command "Wnn". All possible values for "Wnn" command are shown in Table 3.2.

The value of C_{ref} also should be set up with the help of appropriate USTI's command Ei.f, where "i.f" is the fixed point decimal number: "I" is the integer part (1 to 12 decimal digits) and "f" is the fractional part (0 to 12 decimal digits). An example of USTI's commands for capacitance-to-digital conversion for RS232 serial interface, slave mode is shown in Figure 13.3.

In the master communication mode the charging time should be set up with the help of external switches N0, N1, N2. In this mode the result at the converter's output will be presented as C_x/C_{ref}. In order to get the actual capacitance, such result must be multiplied by C_{ref}. Oscillogram on the SMPL pin for capacitance measurement mode for humidity sensor is shown in Fig. 13.4.

```
> M11                    ; Capacitance measurement mode
> E0.00000000012543      ; Reference set up Cref=125.43 pF
> W00                    ; Set up the charging time (100 µs)
> S                      ; Start Measurement
> C                      ; Check the measurement status: 'r'-ready, 'b'-busy
> R                      ; Get the result in F
0.000000000155
```

Fig. 13.3 USTI commands for capacitance measurements (RS232 interface).

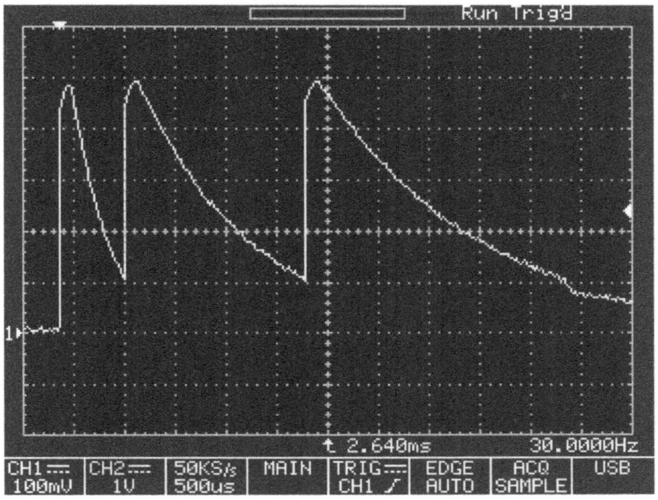

Fig. 13.4 Oscillogram of three-point calibration technique on the USTI's SMPL pin (RH sensor example, $C_x \cong 257$ pF).

Experimental investigation of a capacitance-to-digital converter based on the USTI IC reported in [17] is described below.

The universal capacitance-to-digital converter was applied to the measurement of capacitance in a wide range from 50 pF to 100 µF. Preliminarily, the USTI has been calibrated at laboratory temperature range in order to eliminate an additional systematic error due to a quartz oscillator trimming inaccuracy (calibration tolerance) and a short term temperature instability [18]. The USTI has been connected to a PC, where terminal software Terminal v1.9b [19] was running under Windows operation system. Each of capacitance C_x was measured 100 times in order to compare obtained results with reported in literature, for example, in [13]. The actual capacitance has been measured by the high precision LCR meter GW Instek LCR-819 with the basic accuracy ±0.05 % or better, which also has been preliminary calibrated. The oscillograms for C_x=152.22 pF are shown in Figure 13.5, and measurement results for 152.22 pF and 1.9279 µF are shown in Figures 13.6 and 13.7 respectively. The maximum relative error does not exceed ± 0.7 % in the worst case.

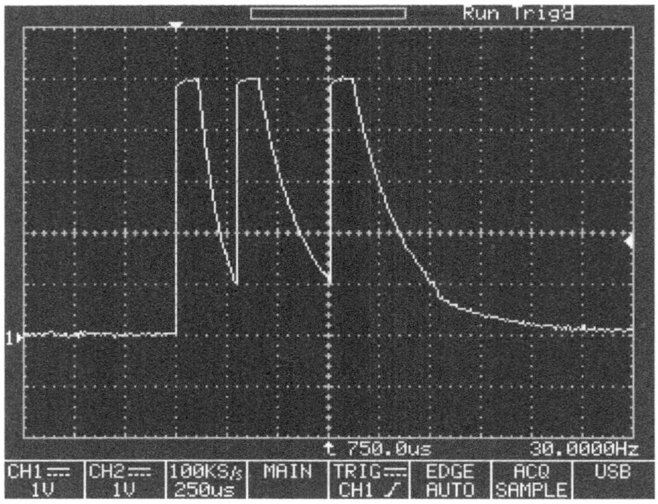

Fig. 13.5 Oscillogram of three-point calibration technique on the USTI's SMPL pin at C_x=152.22 pF.

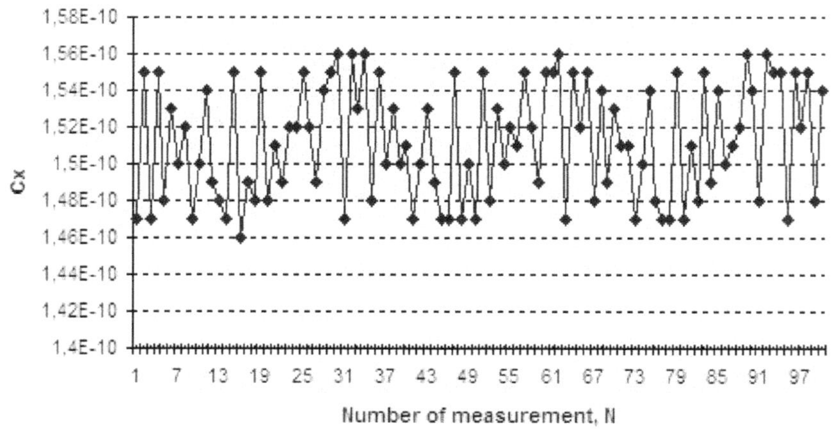

Fig. 13.6 Measurement results for actual capacitance C_x=152.22 pF at C_{ref}=125.43 pF; R=20 MΩ.

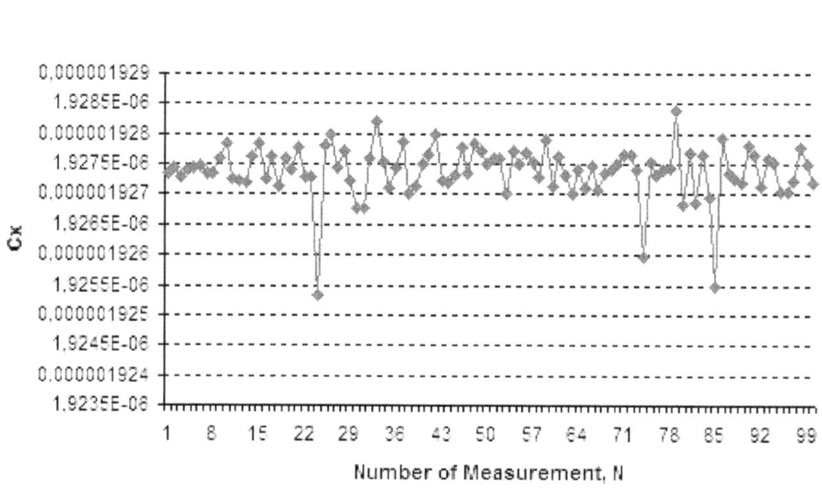

Fig. 13.7 Measurement results for actual capacitance C_x=1.9279 μF at C_{ref}=1.9286 μF; R=99 Ω.

The experimental results have confirmed the average relative error ± 0.036 % (which is in 20-100 times less than the reported in [5, 6]) in the comparative measuring range (Table 13.2, gray color).

Table 13.2 Comparative capacitance measurements results for PICOMETER [5] and USTI.

<table>
<tr><th>Marked Value</th><th>PICOMETER's Error, %</th><th>USTI's Error, %</th><th>Measurement conditions</th></tr>
<tr><td>150 pF</td><td>N/a</td><td>0.6400</td><td>C_{ref}= 125.43 pF, R = 20 MΩ</td></tr>
<tr><td>180 pF</td><td>N/a</td><td>0.2300</td><td rowspan="2">C_{ref}= 174.04 pF, R = 20 MΩ</td></tr>
<tr><td>233 pF</td><td>N/a</td><td>0.6200</td></tr>
<tr><td>470 pF</td><td>N/a</td><td>0.7000</td><td>C_{ref}= 449.25 pF, R = 20 MΩ</td></tr>
<tr><td>2.2 nF</td><td>4.3</td><td>0.0680</td><td rowspan="4">C_{ref}= 22 nF, R = 91 kΩ</td></tr>
<tr><td>20 nF</td><td>1.2</td><td>0.0158</td></tr>
<tr><td>33 nF</td><td>1.7</td><td>0.0011</td></tr>
<tr><td>47 nF</td><td>1.1</td><td>0.0019</td></tr>
<tr><td>100 nF</td><td>6.7</td><td>0.1400</td><td>C_{ref}= 94.63 nF, R = 91 kΩ</td></tr>
<tr><td>470 nF</td><td>2.1</td><td>0.0010</td><td>C_{ref}= 464.89 nF, R = 5 kΩ</td></tr>
<tr><td>940 nF</td><td>6.6</td><td>0.0230</td><td>C_{ref}= 961.11 nF, R = 2.3 kΩ</td></tr>
<tr><td>2 μF</td><td>N/a</td><td>0.0270</td><td>C_{ref}= 1.9286 μF, R = 1 kΩ</td></tr>
<tr><td>47 μF</td><td>N/a</td><td>0.1000</td><td>C_{ref}= 42.206 μF, R = 99 Ω</td></tr>
</table>

In addition, the USTI can work with any known capacitance–to-frequency and capacitance–to-period converters, for example [20-23], in a wide frequency range. In such cases converters' outputs should be directly connected to any of two USTI's channels for frequency-time measurements. The relative error for frequency-to-digital conversion is ≤ ± 0.0005 %.

Some interesting capacitance-to-duty cycle, capacitance-to-period and capacitance-to-frequency converters are also described in [24-26]. The USTI and UFDC ICs can be used with such devices in order to design capacitance-to-digital converters with high metrological performances.

Due to designed USTI integrated circuit the acceptance of capacitive technology will be continuously increasing in various applications during next years.

13.3 References

[1]. Zambov L.M. , Popov C., Plass M. F., Bock A., Jelinek M., Lancok J., Masseli K., Kulisch W., Capacitance Humidity Sensor with Carbon Nitride Detecting Element, *Appl. Phys.*, Vol. A 70, 2000, pp.603–606.

[2]. AD7745/AD7746 24-bit Capacitance-to-Digital Converter with Temperature Sensor, *Analog Devices*, USA, 2005.

[3]. AD7747 24-bit Capacitance-to-Digital Converter with Temperature Sensor, *Analog Devices*, USA, 2007.

[4]. AD7150 Capacitance Converter for Proximity Sensing, *Analog Devices*, USA, 2008.

[5]. Richey R. Resistance and Capacitance Meter using a PIC16C622, AN611. *Microchip Technology, Inc.*, USA, 1995.

[6]. Nikolov G.T., Stoyanov I.I., Nikolova B.M., RLC Measurement using Virtual Counting Techniques, in *Proceedings of the ELECTRONICS' 2006*, September 2006, pp.69-74.

[7]. ZSSC3123 cLite™ Capacitive Sensor Signal Conditioner, Rev.1, *Zentrum Mikroelektronik Dresden AG*, Germany, 2011.

[8]. ZSSC3122 cLite™ Low Voltage Capacitive Sensor Signal Conditioner, Rev.1.20, *Zentrum Mikroelektronik Dresden AG*, Germany, 2011.

[9]. ZSC31210 cLite™ Capacitive Sensor Signal Conditioner, Rev.1.2, *Zentrum Mikroelektronik Dresden AG*, Germany, 2011.

[10].PCapØ1AK-0101xx Single-chip Solution for Humidity Measurement, *acam messelectronic GmbH*, Germany, 2 February 2011.

[11].Universal Transducer Interface (UTI), Datasheet, *Smartec*, 30 November 2010.

[12].Heidary A., Meijer G.C.M., An Integrated Interface Circuit with a Capacitance-to-Voltage Converter as Front-end for Grounded Capacitive Sensors, *Meas.Sci Technol.,* Vol. 20, 2009, pp.1-7.

[13].Reverter F., Pallas-Areny R., Direct Sensor-to-microcontroller Interface Circuits, *Marcombo S. A.,* Barcelona, 2005.

[14].Reverter F., Casas O., Direct Interface Circuit for Capacitive Humidity Sensors, *Sensors and Actuators A,* Vol. 143, 2008, pp. 315-22.

[15].Universal Sensors and Transducers Interface (USTI), Specification and Application Note, *Technology Assistance BCNA 2010, S. L.,* 2011.

[16].Van der Goes F. M. L, Meijer G. C. M., A Novel Low-cost Capacitive Sensor Interface, *IEEE Trans. on Instrum. and Meas.,* Vol.45, Issue 2, 1996, pp.536-540.

[17].Yurish S. Y., Universal Capacitive Sensors and Transducers Interface, in *Proceedings of the 23rd International Conference EUROSENSORS XXIII*, Lausanne, Switzerland, 6-9 September 2009, pp. 441-444.

[18].Yurish S. Y., Advanced Automated Calibration Technique for Universal Sensors and Transducers Interface IC, in *Proceedings of the IEEE International Instrumentation and Measurement Technology Conference (I²MTC 2009)*, Singapore, May 5-7, 2009, pp.402-405.

[19].Terminal v.1.9b: http://www.sensorsportal.com/DOWNLOADS/Terminal.exe

[20].Cheng-Ta Chiang, Chi-Shen Wang, Yu-Chung Huang, A CMOS Integrated Capacitance-to-frequency Converter with Digital Compensation Circuit Designed for Sensor Interface Applications, *IEEE Sensors Journal,* Vol.28, Issue 31, 2007, pp. 954-57.

[21].Chatzandroulis S., Tsoukalas D., Capacitance to Frequency Converter Suitable for Sensor Applications Using Telemetry, *Analog Integrated Circuits and Signal Processing*, Vol. 27, 2001, pp.31-38.

[22].Stoyanov I.I., Nokolov G.T., Nikolova B.M., Design of RLC to Frequency Converter, in *Proc.of the ELECTRONICS*' 2006, September 2006, pp.63-68.

[23].Ioannis Ramfos, Stavros Chatzandroulis, A 16-channel Capacitance-to-Period Converter for Capacitive Sensor Applications, Analog Integrated Circuits and Signal Processing, *Springer*, 2011, pp.1-7.

[24].Svistunov B. L., Frolov I. N., Sensing Device for Capacitance Sensor, Measuring Instruments and Control Systems, No. 12, 1987, pp. 21 (in Russian).

[25].Ménini Ph., Dondon Ph., Blasquez G., Pons P., Modelling of a Capacitance-to-Period Converter Dedicated to Capacitive Sensors, in *Proceedings of the 13th European Conference on Solid-State Transducers*, The Hague, The Netherlands, 12-15 September, 1999, pp.549-552.

[26].N. V. Kirianaki, S. Y. Yurish, N. O. Shpak, V. P. Deynega, Data Acquisition and Signal Processing for Smart Sensors, Chichester, UK, *John Wiley & Sons*, 2001.

Chapter 14

Resistive Sensors Interfacing

14.1 Resistance-to-digital Converters: Introduction and Problem Definition

Resistive sensing elements are widely using in various physical and chemical sensors and transducers for different values as gases concentrations, temperature, humidity, distance, etc.

Various interfaces and readout circuit for resistive gas sensing elements are described in [1-3]. The response of gas sensors, for example, consists of a variation in the resistivity of the active material. Each active material has a different nominal resistance [2]. Their values can be changed from a few tens Ω to several MΩ. This requires an efficient interface circuit.

Many existing resistance-to-digital converters are based on several active external components as analog-to-digital converter (ADC) and/or operational amplifiers and a microcontroller; others based on resistance-to–time or resistance-to-frequency converters and microcontroller report a relatively low accuracy: the average relative error 1 % [4], 1.4 % [5] or more than 3 % [6] for resistance measurements.

Microcontrollers' manufacturers such as *Texas Instruments, Microchip Technology* and others describe direct resistive sensor-to-microcontroller interfaces, for example in [7], however, any metrological performances have not been reported in such application notes.

A good accuracies are reported in microcontroller-based and ADC-based resistance-to-digital converters [8-10], but in the limited ranges of resistances correspond to the rage of platinum resistor Pt1000 at -45 ^{0}C to + 120 ^{0}C (± 0.01 % at 825 Ω … 1470 Ω); 5 kΩ at ± 0.2 % and 600 Ω at ± 0.15 % respectively.

However, till now no universal solution for both resistance-to-digital conversion and resistance-to-frequency (time)-to-digital conversion are available. In addition, many designed converters based on microcontrollers or ASICs have limited resistive range or/and low accuracy.

Hence, a design of simple, universal, single chip direct resistive sensors interfacing circuit with high metrological performances and wide measuring range is an important and timely task.

14.2 Direct Resistive Sensors Interface

The Universal Sensors and Transducers Interface (USTI) integrated circuit has a special mode for resistance measurement (Figure 14.1) [11, 12].

Fig. 14.1 Universal Sensors and Transducers Interfacing circuit (USTI) in miniaturized 32-pad Micro Lead Frame (MLF) package.

It uses a three-signal measurement technique, initially proposed for a capacitance-to-digital conversion [13] and expanded later on a resistance-to-digital conversion [14], at which the offset, reference and measurand values are converted into three time intervals by internal resistive-to-time converter base on the internal comparator. The unknown resistance should be calculated according to the following equation:

$$R_x = \frac{N_x - N_{off}}{N_{ref} - N_{off}} \cdot R_c, \tag{14.1}$$

where N_x, N_{off} and N_{ref} are the numbers of reference frequency pulses counted during the measurand, offset cancellation and reference measurement stages respectively; R_c is the precision reference resistor.

The direct resistive interface needs only a few external components as R_c – the reference resistor; C – the calibrating capacitor; R_0 – the limiting resistor (Figure 3.15).

The USTI has three popular serial interfaces RS232, SPI and I^2C for communication to PC or master microcontroller. The converter can work in a slave communication mode with all mentioned serial interfaces and a master communication mode for RS232 communication. In this case no any master microcontroller neither external PC should be used for the IC controlling. The USTI continuously generates converted results at its output.

The following design consideration should be taken into account for the external components. The capacitance (in Farads) of calibrating capacitor should be calculated according to the equation (3.2), the charging time (in seconds) should be calculated according to equation (3.3) and set up with the help of USTI's command "Wnn". All possible values for "Wnn" command are shown in Table 3.2.

The value of $R_c \leq R_x$ also should be set up with the help of appropriate USTI's command Ei.f, where "i.f" is the fixed point decimal number: "I" is the integer part (1 to 12 decimal digits) and "f" is the fractional part (0 to 12 decimal digits). An example of USTI's commands for resistance-to-digital conversion (RS232 serial interface, slave mode) is shown in Figure 14.2.

>M10	; Set up a resistance R_x measurement mode
>E263000.0	; Set the reference value R_c=263 kΩ
>W1B	; Set the charging time 100 ms
>S	; Start measurement
>C	; Check result status: returns "b" if busy and "r" if ready
>R	; Read conversion result in Ohm

262.208764377042

Fig. 14.2 Commands for resistance conversion at RS232 communication slave mode.

In the master communication mode the charging time should be set up with the help of external switches N_0, N_1, N_2. In this mode the result at the converter's output will be presented as R_x/R_c. In order to get the actual resistance, such result must be multiplied by R_c.

Oscillogram on the SMPL pin for R_x=1006.5 Ω; R_c=604.02 Ω; C= 3 μF and R_0=328.63 Ω is shown in Figure 14.3 (a) and for R_x=10 237 00000 Ω in Figure 14.3 (b). In the first case, the full length of one measuring cycle does not exceed 11 ms (without taking into account the communication and calculation times of IC).

Experimental investigation of a resistance-to-digital converter based on the USTI IC reported in [12] is described below. The USTI was applied to the measurement of resistivity in a wide range from 10 Ω to 10 MΩ, which covers a resistive range of almost all existing resistive sensors and transducers.

Preliminarily, the IC has been calibrated at laboratory temperature range in order to eliminate additional systematic error due to quartz oscillator trimming inaccuracy (calibration tolerance) and a short term temperature instability [15]. The USTI has been connected to a PC, where terminal software Terminal v1.9b was running [16]. Each of resistance R_x was measured 100 times in order to compare obtained results with reported in literature, for example, in [8, 14].

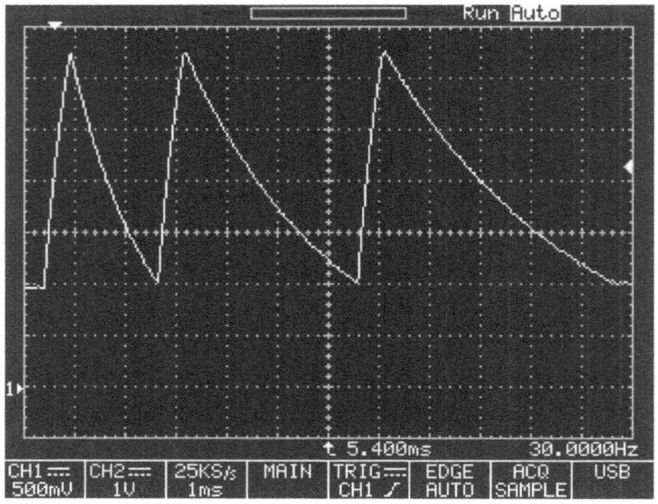

(a)

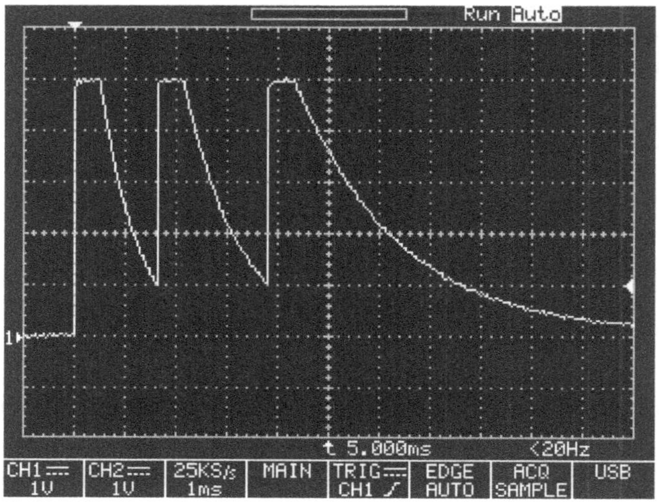

(b)

Fig. 14.3 Oscillogram of three-point calibration at offset, reference and measurand determination charges-discharges stages on the SMPL pin: R_x=1006.5 Ω; R_c=604.02 Ω; C= 3 μF and R_0=328.63 Ω (a) and R_x=10 237 000 Ω (b).

The actual resistance has been measured by the high precision inductance, capacitance and resistance meter series GW Instek LCR-819 with the basic accuracy ±0.05 % or better, which also has been preliminary calibrated.

The measurement results for R_x=1006.5 Ω is shown in Figure 14.4 Statistical characteristics are adduced in Table 14.1. The average relative error does not exceed ± 0.07 %, and the maximum relative error does not exceed ± 0.25 % in the worst case.

The χ^2-test for goodness of fit was applied to investigate the significance of the differences between observed data in the histograms and the theoretical frequency distribution for data from a normal population.

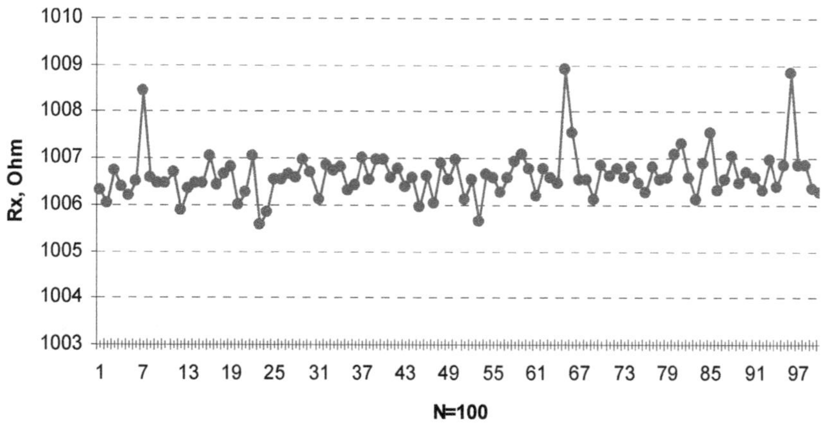

Fig. 14.4 Measurement results for R_x=1006.5 Ω (Trendline: y = 0.0034x + 1006.5; R-squared value: R^2 = 0.0379).

Table 14.1 Statistical characteristics of measuring results for R_x=1006.5 Ω.

Parameter	Value
Number of measurements, *N*	100
Minimum R_x (min), Ω	1004.5897
Maximum R_x (max), Ω	1008.9129
Sampling Range, R_x (max) - R_x (min), Ω	4.3232
Median	0
Arithmetic Mean, Ω	1005.7816
Variance	0.4172
Standard Deviation	0.6459
Coefficient of Variation	1557.1612
Confidence interval for arithmetic mean at *P*=97 %	1005.6414≤ R_x ≤ 1005.9218
Relative error, %	0.07
χ^2 – test (S) at: k=7; *P* = 97 % χ^2_{max} = 10	231.2859
Hypothesis about Gaussian distribution	No acceptable

The number *k* of equiprobable classes should be calculated according to recommendations, described in [17]:

$$k_{\max} = 1.25 \times n^{0.4}, \tag{14.2}$$

and chosen nearest odd value. In our case, at 100 measurements k_{max}=7.89, consequently, we have chosen *k*=7. Taking into account that the sum of deviations between the data set and the assumed distribution $S > \chi^2_{max}$, the hypothesis about Gaussian distribution can not be accepted. The grouping intervals at *k*=7 is shown in Table 14.2. The frequency distribution is shown in Figure 14.5. The hypothesis about uniform distribution is also not accepted.

The dependence of relative error vs. charging time at R_x = 2 043 300 Ω measurement is shown in Figure 14.6. The dependence clear shows that the charging time must be chosen according to the equation (3) and its increasing does not reduce the relative error but even increase it.

Table 14.2 Grouping Intervals at k=7.

No.	Classes	b[i]	w[i]	e[i]	A[i]
1.	1005.20735	15	0.186978	18.7	0.731
2.	1005.82495	45	0.339765	33.98	3.57
3.	1006.44255	30	0.320164	32.02	0.12
4.	1007.06015	6	0.129209	12.92	3.707
5.	1007.67775	1	0.022219	2.22	0.672
6.	1008.29534	2	0.001615	0.16	20.933
7.	1008.91294	1	4.9E-0005	0	201.539
	Amount:	**100**	**0.999999**	**100**	**231.286**

b[i] is the observed frequency distribution
w[i] is the distribution function
e[i] is the expected frequency distribution
A[i] is the deviation

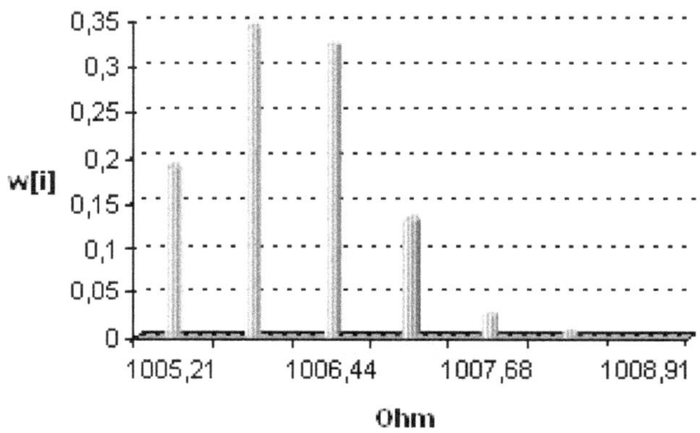

Fig. 14.5 Frequency distribution.

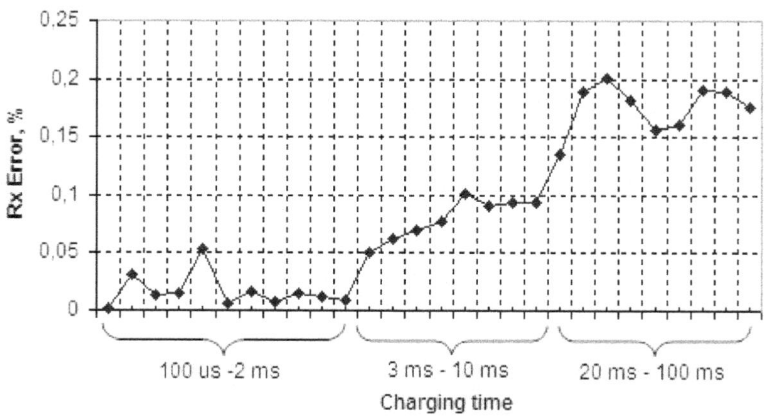

Fig. 14.6 Relative error vs. charging time at R_x = 2 043 300 Ω.

The comparative resistor measurement results in a wide measuring range are shown in Table 14.3. As it is visible from the table, the average relative error of the designed resistance-to-digital converter is ± 0.47 % (which is in 2-3 times less than reported in [4] (grey color) and [5]) in the wide measuring range from 10 Ω to 10 MΩ.

Table 14.3 Comparative resistor measurement results.

<table>
<tr><th colspan="4">Relative Resistance Error</th></tr>
<tr><th>Marked Value</th><th>PICOMETER [4] Error, %</th><th>USTI Error, %</th><th>Measurement Conditions</th></tr>
<tr><td>0.10 K</td><td>N/a</td><td>0.24</td><td>R_c=329.66 Ω; C=2 μF R_0=609.86 Ω</td></tr>
<tr><td>0.30 K</td><td>N/a</td><td>1.58</td><td rowspan="2">R_c=432 260 Ω; C=2 μF
R_0=609.85 Ω</td></tr>
<tr><td>0.82 K</td><td>N/a</td><td>1.31</td></tr>
<tr><td>1.0 K</td><td>N/a</td><td>0.026</td><td>R_c=604.02 Ω; C=3 μF
R_0=328.63 Ω</td></tr>
<tr><td>1.2 K</td><td>1.3</td><td>1.17</td><td rowspan="19">R_c=432 260 Ω
C=2 μF
R_0=609.85 Ω</td></tr>
<tr><td>5.1 K</td><td>1.0</td><td>1.067</td></tr>
<tr><td>8.2 K</td><td>2.0</td><td>0.92</td></tr>
<tr><td>10 K</td><td>2.0</td><td>0.918</td></tr>
<tr><td>15 K</td><td>1.7</td><td>0.860</td></tr>
<tr><td>20 K</td><td>1.5</td><td>0.805</td></tr>
<tr><td>30 K</td><td>1.4</td><td>0.759</td></tr>
<tr><td>51 K</td><td>1.0</td><td>0.641</td></tr>
<tr><td>75 K</td><td>1.0</td><td>0.566</td></tr>
<tr><td>91 K</td><td>0.6</td><td>0.491</td></tr>
<tr><td>150 K</td><td>0.5</td><td>0.392</td></tr>
<tr><td>200 K</td><td>0.3</td><td>0.309</td></tr>
<tr><td>300 K</td><td>0.2</td><td>0.2</td></tr>
<tr><td>430 K</td><td>0.4</td><td>0.137</td></tr>
<tr><td>560 K</td><td>0.6</td><td>0.062</td></tr>
<tr><td>680 K</td><td>0.7</td><td>0.02</td></tr>
<tr><td>820 K</td><td>0.7</td><td>0.0091</td></tr>
<tr><td>910 K</td><td>0.8</td><td>0.0026</td></tr>
<tr><td>1.0 M</td><td>N/a</td><td>0.0493</td></tr>
<tr><td>1.5 M</td><td>N/a</td><td>0.122</td><td>R_0=464 240 Ω
C=2.1 nF
R_c =563 960 Ω</td></tr>
<tr><td>2.0 M</td><td>N/a</td><td>0.063</td><td>R_c=432 280 Ω
C=2.1 nF
R_0=464 240 Ω</td></tr>
</table>

A conditional splitting of the range of measurement into sub ranges with a choice of appropriate values R_c and C has allowed to receive also the relative error ± 0.01 % comparable with the reported in [8, 14] or better.

In addition, the described resistive-to-digital converter can work with any known resistance-to-time, resistance-to-frequency or resistive-to-duty cycle converters, for example, described in [1, 3, 18-22]. In such cases, their outputs should be directly connected to any of two USTI's channels for frequency-time measurements. The converter can measure frequency in a wide measuring range from 0.05 Hz to 9 MHz

(144 MHz with prescaling) with programmable relative error from ±1 to ±0.0005 %. The relative error is constant in the whole frequency range. The converter has non-redundant conversion time for frequency and period measurements, and scalable resolution. The frequency-to-digital error of the USTI can be neglected in comparison with the resistance-to-frequency error. Besides the USTI, the UFDC-1 and UFDC-1M-16 ICs also can be used for any resistance-to-time, resistance-to-frequency or resistive-to-duty cycle converters.

The time interval can be also measured in a wide measuring range from 2 μs to 250 s. The quantization error for time interval measurement can be calculated according to the equation (3.1).

In case when the condition circuit for resistive sensors combines frequency and duty-cycle modulation of the same output signal, for example as it was described in [21], it also can be directly connected to the designed universal resistance-to-digital converter because of the last one can convert both: frequency and duty-cycle into digital in the same channel.

The commands for frequency and duty-cycle measurements at RS232 slave communication mode for a resistive conditioning circuit are shown in Figure 14.7

```
>M00    ; Set up a frequency fx measurement mode
>A09    ; Select the relative error ±0.001 %
>S      ; Start measurement
>C      ; Check if a result is ready: returns 'r' if ready and 'b' if busy
>R      ; Read conversion result for fx

>M04    ; Set up a duty-cycle measurement mode
>S      ; Start measurement
>C      ; Check if a result is ready: returns 'r' if ready and 'b' if busy
>R      ; Read conversion result for duty-cycle
```

Fig. 14.7 Commands for frequency and duty-cycle measurements in the 1st channel at RS232 communication slave mode.

In the common case, two independent resistance-to-frequency and/or resistance-to-time condition circuits can be connected to the designed IC as well as one addition resistive sensing element.

The resistance-to-digital converter based on the USTI IC can be used also with RTD (platinum (Pt100, Pt500, Pt1000) and Cu) temperature detectors and NTC thermistors for temperature measurements. The temperature value should be calculated by the external microcontroller or PC according to the transfer function of the resistance temperature sensing elements.

The RTD can be included into an oscillator circuit to generate a frequency output. As the temperature changes, the resistance of the sensor changes, modifying the oscillating frequency. The frequency is converted to digital with the UFDC-1 IC [23]. In comparison with the solution based on ADC, current source and second-order low-pass filter, this design approach let assures better noise immunity.

The developed resistive-to-digital converter also has several addition advantages such as relatively low cost, universality, small size (32-pad MLF 5 mm × 5 mm × 1 mm package) and high reliability. It is suitable practically for any resistive sensors and transducers and can work with any known resistance-to-time or resistance-to-frequency converters.

The designed IC has been realized in CMOS as a low power ASIP and available in three possible packages: 28-lead PDIP, 32-lead TQFP and 32-pad QFN/MLF packages. It is suitable for the using in industrial temperature range from –40 ^{0}C to +85 ^{0}C. The operating voltage is +4.5 ... +5.5 V. The active supply current at operating voltage +4.5 V and clock frequency 20 MHz is not more than 9.5 mA.

The next step in the future development is an integration of designed converter together with various resistive sensing elements into MEMS, systems-on-chip (SoC) and/or systems-in-package (SiP).

14.3 References

[1]. Balaji Jayaraman, Navakanta Bhat, High Precision 16-bit Readout Gas Sensor Interface in 0.13 μm CMOS, in *Proceedings of the IEEE International Symposium on Circuits and Systems (ISCAS' 2007),* May 2007, pp. 3071-3074.

[2]. J. L. Marino, S. A. Bota, R. Casanova, A. Dieguez, C. Cane, J. Samitier, A Reusable Smart Interface for Gas Sensor Resistance Measurement, *IEEE Transactions on Instrumentation and Measurement*, Vol. 53, Issue 4, August 2004 pp. 1173 - 1178.

[3]. Alejandro Duran Carrillo de Albornoz, Diego Ramirez Munoz, Jime Sanchez Moreno, Silvia Casans Berga, Edith Navarro Anton, A New Gas Sensor Electronic Interface with Generalized Impedance Converter, *Sensors and Actuators B: Chemical*, B134, 2008, pp. 591-596.

[4]. R. Richey, Resistance and Capacitance Meter Using a PIC16C622", AN611, *Microchip Technology, Inc.*, USA, 1995.

[5]. G. T. Nikolov, I. I. Stoyanov, B. M. Nikolova, RLC Measurement Using Virtual Counting Techniques, in *Proceedings of the ELECTRONICS' 2006*, September 2006, pp.69-74.

[6]. I. I. Stoyanov, G. T. Nokolov, B. M. Nikolova, Design of RLC to Frequency Converter, in *Proceedings of the ELECTRONICS' 2006*, September 2006, pp.63-68.

[7]. L. Bierl, Economic Measurement Techniques with the Comparator_A module, Application Report SLAA071, *Texas Instruments*, USA, October 1999.

[8]. F. Reverter, R. Pallas-Areny, Direct Sensor-to-microcontroller Interface Circuits: Design and Characterisation, *Marcombo S. A.*, 2005.

[9]. N. M. Mohan, B. George, V. J. Kumar, Dual Slope Resistance to Digital Converter, in *Proceedings of the. IEEE IMTC' 2007 Conference*, May, 2007, pp.1-5.

[10].N. M. Mohan, B. George, V. J. Kumar, A Sigma-Delta Resistance to Digital Converter Suitable for Differential Resistive Sensors, in *Proceedings of the IEEE I2MTC 2008 Conference*, May 2008, pp.1159-1161.

[11].Universal Sensors and Transducers Interface (USTI), Specification and Application Note, *Technology Assistance BCNA 2010, S. L.*, 2011.

[12].Yurish S. Y., Universal Resistance-to-Digital Converter, in *Proceedings of the 2nd International Conference on Advances in Circuits, Electronics and Micro-electronics (CENICS' 2009)*, Sliema, Malta, 11-16 October 2009, pp.28-33.

[13].F. M. L. Van der Goes, G. C. M. Meijer, A Novel Low-cost Capacitive Sensor Interface, *IEEE Transactions on Instrumentation and Measurement*, Vol.45, Issue 2, April 1996, pp.536-540.

[14] F. Reverter, Microcontroller-based Interfaces for Quasi-digital Sensors, PhD Thesis, *Technical University of Catalonia*, Barcelona, Spain, 2004.

[15].S. Y. Yurish, Advanced Automated Calibration Technique for Universal Sensors and Transducers Interface IC, in *Proceedings of the IEEE International Instrumentation and Measurement Technology Conference (I²MTC 2009)*, May 2009, pp.402-405.

[16].Terminal v.1.9b: http://www.sensorsportal.com/DOWNLOADS/Terminal.exe

[17].P. V. Novitskiy, I. A. Zograf, Error Estimation for Results of Measurements, *Energoatomizdat*, 1991 (in Russian).

[18].S. Kaliyugavaradan, A Linear Resistance-to-time Converter with High Resolution, *IEEE Transactions on Instrumentation and Measurement*, Vol. 49, No.1, , February 2000, pp.151- 153.

[19].S. Kaliyugavaradan, A Simple Resistance-to-time Converter for Signal Conditioning of Resistive Transducers, *Measurement Science and Technology*, Vol.11, Issue 5, May 2000, pp.73-75.

[20].URFC-2501-250 0.25 μm Resistance to Frequency Converter, *Global Unichip Corp*., Taiwan, 2004.

[21].V. Ferrari, C. Ghidini, D. Marioli, A. Taroni, A Conditioning Circuit for Resistive Sensors Combining Frequency and Duty-cycle Modulation of the Same Signal, *Measurement Science and Technology*, Vol.8, Issue 7, July 1997, pp. 827-829.

[22].Sheroz Khan, Mustafa Zeki, Shihab Abdel Hameed, AHM Zahirul Alam, Aisha Hassan Abdalla, A. F. Salami, W. A. Lawal, Signal Generator Circuit Carrying Information as Embedded Features from Multi-Transducer Signals, *Proceedings of World Academy of Science, Engineering and Technology*, Vol. 36, December 2008, pp.635-638.

[23].Edval J. P. Santos, Isabela B. Vasconcelos, RTD-based Smart Temperature Sensor: Process Development and Circuit Design, in *Proceedings of the 26th International Conference on Microelectronics (MIEL' 2008)*, 11-14 May 2008, pp.333-336.

Chapter 15

Resistive-Bridge Sensors Interfacing

15.1 Resistive-Bridge-to-digital Converters: State-of-the-Art Review

A Wheatstone bridge is widely used to convert the resistance variation of sensors and transducers that is related to the physical quantity into a voltage or current. Today many companies produce various resistive-bridge sensors, for example, SX30 Series differential pressure sensors *(Honeywell);* piezoresistive strain gauges; load cells; temperature, force, acceleration, magnetoresistive sensors HMC1052 and AAH002 (*Honeywell*); vacuum, absolute and differential pressure sensors HM, AP301, 3000, 5000, 7000 Series (*Merit Sensor);* Series SW415 barometric pressure sensors (*Sensonor),* Series SPD300 and SPD015 absolute pressure sensors *(Smartec),* etc.

To converter output bridge voltages into digital, analog-to-digital converters are often used, when the resistive sensor bridge is supplied by a voltage or current. The same or another dc reference source is used to perform precise measurement of the bridge-output voltage. Taking into account that the output bridge voltages are small, they must be amplified and filtered before digitization, and a high performance ADC should be used [1, 2]. 18-, 20- and 24-bit precision ADCs for bridge sensors are produced by several manufacturers, for example, ADS1130/31, ADS1230/31 and ADS1232/34 from *Texas Instruments* [3-5] or AD7796/97 and AD7191 from *Analog Devices.* Such classical solution is relatively expensive, because it requires precision components.

A solution with an embedded into a microcontroller ADC is described in [6] and requires analog multiplexer, operation amplifier and at least, six external discreet passive components as resistors and capacitors.

A low cost direct resistive-bridge sensor – to – microcontroller interface and its applications for some sensors and transducers are described in [2, 7, 8]. A microcontroller, external or internal Schmidt trigger and a few passive components are used for direct resistive sensor bridge–to–digital conversion.

However, the microcontroller based resistive-bridge – to – digital conversion needs a careful consideration and reduction additional errors due to a trigger noise, which depends on power supply noise and various program-related effects [9, 10].

ZMDI, Maxim, Texas Instruments, Analog Devices, Smartec, Micro Analog Systems and other manufacturers produce today integrated sensor signal conditioners and universal transducers interface (UTI) circuits with digital and quasi-digital outputs respectively for resistive bridge sensors for various industrial applications including automotive and medical. Main performances of some high performances integrated sensor signal conditioners with digital or quasi-digital output are shown in Table 15.1.

Table 15.1 Main performances of high-performance integrated industrial sensor signal conditioners and sensor signal interfaces.

Type	Resolution, bit	Output	Notices
ZMDI			
ZSSC3008	12	ZACwire serial 1-wire	Piezo-resistive bridge sensor types
ZSI21013	16	SPI	3 channel sensor bridge
ZSC31010	14	ZACwire serial 1-wire	Offset and gain compensation
ZSC31014	14	I^2C and SPI	PC-controlled configuration
ZSC31015	14	ZACwire serial 1-wire	PC-controlled calibration
ZMD31020	12	I^2C	Piezo-resistive bridge sensor types
ZMD31030	12	PWM, LIN compatible	Sample rate 100 Hz
ZMD31035	12	Digital 1-wire, LIN compatible protocol	Piezo-resistive bridge sensor types
ZSC31050	9 to 15	PWM (optional), I^2C, SPI, ZACwire	Additional analog outputs
ZSC31150	12.4	ZACwire or I^2C	Sensor specific correction
ZSSC3017	16	I^2C	Developed for correction of resistive bridge sensors
ZSSC3170	13 to 14	LIN or PWM	Correction of bridge sensor signals
Texas Instruments			
ADS1130	18	SPI	ADC for bridge sensors
ADS1131	18	Two-Wire Serial Digital Interface	ADC for bridge sensors
ADS1230	20	SPI	ADC for bridge sensors
ADS1231	24	SPI	ADC for bridge sensors
ADS1232	24	SPI	ADC for bridge sensors
ADS1234	24	SPI	ADC for bridge sensors
Analog Devices			
AD7796	16	SPI	ADC for bridge sensors
AD7797	24	SPI	ADC for bridge sensors
AD7191	24	SPI	ADC for bridge sensors
Maxim			
MAX1464	12	Two PWM outputs	2 channels
Smartec			
UTI03	13 to 14	Period modulated output	Resistive bridges 250 Ω - 10 Ω with maximum imbalance ± 4 % or ± 0.25 %

The Universal Transducer Interface (UTI) from *Smartec* is a complete analog front end for low frequency measurement applications, based on a period-modulated oscillator. Sensing elements (resistive bridge) can be directly connected to the UTI without the need for extra electronics. The UTI outputs a microcontroller-compatible period-modulated signal [11].

ZMDI provides mixed-signal integrated circuits for highly accurate amplification and sensor specific correction (offset, temperature drift, nonlinearity) of signals from resistive bridge sensor elements. The relative FS error is ±0.1 to ±0.25 %.

MAX1464 form *Maxim* to be tested with a resistive bridge pressure sensor to form a high-precision intelligent pressure monitoring system, the conversion rate 15 times / sec, measurement error is less than ± 0.1 %.

Micro Analog Systems, Oy manufactures MAS6502, a low power 16-bit piezo resistive sensor signal interface IC primarily intended for MEMS based pressure sensor modules with 14-bit ADC, 2-wire I^2C serial interface and selectable input signal and sensor resistance values ranges. A conversion time is 0.8 ms ... 10.6 ms.

Often, for example, in remote sensing applications, the resistance variation is preliminary converted into a frequency [12-17] or duty-cycle signal [14, 16, 17]. An integrated two-wire bridge-to-frequency converter described in [12] is intended to be used as a remote signal conditioner for sensor bridges such as strain-gauge bridges or platinum-wire temperature-sensing bridges. The converter has sensitivity on the order of 1 Hz per 1-μV/V relative bridge output. A center frequency of 10 kHz allows the application of an untrimmed bridge with an unbalance up to ±10 000 μV/V. The instability is less than 10^{-4} per Kelvin and per 1-V supply-voltage variation. The untrimmed transfer inaccuracy is lower than ±1 %. The linearity error is lower than ±0.01 %. Different bridge readout functions can be chosen by different circuit configurations. The converter can be connected to a single supply voltage. The frequency output is modulated on the supply current. The supply voltage is 12-24 V.

A fully integrated bridge-output-to-frequency converter realized as signal-conditioning circuitry for a one chip smart thermal mask flow sensor has described in [13]. The frequency converter reads out four separate resistor bridges of the air-flow sensor in order to realize high sensitivity and directional flow measurement over the full angle range of 360^0. The converter selects the sensor bridges individually by means of a microprocessor controlled analog multiplexer. The flow velocity and direction are also calculated by the microprocessor. The relative measurement cancels all first-order errors. The center frequency is 10 kHz and the sensitivity 1 Hz/(μV/V). A bridge-offset of 10 mV/V is allowed, and the linearity error is ±0.1 %.

The CMOS bridge to frequency converter with gain and offset control is presented in [15]. This circuit incorporates Wheatstone bridge was balanced, the simulated center frequency varied from 10.9 kHz to 220 kHz by changing the external capacitor from 1.5 nF to 0.05 nF. The maximum deviation in frequency from the central frequency increases from 4.5 kHz to 67 kHz as the capacitor decreases.

V. Ferrary *et al* [16] reported a signal conditioning circuit based on a relaxation oscillator proposed for use with resistive bridge sensors. The circuit provides a rectangular-wave output whose frequency is related to the bridge unbalance, and duty-cycle is a function of overall sensor bridge resistance, hence of the sensor operating temperature. In this way, two measurement values are simultaneously and independently carried on the same output signal.

An application specific integrated circuit (ASIC) front-end interface in 0.7 μm CMOS for resistive-bridge sensors is proposed by V. Ferrary *et al* [17]. The circuit is also based on a relaxation oscillator where the frequency is related to the fractional bridge unbalance, and the duty-cycle depends on the overall bridge resistance, which typically is related to temperature. The circuit has been characterized by means of a 1-kΩ reference bridge showing frequency and duty-cycle sensitivities of 60.4 Hz/(1000 ppm) and 0.276%/(mΩ/Ω), respectively, at a central frequency about 6.4 kHz.

No one existing on the modern market integrated converter for resistive bridge sensors (Table 15.1) can work with both: resistive-bridge sensing elements and frequency and duty-cycle signal conditioners' outputs, described above.

The universal solution - a universal interfacing circuit for resistive-bridge sensing elements and bridge-output-to-frequency and/or duty cycle converters designed by the author is described in this article, which is an extension version of the conference paper presented at SENSORDEVICES conference [18] and described in details in [19].

15.2 Direct Resistive-Bridge Sensors Interface

The Universal Sensors and Transducers Interfacing (USTI) integrated circuit has a special resistive bridge-to-digital conversion mode [20]. This fully digital CMOS integrated circuit based on novel patented methods for frequency, period, its ratio, duty-cycle and phase-shift measurements.

It uses the similar conversion method described in [2, 9], which is based on so-called a three-point measuring technique [21]; but instead of the external or internal Schmidt trigger, an embedded analog comparator has been used. The resistive sensor bridge is considered as a resistor network with three inputs and one output [2]. The resistance of each input to the output depends on the measurand. Using each input to turn to charge/discharge a capacitor connected to the bridge output yields three different time intervals. For a full bridge, the ratio between the difference between two time intervals and the third time interval yields the fractional resistance change [2]. This change x for each bridge arm can be obtained by the following way:

$$x = \frac{t_1 - t_3}{t_2} \tag{15.1}$$

The direct resistive-bridge sensors interface needs only a few external components such as a limiting resistor R and capacitor C (Figure 3.17).

The IC has three popular serial interfaces RS232, SPI and I^2C for communication to PC or master microcontroller. It can work in a slave communication mode with all mentioned serial interfaces and a master communication mode for RS232 communication. In this case, no any master microcontroller neither external PC should be used for IC controlling. The USTI continuously generates converted results at its output.

The following design consideration should be taken into account for the external components. The capacitance (in Farads) of capacitor C should be calculated according to the equation (3.5), and charging time (in seconds) – according to formula (3.3). The charging time can be set up with the help of USTI's command "Wnn".

An example of USTI's commands for resistance-bridge – to – digital conversion (RS232 serial interface, slave mode) is shown in Figure 15.1.

```
>M12        ; Set up a resistance-bridge Bx measurement mode
>W13        ; Set the charging time 20 ms
>S          ; Start measurement
>C          ; Check result status: returns "b" if busy and "r" if ready
>R          ; Read conversion result

0.006005379986
```

Fig. 15.1 USTI's commands for resistance-bridge – to – digital conversion at RS232 communication slave mode.

In the master communication mode the charging time should be set up with the help of external switches N_0, N_1, N_2; and the mode - by M_0, M_1, M_2 jumpers at appropriate USTI's inputs (Figure 3.19).

The circuit diagram of USTI working in slave communication mode (RS232) controlled by an external microcontroller or PC is shown in Figure 3.4.

Oscillograms on the USTI's CHRG pin for C=1 μF and charging time 2 ms; and for C=2 μF and charging time 4 ms at measurement for a resistive-bridge low cost differential pressure sensor SX30GD2 (Honeywell) with input/output resistance 4.1 kΩ at atmospheric pressure in laboratory conditions are shown in Figure 15.2 (a, b).

In the first case, the full length of one measuring cycle does not exceed 21.5 ms (without taking into account the communication and calculation times of IC); in the second case – 36 ms.

The relative quantization error for time interval measurement at each of discharging stage can be calculated according to equation (3.1). So, for the minimum discharge time ~5.ms (Figure 15.2 a) the relative quantization error for discharge time measurement will not exceed ± 0.001 % and can be neglected in comparison with a sensor's error.

In addition, the designed universal resistive-bridge – to – digital converter can work with any known bridge – to –frequency or – duty-cycle converters, , for example, described in [10-17] and industrial sensor signal conditioners with PWM and period-modulated outputs, for example, ZMD31030/31050, ZSSC3170 (*ZMDI*), MAX1464 (*Maxim*) and UTI03 (*Smartec*), Table 15.1. In such cases, their outputs of such converters should be directly connected to any of two USTI's channels for frequency-time measurements (pin 4, 6 or 5, 11).

The 2-channel USTI can measure a frequency in a wide measuring range from 0.05 Hz to 9 MHz (144 MHz with prescaling) with programmable relative error from ±1 to ±0.0005 %. The relative error is constant in the whole frequency range. The IC has non-redundant conversion time for frequency and period measurements, and scalable resolution. The frequency-to-digital conversion error of the USTI can be also neglected in comparison with the bridge-to-frequency and sensor error.

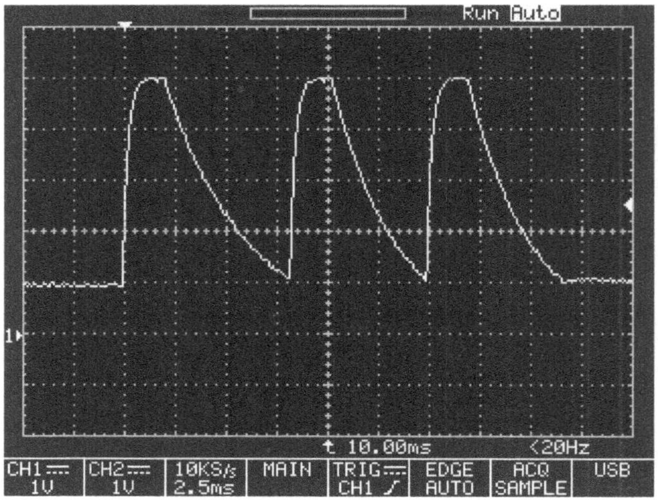

a)

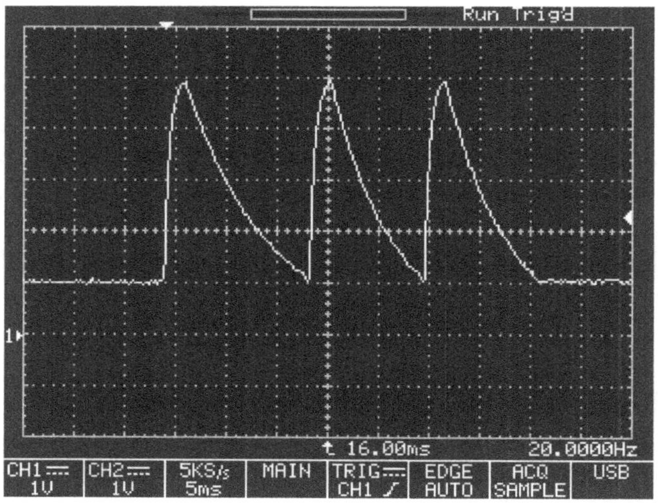

b)

Fig. 15.2 Oscillogram of three-point measurement technique at offset, reference and measurand determination charges-discharges stages on the CHRG pin: C=1 μF , charging time 2 ms (a); C=2 μF , charging time 4 ms (b).

In case when the condition circuit for resistive bridge sensors combines frequency and duty-cycle modulation of the same output signal, as described in [14, 16-17] it also can be directly connected to the designed universal resistive-bridge – to – digital converter because of the last one can convert both: frequency and duty-cycle into digital in the same channel.

The commands for frequency and duty-cycle measurements at RS232 slave communication mode for a resistive conditioning circuit are shown in Figure 7.

```
>M00;              Set up a frequency fx measurement mode
>A09;              Select the relative error 0.001 %
>S;                Start measurement
>C;                Check result status: returns "b" if busy and "r" if ready
>R;                Read conversion result for fx
10000.0045879

>M04;              Set up a duty-cycle measurement mode
>S;                Start measurement
>C;                Check result status: returns "b" if busy and "r" if ready
>R;                Read conversion result for duty-cycle
0.4532
```

Fig. 15.3 USTI's commands for frequency and duty-cycle measurements in the 1st channel at RS232 slave communication mode.

In the common case, two independent resistive-bridges – to – frequency condition circuits can be connected to the USTI as well as one addition resistive sensor bridge. Such advantage is very useful and multichannel sensor systems and data acquisition systems design. Taking into account a small MLF package size (5 mm × 5 mm × 1 mm) such converters can be easily embedded into various sensors' packages to produce low cost universal digital solution.

15.3 Applicaton Examples

The resistive bridge-to-digital converter has been tested with both: a strain gauges bridge emulator and differential pressure resistive bridge sensor SX30GD2.

15.3.1 Strain Gages Emulation

Preliminarily, the USTI has been calibrated at laboratory temperature range (+26.9 ^{0}C…+29.8 ^{0}C) in order to eliminate additional systematic error due to quartz oscillator trimming inaccuracy (calibration tolerance) and a short term temperature instability [22]. The USTI has been connected to a PC (through the serial RS232-to-USB interface) where the terminal software Terminal v1.9b was running under Windows [23].

The full resistive Wheatstone bridge sensor was emulated by the Strain Indicator Calibrator Vishay 1550A - a laboratory standard for verifying the calibration of strain and transducer indicators (Figure 15.4). It has the accuracy ±0.025 % due to high precision ultra-stable Vishay resistors used throughout to ensure excellent stability, repeatability, accuracy and incremental steps required in a laboratory standards instrument. The calibrator reads directly in microstrain. The bridge resistance was 350 Ω. The value of capacitor C was calculated according to the equation (3.5): $C = 0.002/350 = 5.7$ μF. In one's turn, the charging time was calculated according to formula (3.3): $T=2200 \times 5.7$ μF ≅ 13.2 ms. The actual charging time was set up 20 ms with the help of USTI's W13 command (Figure 15.1).

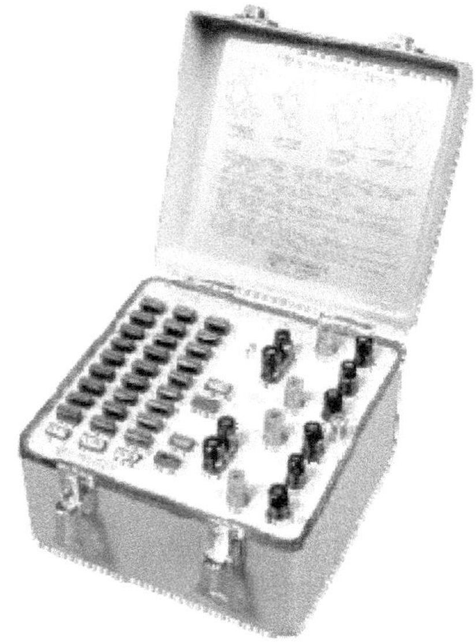

Fig. 15.4 1550A Strain Indicator Calibrator (*VISHAY*).

The strain gage range was changed at the experiment from 0 to ± 99 900 με with appropriate steps. Each of value was measured 60 times in order to compare obtained results with reported in literature It was visible changes for fractional resistance change x form approximately 0.018 (at 0 με) to 0.0995 (at 99 900 με). Statistical characteristics for measuring results are adduced in Table 15.2.The measurement results for emulated strain gage ranges 0 με and 99900 με are shown in Figure 15.5 (a, b).

The χ^2-test for goodness of fit was applied to investigate the significance of the differences between observed data in the histograms and the theoretical frequency distribution for data from a normal (Gaussian) population. Taking into account that the sum of deviations between the data set and the assumed distribution in both cases $S > \chi^2_{max}$, (see Table 15.2), a hypothesis about the Gaussian distribution can be accepted in these two cases.

The dependence of fractional part x vs. emulated values in the whole strain gage range from 0 με to 99900 με is shown in Figure 15.6 It is linear and has two sub ranges: (0 … 1000) με and (2000 … 99900) με.

Table 15.2 Statistical characteristics of measuring results.

Parameter	**Value**	
	x (0 με)	**x (99900 με)**
Number of measurements, N	60	60
Minimum R_x (min)	0.0016	0.0954
Maximum R_x (max)	0.0021	0.0959
Sampling Range, R_x (max) - R_x (min)	0.0005	0.0005
Median	0	0
Arithmetic Mean	0.0019	0.0957
Variance	9.5E-0009	1.4E-0008
Standard Deviation	0.0001	0.0001
Coefficient of Variation	19.5202	803.946
χ^2 – test (S) at: k=6; P = 97 % χ^2_{max} = 8.9	2.2572	5.354
Hypothesis about Gaussian distribution	Accepted	Accepted

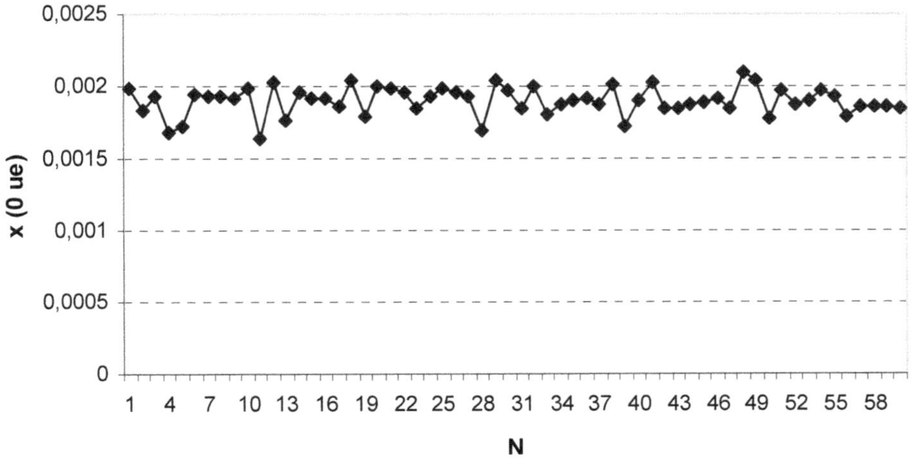

a)

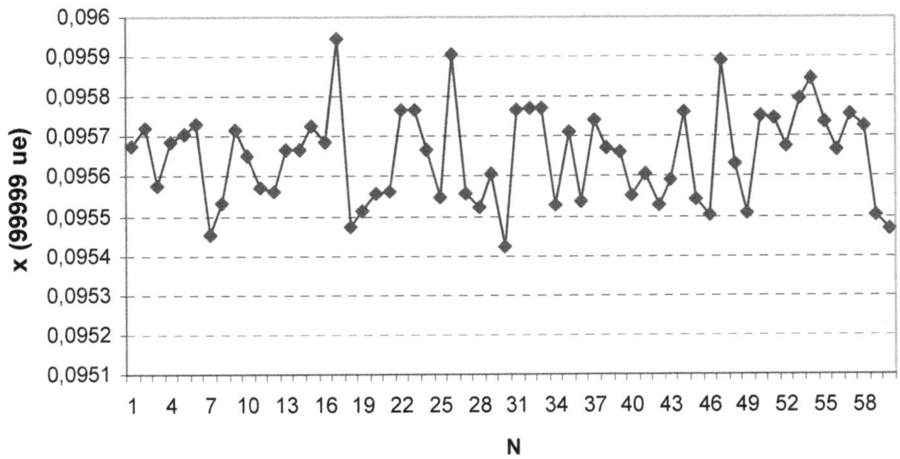

b)

Fig. 15.5 Measurement results for emulated strain gage ranges 0 με (a), and 99900 με (b).

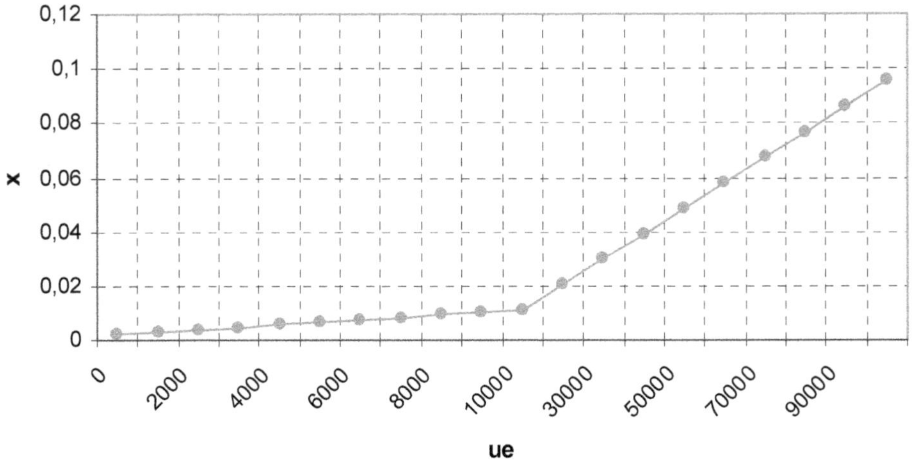

Fig. 15.6 x changes through the strain gage measuring range.

15.3.2 Differential Pressure Sensor Series SX30GD2

The sensor system consisting of USTI IC working in resistive-bridge sensing mode and a resistive differential pressure sensor bridge SX30GD2 (*Honeywell*) without any active component between the IC and bridge also was investigated experimentally.

The SX30GD2 is a low cost silicon differential pressure sensor with four-resistor bridge circuit (Figure 15.7). It has typical ± 0.2 % (± 0.5 % maximal) combined pressure non-linearity and pressure hysteresis FSS error and 0 … 30 psi operating pressure range and 4.1 kΩ bridge resistance.

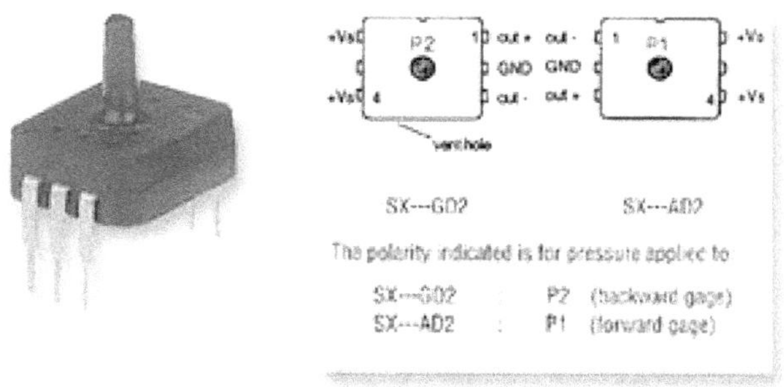

(a) (b)

Fig. 15.7 Differential pressure sensor bridge SX30GD2, *Honeywell* (a), pin-out for SX--GD2 and SX--AD2 (b).

Time intervals on the USTI's CHRG pin where measured 100 times with an external Schmidt trigger 74HC14D and without it (Figure 15.8). C was 2 μF and charging time 4 ms.

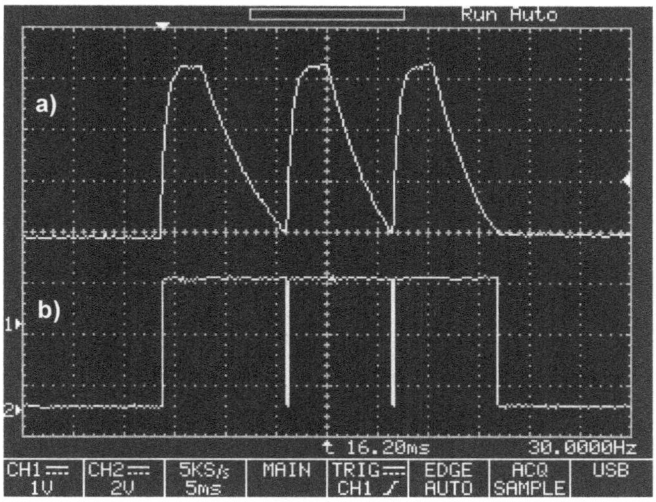

Fig. 15.8 Oscillograms on the USTI's CHRG pin without Schmidt trigger (a), and with external Schmidt trigger (b).

Statistical characteristics and measurement results for both cases at p=0 psi are shown in Table 15.3 and in Figure 15.9 (a, b) respectively. As it visible form the experimental results the relative error in these both cases are practically the same and comparable or less to the sensor's error.

Table 15.3 Statistical characteristics of measuring results.

Parameter	x (p = 0 psi)	
	With Schmidt trigger	**Without Schmidt trigger**
Number of measurements, N	100	100
Minimum R_x (min)	0.0053	0.0058
Maximum R_x (max)	0.0065	0.0062
Sampling Range, R_x (max) - R_x (min)	0.0012	0.0004
Median	0	0
Arithmetic Mean	0.005757267	0.006041671
Variance	1.0E-0007	6.6E-0009
Standard Deviation	0.0003	0.0001
Coefficient of Variation	17.805	74.3148
Confidence interval for arithmetic mean at P=97 %	0.0057 < x < 0.0058	0.006 < x < 0.0061
Relative error, %	0.87	0.83
χ^2 – test (S) at: k=8; P = 97 % χ^2_{max} = 12	57.0539	7.4373
Hypothesis about uniform distribution	Not accepted	Accepted

Experimental results have confirmed high metrological performance of resistive-bridge – to – digital converter based on the Universal Sensors and Transducers Interface circuit and using a simple, cost-effective three-point measuring technique. The converter can work with various resistive bridge sensors that do not include any internal components other than the arms forming a bridge.

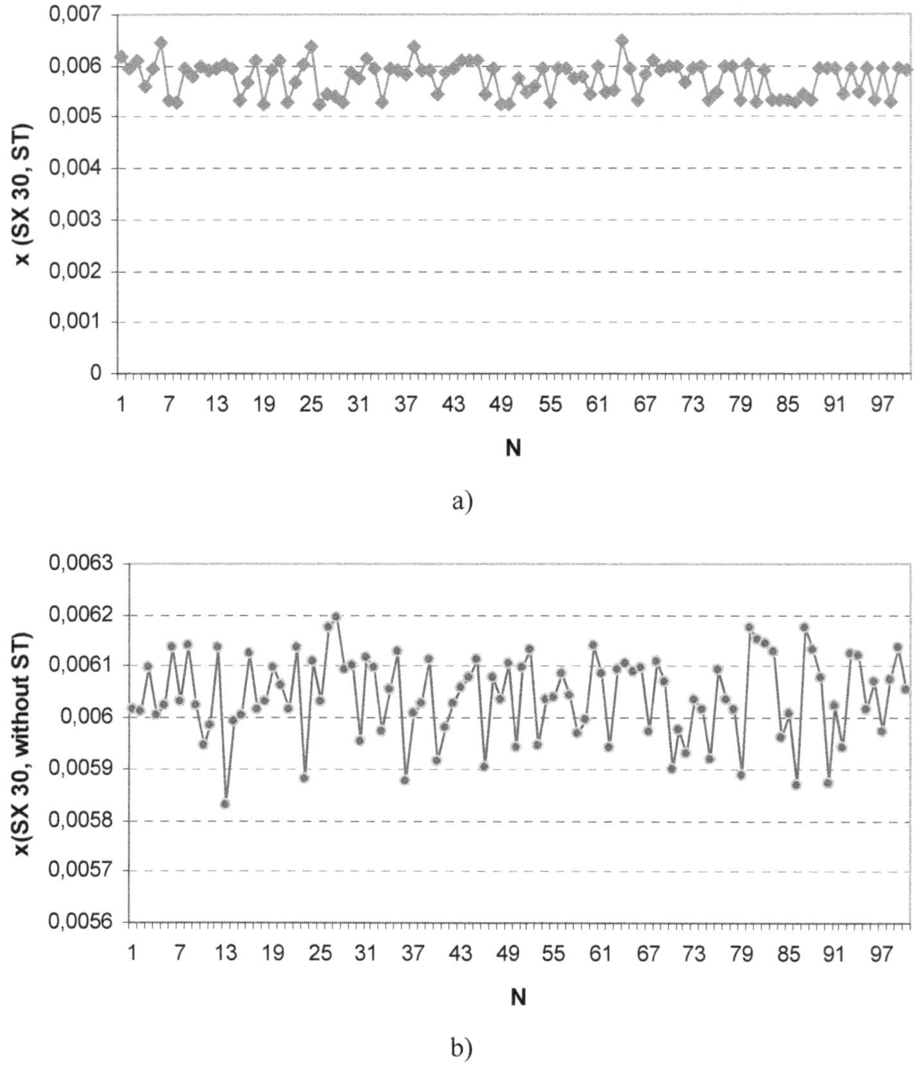

Fig. 15.9 Measurement results at 0 psi for sensor system with external Schmidt trigger (a), and wihtout Schmidt trigger (b).

In addition, the converter can work with any known resistance-bridge – to - frequency or duty-cycle converters and industrial sensor signal conditioners and interfacing circuits with PWM and period-modulated outputs. In such cases, their outputs should be directly connected to any of two USTI's channels for frequency-time measurements.

The USTI is working in industrial operating temperature range from – 40 ^{0}C to +85 ^{0}C. The power supply voltage is +4.5 … +5.5 V. The active supply current at operating voltage +4.5 V and clock frequency 20 MHz is not more than 9.5 mA.

The future research and development will be focused on the further improvement of metrological performance including conversion time, accuracy and extending the operating temperature range up to +150 ^{0}C in the future series of USTI ICs, for example, USTI-EXT and various sensor signal conditioners on its basis.

15.4 References

[1]. R. Pallás-Areny, J. G. Webster, Sensors and Signal Conditioning, 2nd ed., *Wiley-Interscience*, 2000.

[2]. A. Custodio, R. Bragós, R. Pallás-Areny, A Novel Sensor–Bridge-to-Microcontroller Interface, in *Proceedings of the 18th IEEE Instrumentation and Measurement Technology Conference (IMTC' 2001),* Vol. 2, Budapest, Hungary, May 21-23, 2001, pp. 892-895.

[3]. 18-Bit Analog-to-Digital Converter for Bridge Sensors, ADS 1130, Application Note, SBAS458, *Texas Instruments*, June 2009.

[4]. 20-Bit Analog-to-Digital Converter for Bridge Sensors, ADS 1230, Application Note, SBAS366A, *Texas Instruments*, Revised July 2007.

[5]. 24-Bit Analog-to-Digital Converter for Bridge Sensors, ADS1232, ADS 1234, Application Note, SBAS350F, *Texas Instruments*, Revised February 2008.

[6]. B. Baker, Building a 10-bit Bridge Sensing Circuit Using the PIC16CXX and MCP601 Operational Amplifier, Application Note AN717, Chandler (AZ): *Microchip Technology*, 1999.

[7]. J. Jordana, R. Pallás-Areny, A Simple, Efficient Interface Circuit for Piezoresistive Pressure Sensors, *Sensors and Actuators A*, Vol.127, 2006, pp. 69-73.

[8]. E. Sifuentes, O. Casas, F. Reverter, R. Pallás-Areny, Direct Interface Circuit to Linearise Resistive Sensor Bridges, *Sensors and Actuators A*, Vol.147, 2008, pp. 210-215.

[9]. R. Pallás-Areny, Advanced Sensors and Their Interface Circuits: Review and Perspectives, in *Proceedings of the 9th International Conference on Optimization of Electrical and Electronic Equipment (OPTIM' 2004),* on CD, Brasov, Romania, May 20–21, 2004.

[10].F. Reverter, J. Jordana, R. Pallás-Areny, Program-dependent Uncertainty in Period-to-code Converter based on Counters Embedded in Microcontrollers, in *Proceedings of the 19th IEEE Instrumentation and Measurement Technology Conference (IMTC' 2003),* Vail, USA, 20-22 May, 2003, pp. 1078-1082.

[11].Universal Transducer Interface (UTI), Specification, Version 4.12, *Smartec*, The Netherlands, October 2006.

[12].J. H. Huijsing, G. A. van Rossum and A. van der Lee, Two-wire Bridge-to-frequency Converter, *IEEE Journal of Solid-State Circuits*, Vol. SC-22, No. 3, June 1987, pp. 343-349.

[13].G. J. A. van Dijk, J. H. Huijsing, Bridge-output-to-frequency Converter for Smart Thermal Air-flowsensors, *IEEE Transactions on Instrumentation and Measurement*, Vol. 44, Issue 4, August 1995, pp. 881-886.

[14].V. Ferrari, C. Ghidini, D. Marioli, A. Taroni, A Conditioning Circuit for Resistive Sensors Combining Frequency and Duty-cycle Modulation of the Same Output Signal, *Measurement Science and Technology*, 8, 1997, pp. 827-829.

[15].D. McDonagh, K. I. Arshak, CMOS Bridge to Frequency Converter with Gain and Offset Control, *Microelectronics Journal*, Vol.29, 1998, pp. 727-732.

[16].V. Ferrari, D. Marioli, A. Taroni, Oscillatot-based Interface for Measurand-plus-Temperature Readout from Resistive Bridge Sensors, *IEEE Transactions on Instrumentation and Measurement*, Vol. 49, No.3, June 2000, pp. 585-590.

[17].V. Ferrari, A. Ghisla, Zs. K. Vajna, D. Marioli, A. Taroni, ASIC Front-end Interface with Frequency and Duty Cycle Output for Resistive-bridge Sensors, *Sensors and Actuators A*, Vol.138, Issue 1, July 2007, pp. 112-119.

[18].S. Y. Yurish, Universal Interfacing Circuit for Resistive-Bridge Sensors, in *Proceedings of the 1st International Conference on Sensor Device Technologies and Applications (SENSORDEVICES' 2010)*, 18-25 July 2010, Venice, Italy, pp. 211-216.

[19].S. Y. Yurish, A Simple and Universal Resistive-Bridge Sensors Interface, *Sensors & Transducers*, Vol. 10, Special Issue, February 2011, pp. 46-59.

[20].Universal Sensors and Transducers Interface (USTI). Specification and Application Note, *Technology Assistance BCNA 2010, S. L.*, 2011.

[21].G. Meijer (ed.), Smart Sensor Systems, *John Wiley & Sons*, Chichester, UK, 2008.

[22].S. Y. Yurish, Advanced Automated Calibration Technique for Universal Sensors and Transducers Interface IC, in *Proceedings of the IEEE International Instrumentation and Measurement Technology Conference (I^2MTC 2009)*, Victoria, Canada, May 2009, pp. 402-405.

[23].Terminal V1.9b: http://www.sensorsportal.com/DOWNLOADS/Terminal.exe

Chapter 16

DAQ Systems for Quasi-Digital Sensors and Transducers

16.1 Data Acquisition Systems: State-of-the-Art

Technological and manufacturing processes are sources of the initial data for multichannel data acquisition (DAQ) systems. The data are received by a control system as random processes of parameters variations and in the form of random events [1]. The data acquisition in such systems is defined as *"collecting and measuring electrical signals from sensors, transducers, and test probes or fixtures and input them into a computer for processing"* [1]. The last one can include conversion of sensors' characteristics, a joint signal processing of multiple parameters, and statistical processing of measurement results and presents them in a user-friendly form.

Multichannel data acquisition systems are intended for the conversion of initial parameters of processes and events (output signals from one or several sensors and transducers) into equivalent digital signals, suitable for the further processing, transferring and input it into the central computer, which controls the channeling data acquisition and forms data arrays for its further displaying or its use in control systems [1]. Modern data acquisition systems are able to handle practically all physical and chemical, electrical and non-electrical quantities, due to the wide variety of sensors and transducers. Alongside with analog sensors (with voltage, current or charge output), so-called quasi-digital sensors and transducers (with frequency, period, duty-cycle, PWM, phase-shift or pulse number output) are often used in such systems. In comparison with the analog output signals, used as informative parameters of sensors, frequency-time parameters allow to receive many advantages such as high noise immunity and output signal power; wide dynamic range; high accuracy of frequency standards (references); multiparametricity, simplicity of communication, interfacing, integration and coding [2]. Such sensors are rather interesting from a technological and fabrication compatibility point of view, because of their metrology performances and the simplifications of the signal conditioning circuitry. The simplicity yields a reduction of the required hardware and/or of the chip area [2]. Modern quasi-digital sensors have high accuracy (up to

±0.01 % relative error and even better: ±0.003 %) and broad frequency range from part of Hz to some MHz.

Methods of data acquisition depend on solved tasks of control and measurement and directly influence the structure and functionalities of multichannel data acquisition systems. Two traditional methods of data acquisition are widely used in modern automatic control and measuring systems [1]: a method using time-division channeling, based on multiplexing sensors, i.e. on time-shared data acquisition from each of sensor; and a method using space-division channeling, based on simultaneous data acquisition from all the sensors. Data acquisition with the time-division channeling allows to create inexpensive multichannel data acquisition systems for quasi-digital sensors but is not effective at a big number of channels. The method of data acquisition for quasi-digital sensors with the space-division channeling allows to design effective but relatively expensive multichannel systems.

There are many manufacturers who produce DAQ boards for frequency-time parameters of signals (Table 16.1). The maximum frequency and the number of channels meet today's requirements. However, the accuracy till now is not perfect in order to use these DAQ boards with the modern quasi-digital sensors, and/or frequency range and functionality are narrow [3]. Despite of the use of an advanced approach to software and driver design for DAQ systems, all modern boards, modules and systems, as rule, based on the classical methods for frequency (period) measurements with all disadvantages intrinsic to these methods. So, for low frequencies, period measurement (indirect counting method) is used. In this case, the counter typically counts an internal, fixed time base. Quantization error can cause period measurement to be off by ±1 timebase cycle. For high frequencies, the measurement is performing by measuring the time over N periods of the waveform. Then the period of the signal is computed and the reciprocal calculated to obtain the frequency. In this case, quantization error, which can be ±1 timebase cycle, is spread over a number of cycles of the measured signal. As the total time of the measurement increases, the quantization error becomes less significant. For both methods the maximum frequency, f_{xmax} that can be measured with a given error is obtained using the formula [4]:

$$f_{x\max} = \frac{xNf_0}{1+x} \tag{16.1}$$

where x stands for the error, N stands for the number of periods, and f_0 stands for the frequency of the timebase. For example, if we are using an 80 MHz timebase and stay within ± 0.001 or ± 0.1 % relative error, the maximum frequency that can be measured with N=1 is about 79.9 kHz. With N=13, the maximum frequency is 1 MHz.

However such an approach satisfies programmers more than it does mythologists and customers. Such DAQ boards, systems and modules can not be used for all existing on the modern market quasi-digital sensors and transducers [5]. In addition, it is not a low cost solution. Let consider some commercially available DAQ boards, systems and modules in details.

In the Timing I/O boards produced by *National Instruments* (USA), classical methods for frequency (standard counting technique) and period (indirect method) are used. The *National Instruments* also offers a module SCXI-1126 of virtual frequency meter [6].

However, despite the fact that today frequency and time can be measured with the highest possible accuracy in comparison with other informative parameters, the intermediate frequency-to-voltage conversion is used in this device. An analog signal proportional to the input frequency is formed on the converter's output. Then the resulting voltage signal is converted into digital code by a standard analog-digital converter, which is a part of the standard commercial DAQ boards.

Table 16.1 Main Performances of DAQ Boards.

Type	$f_{x\ max}$, MHz	Number of Channels	Error, %
Timing I/O Board, *National Instruments* (USA)			
PC-TIO-10	7	10	0.01
NI 660X	20 … 80 (60 …125)*	4 … 8	0.005 (clock accuracy)
PCI-DAQ Boards, *Kiethley* (USA)			
KPCI-31XX	5 (20)*	3 … 4 (8)	n/a
Frequency-input Card, *IOTECH* (USA)			
DBK7	0.95	4	0.1 (relative error)
Counter and Digital I/O Board, *Mielhaus Electronic* (Germany)			
ME-1400A/B	10	3 … 6	0.01
ME-1800	1	3	n/a
Time-to-Digital Converter, *Agilent-Acqiris* (USA)			
TC840	20 s (time interval)	12	0.0002 (clock accuracy)
DAQ System, *Intelligent Instrumentation* (USA)			
UDAS-1001E	10	1	n/a
Multi-Function Counter/Timer Card, *ADLINK Technology* (Taiwan)			
ACL-8454/X	10	6 … 12	n/a
PCI-8554	10	16	n/a
Timer/Counter Boards, *OMEGA* (USA)			
CIO-CTR10HD	7	10	n/a
CIO-CTR20HD	7	20	n/a
Timer/Counter Boards, *Contec* (Japan)			
CNT16-32S	0.2	32	n/a
CNT24-4	1	4	n/a
TCR-10	7	10	n/a
Frequency Module, *NPP Mera* (Russia)			
MC-451	0.01 Hz … 400 kHz	8	0.001 ÷ 0.01 (FS error)
Digital I/O and Counter Card, *Advantech* (USA)			
PCL-720	2.6	3	n/a
Timer/Counter Boards, *Axiom* (Taiwan)			
AX5216	7	5	n/a
AX5218	7	10	n/a
AX5220	10	3	n/a
Field Module, *MicroControl* (Germany)			
μCAN.4.ci-BOX	0.5	4	0.1 Hz (resolution)
Frequency Measurement and Switching Instruments, *JAQUET Technology Group* (Switzerland)			
DualTach T501	0.05	2	0.002 %
Instrument for QMB and SAW sensors, *JLM Innovation* (Germany)			
FQ4	500	4	n/a

* - with prescalling.

There are also DAQ boards for frequency signals based on the calculation of the spectrum of signal [6]. The frequency of periodical signal is calculated using the software module *Power Spectrum VI*, available in the library of software modules of LabVIEW software, based on a finite number of points obtained by digitized signal with the help of analog I/O boards. The module *Power Spectrum VI* uses the FFT to determine the fundamental frequency signal. This method is used for signals that are not changing in time during the sampling period.

The company *Brilliant Instruments* (USA) recently introduced on the market boards of time interval analyzers BI200 and BI220 with high resolution of time intervals measurement up to 3 ... 8 ps [7]. The time analyzer BI220 also has a mode for frequency measurement in a wide range from 0.15 Hz to 2.5 GHz (with a prescaler (frequency division)), but the relative error for frequency measurement is more than ± 0.002 % at the reference generator error ± 0.005 % and depends on the frequency because of analyzers use the classic method frequency measurement (standard counting technique) with analogue interpolation and the relative error in low frequency range is very intolerable. In addition, the device has a high price ~ US $ 20 000.

The high price more than 2 000 Euro has also the PCI board of Time/Frequency counter T3100(S), manufactured by the *VIGO System S. A.* (Poland) [8]. The device measures the frequency in a wide range from 0.1 Hz to 3.5 GHz by the method of reciprocal counting.

Modern multichannel data acquisition systems for quasi-digital sensors and transducers must be based on advanced methods for frequency (time parameter)-to-digital conversion and optimal design approach [9, 10]. The modified method of the dependent count, used for multichannel broad range measurements of frequency in 4-channel module of impulse signals for input it into PC is described in [11]. However, this modification of the method has led to essential decrease in accuracy, narrowing of a measurement range and dependence of a quantization error on the measured frequency similar to those, which is observed at the use of classical method of measurement of instant frequency (indirect counting technique) [12] and shown in the high frequency range.

The integrated Universal Sensors and Transducers Interface (USTI) and Universal Frequency-to-Digital Converters (UFDC-1 and UFDC-1M-16) are especially designed for applications in various multichannel data acquisition systems for quasi-digital sensors and transducers [13-15]. They lets to built low cost multichannel data acquisition systems with high metrological performances based on the methods of data acquisition for quasi-digital sensors with the space-division, time-division channeling and combined method. Examples and realization of such multichannel data acquisition systems are described in this Chapter.

16.2 DAQ Systems Design

16.2.1 DAQ Channels for Quasi-Digital Signals

The USTI, UFDC-1 and UFDC-1M-16 are well suitable for various intelligent multichannel data acquisition systems based on different DAQ methods due to its high metrological performances, wide functionality, intelligent functions and relatively low price. The designed ICs are based on four novel methods of measurements for frequency (period), its ratios, duty-cycle (duty-off factor) and phase shift.

The latest IC – USTI (Figure 16.1) has programmable relative error from ± 1 to ± 0.0005 %, which is constant in the broad frequency range from 0.05 Hz to 9 MHz (144 MHz with prescaling); non-redundant conversion time; scalable resolution; 29 measuring modes for various frequency-time parameters of electric signals, resistance, capacitance and resistive bridge parameters. The 2-channel ICs have 3 popular interfaces as RS232, SPI and I^2C. The USTI has internal reference frequency f_0=625 kHz, while the USTI-EXT has increased internal reference frequency up to 16 MHz.

Fig. 16.1 Universal Sensors and Transducers Interface (USTI) ICs in 28-lead Plastic Dual Inline Package (PDIP).

The designed multichannel DAQ system with time-division channeling based on the USTI IC is shown in Figure 16.2 [16, 17]. One quasi-digital sensor or transducer, for example, can be directly connected to the 1st USTI's channel, the rest sensors (or all of them) can be connected to the 2nd channel with the help of digital multiplexer (MUX). Such flexible DAQ architecture lets easy realize an advanced DAQ algorithm, when sensors of critical parameters (for example, in the 1st channel) must be served as soon as possible [1].

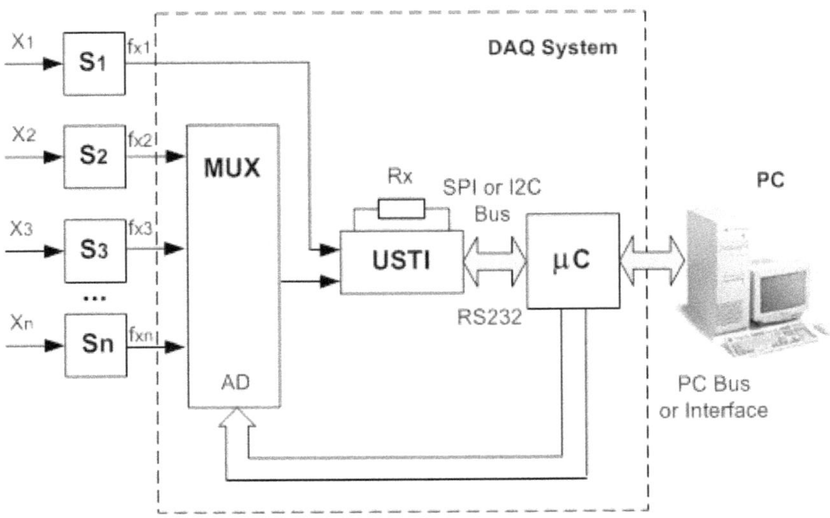

Fig. 16.2 Multichannel DAQ system with time-division channeling.

A microcontroller or microprocessor realizes communication functions between the IC and PC as well as control functions for both: MUX and USTI. In addition, one resistive, capacitive of resistive bridge sensing element can be connected directly to the USTI.

For the time-division channeling, the cycle polling time τ can be calculated according to the following equation:

$$\tau = n \cdot (T_{meas} + \tau_{delay1} + \tau_{delay2}), \tag{16.2}$$

where τ_{delay1} is the time delay between the ending of the conversion in the previous sensor and the command to poll the next sensor; τ_{delay2} is the time delay of the frequency conversion starting after the sensor connection; n is the number of sensors in the multichannel DAQ system; T_{meas} is the measurement time.

In a similar to the UFDC IC, the measurement time T_{meas} for the USTI IC includes three main components: conversion rate (t_{conv}), communication time (t_{comm}) and calculations time (t_{calc}):

$$T_{meas} = t_{conv} + t_{comm} + t_{calc}, \tag{16.3}$$

The conversion rate t_{conv} can be calculated according to the following equation:

$$\begin{cases} t_{conv} = \dfrac{1}{f_x} \quad if \quad \dfrac{N_\delta}{f_0} \prec T_x \\ t_{conv} = \dfrac{N_\delta}{f_0} + (0 \div T_x) \quad if \quad \dfrac{N_\delta}{f_0} \geq T_x \end{cases} \tag{16.4}$$

where $N_\delta = 1/\delta$ is the number proportional to the required programmable relative error δ; $T_x = 1/f_x$ is the period of unknown frequency f_x. The communication time for a slave communication mode (RS232 interface) can be calculated as:

$$t_{comm} = 10 \cdot k \cdot t_{bit}, \tag{16.5}$$

where t_{bit} = 1/300, 1/600, 1/1200, 1/2400, 1/4800, 1/9600, 1/14400, 1/19200, 1/28800 or 1/38400 is the time for one bit transmitting; k is the number of bytes (k=13÷24 for ASCII format). As usually, at the right chosen of baud rate (maximum possible for a certain application) the $t_{comm} \leq t_{conv}$. The communication time for SPI interface should be calculated according to the next formula:

$$t_{comm} = 8 \cdot k \cdot \frac{1}{f_{SCLK}}, \tag{16.6}$$

where f_{SCLK} is the serial clock frequency, which should be chosen for the USTI in the range from 100 to 500 kHz (5 MHz); k=12÷13 is the number of bytes. The number k is dependent on measurement result format: BCD (k=13) or binary (k=12). In the communication standard mode the speed for I^2C interfaces can be determined according to the following equation:

$$t_{comm} = 8 \cdot k \cdot \frac{1}{f_{SCL}}, \tag{16.7}$$

where f_{SCL} is the serial clock frequency, which should be equals to 100 kHz for the USTI; k=12÷13 is the number of bytes for measurement result: BCD (k=13) or binary (k=12). The calculation time t_{calc} depends on operands and as usually is not more than some ms.

The multichannel DAQ system with combined: space- and time-division channeling based on the USTI IC is shown in Figure 16.3. In such system instead of one USTI and an n-channel multiplexer, n ICs (according to the number of channels) and a microprocessor system with n inputs are used. That is, for simultaneous measurement of several frequencies, there is an independent channel with the USTI. The microprocessor simultaneously starts all converters and at the end of the measurement processes reads results. Quasi-digital sensors and transducers can be also connected in pairs to one USTI. In addition, one resistive, capacitive of resistive bridge sensing element can be also connected directly to the USTI. For one's turn, all USTI can be connected to a master microcontroller or microprocessor with the help of SPI or I^2C bus. Each USTI IC can serves up to 3 channels by itself in such DAQ systems.

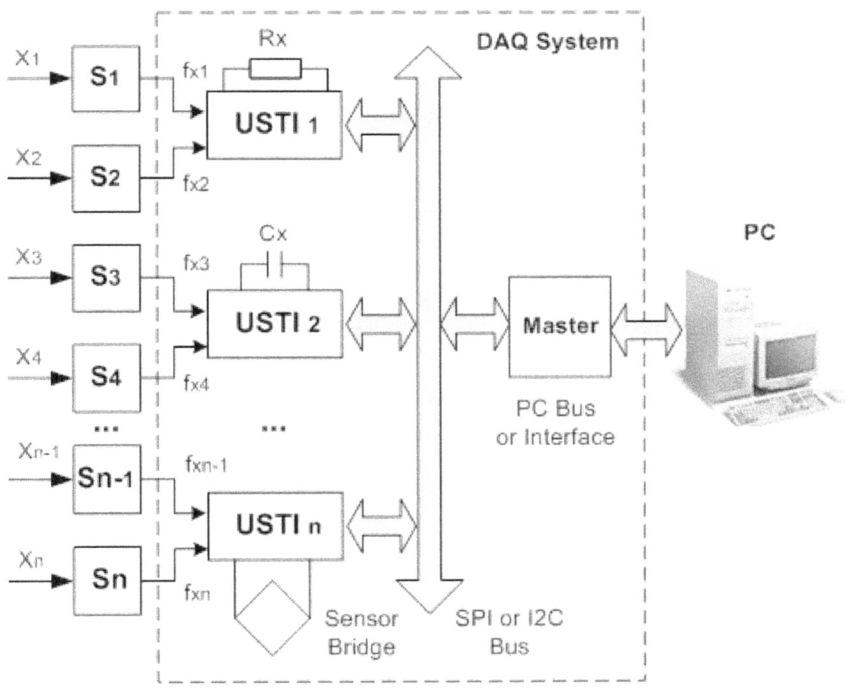

Fig. 16.3 DAQ system with combined space- and time-division channeling.

The cycle polling time τ for the space-division channeling is increased approximately in k times in comparison with the time-division channeling and should be calculated as $\tau = T_{meas} + t_{readout}$.

The commands for USTI working with RS232 interface in the slave communication mode at frequency measurements in both channels are shown in Figure 16.4.

```
> A0A            ; Set up the relative error ± 0.0005 %
> M00            ; Frequency measurement in the 1st channel
> S              ; Start measurement
> C              ; Check in the result is ready
r                ; Returns'b'-if in progress; 'r'-if ready
> R              ; Read result
100000.07507

> M0E            ; Frequency measurement in the 2nd channel
> S              ; Start measurement
> C              ; Check in the result is ready
r                ; returns'b'-if in progress; 'r'-if ready
> R              ; Read result
0.0500085968
```

Fig. 16.4 USTI commands for frequency measurements in two channels (RS232, slave communication mode).

If in the 2nd channel other than ± 0.0005 % relative error is necessary, it must be set up in advance for the 2nd channel by appropriate "A" command.

Circuit diagrams of capacitive, resistive sensors and passive resistive sensor bridges interfacing based on the USTI are shown in Figure 3.16, 3.15 and 3.17 respectively. The commands for capacitance, resistance and resistive bridge measurements are shown in Figure 13.3, 14.2 and 15.1 accordingly. In these three cases a three-signal measurement technique and preliminary parameter-to-time interval conversion are used in order to measure appropriate parameters [18-21].

The range of capacitance measurements is from 50 pF to 100 μF with an average relative error ± 0.036 % in a wide capacitance range. The range of resistance measurements is from 10 Ω to 10 MΩ with an average relative error 0.47 % and ± 0.01 % relative error at conditional splitting of the range of measurement into sub ranges with a choice of appropriate values of the reference resistor and charging-discharging capacitor. The full sensor bridges (four sensing arms) input resistances are from some hundred Ω to some kΩ.

Any capacitance-, resistive- or resistive bridge-to-frequency (period, time interval, duty-cycle, etc.) converters can be also connected to the USTI. The relative error for the further frequency-to-digital conversion is constant in the whole frequency range and not exceeds ± 0.0005 %.

16.2.2 DAQ Channels for Analog Signals

Analog sensors with voltage output can be also connected to the USTI and UFDC ICs by using an intermediate voltage-to-frequency conversion based on appropriate voltage-to-frequency converter (VFC). One channel of DAQ system for the analog signal is shown in Figure 16.5.

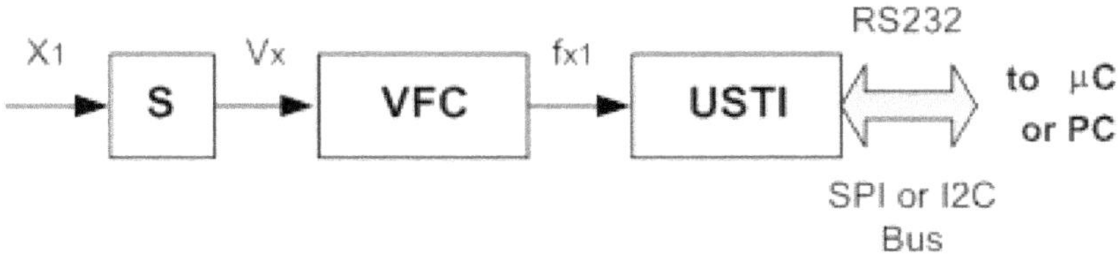

Fig. 16.5 DAQ system's channel for analog sensor.

Such approach is also an excellent solution to transmit an analog signal from remote sensors over a long distance without losing accuracy [22]. It is relatively easy to send a frequency signal over a long transmission path without interference via optical isolators, optical fiber links, telephone lines, twisted-pair or co-axial lines, or radio links. If the data must be digital, the receiver will consist of a frequency-to-digital converter. Such approach is used in different telemetry systems: VFC converts the sensor's analog output into frequencies, which can be used to modulate an RF transmitter. At the other end, a receiver pick up the RF signal and demodulates it back with the purpose of further frequency-to-digital conversion [23]. By convert in an analog voltage to a linearly-proportional pulse train of short duration, it is possible to transmit this data on the same wires used to energize the VFC.

Alongside with such main ADC architectures as successive approximation type, flash, integrating (dual slope) and Σ-Δ, the VFC can be successfully used for analog-to-digital conversion [24]. The VFC performs an indirect conversion by first converting the input analog signal to frequency and then using a frequency-to-digital converter to convert the frequency to a digital world. There are a number of advantages to using the VFC in an ADC conversion scheme. Fist, unlike converters based on binary-weighted network, monotonicity is inherent under all supply and temperature conditions. Second, the fact that the signal is converted to an easily-transmitted serial bit stream allows the analog circuitry (the VFC and analog signal conditioning circuits) to be located close to the signal source and the digital circuitry (frequency-to-digital converter) to be located elsewhere. This is especially advantageous when a large number of channels are required: the remote VFCs can be used to provide "converter-per-channel" data acquisition. Finally, since the digital number is accumulated over a large number of cycles, integration and therefore reduction of unwanted signals are inherent [25]. Resolution can, in theory be increased almost indefinitely.

Numerous of integrated VFC and modules are commercially available and overall performance is adequate to meet many requirements (for example, a new generation of CMOS synchronous charge-balance and current balance VFC with improved linearity and stability) [26, 27]. As distinct from the asynchronous VFC, which relay on the stability of an external capacitor to set their full-scale frequency, the charge-balance VFC AD7741/AD7742 from *Analog Devices* uses an external clock to define the full-scale output frequency [28]. The result is more stable, more linear transfer function (± 0.012 % integral nonlinearity) and allows the designer to determine the system stability and drift based upon the external clock selected. The interactive design tool for VFC AD650 is shown in Figure 16.6.

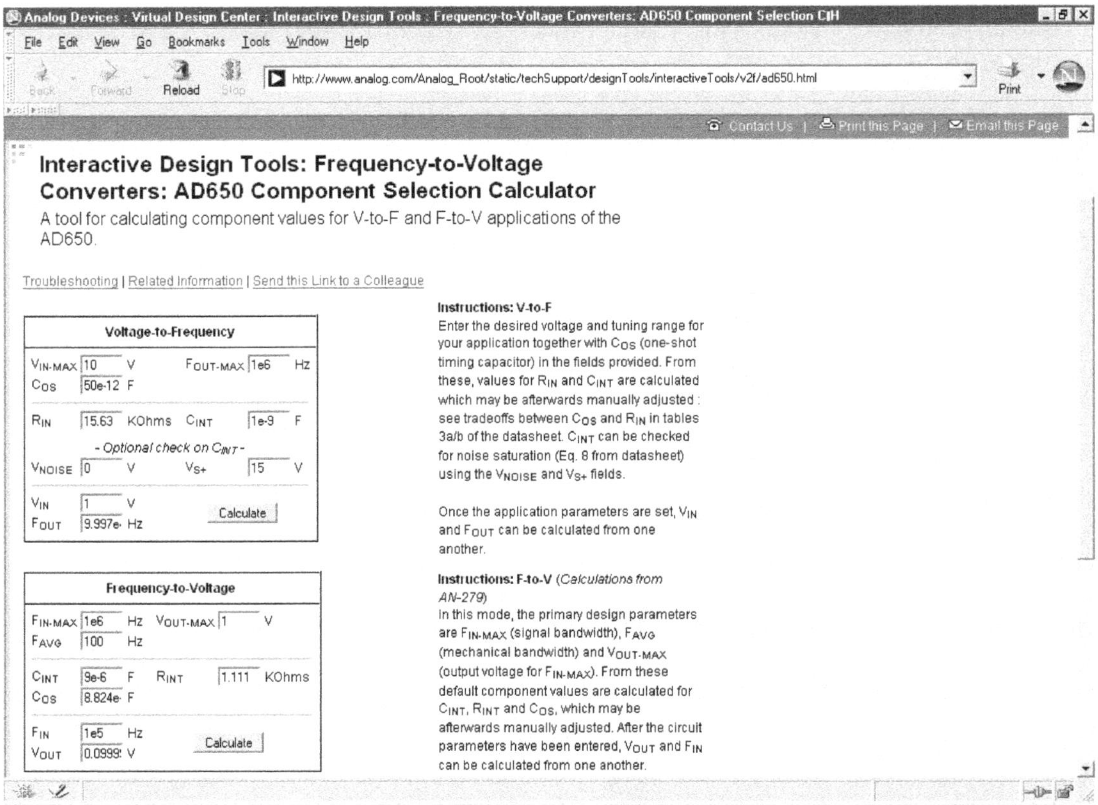

Fig. 16.6 Interactive Design Tool for VFC AD650 (http://designtools.analog.com/dt/v2f/ad650.html).

In many cases, however, very high performance or special characteristics are required and available ICs will not work. In these instances VFCs specially optimized for the desired parameters are required in order to improve speed, dynamic range, stability and linearity [29, 30].

In spite of the fact that the design methodology and technique for high performance VFCs (for example, ultra-high speed 1 Hz-100 MHz VFC with 0.06 % linearity; fast response (3 μs) 1 Hz-2.5 MHz VFC with 0.05 % linearity; high stability quartz stabilized 10 kHz – 100 kHz VFC with 0.005 % linearity and ultra-linear 100 kHz – 1 MHz VFC with linearity inside 7 ppm (0.000 7%) and 1 ppm resolution for 17-bit accuracy applications) exists over long time [29], the task for further frequency-to-digital conversion for such VFC was a "bottleneck" in such promised ADC scheme. Simple pulse or period counting techniques based on classical frequency-to-digital conversion methods essentially limited the conversion rate or dynamic range. Accuracy was decreased the conversion rate was increased. Typical accuracy at 100-300 conversions per second was 12 bits [24].

The solution of the mentioned problem is based on the novel integrated universal frequency-to-digital converters (UFDC-1M-16) and universal sensors and transducers interface (USTI) based on advanced conversion method, with the following advantages: constant quantization error δ_q in all frequency range (from f_{xmin} to f_{xmax}), programmable relative error δ, non-redundant conversion rate T_{conv} and a capability to measure frequency $f_x >> f_0$ (reference frequency). The conversion rate for such converters can be calculated according to the equation (16.4).

An example of data acquisition system for remote sensor using the ADC based on VFC and UFDC-1M-16 is shown in Figure 16.7. The conversion rate of IC UFDC-1M-16 is shown in Figure 2.28 (b). The IC has a wide frequency range from 0.05 Hz to 7.5 MHz (120 MHz with prescaling), programmable relative error from ±1 to ±0.001 % at appropriate conversion rate from 6.25 μs to 6.25 ms, I^2C, SPI and RS232 communication interfaces. Due to the programmable relative error the conversion time and accuracy can be optimized for specific applications.

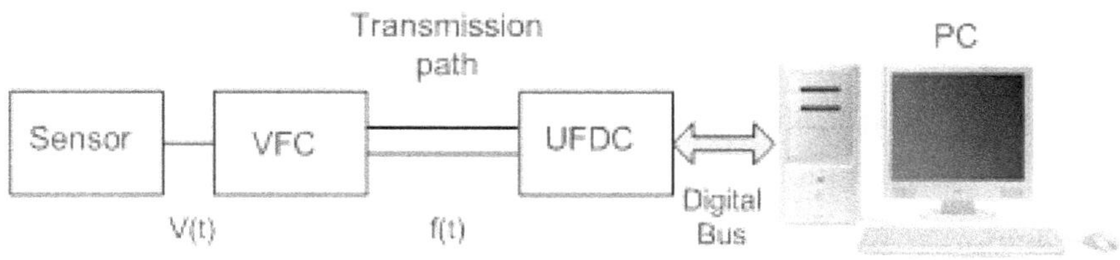

Fig. 16.7 Data acquisition system for remote sensors.

One of the more important advantages is that the UFDC-1, UFDC-1M-16 and USTI ICs can be easily fabricated with the VFC on one digital CMOS chip, for example, system-on-chip developed by *Cypress Microsystem* [31, 32] to provide low cost, high accuracy, multichannel signal acquisition and adaptive conversion to digital on a single chip. From other side, these ICs can be fabricated together with a CMOS MEMS quartz crystal oscillator to provide both: voltage-to-frequency and frequency-to-digital conversions with internal, stable, precision frequency base [33].

This technique will give a wide dynamic range and faster conversion time. The last one is non-redundant, shorter than for the pulse counting and comparable with the successive-approximation and Σ-Δ ADC. This will be especially important in systems where a number of channels are being converted. Table 16.2 briefly shows some of the main comparative characteristics and tradeoffs of mainstream analog-to-digital converter technologies [34].

Table 16.2 Analog-to-digital converter types for data acquisition, measuring and sensors systems.

Converter Type	Maximum speed	Typical resolution, bit	Noise Immunity	Relative Cost
Successive Approximation	Medium (10 kHz to 1 MHz)	6-16	Little	Low
Integrating	Slow (10 Hz to 30 Hz)	12-24	Good	Low
VFC-based	Medium (160 kHz to 1 MHz)	16-24 or more	Excellent	Low
Sigma-Delta	Slow to Medium (Up to 1 MHz or higher)	16 or more	High	Low
Flash	Very Fast (1 MHz to 500 MHz)	4-8	None	High

The mainstream analog-to-digital conversion methods for data acquisition and measurement instruments include successive approximation, integrating, sigma-delta, flash and VFC-base converters. Each conversion method offers a different combination of performance and price that makes it suitable for specific set of data acquisition applications. It is expediently to use the VFC-based analog-to-digital conversion in such applications as remote sensors, telemetry applications and multichannel data acquisition systems.

The advanced analog-to-digital conversion based on the voltage-to-frequency-to-digital conversion lets receive many advantages such as high accuracy, relatively low power consumption relatively low cost solution, great stability, wide dynamic range and faster conversion time in comparison with existing ADC techniques and comparable with the successive-approximation and Σ-Δ ADC. The VFCs are most cheap solutions for multichannel DAQ systems with high noise immunity and simple galvanic isolation [35].

The prototype of DAQ board for quasi-digital and analog sensors and transducers is shown in Figure 16.8. It was based on the USTI evaluation board and the evaluation board for single and multi-channel synchronous voltage-to-frequency converters [35]. The last one is based on the integrated VFCs AD7741/AD7742 [28]. The experimental dependence of measured by the USTI frequency and input voltage is shown in Figure 16.9 Due to high programmable accuracy, the USTI's relative error can be neglected in comparison with the VFC's error, because of it can be selected in one order less than the VFC's conversion error.

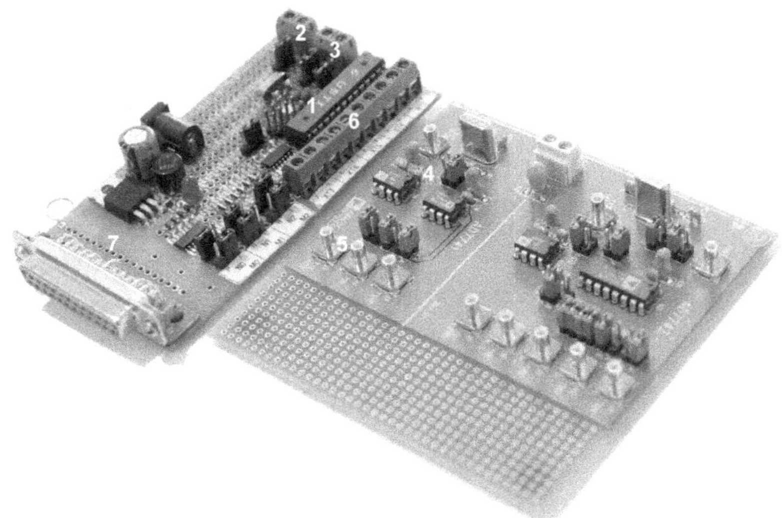

Fig. 16.8 DAQ board prototype: 1- USTI; 2,3 – inputs for quasi-digital sensors; 4-VFC AD7741; 5- input for analog sensor; 6 – inputs for sensing elements and external components; 7 – RS232 connector.

In addition to high metrological performances, wide functionality and parallelism, the designed multichannel DAQ systems for analog and quasi-digital sensors and transducers have several intelligent (smart) functions: parametric adaptation and self-identification functions [36]. An adaptation in DAQ systems can be used for increasing of measurement accuracy and /or decreasing of measuring time. Based on an appropriate adaptive measuring algorithm, a trade-off between required relative error and time of measurement can be chosen. The parametrical adaptation consists in production of control signals supporting the DAQ system in the required state by means of adopted parameters depending on measuring conditions and/or measuring algorithm.

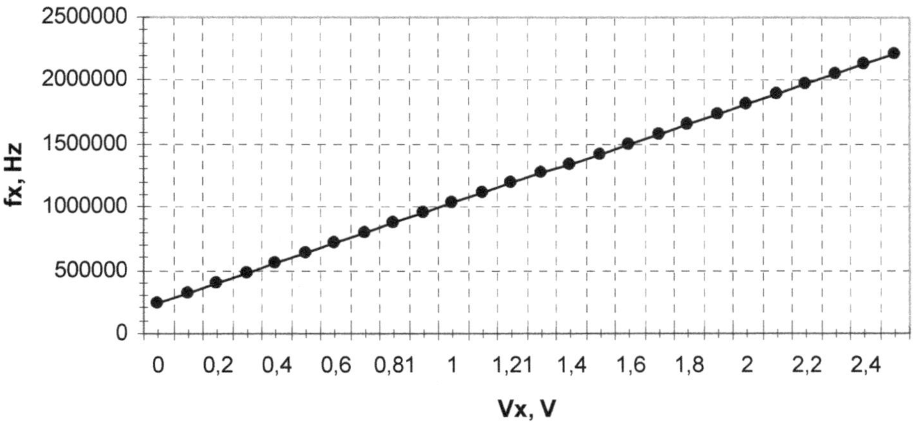

Fig. 16.9 Measured frequency by USTI vs. input voltage V_x.

The USTI can also store the Transducer Electronic Data Sheet (TEDS) according to the IEEE 1451 standard with appropriate sensors calibration constants and frequency-to-digital programmable quantization error in its nonvolatile flash memory [37].

Taking into account the programmable relative error of the mentioned ICs, it is possible to set an appropriate conversion error at non-redundant time of measurement and flexibly change it according to the advanced measuring algorithm or current measuring conditions. This possibility supplies an intelligence extension due to the self-adaptation capability together with the self-identification of smart transducers. An intelligent sensor systems design based on USTI and UFDC ICs are described in details in Chapter 17.

The designed intelligent multichannel DAQ systems based on the novel ICs of USTI and UFDC-1M-16 have demonstrated high metrological performances. It can be used with any existing quasi-digital as well as analog sensors and transducers. A parallelism in such systems was realized as a time-division channeling and a combined channeling, included both: the time- and space- division channeling. In addition to the self-adaptation capability the sensors in such systems will have self-identification capability due to the TEDS according to the IEEE 1451 standard in the ICs' memory. The described design approach can be used also for creation low cost, high performances multichannel, multifunctional DAQ systems for basic environmental sensing parameters as well as many other custom sensing applications. In this case, only one IC of USTI can deal with three different sensing parameters and DAQ system should use the time-division channeling.

16.3 Multifunctional and Multiparametric Sensors, Transducers and Sensor Systems

Modern micro-electro-mechanical (MEMS), thin film and semiconductor technologies enable the integration of microelectronics circuits and multifunctional sensor arrays fabricated on silicon substrates for detecting of different kinds of chemical and physical parameters [38]. Such sensors are called *multfunctional sensors* and represent the single construction, allowing the perceiving of the set of physical or chemical quantities, which action is concentrated in a small, local area. It is not a simple unit of one-functional sensors, but rather its structurally advanced combination is created with the aim to reduce the chip area and share the usage of the digital or quasi-digital output, etc. [39]. Modern multifunctional sensors and transducers can be created either within one signal sensor domain, such as radiation or a combination of several signaling domains, namely, first of all, thermal, mechanical, chemical and electrical.

Depending on the number of informative parameters in one of the outputs of the multifunctional quasi-digital sensor, such devices can be divided into *single-* or *multiparametric sensors*. So multifunctional, single-parametric sensors and transducers can have some outputs with measurands proportional to a single frequency-time parameter: frequency, period, duty-cycle, etc. on each of outputs. In turn, the multifunctional, multiparametric sensors and transducers can contain one or more outputs, the measured values on which are proportional to the multiple time-frequency parameters for each of the outputs. For example, the frequency of one of the output is proportional to a measured value and the duty cycle on the same output – to another measured quantity.

Advanced sensors for detecting simultaneously of various parameters such as temperature, pressure, gas and vapour concentration, odour, acceleration, inertia, electric and magnetic fields, etc. can generate frequency-time signals on its outputs carrying the information provided by sensing elements. It must provide not only high signal-to-noise ratio in a wide dynamic range but also good cross-sensitivity [38].

The first multifunctional sensors were a natural combination of temperature and pressure or temperature and humidity sensors in the single device. The temperature sensors have been used also for temperature error compensation. Late, this set of quantities has been added by conductivity, dew-point, pH, oxygen, etc.

Interesting examples are the integrated smart optical sensors developed in Delft University of Technology [40, 41]. Integrated on-chip colour sensors have been designed and fabricated to provide a digital output in the IS2 bus format. The readout of photodiodes in the silicon takes place in such a way that pulse series are generated with the pulse frequency proportional to the optical intensity (luminance) and the duty cycle to the colour (chrominance). The colour information is obtained using the wavelength dependence of the absorption coefficient in the silicon in the optical part of the spectrum, so no filters are required. The counters and the bus interface have been realized in a bipolar and CMOS version with the enhanced resolution, which is being investigated. An example of sensor system based on such sensor and UFDC-1 IC is shown in Figure 16.10 [42].

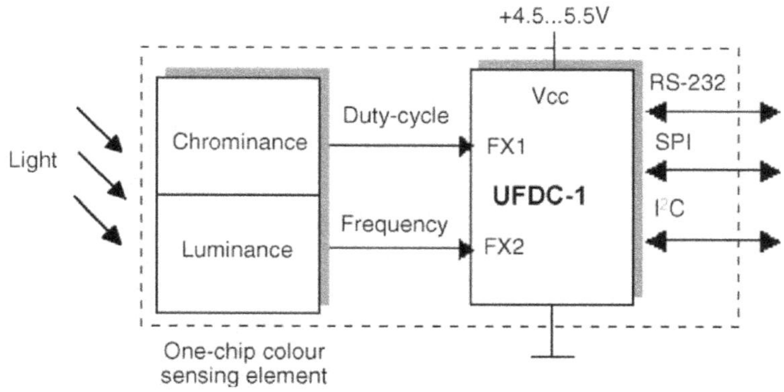

Fig. 16.10 The UFDC-1 application in sensor system with multiparametric opto sensor [42].

Another example of quasi-digital, multifunctional, multiparametric sensor is described in [43]. This piezoresistive pressure sensor has one output with frequency proportional to pressure and duty-cycle proportional to temperature.

The multifunctional, single-parametric sensor system for measuring humidity (based on the principle of hygrometric temperature difference between dry and wet quartz thermometers), pressure and temperature is described in [44]. The system has three frequency outputs with a single informative parameter for each of the outputs.

Some quasi-digital multifunctional sensors and transducers are available on the market. For example, all series of multifunctional, single-parametric frequency output pressure and temperature transducers with two square waves outputs from *Parascientific* (USA): (30…42) kHz for pressure and (168…172) kHz for temperature [45]; pressure transducers QHB, SPB, QMB from *Quartzdyne* [45] with two frequency outputs: (10 … 70) kHz for pressure and (10 … 90) kHz for temperature, and other sensors.

The UFDC and USTI ICs can be used with all mentioned above multifunctional, mustiparametric sensors. By this way, for example, only one USTI IC can be used in sensor systems for five various quantities: two quasi-digital, multifunctional, multiparametric sensors or transducers should be connected to F_{x1} and F_{x2} inputs, and one capacitive, resistive or resistive bridge sensing element should be connected to appropriate IC's inputs. In such sensor system the USTI will measure frequency and duty-cycle in each of channel, and convert capacitance, resistance or bridge to digital.

16.4 References

[1]. N. V. Kirianaki, S. Y. Yurish, N. O. Shpak and V. P. Deynega, Data Acquisition and Signal Processing for Smart Sensors, Chichester, UK, *John Wiley & Sons*, 2002.

[2]. S. Y. Yurish, Data Acquisition for Frequency- and Time Domain Sensors, in Smart Sensor Systems, ed. by G. C. M. Meijer, Chichester, UK, *John Wiley & Sons*, 2008.

[3]. Manufacturers of Data Acquisition (DAQ) Systems at Sensors Web Portal (http://www.sensorsportal.com): http://www.sensorsportal.com/HTML/SENSORS/DAQ_Systems_Manufacturers.htm

[4]. The Measurement and Automation Catalog, *National Instruments*, USA, 2000.

[5]. Sensors Web Portal: http://www.sensorsportal.com/HTML/Sensor.htm

[6]. Ghercioiu M., Hubbard E., Frequency Measurement with SCXI-1126, Application Note 138, *National Instruments*, USA, 2000.

[7]. Time Interval Analyzers, Brilliant Instruments webs site: http://www.b-i-inc.com/

[8]. Time/Frequency Counter Model T3100(S), *Vigo System S. A.*, 2011.

[9]. N. V. Kirianaki, Yurish S. Y., Advanced DAQ Board for Frequency-Time Parameters of Electric Signals, in *Proceedings of the IEEE International Workshop on Intelligent Data Acquisition and Advanced Computing Systems: Technology and Applications (IDAACS' 2003)*, 8-10 September 2003, Lviv, Ukraine, pp.150-153.

[10].N. V. Kirianaki, S. Y. Yurish, Measuring System for Acquisition and Signal Processing of Frequency-time Domain Sensors Signals, *Sensors and Systems*, No.12, December, 2004, pp. 2-5 (in Russian).

[11].V. V. Makarov, S. M. Volodin, Multichannel measurements of frequency pulse signals by the modified method of the dependent count, *Sensors and Systems*, Vol. 9, pp.34-39, September 2004 (in Russian).

[12].R. S. Yermolov, Digital Frequency Counters, Moscow, USSR, *Energija*, 1973 (in Russian).

[13].Universal Frequency-to-Digital Converter (UFDC-1), Specification and Application Note, *Technology Assistance BCNA 2010, S. L.*, 2011.

[14].Universal Sensors and Transducers Interface (USTI), Specification and Application Note, *Technology Assistance BCNA 2010, S. L.*, 2011.

[15].High-Speed Universal Frequency-to-Digital Converter UFDC-1M-16, Data Sheet, *Technology Assistance BCNA 2010, S. L.*, 2011.

[16].S. Y. Yurish, Intelligent Data Acquisition System for Quartz Crystal Microbalance and Other Resonator-Based Sensors, in *Proceedings of the IEEE International Workshop on Intelligent Data Acquisition and Advanced Computing Systems: Technology and Applications (IDAACS' 2007)*, 6-8 September 2007, Dortmund, Germany, pp. 2-7.

[17].S. Y. Yurish, Multichannel Data Acquisition System for Smart Sensors and Transducers, *Sensors & Transducers*, Special Issue, Vol.8, February 2010, pp. 1-12.

[18].S. Y. Yurish, Universal Capacitive Sensors and Transducers Interface, in *Proceedings of the 23rd International Conference EUROSENSORS XXIII*, Procedia Chemistry 1, Lausanne, Switzerland, 6-9 September 2009, pp.441-444.

[19].S. Y. Yurish, Universal Resistance-to-Digital Converter, in *Proceedings of the 2nd International Conference on Advances in Circuits, Electronics and Micro-electronics (CENICS' 2009)*, Sliema, Malta, 11-16 October 2009, pp.28-33.

[20].S. Y. Yurish, Universal Interfacing Circuit for Resistive-Bridge Sensors, in *Proceedings of the 1st International Conference on Sensor Device Technologies and Applications (SENSORDEVICES' 2010)*, 18-25 July 2010, Venice, Italy, pp. 211-216.

[21].S. Y. Yurish, A Simple and Universal Resistive-Bridge Sensors Interface, *Sensors & Transducers*, Vol. 10, Special Issue, February 2011, pp. 46-59.

[22].S. Y. Yurish, Advanced Analog-to-Digital Conversion Using Voltage-to-Frequency Converters for Remote Sensors, *Key Engineering Materials: Measurement Technology and Intelligent Instruments VIII*, Vol. 381 – 382, 2008, pp.623-626.

[23].M. O. Paiva, Voltage-to-Frequency/Frequency-to-Voltage Converter, AN795, *Microchip Technology, Inc.*, 2002.

[24].A. Collins, Solid State Solutions for Electricity Metrology, Metering and Tariffs for Energy Supply, *Conference Publication*, No. 462, 1999.

[25].P. Klonowski, Analog-to-Digital Conversion Using Voltage-to-Frequency Converters, AN-276 Application Note, *Analog Devices*.

[26].Voltage-to-Frequency Converters (VFC) and Current-to-Frequency Converters (CFC): www.sensorsportal.com/HTML/SENSORS/Voltage_to_Frequency_Converters.htm
[27].Manufacturers of Voltage-to-Frequency Converters at Sensors Web Portal: http://www.sensorsportal.com/HTML/SENSORS/VFC_Manufacturers.htm
[28].Single and Multichannel, Synchronous Voltage-to-Frequency Converters, AD7741/AD7742, *Analog Devices, Inc.*, USA, 1999.
[29].Williams J. Designs for High Performance Voltage-to-Frequency Converters, Application Note 14, *Linear Technology*, March 1986.
[30].V. Deryabin, Voltage-to-frequency Converters, *Components and Technologies*, No.7, 2000 (in Russian).
[31].R. Aragonis, C. Ferrer, J. Oliver, Use of PSoCTM Component in a Frequency Based Acquisition Method for Smart Sensors Applications, in *Proceedings of the IEEE International Conference on Industrial Technology (ICIT)*, 2004, pp.1357-1362.
[32].R. Aragonis, J. Oliver, Carles Ferrer, Self-Adaptive A/D Converter Based on a Frequency Based Acquisition Method for Smart Sensors Applications, in *Proceedings of the IEEE ISIE Conference*, 20-23 June, 2005, Dubrovnik, Croatia, pp. 1055-1056.
[33].S. Y. Yurish, MEMS Based System-on-Chip and System in Package: New Perspectives, *Sensors & Transducers*, Vol. 78, Issue 4, April 2007, pp. I-IV.
[34].Data Acquisition and Control Hand Book, *Keithley*, USA 2001.
[35].Evaluation Board for Single & Multi-Channel Synchronous Voltage-to-Frequency Converters EVAL-AD7741EB/EVAL-AD7742EB, *Analog Devices*, USA.
[36] S. Y. Yurish, Self-Adaptive Smart Sensors and Sensor Systems, *Sensors & Transducers*, Vol.94, Issue 7, July 2008, pp.1-14.
[37] S. Y. Yurish, Extension of IEEE 1451 Standard to Quasi-Digital Sensors, in *Proceedings of the IEEE Sensors Applications Symposium 2007 (SAS-2007)*, San Diego, California, USA, February 6-8, 2007.
[38].D. V. Ivanov, Advanced Sensors for Multifunctional Applications, *JOM*, October 2000.
[39].Smart Sensors and MEMS, ed. by S. Y. Yurish and M. T. Gomes, Dordrecht, Boston, London, *Kluwer Academic Publishers*, 2004.
[40].G. de Graaf, R. F. Wolffenbuttel, Light-to-Frequency Converter Using Integrated Mode Photodiodes, in *Proceedings of the IMTC'96 Conference*, 4-6 June, Brussels, Belgium, 1996, pp.1072-1075.
[41].G. de Graaf, F. Riedijk, R. F. Wolffenbuttel, Colour Sensor System with a Frequency Output and an ISS or 1^2C Bus Interface, *Sensors and Actuators A*: *Physical*, Vol.61, Issue 1, 1997, pp. 441-445.
[42].S. Y. Yurish, Digital Sensors Design based on Universal Frequency Sensors Interfacing IC, *Sensors and Actuators A: Pysical*, Vol.132, 2006, pp.265–270.
[43].Crescini, V. Ferrari, Z. K. Vajna, D. Marioli, A. Taroni, A. Borgese, M. Marinelli, E. Milani, A. Paoletti, A. Tucciarone, G. Verona-Rinati, Design and Development of a Piezoresistive Pressure Sensor on Micromachined Silicon for High-temperature Applications and of a Signal-conditioning Electronic Circuit, *Microsystem Technologies*, Vol. 9, 2003 pp. 431–435.
[44].T. T. Lang, Computerized Instrumentation, Chichester, *John Wiley & Sons*, 1991.
[45].Manufacturers of Pressure Sensors at Sensors Web Portal (www.sensorsportal.com): http://www.sensorsportal.com/HTML/SENSORS/PresSens_Manuf.htm

Chapter 17

Intelligent Sensor Systems

17.1 Smart vs. Intelligent: Modern Definitions and Trends

The world smart sensors and intelligent sensor systems market is projected to be growth strongly [1] (See Chapter 1 for more details). But what does it really mean a "smart sensor" ? In some languages, for example, German and Russian, "smart" and "intelligent" are translated from English by the same word that normally means "intelligent". It introduces some misunderstanding in recent sensors related publications. But in English there is a difference between "smart" and "intelligent". It is a reason of existing of two widely used smart and intelligent sensor definitions. The first one is more related to technological or so-called "smart" aspects; the second one is more related to functional or so-called "intelligent" aspects. The first definition was formulated by the following way: ***"Smart sensor*** *is a combination of a sensor, an analog interface circuit, an analog to digital converter (ADC) and a bus interface in one housing"* [2, 3]. It was the first smart sensor definition and is used widely till now. A smart sensor can be made as integrated (if all elements of smart sensors are integrated into one chip), hybrid, system-on-chip, system-in-package and IC-compatible 3D micro-structuring [3]. The intelligent sensor definition was formulated by the following manner: ***"Intelligent sensor*** *is the sensor that has one or several intelligent functions,* and is used often in various references. There was also an attempt to merge both "technological" and "intelligent" aspects of smart sensors in a single definition [4]: ***"Smart sensor*** *is one chip, without external components, including the sensing, interfacing, signal processing and intelligence (self-testing, self-identification, self-validation or self-adaptation) functions.* This definition reflects modern technological trends and challenges: smart sensors must be intelligent, and intelligent sensors must be smart. Some interesting smart and intelligent sensors definitions in its developments were also described in [5].

What does it make a sensor to be "intelligent" ? Very often it means a presence of microcontroller or microprocessor. However, it is a necessary but not enough condition. Sometimes a microcontroller is used only for calculation according to predetermined equations. Speaking about intelligent sensors various intelligent functions mean a self-testing, self-validation, self-checking, self-diagnosis, self-adaptation, self-identification, self-calibration, self-compensation, self-etc.

In common case, a smart sensor can not have any intelligent function, as well as an intelligent sensor can not be a smart. In other turn, *an intelligent sensor system* can be based on any sensors, not only on smart or intelligent.

In 2009 a mini survey "What does it mean "smart sensor"?" was set up by the IFSA at Sensors Web Portal. Very good feedback from 227 participants from both: academia and industry has been obtained. The survey results are shown in Figure17.1.

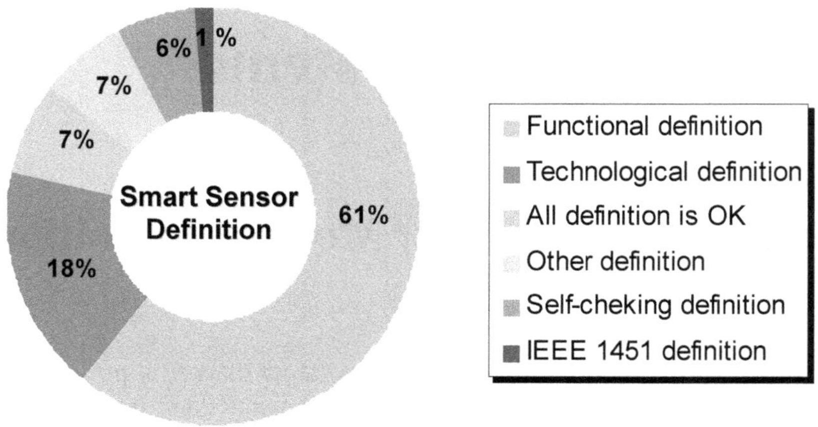

- Sensor with any intelligent function as self-identification, self-validation, self-testing or self-adaptation **(138) 61%**
- Combination of sensing element, analog interface circuit, ADC and bus interface **(40) 18 %**
- Sensor with only self-checking (self-calibration, self-validation) function **(14) 6 %**
- IEEE 1451 compatible sensor **(3) 1 %**
- Any definition is OK **(16) 7 %**
- Other suggestions **(16) 7 %**

Total Votes: 227

Fig. 17.1 Survey results 2009: "What does it mean "Smart Sensor" ? " (IFSA, 2009).

Speaking about other suggestions made by survey's participants, first of all it should be a sensor with extended network capabilities.

The survey's results justify that modern sensors and sensor systems include more and more intelligent functions and are based on modern microelectronics and MEMS technologies.

Modern microelectronics and MEMS technologies and its rapid developments bring new possibilities for integration of all smart sensors elements, in other turn, novel methods of measurements and advanced algorithms introduce more intelligence into smart, intelligent sensors and systems.

From technological, "smart", point of view an IC-compatible 3D micro-structuring technology is being developed [3]. An integrated smart sensor should contain all elements including wireless communication and power management. Without doubt, innovative technology concept such as smart systems integration will play a crucial role in this respect. Smart systems, defined as intelligent, often miniaturized, technical subsystem with their own and independent functionality evolving from microsystems technology. Smart systems are able to sense and diagnose complex situations. They are "predictive", they have the capability to decide and help to decide as well as to interact with the environment [6].

The greatest progress in innovation, however, will happen where MEMS technologies overlap with smart technologies. For intelligent sensor systems this means a trend from miniaturized sensor to the smart and miniaturized sensor systems, which can integrate processing functions in a minimal space. The main goal of smart sensor development is to improve the reliability and durability of these sensors and make them more easily adaptable to new functions and conditions during the operating phase. In addition to self-diagnosis, self-calibration and self-validation capabilities, smart sensors must have the self-adaptation function. Self-adaptive intelligent sensors can adapt themselves automatically to the measurement task according to current measurement conditions and algorithms. The main goal in bus sensor development is to make them more easily adaptable to communication networks (including wireless networks) and control systems. According to the 'plug and play' principle, installation in complex measuring and automation systems can be made much easier; the intelligence of the system as a whole can be decentralized, and the sensors can be used in a multiple way by the various control systems.

Most effort in recent years has focused upon the advancement of the sensing devices and associated hardware, increasing the sensitivity and reactivity of these systems. These capabilities however cannot be fully used in any cost-sensitive application, as this traditional solution to the measurement challenge is several orders of magnitude more expensive than a smart-sensor should cost or can be afforded in most applications (high performance ADCs cost several dollars). However there is a need for improved sensors with increased accuracy, reliability and acquisition speed.

The ability of modern intelligent sensors systems to process information is not enough: efficient coupling this ability with decision making based on data processing, on advances in data structures and upon the knowledge of each application environment in order to learn and adapt will be required.

There are a lot of new technologies, suitable for smart sensors creation. But still there is a problem how to joint them and use in a frame of intelligent sensor systems on chip.

Advanced packaging technologies, such as System-in-Package (SiP) technology, is gaining importance in the smart sensors systems integration. With its significantly higher flexibility compared to System-on-Chip (SoC), the SiP can combine multiple technologies and reuse intellectual property from numerous sources, allowing designers to overcome integration difficulties without compromising on individual chip technologies. In other words, SiP-based smart sensors systems offer numerous advantages – more functions in less space, reduced design cycle times and time-to-market.

Fully integration of all blocks and functions into a SiP is expensive, and only a revolution approach will make the cost per integrated smart, intelligent sensor reasonable. How to achieve this important goal ? Nevertheless the Strategic Research Agenda of the European Technology Platform on Smart System Integration [6] describes well "what" should be done, but "how" and in "which way" are till not determined, especially in smart sensors and systems area. Even the usage of more advanced technological processes, for example, 45 nm CMOS for SoC design, it will be slow evolution but no revolution in design because of it is related only on a further scaling, and as usually, bring initially many problems at production: low yield rate, high price, etc., what we have in a few last decades. Where is a solution ? It is in a revolution approach to smart and intelligent sensors design.

The innovation, revolution approach for intelligent sensors and sensors systems creation must:

1) Take advantage of both modern technologies and advanced methods of measurements. The most effective method to achieve this purpose will be a combination of technological and structural-algorithmic methods for metrological performance enhancement that can increase essentially the intelligent sensors systems development productivity;
2) Move from traditional information analog signal domain to frequency signal domain for sensors' and transducers' outputs.

All these possible if smart and intelligent sensors, transducers and sensor systems design will be based on novel microelectronics components such as Universal Frequency-to-Digital Converter (UFDC-1, UFDC-1M-16) and Universal Sensors and Transducers Interface (USTI) [7-9].

In addition such innovation approach will open up new horizons also in nano-electromechanical systems e.g. sensors/resonators/arrays; high-performance nanomechanical oscillators fabricated by bottom-up integration of silicon nanowires for applications such as sensitive mass detection and radio-frequency signal processing.

Without doubt, innovative, revolution technology concept will play a crucial role in the coming future. Smart sensors will be smarter and intelligent sensors will have more artificial intelligence, and finally, we will get real **smart, intelligent sensors.**

Intelligent sensor systems design with self-identification and self-adaptation functions will be considered below.

17.2 IEEE 1451 Standard Extension and Adaptation for Quasi-Digital Transducers

The UFDC and USTI ICs are based on the advanced method for frequency-to-digital conversion with programmable relative error and non redundant conversion time. It introduces a unique opportunity to build various intelligent sensor systems with self-adaptation very easy. In other turn, this function must be also reflected in the IEEE 1451 standard. The next parts of this Chapter described these two important intelligent functions in details.

17.2.1 Smart IEEE 1451 Compatible Transducers and Modern Smart Sensor Definition

The self-identification function of intelligent sensors, transducers and sensor systems is connected directly with the IEEE 1451 Smart Transducer Interface standard which is working out by the Institute of Electrical and Electronics Engineers (IEEE) Instrumentation and Measurement Society's Technical Committee on Sensors Technology [10]. Due to a wide spread of smart sensors, transducers and sensor systems this standard becomes more and more popular. The standard defines the concept of plug-and-play sensors with analog outputs, maintaining compatibility with the large existing base of analog instrumentation and interfaces. More specifically, the standard spells out how an analog sensor can be augmented with an embedded Transducer Electronic Data Sheet (TEDS), which contains technical information that identifies the sensor, specifies the sensor's analog interface, and describes the sensor's use [11].

It is expected that any successful standard must benefit both manufacturers and customers. Today, the IEEE 1451 standard is an example of one such standard that is becoming increasingly popular. The standard IEEE 1451.4 was approved in 2004 and there were manufactured more than 3200 different models of sensors according to this standard till now [12].

Concerning this standard, quasi-digital sensors with frequency output also mentioned in some documents, articles and papers but real results are not observed in this area. However, today exist a lot of quasi-digital sensors with frequency, period, duty-cycle; PWM, phase-shift or pulse number output signal practically for any physical and chemical, electrical or non-electrical quantities [13]. They have a high accuracy (up to ± 0.01 % relative error and better) and a wide frequency range from parts of Hz up to some MHz. According to the International Frequency Association (IFSA) study (2010), such quasi-digital sensors take 15 % among all sensors in the global modern sensor market. This is a reason why they also must be covered by the IEEE 1451 standardization process.

If a TEDS example for any frequency sensor exists, it does not reflect self-adaptation possibilities of modern smart sensor systems based on quasi-digital sensors and novel frequency-time parameters – to – digital conversion methods. The main problem consists in the following. Taking into account that quantization error of frequency-to-digital conversion depends on the method for frequency measurement, there are not any standard frequency-to-digital converters like standard analog-to-digital converters (ADC). Some years ago there was an attempt to standardize the frequency output interface signal from 2 kHz to 22 kHz or unify it from 4 kHz to 8 kHz as it was done for current and voltage output sensor signals. In the 70th and late there were proposed some interesting devices for frequency sensors output unification, for example, described in [14, 15]. However, all these strongly restricted frequency output sensors and transducers manufactures and fortunately the frequency output interface signal was not standardized.

Today, *HONSBERG Instruments GmbH* (Germany) manufactures the EFFF Frequency/ Frequency-Converter [16]. It is intended for sensors with variable frequency output. The EFFF has input frequency range from 4 Hz to 10 kHz, and output frequency range from 10 Hz to 2 kHz. The input frequency is adjustable via program pin.

The Proposal for Analogue Interface Standardization in Avionics [17] determines the standard variable frequency inputs (the frequencies of these inputs are proportional to the parameter being measured and are in widespread use by some civil avionics suppliers) in broad ranges: from 0 to 50 kHz and from 50 kHz to 1 MHz. The Table 17.1 summarizes the proposed parameters of Variable Frequency Inputs (VFIs) for standardized interfaces.

Table 17.1 Parameters of Variable Frequency Inputs (VFIs) for standardized interfaces [17].

Unique Reference	Frequency Range, kHz	Accuracy, %	Signal Level
VFI-1	0-1000	± 0.1	0-20 V
VFI-2	0.001 -1	± 1	1- 74 mV
VFI-3	0.0009 - 16	TBD	Square wave, 50 % duty cycle, 0-8 V
VFI-4	0-20	± 1	Selectable for 0 to 0.5, 1, 5, 10, 15, 20, 40, 50, 75 V_{pp}
VFI-5	0.1 - 15	± 0.1	0.75 ± 0.25 V
VFI-6	0 – 0.4	TBD	Square wave, 50 % duty cycle, 0 ... 10 V ± 3 V
VFI-7	0.083 to 2 pulses/s	± 1	Square wave, 5.1V_{pp}

TBD- to be defined by the system designer.

Nevertheless that the required relative error is big enough ± (0.1 ... 1) % it must be constant in broad frequency range from 0 to 1 MHz. Therefore it will be good to have a universal frequency-to-digital converter working in such frequency range and having programmable relative error from 0.01 % to 0.1 % in order to be neglected in comparison with sensor's error. The UFDC-1, UFDC-1M-16 and USTI ICs satisfy to such requirements [7-9].

Some papers and articles published in the last few years examine the IEEE 1451 standard and existing quasi-digital sensors in order to add smart, plug-and-play and self-adaptation capabilities to systems based on different quasi-digital sensors and find entry points that can provide a better adoption of this family of standards in industry [18-20].

As it was written by R. N. Johnson in [21] "Some organizations apply the label to any sensor that generates digital output, but a real smart sensor must do something with the data, not just pass them on in a digital format... The sensor is capable of making logical decisions at the source of the information". The modern smart and intelligent sensors definition is given above.

The IEEE 1451 is a developing family of standards that brings the important element of intelligence – the self-identification (or self-describing), to the traditional analog-mode sensors in order to reduce the time and challenges associated with sensor configuration. The standard defines a mechanism for adding self-describing behavior via the TEDS including calibration information to sensors with an analog signal interface. The IEEE 1451 smart transducer definition is well agrees with the modern smart and intelligent sensor definitions.

17.2.2 Family of IEEE 1451 Standards and Quasi-Digital Sensors

Recognizing that smart sensors will inevitably take a number of different forms with different levels of integration, the IEEE 1451 standards define a set of complementary interfaces designed to work together or independently [11]. The IEEE 1451 family of smart transducer interface standards in term of quasi-digital sensors is shown in Figure 17.2. Instead of ADC the Frequency-to-Digital Converter (FDC) must be used. It is naturally that for working according to IEEE 1451 standards advanced requirements are imposed to the FDC. It should have a programmable relative conversion error, scalable resolution, high accuracy, non redundant conversion time, wide frequency range and multifunctionality. In other words the FDC should be based on modern, advanced methods for conversion of different frequency-time parameters to digital.

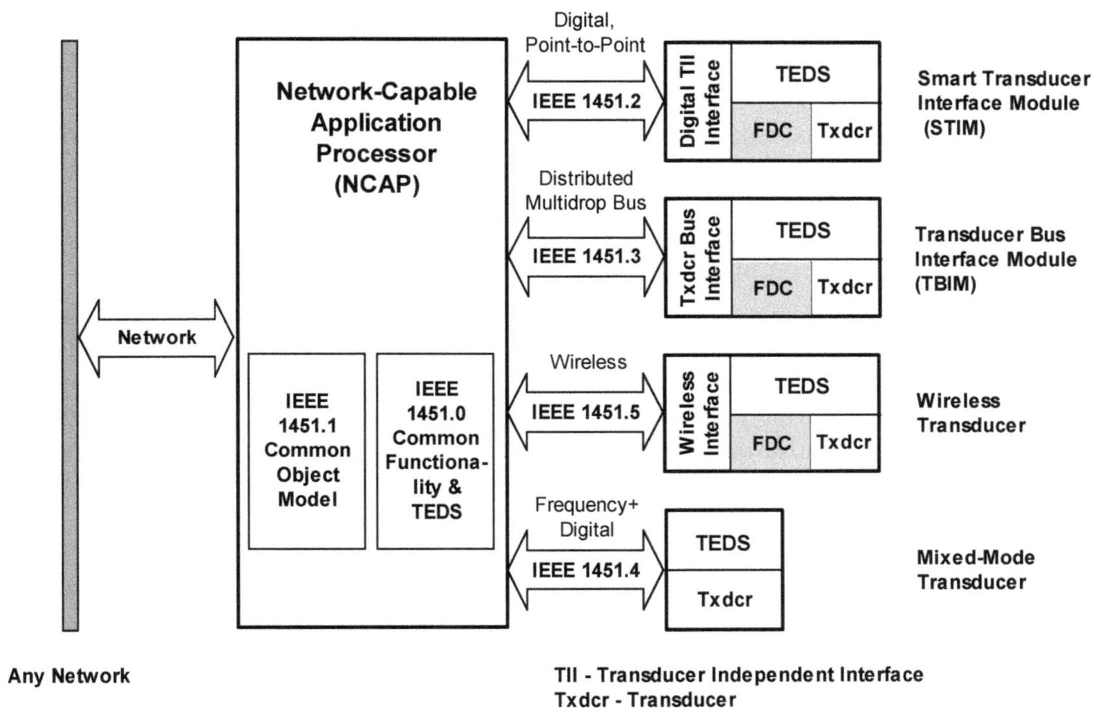

Fig. 17.2 IEEE 1451 family of smart transducer interface standards in term of quasi-digital sensors.

All of the current IEEE 1451 standards are designed to be complementary; they can be used independently or together. For example, an IEEE 1451.4 mixed-mode sensor could plug into an IEEE 1451.3 transducer bus interface module that digitizes the transducer signal and communicates the data over the IEEE 1451.3 sensor network [11].

A TEDS for quasi-digital sensor should contain information about sensor's error and it was difficult to standardize the FDC because of its quantization error dependent on the conversion frequency-to-digital method that also cannot be standardized.

For any of traditional, classical methods for frequency (period) measurement (standard counting technique or indirect counting technique), the quantization error dependent on the converted frequency f_x [22]. Any of convenient methods cannot work with modern high accurate frequency output sensors working in the wide frequency range. Therefore, an advanced frequency-to-digital conversion method with constant quantization error in all frequency range should be used in modern FDCs. However, such methods, as rule, are patented and in order to use it sensors manufacturers must buy a license. Up to now, such situation was a very strong restriction for the global standardization process because of it needs introduction of new member of IEEE 1451 standards family or its extension to quasi-digital sensors.

Fortunately, this problem has been solved due to development and introduction to the market the family of integrated Universal Sensors and Transducers Interface (USTI) IC [9]. This converter is based on four novel patented conversion methods for frequency (period), its ratio, duty-cycle and phase-shift. It has a programmable relative error of measurement (from 1 % to 0.0005 %), which is constant in all frequency range from 0.05 Hz to 9 MHz (144 MHz with prescaling), scalable resolution and non redundant conversion time. Thereby, converter have a self-adaptive possibility, i.e. a possibility to change the relative conversion error, which influences to the conversion time and conversely, to change the conversion time due to conversion error variation practically for each of measurement according to an adaptive conversion algorithm. The quantization error does not dependent on the converted frequency and less than the relative programmable error $\delta_x < \delta$. The integrated circuit has a quartz-accurate automated calibration; RS232 (422), SPI and I^2C interfaces; master and slave communication modes. There are one generating (with a calibration purpose) and 29 measuring modes designed for conversion to digital of any frequency-time parameters of sensor's and transducer's output: frequency, period, duty-cycle, duty-off factor, frequency and period ratios and differences, time interval, pulse space and width, phase shift, absolute and relative frequency deviation, pulse number count. It addition, the USTI can directly convert to digital a resistance, capacitance and resistive bridge parameters of sensing elements. In this case the number of external components are minimum possible.

Smart Transducer Interface Module (STIM) for analog signals (standard IEEE 1451.2) can be based, for example, on the MicroConverterTM microcontrollers family (ADuC812/816) from *Analog Devices* [23, 24]. The devices include all STIM's components on the one chip: ADC, embedded microcontroller and flash memory. In contradistinction to the MicroConverter™, the USTI IC instead of ADC has a two-channel multifunctional FDC for all possible frequency-time parameters. Furthermore, the USTI is available in small 28-lead, 32-lead and 32-pad packages in comparison with 52-lead package for MicroConverter™ microcontrollers family.

The custom designed versions of the USTI IC can be easy used in any member of IEEE standards family and brings a lot of benefits to customers and smart transducer manufacturers.

17.2.3 Physical Representation of IEEE 1451.2 for Quasi-Digital Sensors

The IEEE 1451.2 introduces the Transducer-Independent Interface (TII) [21]. This standard was defined to provide an industry-standard interface to connect sensors to microprocessor. The objective of the standard is to provide a common digital interface between transducers and the network communications nodes, and consequently facilitate the transducer manufactures to develop smart devices and to interface those devices to networks, systems, and instruments by incorporating existing and emerging sensor and networking technologies. The physical representation of IEEE 1451.2 standard in terms of quasi-digital sensors and USTI IC place is shown in Figure 17.3.

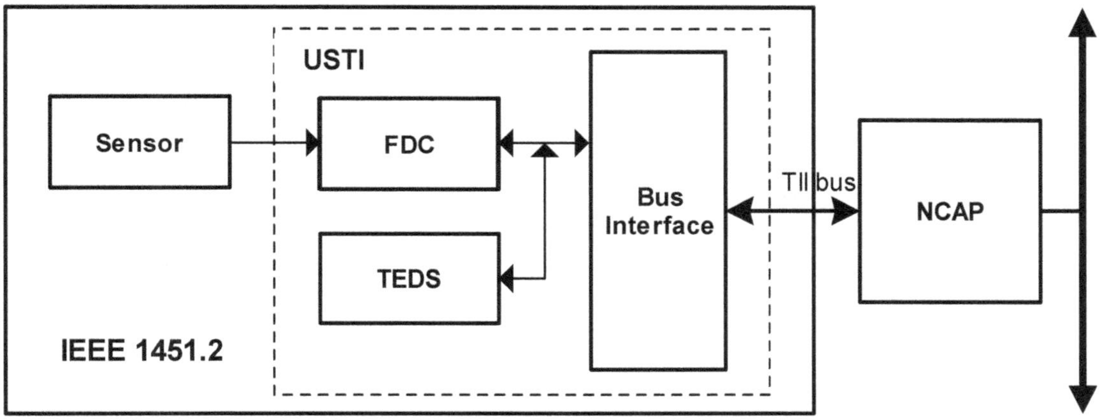

Fig. 17.3 Physical representation of IEEE 1451.2 standard for quasi-digital sensors.

The STIM can include sensors (Sensor STIM), actuators (Actuator STIM) or both in the same time (Sensor/Actuator STIM). The STIM contains a TEDS, sensor interface logic, converter or signal conditioner. It also contains address logic and digital readout circuits for signal conversion (in case of quasi-digital sensors it should be the FDC).

The address logic is used for communication between the STIM and NCAP. The network processor NCAP is situated between a Network and STIM and is intended for network communication and data exchange with the STIM, information processing and application tasks execution. It also transmits a power supply signal to the STIM.

The network processor NCAP, as rule contains controller and extended network interface to support other devices. After information was read form a TEDS, the processor knows about data exchange speed for STIM, how many channels are in the system and data format for each of converters in the interface module as well as units for measurements.

The NCAP starts a measurement or control by the STIM initialization and manage control signals after the measurement finished. The NCAP can be used as a node controller in multinode network as well as to support data exchange with other NCAP.

TII is a 10-wire digital bus. It is built around synchronous serial communication based on the SPI protocol. The TII is an interface that the sensor manufactures can use which is independent of the network being used to interconnect the sensors. It provides trigger, interrupt and plug-and-play functions through dedicated signal lines. The data transport is the same as SPI with 8-bit data words [25]. The USTI IC also supports the SPI interface, convert frequency-to-digital and can contain the TEDS in its own flash memory.

Such embedded system can be used with any quasi-digital sensor and lets to build smart transducers or sensor systems with minimum possible hardware or chip area. Integrated or hybrid technologies can be used to embedded these universal modules to smart transducers. Taking into account the programmable relative error of FDC based on the USTI it is possible to set an appropriate conversion error at non redundant conversion time and flexibly change it according to the advanced measuring algorithm. This possibility supplies an intelligence extension due to the self-adaptation capability of smart transducers together with the self-identification.

An alternative architecture, which has become quite popular in many applications it to employ "smart sensors" that can be connected on a network (typically hardwired) to a master (host) control system [25]. In this approach, each sensor module must also contain one USTI IC. The advantage of this architecture is that significant signal processing can take place within the sensor module and, by using a standard network bus, sensors can share information. The I^2C bus, which also supported by the developed converters, is well suited for this architecture, and the TII (IEEE 1451.2) bus can be used when one sensor module will be connected to the host system. This structure also lends itself well to wireless network applications.

17.2.4 TEDS for Quasi-Digital Sensors and Transducers

The TEDS data structure for quasi-digital sensors must describe specific parameters as a frequency output range and relative sensor's error into the extended TEDS section. An example of TEDS for pressure frequency output sensor is shown in Table 17.2.

Table 17.2 A possible structure of IEEE 1451.4 TEDS for pressure frequency output sensor.

TEDS Structure	Example of Pressure Frequency Output Sensors	
Basic TEDS	Manufacturer ID	19
	Model ID	5300
	Version letter	A
	Serial number	00639F
Standard and Extended TEDS (fields will vary according to transducer type)	Calibration date	14 October 2011
	Measurement range	1000 psia
	Frequency output minimal	30 kHz
	Frequency output maximal	42 kHz
	Relative FS error	± 0.01 %
	FDC quantization error	± 0.001 %
	Thermal Sensitivity	± 0.005 %/°C
	Response time	3 ms
User Area	Sensor location	B35-2
	Calibration due date	15 October 2011

The first 64 bits of the transducer TEDS is the Basic TEDS. The Basic TEDS uniquely identifies the transducer and includes the Manufacturer ID (14 bits), model number (15 bits), version letter (5-bit character code), version number (6 bits) and serial number of the device (24 bits). This data shall be organized according to the format described Table a in a non-volatile memory.

A frequency output sensors template is intended for any quasi-digital sensors whose output informative parameter is a frequency of pulse signal. The first 6-bit Select Case will indicate the physical measured and the corresponding units (i.e. PSI, N, m/s^2, mm, %, etc.). The TemplateID has no value because is an undefined template but it can be added as an integer of 8 bits. The frequency output sensors template is shown in Table 17.3.

Table 17.3 Frequency output sensors template (ID = XX) summary.

Select	Property	Description	Access	Bits	Data Type (and Range)	Units
–	TEMPLATE	Template ID	–	8	Integer (value = xx)	–
Select Case – Selects Type of Physical Measurand (Units)				6	Select Case	–
Cases 0 - 45	%MinPhysVal	Minimum physical value	CAL	32	Single	Various*
	%MaxPhysVal	Maximum physical value	CAL	32	Single	Various*
-	%MinElecVal	Minimum electrical output	CAL	32	Single	Hz
	%MaxElecVal	Maximum electrical output	CAL	32	Single	Hz
–	%MapMeth	Mapping Method	ID	-	Assign = 0, "Linear"	–
–	%RespTime	Response time	ID	6	ConRelRes (1E-6 to 7.9, ±15%)	Seconds
–	%RelativeError	Relative Error	ID	16	ConRelRes (1E-3 to 9.64, ±0.007%)	%
–	%QuantizationError	FDC Quantization Error	ID	16	ConRelRes (1E-4 to 0.96, ±0.007%)	%
–	%TempSensitivity	Temperature sensitivity.	ID	16	ConRelRes (1E-3 to 9.64, ±0.007%)	%/°C
–	%CalDate	Calibration date	CAL	16	DATE	–
–	%CalInitials	Calibration initials	CAL	15	CHR5	–
–	%CalPeriod	Calibration period	CAL	12	UNINT	days
–	%MeasID	Measurement location ID	USR	11	UNINT	–

* Units for %MinPhysVal and %MaxPhysVal are determined by value of Select Case "Physical Measurand (Units)" as summarized in Table 17.4.

Taking into account a self-adaptation capability, the TEDS for quasi-digital sensors should includes such parameter as a quantization error of frequency-to-digital conversion. Due to the USTI IC, it is programmable and can be chosen, for example, higher than sensor's FS error in order to have very short conversion time in critical situations, for example, or less than the FS error to have a maximum possible accuracy.

All described above are also true for so-called virtual TEDS, which is saved not in a sensor but exist in a separate file for a particular model of sensor. A measuring system can read information from such files at its measuring channels configuration or estimation of common error for measuring system.

Table 17.4 Enumeration of select case values for "Physical Measurand (units)".

Case	Physical Units	Case	Physical Units	Case	Physical Units	Case	Physical Units
0	K	12	Pa	24	radian/s	36	gpm
1	°C	13	psi	25	rpm	37	cfm
2	strain	14	Kg	26	Hz	38	l/min
3	microstrain	15	G	27	g/l	39	RH
4	N	16	m	28	kg/m^3	40	%
5	lb	17	mm	29	$mole/m^3$	41	Volts
6	kgf	18	in	30	mole/l	42	Volts rms
7	m/s^2	19	m/s	31	m^3/m^3	43	Amperes
8	ga	20	mph	32	l/l	44	Amperes rms
9	Nm/radian	21	fps	33	kg/s	45	Watts
10	Nm	22	radians	34	m^3/s		
11	oz-in	23	degrees	35	m^3/hr		

The data types used in standard templates are summarized in Table 17.5.

Table 17.5 Data types used in IEEE standard templates.

Data type	Description
UNINT	Unsigned integer
Chr5	5-bit character
ASCII	Standard 7-bit ASCII
Date	Number of days since January 1, 1998
Single	Single-precision floating point
ConRes	Constant resolution. This is a custom data type for compressed floating point values that provides a linear mapping of a defined interval
ConRelRes	Constant relative resolution. This is a custom data type for compressed floating point values that provides a logarithmic mapping of a defined interval
Enumeration	References a defined enumerated data type defined in the template

ConRes and ConRelRes provide a method of coding non-integer values with minimal bit usage by defining a limited range over which the bits are mapped. The range and resolution for ConRes and ConRelRes properties are listed in the template.

17.2.5 IEEE 1451.4 Mixed-Mode Interface for Quasi-Digital Sensors and Transducers

A smart TEDS sensor, as defined by IEEE 1451.4, includes a mixed-mode interface that accommodates both an analog signal (for the measurement signal) and a serial digital channel (for accessing the digital TEDS information). Speaking about quasi-digital sensors, the Class II of multiwire interface should be used in this case (Figure 17.4).

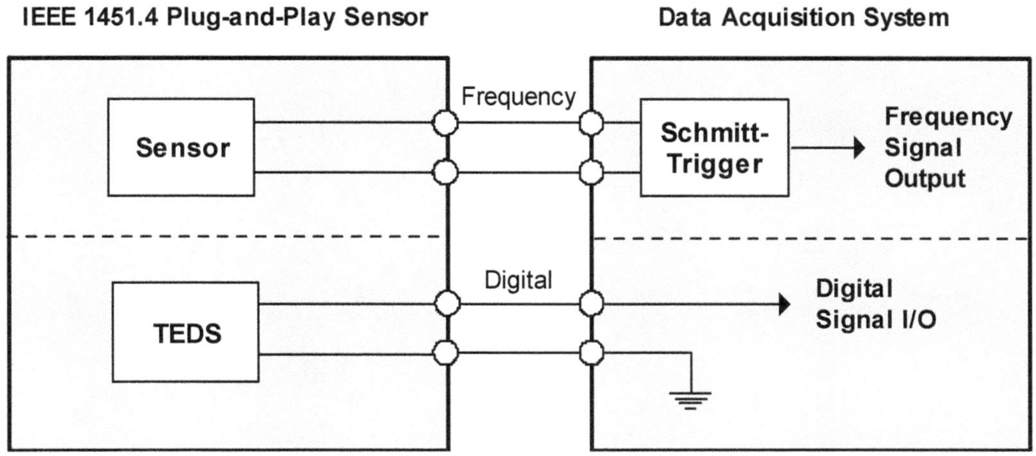

Fig. 17.4 Class II multiwire interface, shown with a frequency output sensor.

The digital portion of the mixed-mode interface is the same as in IEEE 1451.4. The analog portion must be changed by the quasi-digital portion for quasi-digital output sensor signal. In this case, we have the IEEE 1451.4 compatible smart transducer and we should have the IEEE 1451.4 compatible data acquisition system or measurement instrument.

The application for read and write the IEEE 1451.4 TEDS can be developed, for example, in LabVIEW and is indented to be used to read and write hexadecimal data that represents TEDS. The hexadecimal format to read and write TEDS was used to let the users to write and read data from USTI's internal flash memory according to the mixed mode interface format as can be seen in the example below in Figure 17.5.

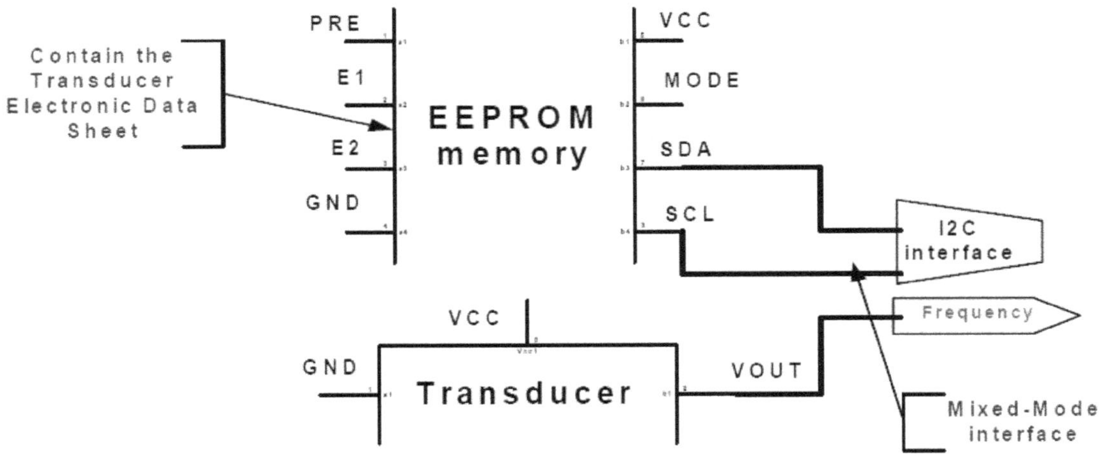

Fig. 17.5 Mixed mode interface.

The graphical interface for TEDS writing is shown in Figure 17.6 and for TEDS reading in Figure 17.7. The software uses a text file in which the hexadecimal format of the TEDS is written. The stream of data written in this file for a TEDS configuration.

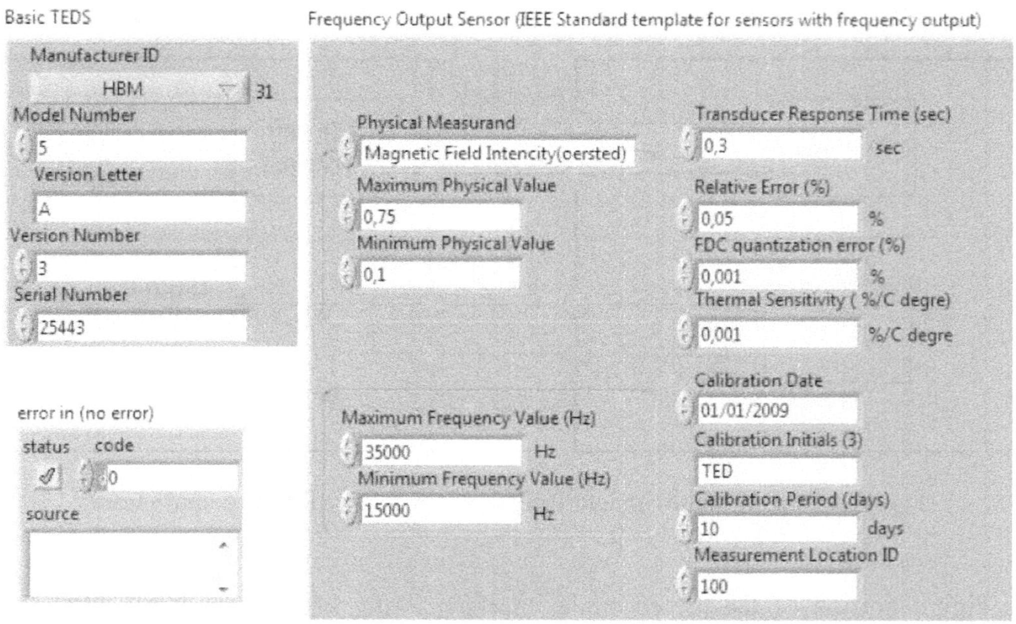

Fig. 17.6 Graphical interface for TEDS writing.

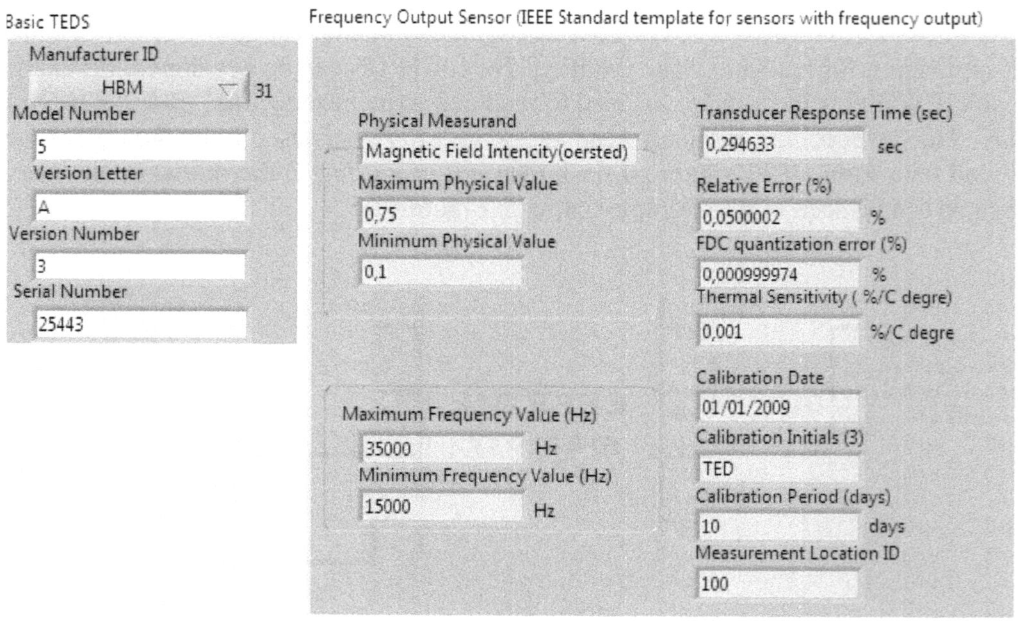

Fig. 17.7 Graphical interface for TEDS writing.

The TEDS data content is shown in Figure 17.8. The first line in figure represents the basic TEDS and data are starting with the LSB. The second line is for frequency output sensor description according to the special IEEE 1451.4 s template used for such sensors.

```
0063630C2001401F
03200510B40FB2000040406D29C119A980011C22E000F7333334FD000002E
```

Fig. 17.8 TEDS data content.

The write and read TEDS application contain two blocks, one for basic TEDS and one for reading and writing the special frequency template. The diagrams of these two applications are shown in the Figures 17.9 and 17.10 respectively.

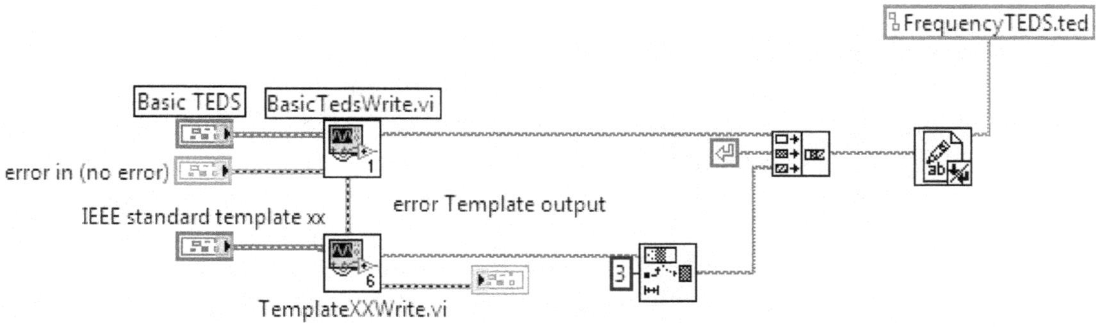

Fig. 17.9 TEDS write block diagram.

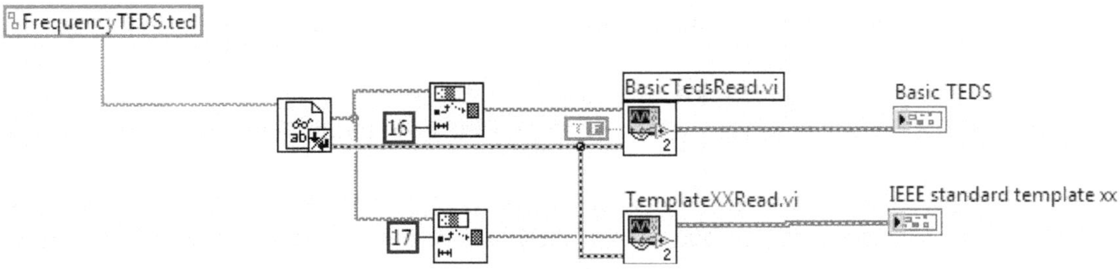

Fig. 17.10 TEDS read block diagram.

Setting up sensor-based data-acquisition systems is tedious and time consuming. Data-acquisition systems can monitor thousands of sensor outputs. Nearly 20 percent of the total cost of most data acquisition applications is spent on hardware setup and configuration. As it was marked in [26], today there are not any data acquisition board, which can work well with any modern wide range and precision quasi-digital sensors. Hence, the IEEE 1451.4 compatible data acquisition system should be also based on new technologies. In this case, the USTI IC can be used in such systems in order to minimize hardware and guaranty high metrological performances including the programmable quantization error (up to ± 0.0005 %) and scalable resolution at non redundant conversion time.

The converter has two channels and can work in 29 measuring modes and one generating and one calibration mode. Besides frequency, period, duty-cycle and pulse number integrated circuits can convert also duty-off factor, frequency (period) ratio and difference, time interval, pulse space and pulse width, phase-shift, absolute and relative frequency deviation and rotational speed.

Very simple design approach based on the FDC chips lets expand the use of IEEE 1451 standard for different frequency (period, duty-cycle, PWM, time interval, phase shift, etc.) output sensors. Even more, each existing quasi-digital sensor can be used according to IEEE 1451.4. The designed FDCs can convert all frequency-time parameters to digital and also can store TEDS and realize communication functions.

What benefits will receive customers ? Together with all IEEE 1451 standards family benefits (including self-identification, simple installation, upgrade and maintenance) customers will receive the self-adaptation capability and digital plug-and-play sensors, using all advantages of frequency-time parameters as an informative parameters of sensors' outputs.

Smart transducer manufactures will receive a unique opportunity to produce low-cost IEEE 1451 compatible sensors with minimum possible hardware. They should not think now about the FDC's accuracy. Only one component – the USTI IC covers three functions of smart transducers: frequency (time)-to-digital conversion with high metrological performances, TEDS storage in the flash memory and communications.

The FDC chips can be embedded into any existing quasi-digital sensors (with the help of integrated or hybrid technologies) to produce IEEE 1451.4 compatible sensors as well as into IEEE 1451.4 compatible data acquisition systems. Due to all these, smart transducer manufactures can receive 15 % of addition global sensors market, which was not involved before into the IEEE 1451 standardization process. Design and development of IEEE 1451 compatible sensors will be quick, low-cost and effective. Such approach will improve the level of commercial adoption of IEEE 1451 standards family in industry.

In addition, manufactures can use the USTI IC to produce traditional digital output sensors with RS232 (422) serial interface, SPI and I^2C communication buses.

The simple extension of IEEE 1451 standards family for quasi-digital sensors will provide smart transducer manufactures with the ability to produce quickly and with low-cost different IEEE 1451 compatible sensors, transducers, measuring and data acquisition systems. The approach needs only one of universal component - the USTI IC, which covers three functions of smart transducers: high accurate frequency (time)-to-digital conversion, TEDS storage in the flash memory and communications. The USTI can be embedded into any existing frequency-time domain output sensors or transducers (with the help of integrated or hybrid technologies) to produce IEEE 1451.4 compatible sensors as well as into IEEE 1451.4 compatible data acquisition systems.

Design and development of IEEE 1451 compatible sensors and transducers will be quick, low-cost and effective. Such approach will improve the level of commercial adoption of IEEE 1451 standards family in industry.

17.3 Self-Adaptive Sensors and Sensor Systems

The self-adaptation is relatively new function of smart and intelligent sensors. Novel designed self-adaptation intelligent sensor systems are based on so-called adaptive algorithms, which were used at the fist time in various digital measuring systems. Let consider such intelligent sensor systems and its algorithms below in details.

17.3.1 Adaptive Algorithms and Parametric Adaptation

Adaptive measurements are measurements in which measuring procedures can be changed at change of properties of a signal or measuring conditions [27]. An adaptation in intelligent sensors and sensor systems can be used for increasing of measurement accuracy, decreasing of measuring time, power consumption reduction, etc. A typical task of algorithmic adaptation is a control of inclusion in the measuring procedure of operations for increase of accuracy (improvement of metrological performance) at appropriate alterations of measuring conditions or properties of measurand. The necessity to control of operation usage for increase accuracy is caused by an opportunity of occurrence of situations when the application of this operation is inexpedient.

There are parametrical and algorithmic adaptations. The adaptation in intelligent sensors, transducers and sensor systems can be determined as a process of purposeful change of system's parameters, which means the determination of criterion of functioning and its fulfillment. If an overall performance of system to estimate by means of criterion of optimality, the adaptation will be a process to change of parameters by means of control influences on the basis of the current information with the purpose to achieve the optimal or required state of sensor system, according to a measuring algorithm at operating conditions changing. Hence, the process of adaptation consists in maintenance of the required (specified) quality of functional for a sensor system at change of operating factor U [28]:

$$U \in Y_i, \tag{17.1}$$

where Y_i is the ith number of feasible controls.

In the case of parametrical adaptation

$$U = \langle P_1, P_2, ..., P_n \rangle, \tag{17.2}$$

where P_n are parameters of adaptation.

Parametrical adaptation consists in production of control signals supporting a smart sensor or sensor system in the required state by means of adopted parameters depending on measuring conditions and/or measuring algorithm. Here the parameter has a final number of values [28]:

$$U \in Y_i = D_i, \tag{17.3}$$

where D_i is the discrete number of control values.

Naturally, self-adaptive intelligent sensor systems should be based on novel methods of measurements and use of quasi-digital sensors (or sensing elements) with frequency, period, duty-cycle, PWM, phase-shift and pulse number outputs [29]. First of all it means a novel patented method with programmable relative error and non-redundant time of measurement. This method has so-called self-adaptive possibilities for change accuracy on speed of measurement due to non-redundant time of measurement.

Being based on the approach described in [27] we shall result the equation of measurements for modified method of the dependent count in the operator form (for two possible algorithms of measurements: with maximal accuracy and maximal speed of measurement):

$$\lambda^*_j = T_s L\gamma_j(t) \vee \delta_s L\gamma_j(t), \tag{17.4}$$

where T_s and δ_s are operations of increase of speed and accuracy respectively, the introduction of which is made according to the established decision rule; L is the operator representing an algorithm of measurement; γ_j is the multivariate input action, generally, time-dependent [30, 31]. At ordinary direct measurements the input action correlated with that moment of time for which the result of measurement is fixed is used; $t \in [t_j, t_{j+T}]$, where T is the time of measurement cycle.

In turn, the equation of measurements for the method with non-redundant reference frequency in the operator form (at presence of two possible algorithms of measurements: with maximal accuracy and minimal power consumption) can be written by the following way:

$$\lambda^*_j = P_s L\gamma_j(t) \vee \delta_s L\gamma_j(t), \tag{17.5}$$

where P_s is the operation of power consumption reduction.

The choice of variants is made according to the current result of measurement β. So, for the modified method of the dependent count:

$$\begin{cases} \lambda^*_j = T_s L\gamma_j(t), \; if \; F_x(\beta^*) \in I_f \\ \lambda^*_j = \delta_s L\gamma_j(t), \; if \; F_x(\beta^*) \notin I_f \end{cases} \text{at } I_f \in I, \tag{17.6}$$

where $F_x(\beta^*)$ is the characteristic of input action or measuring conditions, which value defines the decision for change of measuring algorithm (parameters of system); I_f is the subset of certain area I of possible values of characteristic $F_x(\beta^*)$, the set membership to which defines the necessity to change of system's parameters. Generally, the task of determination of area I_f for the resulted parameter arises.

In turn, for the method with non-redundant reference frequency we will have:

$$\begin{cases} \lambda^*_j = P_s L\gamma_j(t), \; if \; F_x(\beta^*) \in I_f \\ \lambda^*_j = \delta_s L\gamma_j(t), \; if \; F_x(\beta^*) \notin I_f \end{cases} \text{at } I_f \in I, \tag{17.7}$$

The decision about inclusion in the measuring procedure of operation of accuracy or speed increase; or accuracy increase or decrease power consumption, is accepted on the basis of the set-up decision rule. The decision rule is built on the basis of comparison of value of measuring results with the liminal value set in advance at which there is a change of system's parameters depending on the required measuring algorithm.

The common flowchart of adaptive measurement algorithm for smart sensor system based on the modified method of the dependent count at presence of two possible algorithms with parametrical adaptation is shown in Figure 17.11 [32].

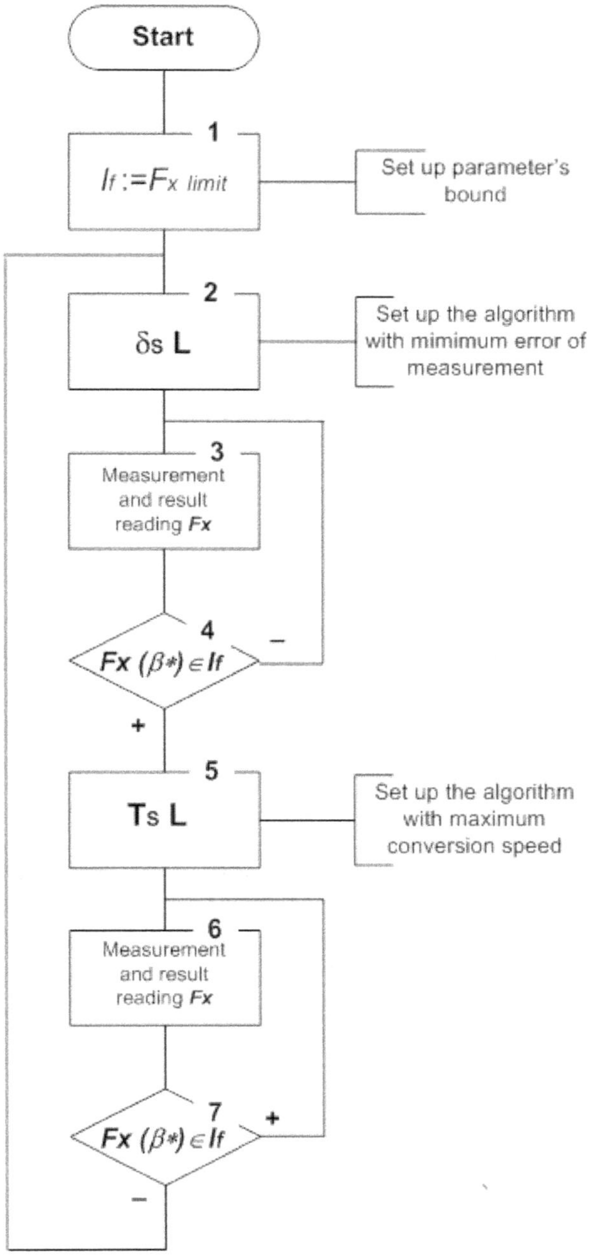

Fig. 17.11 Flowchart of adaptive measurement algorithm for intelligent sensor system based on the modified method of the dependent count [32].

The algorithm of adaptive measurement is realized by the following way. At the beginning, the boundary value of frequency $F_{x\ limit}$ is set-up at which the algorithm change should be made (Block 1). The subset I_f consists of only one value of boundary frequency belonging to the set I of all values of frequencies from the frequency measuring range. Then the programming of maximal accuracy for sensor system by means of the corresponding parameter is carried out (Block 2). After, the frequency measurement and reading of result of measurement F_x (Block 3) is carried out, which is compared to the boundary value $F_{x\ limit}$ (Block 4). If the result of current measurement does not exceed the set-up in advance the boundary value $F_{x\ limit}$, the measurement is proceeded with the greatest possible accuracy (Block 3). Otherwise, the exchange of accuracy on speed (execution of operation for increase of speed due to accuracy decrease) should be made

(Block 5), and system continues to function with the maximal speed, but with the lowered accuracy of measurements (Block 6), whether the checking measured parameter has returned to admissible limits (Block 7). In case $F_x \le F_{x\,limit}$, the measuring system returns back again in the mode of measurement with the increased accuracy by reprogramming (set up) of required error of measurement (Block 2).

17.3.2 Examples of Self-Adaptive Intelligent Sensor Systems and Its Realizations

Let's consider some examples of sensors systems where adaptive measurements are necessary, for example, antilock braking system (ABS) for automobiles; pressure sensor system for gas pipelines; sensors systems based on different sensing elements, having a various error of measurement in different parts of a measuring range; and sensor system for fail-safe cooling with fan speed control based on current temperature measurements.

17.3.2.1 Self-Adaptive Antilock Braking System (ABS)

According to the self-adaptive ABS algorithm based on the modified method of the dependent count for rotation speed, at the beginning, the critical value of rotation speed for wheels at its blocking is set-up. When wheels are not blocked, the measurement of rotation speed is made with the maximal accuracy with the aim to receive information for engine, gear set and other car's systems optimization. At the moment when wheels are blocked, the sensor system is adopted for the changed measuring conditions by an exchange of accuracy on speed with the aim to receive information necessary for wheels unblocking and the prompt generating of corresponding control signals. At the moment when wheels are unblocked again and rotation speed of wheels lays outside of critical values, the sensor system adapts again for measurement of rotation speed with the increased accuracy by reprogramming the required relative error for rotation speed - to - digital conversion.

Naturally, the use of such adaptive algorithms means the presence of computing power (microcontroller or embedded computer) in a sensor system. However, especially developed for such self-adaptive sensors systems the Universal Frequency-to-Digital Converters (UFDC-1, UFDC-1M-16) [7, 8] and Universal Sensors and Transducers Interfaces (USTI and USTI-EXT) [9] integrated circuits essentially simplify realization of such intelligent sensors and sensor systems at minimal possible hardware and low cost.

All ICs are based on advanced conversion method, working in a broad frequency range from 0.05 Hz to 7.5 MHz (9 MHz USTI) without prescalling, have constant programmable relative error from ±1 to ±0.001 % (±0.0005 % USTI) in all frequency range and non-redundant conversion time. These ICs are ideally suited for rotation speed measurements. They can measure a rotation speed and indicate the results in *rpm* units. The number of gear teeth *Z* (1...255) must be set up in advance [33]. Any rotation speed sensor with frequency output can be connected directly to the UFDC or USTI.

Relative errors and appropriate conversion times for the mentioned ICs are shown in Table 17.6 and Figure 17.12. Based on an appropriate adaptive measuring algorithm, a trade-off between required relative error and time for rotation speed-to-digital conversion or frequency (period)-to-digital conversion can be chosen.

Table 17.6 Relative errors and appropriate conversion times.

Relative error, δ_x %	**UFDC-1** (at f_0=500 kHz)	**UFDC-1M-16** (at f_0=16 MHz)	**USTI** (at f_0=625 kHz)
	t_{conv}, s		
1	0.0002	0.00000625	0.00016
0.5	0.0004	0.0000125	0.00032
0.25	0.0008	0.000025	0.00064
0.1	0.002	0.0000625	0.0016
0.05	0.004	0.00125	0.0032
0.025	0.008	0.0025	0.0064
0.01	0.02	0.00625	0.016
0.005	0.04	0.0125	0.032
0.0025	0.08	0.025	0.064
0.001	0.2	0.0625	0.16
0.0005	-	-	0.32

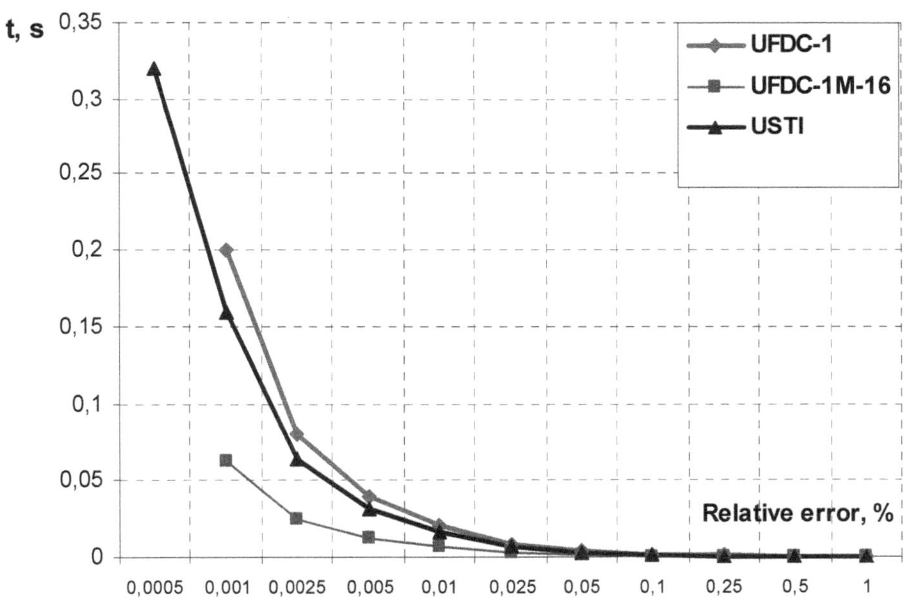

Fig. 17.12 Conversion times vs. relative error.

A self-adaptive smart sensor system for rotation speed based on the mentioned ICs should contain an appropriate frequency output sensor for rotation speed, for example [34]. A smart, intelligent sensor system example is shown in Figure 17.13. Any from UFDC or USTI ICs have three popular sensor systems interfaces such as SPI, I^2C and RS232, and can be controlled by an external microcontroller or computer (slave communication mode).

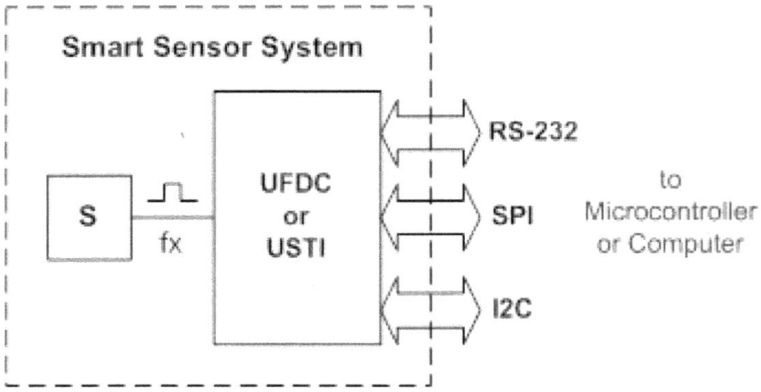

Fig. 17.13 Block diagram of self-adaptive smart sensor system based on UFDC or USTI integrated circuit (S - frequency (period) output sensor).

A microcontroller or computer can change accuracy or conversion time by sending an appropriate command to the UFDC or USTI based on the current measurement result and measuring algorithm. An example of such commands for RS232 communication mode is shown in Figure 17.14.

>M0A ; Rotation speed measurement initialization in the 1st channel
>Z30 ; Set up the modulation rotor teeth number $Z=48_{(10)}=30_{(16)}$
>A09 ; Choose the relative error of frequency measurement ± 0.0005 %
>S ; Start a measurement
>C ; Check result status: return 'r' if ready or 'b' if busy
>R ; Read a result of measurement in *rpm*

; Here microcontroller or computer should check the condition for an algorithm changing and prepare the USTI to measure with highest speed (maximum relative error) if a critical rotation speed has been achieved:

>A00 ; Choose the relative error of measurement ± 1 %
>S ; Start a measurement
>C ; Check result status: return 'r' if ready or 'b' if busy
>R ; Read a result of measurement in *rpm*

Fig. 17.14 Commands for RS232 communication mode at adaptive rotation speed measurements by the USTI.

It is also expediently to use the additional command "C" between "S" and "R" commands to check the measurement status especially at near zero rotation speed. It returns "r" if the result is ready and "b" if the measurement is in a progress.

The described above low cost, intelligent adaptive sensor system lets choose the minimum possible conversion time at critical rotation speed of wheels and maximum possible accuracy for the best engine optimization at nominal rotation speed of wheels.

17.3.2.2 Self-Adaptive Smart Pressure Sensor System for Gas Pipeline

The UFDC and USTI integrated converters works well with all frequency ranges of modern quasi-digital pressure sensors and transducers [35]. The ICs based sensor systems are extremely cost-effective. The level of sensor intelligence that can be obtained for only a few percents of the total cost has made the UFDC-1 or USTI the element of choice for MEMS based pressure sensors and transducers. The UFDC-1 interfacing with various quasi-digital pressure sensors is described in [36]. The self-adaptive intelligent sensor system block diagram is the same as adduced in Figure 17.11. Two frequency output sensors can be connected to each of ICs.

At quick pressure drop and achievements of critical values of pressure in a gas pipeline, the smart adaptive pressure sensor system automatically increases a relative error of measurement of frequency in UFDC-1 or USTI, providing with that a high speed and reaction of sensor system for monitoring of current situation and produce appropriate control signals. At the returning of values of pressure in a nominal range, the accuracy of sensor system is increased up to the level, at which it is possible to neglect this error in comparison with a relative error of pressure sensor.

The commands for RS232 communication mode at adaptive measurement of pressure by the UFDC-1 are shown in Figure 17.15.

```
>M0   ; Frequency measurement initialization in the 1st channel
>A9   ; Choose the relative error of frequency measurement ± 0.001 %
>S    ; Start a measurement
>C    ; Check result status: return 'r' if ready or 'b' if busy
>R    ; Read a result of measurement
```

; Here microcontroller or computer should check the condition for an algorithm changing and prepare the UFDC-1 to measure with highest speed (maximum relative error) if a critical value of pressure has been reached:

```
>A0   ; Choose the relative error of measurement ± 1 %
>S    ; Start a measurement
>C    ; Check result status: return 'r' if ready or 'b' if busy
>R    ; Read a result of measurement
```

Fig. 17.15 Commands for RS232 communication mode at adaptive measurement of pressure by the UFDC-1.

Thus, the parametrical adaptation allows to reduce the measuring time for the UFDC-1 in 3 orders, and in view of communication and calculation times [16], the frequency-to-digital conversion time can be reduced in 23 times in average for critical values of pressure.

17.3.2.3 Temperature and Humidity Self-Adaptive Intelligent Sensor Systems

Various quasi-digital sensors, for example, temperature and humidity sensors with frequency (period) outputs have different errors in various parts of measuring range [37, 38]. For example, temperature sensors with period (MAX6576) and frequency (MAX6577) output from *Maxim* have error, which is changed in different part of working temperature range (-40 ^{0}C to +125 ^{0}C) in 2.2 times [39], Figure 17.16.

Temperature Sensor Error (Note 1)		MAX6576	T_A = -20°C	-7.5	±1.1	+7.5	°C
			T_A = 0°C	-5.5	±0.9	+5.5	
			T_A = +25°C	-3.0	±0.8	+3.0	
			T_A = +85°C	-4.5	±0.5	+4.5	
			T_A = +125°C	-5.0	±0.5	+5.0	
		MAX6577	T_A = -20°C	-7.5	±1.1	+7.5	°C
			T_A = 0°C	-6.5	±0.9	+6.5	
			T_A = +25°C	-3.0	±0.8	+3.0	
			T_A = +85°C	-3.5	±0.5	+3.5	
			T_A = +125°C	-4.5	±0.5	+4.5	

(a)

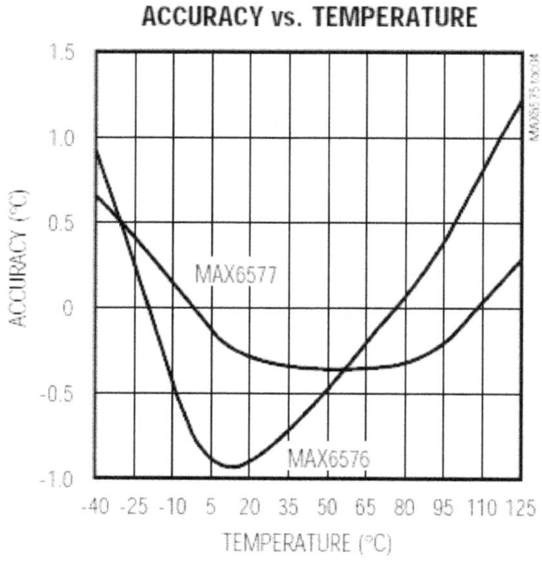

(b)

Fig. 17.16 Absolute errors for MAX6576 and MAX6577 quasi-digital temperature sensors [39].

Based on sensors systems design considerations, in order to be neglected, the relative error of frequency (period) – to – digital conversion should be chosen in one order (or at the least in 5 times) less than the relative error of sensor. Based on the parametric adaptation, it is possible to reduce the conversion time in self-adaptive intelligent temperature sensors systems [40]. In this case, the programmable relative error for frequency-to-digital conversion must be changed dependent on the current measuring range of temperature. According to considerations about the parametric adaptation described above, the lowest and highest parts of working measuring range (- 20 ^{0}C to 0 ^{0}C) and (+ 85 ^{0}C to + 125 ^{0}C) will represent the subset If of I set of all possible values (- 40 ^{0}C to + 125 ^{0}C) of characteristic $F_x(\beta^*)$. The commands for RS232 communication mode at adaptive measurement of temperature by the UFDC-1 are shown in Figure 17.17.

```
>M0   ; Start frequency measurement in the 1st channel
>A4   ; Set the relative error ± 0.05 %
>S    ; Start a measurement
>C    ; Check result status: return 'r' if ready or 'b' if busy
>R    ; Read a result of measurement
```

; Here microcontroller or computer should check the condition for an algorithm changing and prepare the UFDC-1 to measure frequency with ± 1 % relative error if a value of temperature is in the -20^{0}C ... 0^{0}C range.

```
>A0   ; Choose the relative error of measurement ± 1 %
>S    ; Start a measurement
>C    ; Check result status: return 'r' if ready or 'b' if busy
>R    ; Read a result of measurement
```

Fig. 17.17 Commands for RS232 communication mode at adaptive measurement of temperature by the UFDC-1.

At this, the frequency (period)-to-digital conversion time can be decreased in 20 times at - 20 ^{0}C ... 0 ^{0}C temperature range, and the total conversion time in this range in view of communication and calculation times will be decreased on 3 ms. Obviously, if the range of sensor's error is changing in wide limits, the effect of measurement time reduction will be greater.

The temperature and humidity module HTF3130 from *Humirel* with frequency output proportional to a relative humidity has an error that is changed in 1.7 times in the whole working range (from 0 to 99 % RH) [41], Figure 17.18. The interfacing of such sensors with the UFDC-1 is described in [42, 43].

Based on sensors systems design considerations, described above, in the case of maximum possible ± 5 % full scale relative error for HTF3130 humidity module, the programmable relative error for UFDC-1 should be chosen equal to ± 0.5 %. But from 0 to 10 % RH and from 90 to 100 RH % the module has maximum error only ± 3 %. It means ± 0.25 % programmable relative error for the UFDC-1. If we use this relative error for all humidity range (0 – 90 % RH), the time for frequency measurement in such sensor system will be redundant at the beginning (0 – 10 % RH) and at the end (90 – 100 % RH) of this range.

Based on the parametric adaptation, it is possible to reduce the conversion time in a self-adaptive smart humidity sensor system. In this case, the programmable relative error for frequency-to-digital conversion should be changed dependent on the current measuring range of humidity. According to considerations about the parametric adaptation described above, the lowest and highest parts of working measuring range (0 – 10 % RH) and

(90 – 100 % RH) will represent the subset I_f of I set of all possible values (0 – 90 % RH) of characteristic $F_x(\beta^*)$.

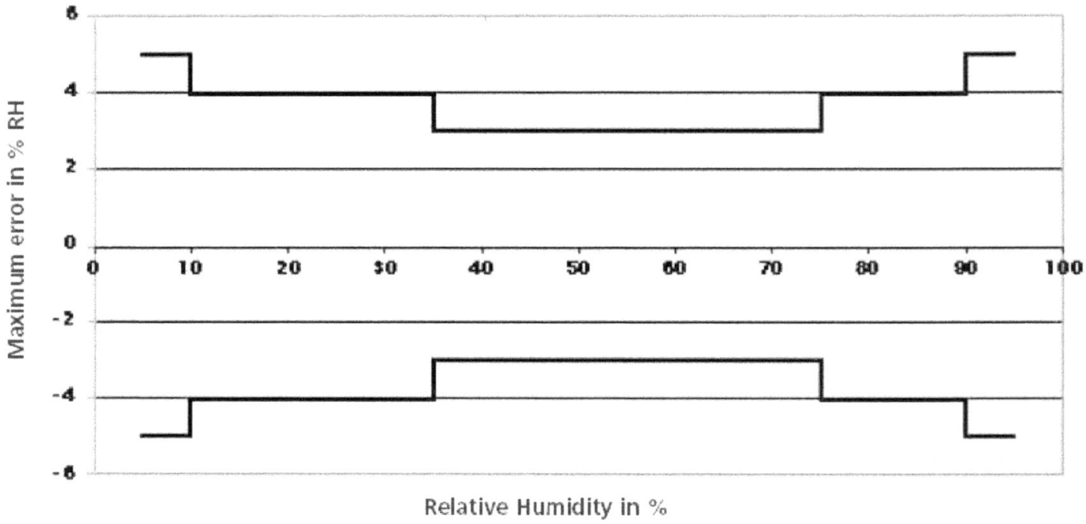

Fig. 17.18 Maximum relative error for HTF3130 humidity module [41].

The commands for RS232 communication mode at adaptive measurement of humidity by the UFDC-1 are shown in Figure 17.19. At this, the total conversion time in view of communication and calculation times will be decreased on 0.5 ms. Obviously, if the range of sensor's error is changing in wide limits, the effect of measurement time reduction will be greater.

```
>M0   ; Frequency measurement initialization in the 1st channel
>A2   ; Choose the relative error of frequency measurement ± 0.25 %
>S    ; Start a measurement
>C    ; Check result status: return 'r' if ready or 'b' if busy
>R    ; Read a result of measurement
```

; Here microcontroller or computer should check the condition for an algorithm changing and prepare the UFDC-1 to measure frequency with 0.5 % relative error if a value of humidity is in the 0 – 10 % RH or 90-100 % RH relative humidity range.

```
>A1   ; Choose the relative error of measurement ± 0.5 %
>S    ; Start a measurement
>C    ; Check result status: return 'r' if ready or 'b' if busy
>R    ; Read a result of measurement
```

Fig. 17.19 Commands for RS-232 communication mode at adaptive measurement of humidity by the UFDC-1.

The same design consideration must be taken in account at design of intelligent humidity sensor systems based on the USTI IC [44]. The circuit diagram for the intelligent sensor system for humidity and temperature measurement based on the HTF3130 humidity module is shown in Figure 8.14 and commands for RS232 communication mode at adaptive measurement of humidity and temperature by the USTI – in Figure 17.20.

```
>M00          ; Set a frequency measurement mode in the 1st channel
>A02          ; Set the relative error ± 0.25 %
>S            ; Start a frequency measurement (humidity)
>C            ; Check result status: return 'r' if ready or 'b' if busy
>R            ; Read a result of frequency measurement (humidity)
>M10          ; Set a resistance Rx measurement mode
>E263000.0    ; Set the reference value of Rc=263 kΩ
>W1B          ; Set the charging time 100 ms
>S            ; Start a resistive measurement (temperature)
>C            ; Check result status: return 'r' if ready or 'b' if busy
>R            ; Read a result of resistive measurement (temperature)
```

;Here microcontroller or computer should check the condition for an algorithm changing and prepare the ;USTI to measure frequency with ± 0.5 % relative error if a value of humidity is in the 0 – 10 % RH or ;90-100 % RH relative humidity range.

```
>M00          ; Set a frequency measurement mode in the 1st channel
>A00          ; Choose the relative error of measurement ± 1 %
>S            ; Start a frequency measurement (humidity)
>C            ; Check result status: return 'r' if ready or 'b' if busy
>R            ; Read a result of frequency measurement (humidity)
>M10          ; Set a resistance Rx measurement mode
>E263000.0    ; Set the reference value of Rc=263 kOhm
>W1B          ; Set the charging time 100 ms
>S            ; Start a resistive measurement (temperature)
>C            ; Check result status: return 'r' if ready or 'b' if busy
>R            ; Read a result of resistive measurement (temperature)
```

Fig. 17.20 Commands for RS232 communication mode at adaptive measurement of humidity and temperature by the USTI.

17.3.2.4 Self-Adaptive Smart Sensor System for Fail-Safe Cooling with Fan Speed Control

In various fail-safe cooling systems the UFDC and USTI integrated circuits families can be used for low-cost realization of adaptive temperature and fan rotation speed measurements. For example, any frequency output rotation speed sensor can be connected to the 1st channel of USTI and temperature sensor with frequency, period, duty-cycle or PWM output – to the 2nd channel of IC.

At the optimal conditions, the relative errors for each of channel should be chosen (programmed) in one order less then the sensors' relative errors. In critical conditions, for example, increased temperature or sudden decreasing of fan rotation speed (this can be an evidence of fan fail), the self-adaptive sensors system will automatically to decrease time of rotation speed and/or temperature measurement due to temporary accuracy reduction in appropriate channel in order to produce appropriate control signals as soon as possible by controller (master). The master initiates communication with the USTI (slave) according to I2C, SPI and RS-232 serial interfaces in order to send appropriate commands, read results and produce control signals. All these make such sensors systems ideal for intelligent fan control in a wide range of cooling applications with fan failure detection in communication, networking, high-end consumer and test equipments.

Self-adaptive intelligent sensors and sensor systems based on novel method of for frequency (period)-to-digital conversion of sensor's outputs can be easy realized on the UFDC and USTI ICs well-suited for such kind of applications.

All described self-adaptive intelligent sensors systems are based on parametric adaptation. Simple decision rules lets chose one of two possible adaptive algorithms: with a high accuracy or speed of measurement dependent on measuring algorithm. The use of such sensors systems allows to change flexibly accuracy on speed and contrary when it is necessary to receive information about measurand in critical points.

17.4 References

[1]. S. Y. Yurish, Sensors: Smart vs. Intelligent, *Sensors & Transducers*, Vol. 114, Issue 3, March 2010, pp. I-VI.

[2]. J., H. Huijsing, F. R. Riedijk, van der Horn G., Developments in Integrated Smart Sensors, *Sensors and Actuators, A: Physical*, Vol.43, 1994, pp.276-288.

[3]. J. H. Huijsing, Smart Sensor Systems: Why ? Where ? How ?, in Smart Sensor Systems, ed. by Gerard C. M. Meijer, *John Wiley & Sons*, Chichester, UK, 2008.

[4]. N. V. Kirianaki, S. Y. Yurish, N. O. Shpak, V. P. Deynega, Data Acquisition and Signal Processing for Smart Sensors, *John Wiley & Sons*, Chichester, UK, 2001.

[5]. R. Taymanov, K. Sapozhnikova, Problems of Terminology in the Field of Measuring Instruments with Elements of Artificial Intelligence, *Sensors & Transducers*, Vol. 102, Issue 3, March 2009, pp.51-61.

[6]. Strategic Research Agenda of the European Technology Platform on Smart System Integration, Version 2, March 2009.

[7]. Universal Frequency-to-Digital Converter (UFDC-1), Specification and Application Note, *Technology Assistance BCNA 2010, S. L.*, 2011.

[8]. High-Speed Universal Frequency-to-Digital Converter UFDC-1M-16, Data Sheet, *Technology Assistance BCNA 2010, S. L.*, 2011.

[9]. Universal Sensors and Transducers Interface (USTI), Specification and Application Note, *Technology Assistance BCNA 2010, S. L.*, 2011.

[10]. Standardization. IEEE 1451 Smart Transducer Interface at Sensors Web Portal: http://www.sensorsportal.com/HTML/standard.htm

[11]. D. Potter, IEEE P1451.4's Plug-and-play Sensors, *Sensors Magazine*, Vol.19, No.2, December 2002.

[12]. National Instruments and Leading Sensor Vendors Release More than 3,200 TEDS Smart Sensor Products, Press Release, 10 October 2005: http://news.thomasnet.com/companystory/468384

[13]. Sensors Web Portal: http://www.sensorsportal.com/HTML/Sensor.htm

[14]. N. V. Kirianaki, R. M. Gayduchok, Digital Measurements of Frequency-Time Parameters of Signal, *Vyscha Shkola*, Lviv, 1978 (in Russian).

[15]. N. V. Kirianaki, S. Y. Yurish, N. O. Shpak, V. P. Deynega, Data Acquisition and Signal Processing for Smart Sensors, *John Wiley & Sons*, Chichester, UK, 2001.

[16].EFFF Frequency/ Frequency-Converter, 85.1 EFFF, *HONSBERG Instruments GmbH*, 9 March 2005.

[17].Proposal for Analogue Interface Standardization in Avionics, Avionic Systems Standardization Committee (ASSC), Avionic Interfaces Working Group, Doc. No: ASSC/110/5/1, Issue 3, July 2003.

[18].S. Y. Yurish, IEEE 1451 Standard and Frequency Output Sensors: How to Obtain a Broad-Based Industry Adoption ? *Sensors & Transducers*, Vol.59, Issue 9, September 2005, pp.412-418.

[19].S. Y. Yurish, Extension of IEEE 1451 Standard to Quasi-Digital Sensors, in *Proceedings of the IEEE Sensors Applications Symposium (SAS' 2007)*, San Diego, California, USA, February 6-8, 2007.

[20].S. Y. Yurish, IEEE 1451 Standard Adaptation for Frequency Output Sensors, *Sensor Electronics and Microsystems Technologies*, No.1, 2006, pp. 71-79.

[21].R. N. Johnson, Building Plug-and-play Networked Smart Transducers, *Sensors Magazine*, October 1997, pp. 40-61.

[22]. R. S. Yermolov, Digital Frequency Counters, *Energija*, Moscow, 1973 (in Russian).

[23] MicroConverter®, Multichannel 12-Bit ADC with Embedded Flash MCU, ADuC812, Rev. E, *Analog Devices, Inc.*, USA, 2003.

[24].MicroConverter®, Dual-Channel 16-Bit ADCs with Embedded Flash MCU, ADuC816, Rev. E, *Analog Devices, Inc.*, USA, 2010.

[25].J. Zhou, A. Mason, Communication Buses and Protocols for Sensor Networks, *Sensors*, Vol.2, February 2002, pp. 244-257.

[26].N. V. Kirianaki, S. Y. Yurish, Advanced DAQ Board for Frequency-time Parameters of Electric Signals, in *Proceedings of the 2nd IEEE International Workshop on Intelligent Data Acquisition and Advanced Computing Systems: Technology and Applications (IDAACS' 2003)*, Lviv, Ukraine, 2003, pp. 150-153.

[27].S. A. Korobeynikov, Determination of Characteristics of Decision Rule at Adaptive Measurement, *Investigated in Russia*, 2004, pp. 1851-1855 (in Russian).

[28].I. N. Pystynskiy, V. C. Titov, T. A. Shyrabakina, Adaptive Photo-Electric Converters with Microprocessors, *Energoatomizdat*, Moscow, 1990 (in Russian).

[29].S. Y. Yurish, Sensors and Transducers: Frequency Output Versus Voltage Output, *Sensors & Transducers*, Vol.49, Issue 11, November 2004, pp. 302-305.

[30].E. I. Tsvetkov, Processor Measuring Devices, *Energoatomizdat,* Leningrad, 1989 (in Russian).

[31].V. Ya. Rozenberg, Introduction to Theory of Accuracy of Measuring Systems, *Sovetskoe Radio*, Moscow, 1975 (in Russian).

[32].S. Y. Yurish, Self-Adaptive Smart Sensors and Sensor Systems, *Sensors & Transducers*, Vol.94, Issue 7, July 2008, pp.1-14.

[33].N. V. Kirianaki, S. Y. Yurish, High Precision, Wide Speed Range Rotation Sensing with UFDC-1, *Sensors & Transducers*, Vol. 59, Issue 9, September 2005, pp. 426-431.

[34].Manufacturers of Rotation Speed and Angular Position Sensors http://www.sensorsportal.com/HTML/SENSORS/RotationSens_Manuf.htm

[35].List of Pressure Sensors Manufacturers at Sensors Web Portal: http://www.sensorsportal.com/HTML/SENSORS/PresSens_Manuf.htm

[36].S. Y. Yurish, Intelligent Digital Pressure Sensors and Transducers Based on Universal Frequency-to-Digital Converter (UFDC-1), *Sensors & Transducers*, Vol. 60, Issue 10, October 2005, pp. 432-438.

[37].List of Temperature Sensors Manufacturers at Sensors Web Portal: http://www.sensorsportal.com/HTML/SENSORS/TempSens_Manufacturers.htm

[38].List of Humidity and Moisture Sensors Manufacturers at Sensors Web Portal:

http://www.sensorsportal.com/HTML/SENSORS/HumiditySens_Manufacturers.htm
[39].MAX6576/MAX6577 Maxim SOT Temperature Sensors with Period/Frequency Output, *Maxim*, USA, 1999.
[40].S. Y. Yurish, Low-Cost, Smart Temperature Sensors Systems Based on Universal Frequency-to-Digital Converter, in *Proceedings of the IEEE Sensors Applications Symposium (SAS' 2009)*, 17-19 February 2009, New Orleans, LA, USA, pp.287-292.
[41].Temperature and Humidity Module HTF3130, Technical Data, Humirel Sensor Device Data, HPC050V0, January 2004.
[42].S. Y. Yurish, Data Acquisition Systems for Quasi-Digital Temperature Sensors Based on Universal Frequency-to-Digital Converter, *Sensors & Transducers*, Vol. 57, Issue 7, July 2005, pp. 341-351.
[43].S. Y. Yurish, Digital Humidity Sensors and Data Logger Design Based on Modern Frequency-to-Digital Converter, *Sensors & Transducers*, Vol. 59, Issue 9, September 2005, pp. 419-425.
[44].S. Y. Yurish, Low-Cost Smart Self-Adaptive Humidity Sensors with Parametric Adaptation, in *Proceedings of SENSOR+TEST Conference 2009*, Nurnberg, Germany, 26-28 May 2009, Vol. 2, pp. 117-122.

Chapter 18

Other Sensors, Sensor Systems and UFDC IC Applications

18.1 Sensors from Mechanical Signal Domain

In addition to described early accelerometers, inclinometers, pressure and displacement sensors, there are many other quasi-digital sensors from mechanical signal domain. First of all, it is load cells, flow, level, position, proximity and torque sensors and transducers [1]. Such most interesting quasi-digital sensors will be considered below in order to be used in various digital sensors and sensor systems designs.

18.1.1 Flow Sensors

Flow meters are widely used in various end-user industries such as oil and gas; chemical and petrochemical; water and wastewater; food and beverage; pulp and paper; pharmaceuticals; power; steel; mining; semiconductors, etc. According to new report by *Global Industry Analysts, Inc*., global flow meters market to reach US $ 4.6 Billion by 2015 [2]. Future growth in the market will be driven by restoration of health in end-use industries post recession, and subsequent re-investment of manufacturing majors in plant renovation, modernization, capacity expansions, technology developments and entry of new players into the market [2].

The following type of flow meters are used in different applications: open channel; differential pressure; variable area; positive displacement; Vortex; electromagnetic; Coriolis; insertion thermal mass; ultrasonic; turbine; MEMS. As usually, almost all turbine (paddlewheel, propeller, etc.), electromagnetic and Vortex types flow meters for liquids and gases have frequency output [3-6]. Such sensors offer the advantages of an attractive performance-to-cost ratio and user-friendly installation and maintenance, from small to large pipes, and low to high flow rates.

Both types of ultrasonic flowmeters: Doppler flowmeters and transit-time flowmeters are based on Doppler frequency shift and the difference in travel time measurements accordingly [7].

Main metrological performances for selected flowmeters [6] are shown in Table 18.1. Some flow meters have also current output (4-20 mA), some other – voltage output.

Table 18.1 Metrological performances of flow meters.

Model	Type	Range	Error, %	Output
		+GF+ SIGNET		
7000	Vortex	0.3 to 4 m/s	± 1	Frequency
7002	Vortex	N/a	± 1	Frequency
2550/2560	Electro-magnetic	0.1 to 7 m/s	± 2	Frequency
		KOBOLD Instruments, Inc.		
DF	Paddlewheel	0.02 - 0.14 GPM to 1.5 - 36 GPM	± 2.5	Frequency
		Badger Meter		
SDI	N/a	0.3 ft/sec to over 20 ft/sec	± 1	Frequency
		Newport		
FP7000	Paddlewheel	N/a	± 2	1 Hz/fps
		Clark Solutions		
CFT110	Turbine	0.1 GPM to 8 GPM	± 3	33 - 1150 Hz
		Omega		
FP7001	Paddlewheel	N/a	± 2	1 Hz/fps
FMG3000 FMG3100	Magnetic	0.05 to 5 m/s	N/a	Frequency
		Dwyer Instruments, Inc.		
SFI-100T	N/a	N/a	± 5	0 to 100 Hz
		INTEC Controls		
SDI	Turbine	0.3 ft/sec to over 20 ft/sec	± 1	800 Hz max
		Proteus Industries, Inc.		
PS600	Turbine	0.1 – 60 GPM	N/a	5 -120 Hz
		ERDCO		
3600	N/a	N/a	N/a	0-1000 Hz
3700	N/a	N/a	N/a	Frequency

N/a – information is not available.

All quasi-digital flow meter can be used with UFDC-1, UFDC-1M-16 or USTI ICs to obtain digital output according to one of popular serial interfaces and buses: RS232, SPI and I^2C [8-10]. Two flow meters can be connected directly to each of IC. The programmable relative error must be selected in one order less, than flow meter's error. For example, if a flow meter has ± 1 % relative error, the programmable relative error for ICs must be ± 0.1 % in order to be neglected.

The ICs are also very suitable for MEMS based flow sensors with duty-cycle output [11]. The signal conditioning circuit is based on a relaxation oscillator with duty-cycle modulation. The implemented resistive bridge-to-duly-cycle converter produces an output signal with due-cycle highly linear in the change in resistance of the flow sensor's elements [11].

18.1.2 Level Sensors

Rapid growth in water and wastewater management market spurred development in level sensors. Buoyed by the favorable trends, the global level sensors market is poised to touch US $ 2.4 billion by 2012 [12]. Level sensor industry, following the trends in electronics industry, is also inclined towards compactness and micro designs. Non-contact sensors gave new dimension to the industry, as they do not interact and interfere with the matter to be sensed. With level sensors market moving towards non-contact or electronic type of measurement from contact or mechanical type of measurements, non-contact level sensor is becoming a predominant technology. Most of the industries are focusing on technologies such as radar and ultrasonic for accurate, reliable and cost effective means for level measurement [12]. RF capacitance, conductance, hydrostatic tank gauging, radar, and ultrasonic are the leading sensor technologies in liquid level tank measurement and control operations [13].

Some manufacturers also supply various quasi-digital level sensors with frequency output [14]. Main metrological performances for selected quasi-digital level sensors are shown in Table 18.2.

Table 18.2 Metrological performances of level sensors.

Model	Type	Range	Error	Output, Hz
		Flowline, Inc.		
EchoPod DX10	Ultrasonic	1.25 m	± 3 mm	976 ... 2000
		CKPT		
DUT-E-F	Capacitive	Length 180 - 2000 mm	± 1 %	500 … 1500
		Oceanscan, Ltd.		
DIGIQUARTZ	Pressure	0 ... 7000 m	± 0.01 %	Frequency output
		Strela		
F	Capacitive	Max. Length up to 4000 mm	± 0.1 %	500 ... 1500

As it is visible from the table, some quasi-digital sensors have very low relative error. For example, the F Series of capacitive fuel level sensor with frequency output has full scale error ± 0.1 % at 500 Hz (empty) and 1500 Hz (full) frequency range. The depth sensor DIGIQUARTZ has relative error ± 0.01 % and provides the ultimate in precision water level measurements up to 7000 m under difficult environmental conditions. The remarkable performance of these depth sensors is achieved through the use of a precision quartz crystal resonator whose frequency of oscillation varies with pressure induced stress. A quartz crystal temperature signal is provided to thermally compensate the calculated pressure and achieve high accuracy over a broad range of temperatures.

Such sensors are easy to interface with the UFDC and USTI ICs [8-10]. Programmable relative errors of ICs in such applications must be selected ± 0.01 % and ± 0.001 % respectively.

The ultrasonic level frequency transmitter DX10-DS EchoPod from *Flowline, Inc.* (Figure 18.1) has the full scale relative error ± 0.24 %.

Fig. 18.1 Ultrasonic, frequency output level transmitter DX10-DS EchoPod from *Flowline, Inc.*

In this case the programmable relative error of UFDC or USTI ICs must be selected in one order less (or in 5 times less, at least) and equals to ± 0.02 % (± 0.05 %). The DX10-DS EchoPod provides also a 0-5 V, 0 – 10 V voltage outputs.

18.1.3 Load Cells

Most load cells which are sold today as "digital load cells" are strain gauge based analogue load cells with built-in electronics to convert the analogue output into a digital output signal [15]. A digital load cell comprises a measurement system illustrated in Figure 18.2 [16].

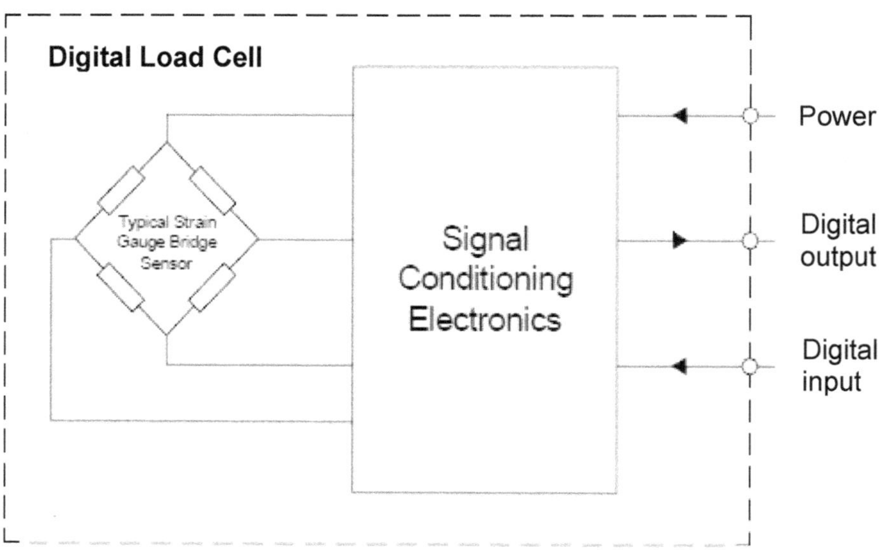

Fig. 18.2 Digital Load Cell [16].

Optimized performance is the result of high-dynamic range A/D converters and optimal filtering [17, 18]. Another, modern, technology driven design approach means the use of quasi-digital load cells with frequency output, for example, and UFDC or USTI ICs

[8-10]. A miniature, capacitive type load cell - iLoad Mini, with a square wave output whose frequency is proportional to the applied load is manufactured by *Loadstar Sensors* [15]. It has a full scale relative error from ±1 % to ±0.5 % and frequency output range from 250 Hz to 150 kHz, which corresponds to 10-200 lb capacity.

Unlike conventional resistive sensors based on either strain gauges or piezo-resistive techniques, Loadstar's breakthrough patented technology harnesses changes in capacitance to measure loads quickly and accurately. In the Mini, the change in capacitance is converted into a change in frequency of the output signal [19].

The load cell accepts a 5 V DC input and outputs a TTL square wave whose frequency is proportional to the applied load. The UFDC or USTI ICs have the capability to measure the frequency of the signal. In dependent on control signal, the iLoad Mini load cell has sensor frequency f_{sensor} or reference frequency f_{ref} on its output:

$$f_{comp} = f_{sensor} - K \cdot f_{ref}, \quad (18.1)$$

where K is the compensation factor.

The connection of two iLoad Mini load cells to development board EVAL UFDC-1 is shown in Figure 18.3.

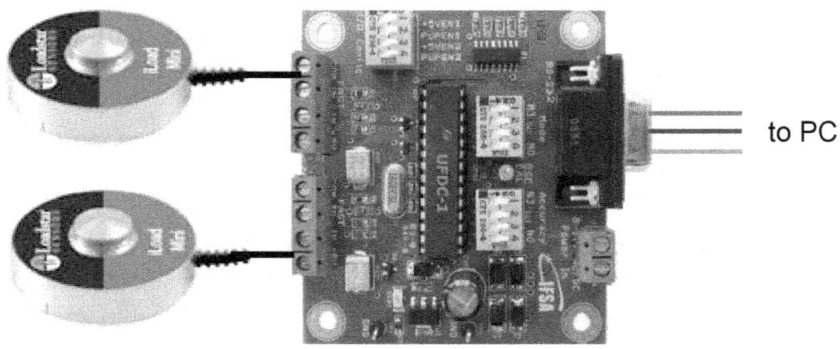

Fig. 18.3 Connection of two iLoad Mini load cells to development board EVAL UFDC-1 (Not for scale).

A stiff load cell with high overload capability and direct frequency output (6.6 - 7.4 kHz) is described in [20]. This novel load cell having stiffness and potential overload capability some 10 times that of a resistive strain gauge load cell. The experimental cell loaded up to 600 kg gives a direct output frequency change of 870 Hz (safety factor of 5). The new approach to design load cells should therefore result in far more robust weighing systems than those using resistive strain gauges. Resonant sensing also offers many other advantages, such as direct digital frequency output compatible with digital microelectronics, low power consumption, better long-term stability and greater resolution [20].

18.1.4 Position and Proximity Sensors

Position measurements answers the question, "Where is the target?" while a proximity sensing answers the question "Is it present or not?" Devices used to measure a target's true position, whether contact or noncontact, provide a continuously varying response (analog voltage or current, quasi-digital or digital signal) that is proportional to target position, anywhere within the calibrated measuring range of the device [21]. There are absolute position sensors or a relative one (displacement sensor); linear or angular (rotary encoders). The following types of position sensors are used widely: inductive, capacitive, Eddy-current, Hall effect, optical, laser and piezo-electric. The Hall effect based sensors (for example, duty-cycle output MLX90291, MLX90292 programmable linear Hall sensors) and quasi-digital optical sensors, which can be used for position and/or proximity sensing have been considered in appropriate book's chapters.

The market pressures to create less expensive, competitive sensor designs in shorter time are forcing developers to search out better design technologies. Good position sensors must maximize a system sensitivity to target movement and minimize sensitivity to all other effects and variables. They also must be inherently immune to noise and capable of performing fast, accurate measurements in inhospitable environments [21]. The failure of position or speed sensors in the field can have a massive technical or commercial impact [22]. One option is to use quasi-digital position sensors and digital sensor systems on its basis, which are well suitable for applications where reliable and accurate operation is essential.

With the goal of designing an Eddy current position sensor of the lowest possible cost, the self-oscillating circuit should be considered [23, 24]. Such sensor has a square wave output whose frequency is a function of displacement. The output can be connected directly to the UFDC or USTI ICs in order to obtain one of the popular serial interface and sensor buses such as RS232, SPI and I^2C [8-10].

An interesting quasi-digital Temposonics LD absolute linear position sensor is produced by the *MTS Systems Corporation* [25, 26]. This sensor is available with start/stop and pulse-width-modulated (PWM) output and working in 50-7620 mm measuring range in comparison with 50-2540 mm range for analog output sensor's version [25]. All outputs are absolute rather than incremental. The sensor photo and output diagrams are shown in Figure 18.4 (a, b). The Temposonics L Series position sensors provide direct Start/Stop and PWM outputs, and no signal-conditioning electronics are needed when interfacing with the UFDC and USTI ICs. The ICs can measure both: time interval between Start and Stop pulses and pulse width in one channel (for UFDC-1, UFDC-1M-16 ICs) or two channels (for USTI IC at pulse width measuring mode). Hence, one Temposonics LD position sensor can be connected to the UFDC, and two sensors to the USTI IC. In case of time interval measurement between Start and Stop pulses by the USTI IC, the Start pulse must be connected to the first F_{x1} input, and Stop pulse – to the second F_{x2} input. Appropriate commands for measuring time interval between Start and Stop pulses and pulse width by USTI IC working in RS232 slave communication mode are shown in Figure 18.5. It is not necessary to set up the relative error, because of the accuracy for such measuring modes are selected automatically as high as possible. The relative error can be calculated according to equation (2.1) for the UFDC-1 and (3.1) for the USTI IC. Resolution is 0.18 mm (when using the UFDC-1 or UFDC-1M-16 ICs) and 0.14 mm (when using the USTI IC).

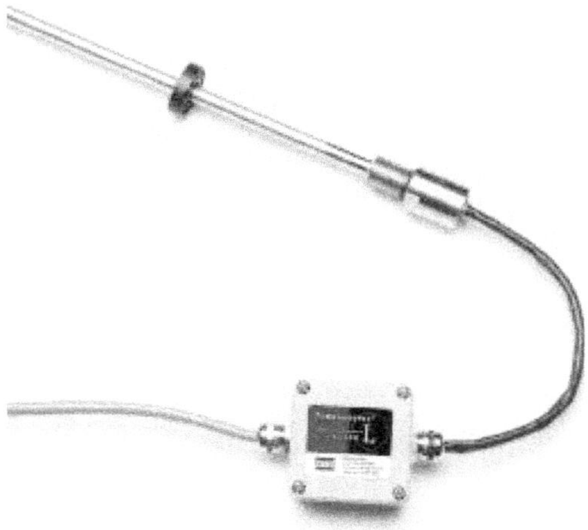

(a)

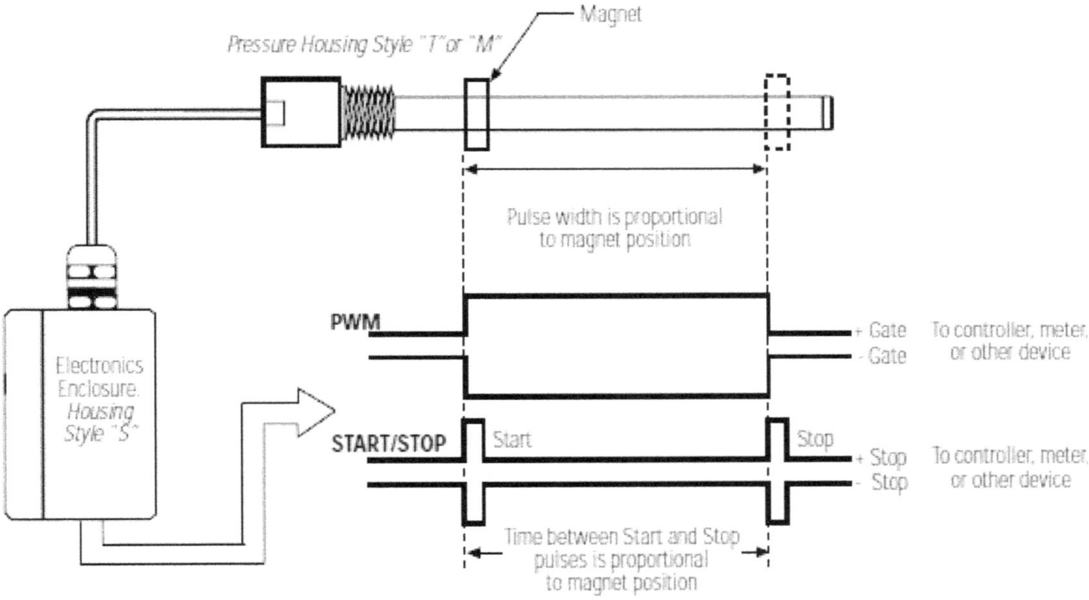

(b)

Fig. 18.4 Temposonics LD absolute linear position sensor (a), output diagrams (b).

```
>M03    ; Time interval measurement between Start and Stop pulses
>S      ; Start a measurement
>C      ; Check result status: returns 'r' if ready or 'b' if busy
>R      ; Read a result

>M0B    Pulse width measurement in the 1st2 channel
>S      ; Start a measurement
>C      ; Check result status: returns 'r' if ready or 'b' if busy
>R      ; Read a result
```

Fig. 18.5 Appropriate USTI commands (RS232) for time interval measurement between Start and Stop pulses and pulse width measurements.

Proximity sensors are using the same technologies as position sensors. The following types of position sensors are used widely in various applications: capacitive, inductive, Eddy-current, magnetic, laser, optical, Doppler effect, radar and sonar [27]. Different proximity sensor targets demand different sensors. For example, a capacitive or photoelectric sensor might be suitable for a plastic target; an inductive proximity sensor requires a metal target. A proximity sensor adjusted to a very short range is often used as a touch sensor.

As usually, several manufacturers produce quasi-digital proximity sensors [27]. For example, the inductive sensor Bi8-M18-LF10 from *Hans Turck GmbH & Co* has 1 ... 10 kHz frequency output and measuring range (air gap) 1 ... 5 mm. The frequency of output signal is proportional to the target's distance (Figure 18.6).

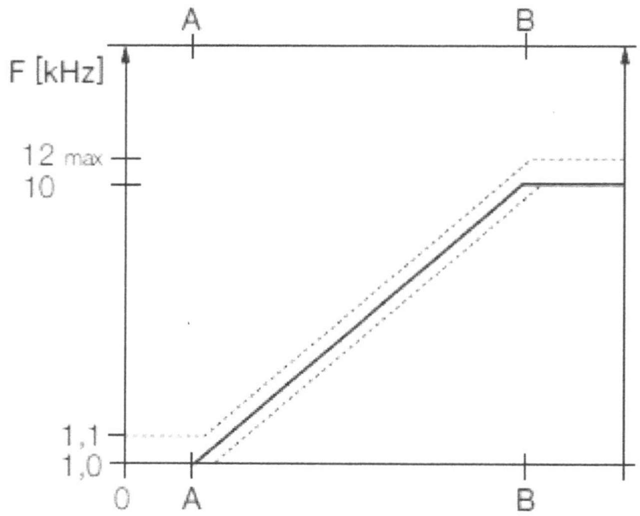

Fig. 18.6 Measuring range of Bi8-M18-LF10 proximity sensor.

The SIMATIC PXS240 ultrasonic and SIMATIC PXS200 sonar proximity sensors (switches) from *Siemens* also have frequency outputs [27]. The ultrasonic diffuse, frequency output sensor UC 80 CND 40 from *Carlo Gavazzi* has the 400-4000 mm operating distance and frequency output in a range from 132 Hz to 1.312 kHz [27]. Such proximity sensors also can be easy interfacing with the UFDC and USTI ICs.

18.1.5 Torque Sensors

According to new World Torque Sensors Market report from Frost & Sullivan, the US $ 647.1 Million this market to grow to US $ 1.06 Billion by 2017 [28]. Torque sensors are used in different applications: telemetry, prosthetics in medical applications, turbo-engine applications in aerospace, and engines in electric and hybrid vehicles.

A torque sensor or transducer is a device for measuring the torque (torque is just a rotational force, or a force through a distance) on a rotating system, such as an engine, crankshaft, gearbox, transmission, rotor, etc. Such sensors can be based on various principles [29, 30]: inductive, Surface Acoustic Wave (SAW), magnetoelastic, infrared, etc. Till now many torque sensors have analog output. Commonly, torque sensors or torque transducers use strain gauges applied to a rotating shaft or axle. Such types of torque transducers add conditioning electronics and an A/D converter to the rotating shaft. However, for many applications the interference immunity is not sufficient [31]. The solution is a quasi-digital torque sensors or digital transducers, based on frequency output (quasi-digital) torque sensors. Today many torque sensors manufacturers produce quasi-digital torque sensors with frequency output [29]. The main metrological performances of quasi-digital torque sensors are adduced in Table 18.3.

Table 18.3 Metrological performances of quasi-digital torque sensors.

Model		Range, N·m	Accuracy Class, %	Output, kHz
		Kistler		
4504B	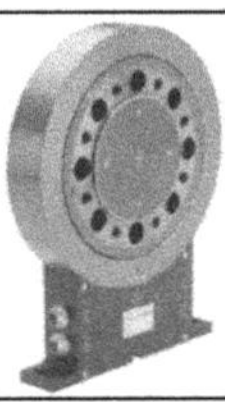	50 - 5000	± 0.1 (± 0.05 optional)	10 ± 5 60 ± 20 100 ± 40
4510B		100 - 4000	± 0.2	100 ± 40
		HBM		
T10F		50 - 10000	± 0.1	10 ± 5
T12		100- 10000	*	10 ± 5 60 ± 30
T40FM		50-80	± 0.1 (± 0.05 optional)	10, 60, 240
		Honeywell		
TMS 9000		200000 lb-ft	± 0.1	10 ± 5 60 ± 20

The block diagram of ultra high precision torque transducer from *HBM* is shown in Figure 18.7.

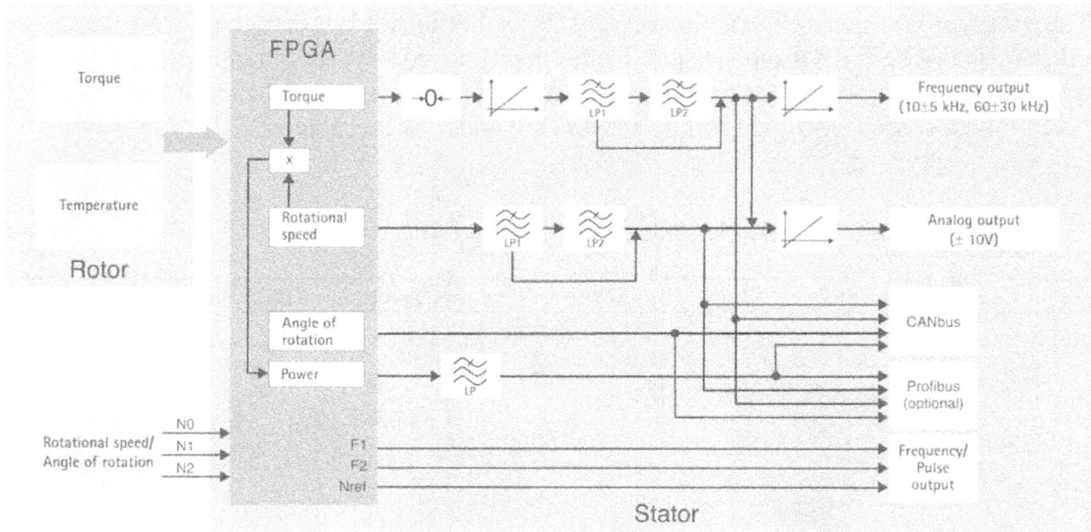

Fig. 18.7 Block diagram of T12 torque sensor.

Some other torque sensors, for example, non-contact rotary torque sensor Series 3000 from *NCT Engineering GmbH* is available with frequency and PWM output.

A new torque transducer with high overload capability and direct 5.8 … 6.6 kHz frequency output is reported in [32]. The transducer is based on a thick-film PZT printed metallic triple-beam resonators. It has been tested in a torque range of up to 20 Nm, resulting in a strain level of 400 microstrains on the measurement shaft surface and giving a large frequency-change output of 800 Hz [32].

All mentioned above torque sensors and transducers can be directly connected to UFDC or USTI ICs in order to convert output frequency in digital according to RS232, SPI or I^2C serial interfaces. Each of ICs can serves two frequency output torque transducers. In addition, the USTI can directly interface a resistive output torque transducer.

18.1.6 Other Sensors and Transducers

Other quasi-digital sensors and transducers with frequency outputs from different signal domains, for example, rain, wind speed, conductivity sensors, true power transducer and other are described in the List of Sensors Manufacturers on Sensors Web Portal at: http://www.sensorsportal.com/HTML/SENSORS/Other_Manufacturers.htm

18.2 Wireless Sensors

The total revenue for wireless sensors and transmitters in industrial applications is likely to reach $1.8 billion in the next 3-4 years [33]. Wireless sensors can be deployed almost anywhere at a far lower cost than can a wired system. With the recent advances in embedded systems and wireless technology, the hardware used is becoming more inexpensive and more widely available. Because wireless sensor networking is built around low-power radios, the nodes that make up the network play a key role in wireless communication. From a physical perspective, the deployment of nodes may take several forms depending on the sensor application and the desired pattern of communication. Deployment may also be a one-time activity, where the installation and use of a sensor network are strictly separate activities. It can also be a continuous process where more nodes are deployed over the lifetime of the network [33]. The application needs determine the actual size of the network. The application can vary from a single sensor node to multiple sensor nodes.

A wireless sensor net is made up of a group of sensor nodes. Each sensor node possesses the ability to monitor some aspect of its environment, and each is able to communicate its observations through other nodes to a destination where data from the network is gathered and processed. Recent developments in wireless technologies and the semiconductor fabrication of miniature sensors are making WSNs smaller and more cost-effective for a growing number of uses [34].

Cost reduction in wireless sensor networks becomes a requirement to extend their application in fields where a great amount of sensors is needed [35]. In this case it should be a good solution to connect many existing low-cost sensors both: analog and quasi-digital to one sensor node to reduce the cost of nodes. Traditional approach to use multichannel ADC and/or analog multiplexers for analog sensors will not give any reduction in price. In addition, the analog multiplexer introduces additional measuring error. Hence, the analog signal must be preliminary converted to the quasi-digital signal (frequency or PWM output).

The described in [35] sensor interface (Figure 18.8) transforms the voltage provided by different sensors with different output ranges to a pulse signal which frequency will depend proportionally on the input voltage. The conversion of the sensor signal to a frequency value will bear much less sensitivity to interferences. In order to achieve the best performance in the conversion to digital values of the frequency signal, it is necessary to the different sensor output ranges are converted into the same frequency range. For this, the conditioning electronics change sensor's gain and offset voltage depending on the sensor signal characteristics.

The circuit consists of two operational amplifiers (OA), analog multiplexer, two programmable potentiometers and voltage-controlled oscillator (VCO), which performs the transformation to the frequency range. The low resolution serial DAC is used for self-calibration purposes. The expression that gives the output frequency is:

$$F_{OUT} = 0.1 \cdot f_{clk} + 0.8 \cdot \frac{V_{out}}{V_{cc}} \cdot f_{clk} \tag{18.2}$$

When a node of the network is activated, the microcontroller selects the sensor to be read by means of the control lines of the analogue multiplexer and carry out the conditioning and the conversion. The frequency-to-digital conversion is performed in this signal conditioning circuit using the classical direct counting method. However, such sensor interface has several disadvantages. It introduces additional error, because of mainly based on analog electronics components. The standard counting technique, which is used for frequency-to-digital conversion has also several known disadvantages, such as dependence of relative error on frequency range and redundant conversion time T_0. This sensor interface can not be used with various quasi-digital sensors and transducers, which as rule have very broad dynamic frequency range: form part of Hz to some tens MHz. In order to get the optimal trade-off between metrological performances and price for the sensor node it is expediently to use other, universal, advanced solutions based on the novel frequency-to-digital conversion method.

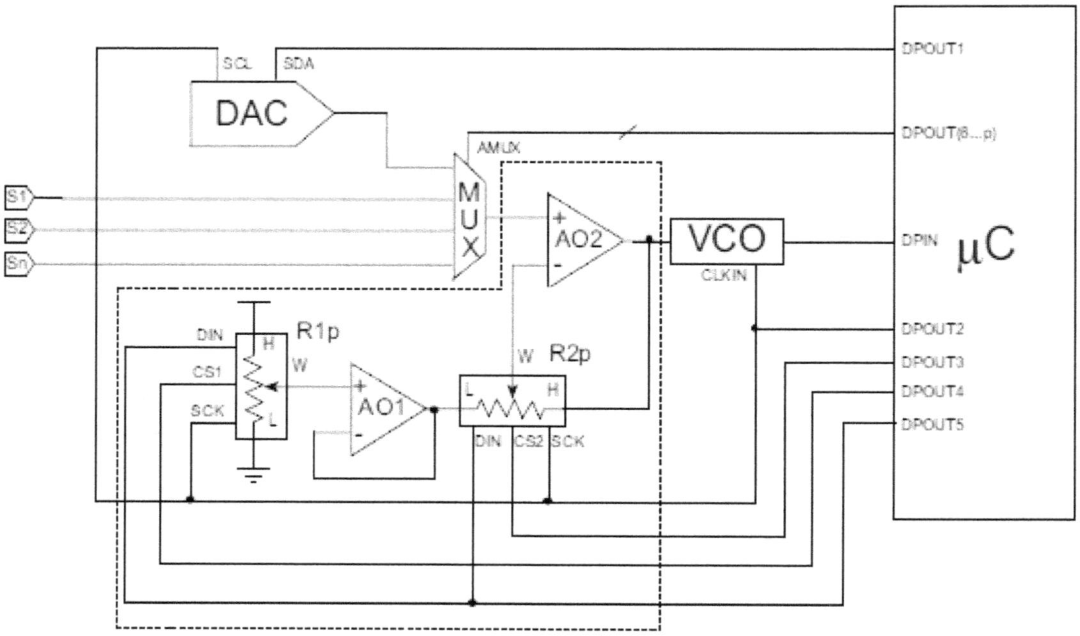

Fig. 18.8 Electronic sensor interface [35].

Let consider several advanced electronic sensor interface for analog and quasi-digital sensors and transducers. The simple, low cost sensor interface for analog and quasi-digital sensors with analog multiplexer and time division channeling is shown in Figure 18.9.

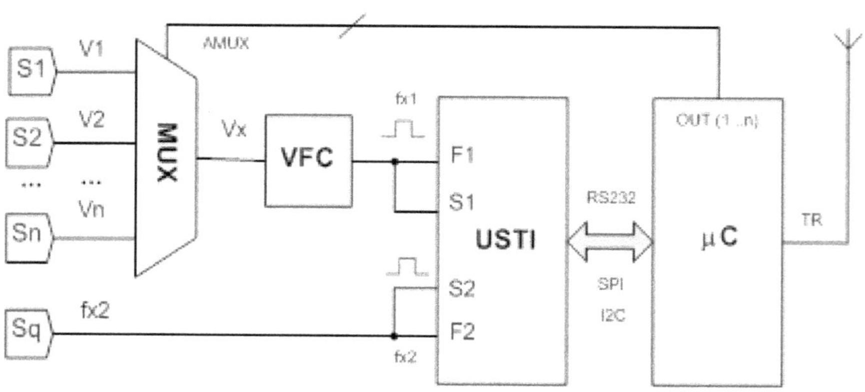

Fig. 18.9 Low-cost electronic sensor interface for analog and quasi-digital sensors with analog multiplexer.

The interfacing circuit consists of the USTI IC [10], which is based on advanced frequency-to-digital conversion method with a constant quantization error in all broad frequency range and non-redundant conversion time. In this case it is not necessary to convert the different sensor output ranges into the same frequency range. Only one voltage-to-frequency converter (VFC) used to convert sensor's output to frequency. As usually, integrated VFC have broad frequency ranges [36, 37]. In addition, one quasi-digital sensor can be directly connected to the second channel of USTI IC, and one sensing element (resistive, capacitance or resistive bridge) can be also connected to this IC. However, the analog multiplexer and VFC introduces additional errors. To eliminate the error due to analog multiplexer, it is possible to convert voltage-to-frequency for each of analog sensors before the multiplexer, and use a digital one, instead of analog multiplexer. Such electronic sensor interface with the digital multiplexer and combined (time for digital signal domain and space for analog signal domain) channeling is shown in Figure 18.10.

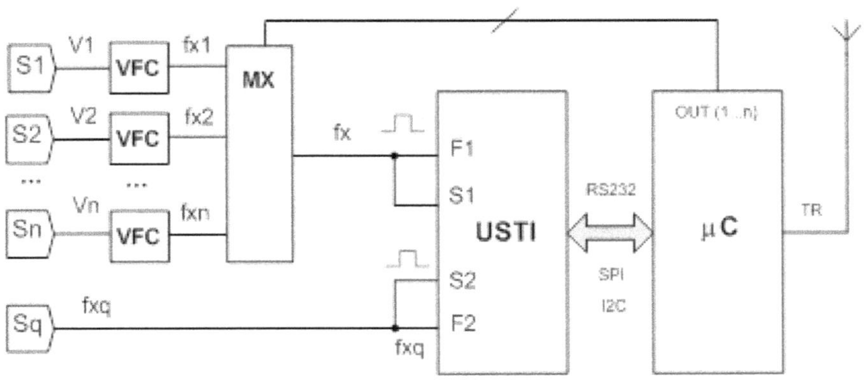

Fig. 18.10 Electronic sensor interface for analog and quasi-digital sensors with digital multiplexer.

The digital multiplexer (MX) does not introduce any additional error. However, the VFCs still do it. So, the solution with minimum possible hardware and high metrological performance is available if instead analog sensors and transducers to use quasi-digital [1]. Such low-cost electronic sensor interface with high metrological performance for sensor nodes is shown in Figure 18.11.

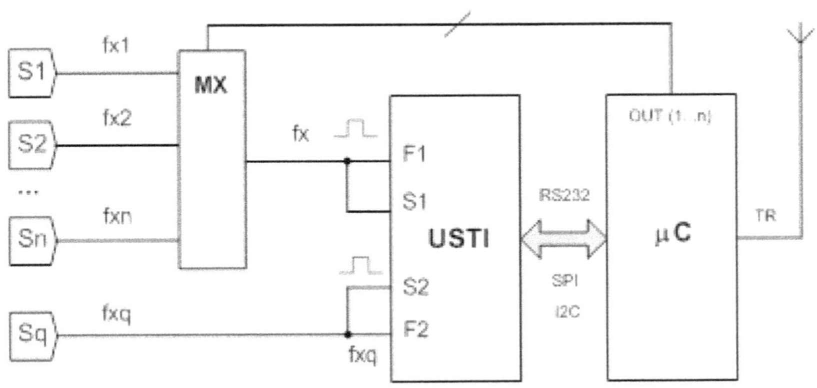

Fig. 18.11 Low-cost electronic sensor interface for quasi-digital sensors and transducers.

The next step to decrease hardware is to use the IC USTI-WSN instead of standard USTI (see Chapter 3). The USTI-WSN has a high speed embedded transmitter with high data rates from 250 kb/s up to 2 Mb/s for wireless communications and provides frame handling, outstanding receiver sensitivity and high transmit output power enabling a very robust wireless communication.

Another benefit to use quasi-digital sensors and transducers instead of analog sensors in WSN is a possibility to transmit frequency-time signals without preliminary conversion to digital. Some examples are described below.

The RF transmitter using pulse width modulation (PWM) method is reported in [38]. It does not use an analog to digital converter (A/D converter) for the RF transmission of analog data. The transmitter consists of a pulse width modulator, a voltage-controlled oscillator (VCO) and on-chip antenna. The PWM method digitally encodes analog signal levels, but the PWM signal remains quasi-digital. By keeping the signal quasi--digital, noise effects are minimized. The modulated signals are inputted into the VCO using PWM. If the voltage of the modulated signal is the high level, the VCO is oscillated, and the RF carrier waves for transmission can be obtained. Then, the output of the VCO is transmitted by the on-chip antenna [38]. The transmitter was possible to transmit with low power dissipation of 0.75 mW at the carrier of 315 MHz.

An interdigital capacitor based battery-free wireless pressure sensor is described in [39]. It consist of an interdigital capacitor (IDC) that serves as a pressure sensing element and an inductor, which works as a passive power source and data communication element (Figure 18.12). These two components work together as an LC resonator to realize the wireless pressure sensing and remote power to eliminate the need for wire connection in conventional pressure sensor. The sensing element is comprised of a set of linear parallel electrodes coated with Polyvinylidene Fluoride (PVDF) pressure sensing material on the top. The change of capacitance in the IDC is a function of the geometry of the electrodes and the electric properties of the sensitive layer. The sensor prototype demonstrated that the sensor prototype performs well in the range of 0 psi to 60 psi with an average pressure sensitivity of 25 kHz/psi [39].

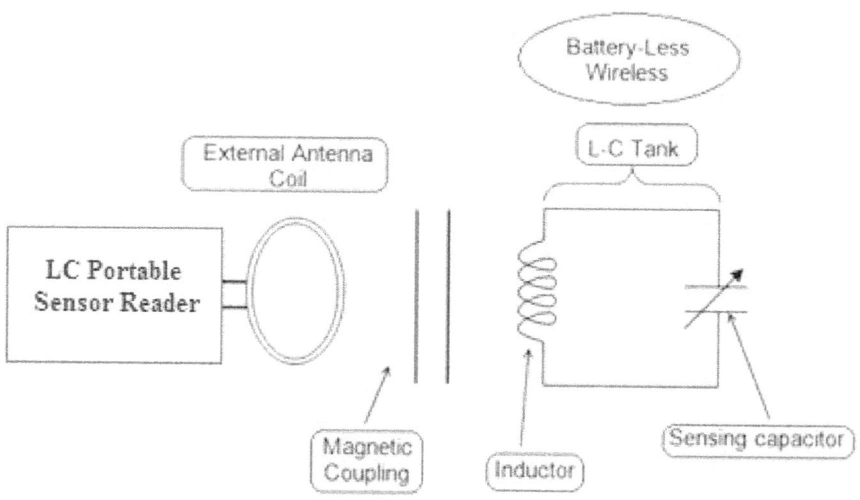

Fig. 18.12 Battery-free passive wireless sensing system [39].

Principle of the described pressure wireless sensor is based on the operation of an inductor-capacitor (L-C) tuned electric oscillator. The L-C oscillator works on the storage of energy in the form of electric charge in the capacitor and in the form of magnetic field in the inductor. The property to be sensed, such as pressure in this case, induced the variation of sensor capacitance due to electrical properity change.of the sensing material. The remote reader detected this information by monitoring resonant frequency shifts due to impedance changes in response to property change. In the L-C tank circuit, the velocity of the current flows from the capacitor to the inductor or vice versa, produces resonant frequency F that depends on the values of the capacitor C and the inductor L, and is expressed by the equation (18.3):

$$F = \frac{1}{2\pi\sqrt{LC}} \tag{18.3}$$

Figure18.13 shows the experimental relationship between the pressure applied and the resonant frequency of the fabricated sensor.

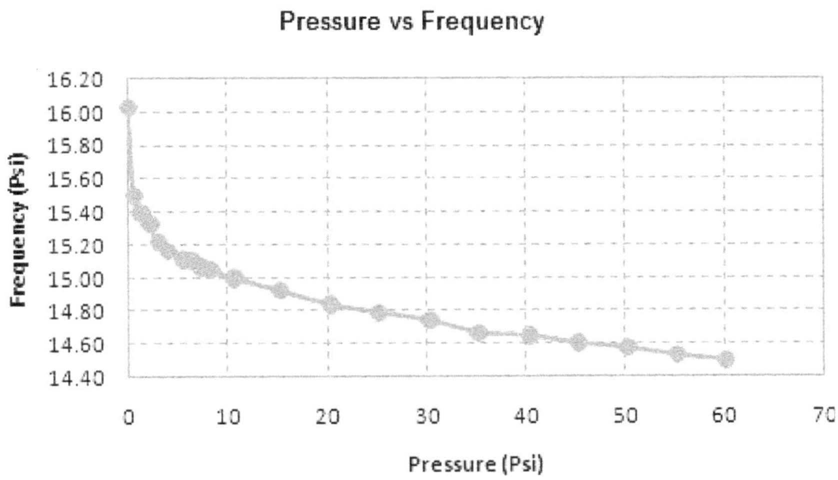

Fig. 18.13 Sensor's resonant frequency vs. pressure measured [39].

Battery-free and wireless pressure sensing can be realized without requiring physical connection, power supplies and active elements in the sensor circuit [39].

18.3 Sensor Systems based on Universal Frequency-to-Digital Converter (UFDC)

In addition to various digital sensors and sensor systems applications, the UFDC and USTI ICs can be used in different measuring instruments (such as tachometers, frequency counters, multimeters, etc.), measuring systems and data acquisition systems. Some recently published UFDC application examples are described below.

18.3.1 Low-Cost Tide Measurement System for Water Quality Assessment

A low-cost tide measurement system for water quality assessment based on two inclinometers is described in [40]. The inclination sensors deliver a DC output voltage, which varies linearly with the angle of its working direction. Thus, with two inclinometers assembled in orthogonal directions, it is possible to provide the tide direction and intensity. The system also includes additional measuring channels, for temperature, electrical conductivity, turbidity and pH measurements. The tide measurement system consists of two voltage-to-frequency converters and a universal frequency-to-digital converter (UFDC-1) [8]. Wi-Fi data communication is also provided including an Ethernet-wireless to the FP-2000 Ethernet port [40]. The block diagram of tide sensor is shown in Figure 18.14.

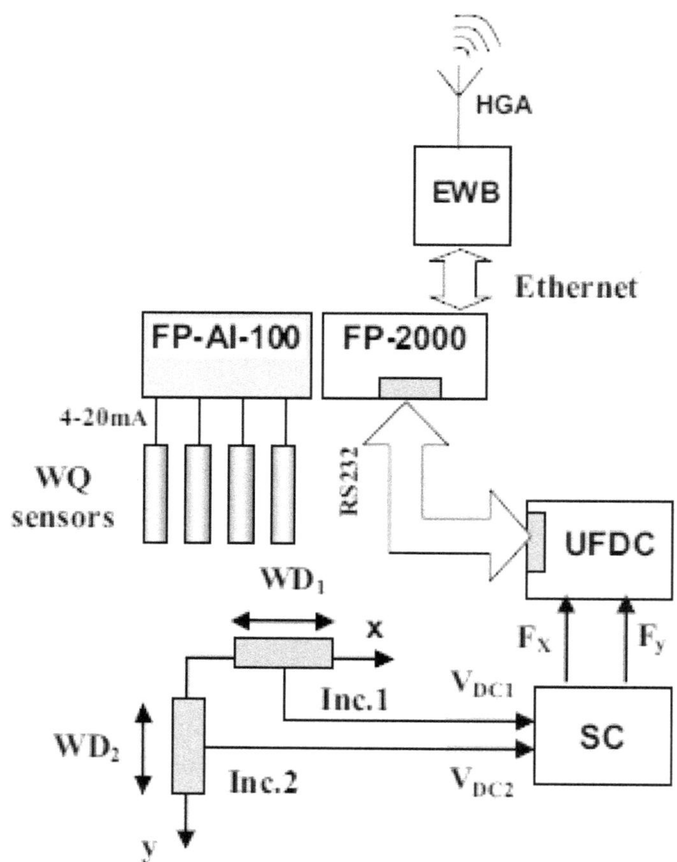

Fig. 18.14 Block diagram of the tide sensor (SC- signal conditioning unit; FP- FieldPoint unit; EWB- Ethernet-wireless bridge; HGA- high gain antenna).

Referring to the water quality metrological characteristics, the implemented system has ±0.1°C temperature absolute measurement error, ±2 % of full scale (0-14) pH measurement relative, ± 1% of full scale (0-20000 uS/cm) conductivity relative error and ±2 % of full scale (0-50NTU) turbidity relative error. The linearity error associated with the inclinometer sensing unit when the working direction angle varies between -60° and

+60° (2800–1300 Hz output frequency range) is lower than ±1 % FS. These measurement characteristics are very attractive specially when compared with those provided by expensive commercial tide measurement system [40].

18.3.2 Sensor System for Energy and Power Consumption in One-phase AC Line

A mixed electronic system for measurement of active, apparent and reactive energies delivered to a load in a single-phase AC voltage line is described in [41]. For this purpose a smart sensor ADE7753 from *Analog Devices* was used [42]. ADE7753 circuit is a mixed processor capable to measure active, apparent and reactive energies and transmitting the information through serial interface. It also has a CF pulse output which frequency is proportional to the active power. For this purpose, the UFDC-1 IC has been used [8]. Additionally, a LabView procedure installed in a PC reads and acquires the frequency value using the serial protocol. The block diagram of the designed system is shown in Figure 18.15.

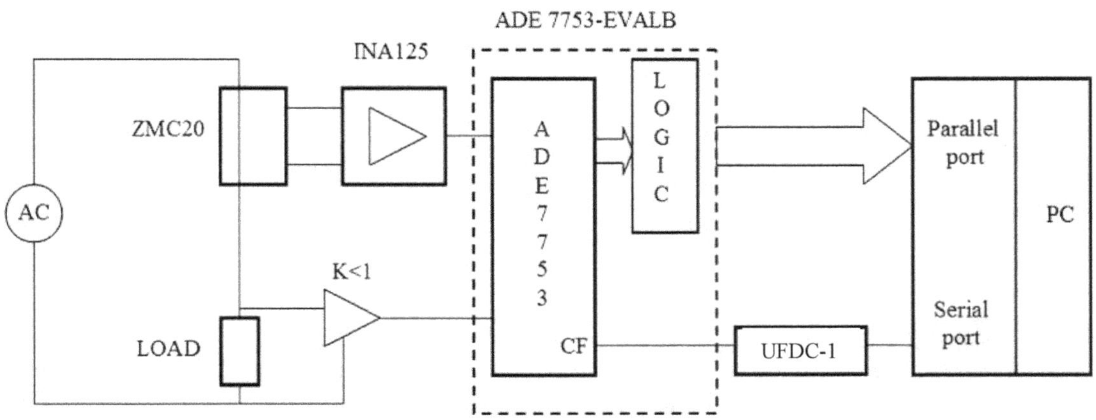

Fig. 18.15 Block diagram of the designed system [41].

The relationship between CF frequency and active power is given by the following equation:

$$f_{CF} = S \cdot P_{ac}; \quad S \equiv 4\frac{Hz}{W} \tag{18.4}$$

The CF output is used to calibrate the ADE7753 sensor interface, but in this application the frequency of CF output will be digitalized by the UFDC-1, and once its TTL output voltage levels are adapted to RS232, its value will be read by the computer serial port to calculate the active power P_{ac}.

18.3.3 Intelligent Stepper Motor Control

For applications requiring critical rotor positioning, usually stepper motors are the best choice. An intelligent controlling device based on a PIC16F84 microcontroller and

UFDC-1 interfacing circuit is described in [43]. The block diagram of the controlling device is shown in Figure 18.6. A four-phase unipolar stepper could use the ULN 2003 driver and a current sensing circuitry built with operational amplifiers, which output in the end a frequency signal proportional to the measured winding current. This signal is fed into the UFDC-1 IC, which communicates with the PIC16F84 microcontroller via the I^2C protocol [8].

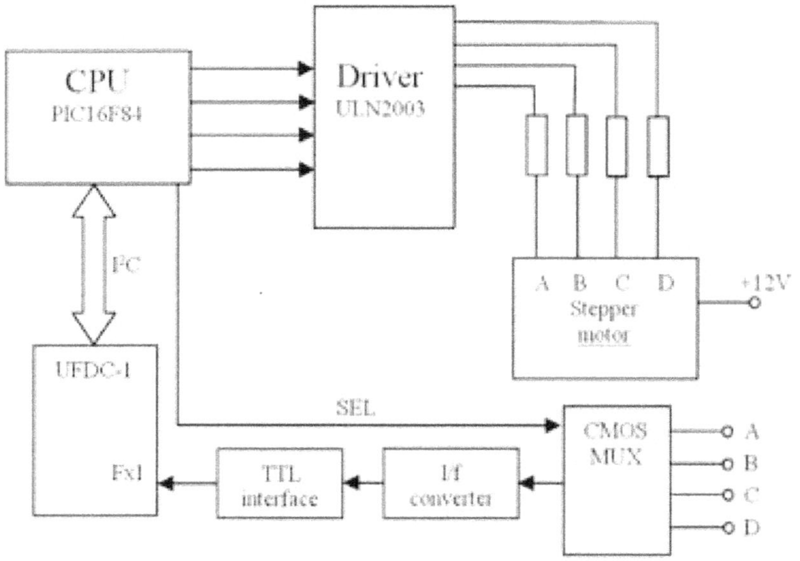

Fig. 18.16 Block diagram of controlling device [43].

Since the stepper motor has four terminals for the windings an analog CMOS multiplexer should be used to feed the sensed voltages into the converter. The TTL interface will provide the rectangular pulses for the UFDC-1's input.

In comparison to the classic linear constant current control a digital sensing system for the winding's current, using current to frequency conversion and a powerful interface like UFDC-1 has been designed.

18.3.4 Remote Laboratory for Smart Sensor Systems Design

Modern concepts in microelectronics education brings many new challenges from the pedagogical and technical aspects. One of the more significant is the integration of project based learning (PBL) into the long-distance educational process [44]. This is very important in digital, smart sensor and sensor systems design where the requests of the industry for faster and more efficient acquisition of knowledge and practical skills lead to a constant search for new solutions in the learning process, which would attract and well motivate students as well as to be friendly. Students are involved into a new learning scenario quite different from the traditional well-determined environments as a classroom and laboratory. Accessing remote laboratory instrumentation and performing experiments with the designed smart, intelligent sensors system individually, have become key elements in distance learning and training.

At the last decade many remote laboratories for distance and on-site microelectronics education and research including sensors design and microelectronics fabrication, where run in various universities around the world [45-52]. Most distance learning programs in microelectronics are focused on online experiments in virtual laboratories with simulators or/and remote laboratories with real equipment. However, a project based e-learning (PBeL) methodology needs some specific approach, first of all, from the hardware and teaching strategy (pedagogical methodology) points of view and must take into account special considerations. In order to be more attractive and useful for a modern "net generation" of students, a remote laboratory for the long-distance project based global e-learning must be built on the latest technology in appropriate subjective area and has a reasonable cost. Such remote laboratory, which satisfies to specific requirements at a reasonable cost, is described below.

The PBeL methodology means that students will propose and create many various projects with different sensors. So, the remote laboratory for such purpose should be based on a unified microelectronics component, which will let to design various sensors systems on the same flexible hardware platform. In case of sensors and measurement related courses such unified element should be the Universal Frequency-to-Digital Converter (UFDC-1, UFDC-1M-16) [8, 9] and its especially designed evaluation board EVAL UFDC-1/ UFDC-1M-16 (Figures 18.17) [53]. Dozens laboratories activities and projects are possible based on this development kit for both: online and on site access [44, 54-58].

Fig. 18.17 Evaluation board EVAL UFDC-1/ UFDC-1M-16.

Remote laboratories are a rapidly growing innovation trend in distance engineering education. The mass proliferation of the Internet, together with the ever-increasing user-friendly computer-controlled instrumentation, have favored the development of such remote laboratories with interactive experiments in the frame of PBeL conception.

The remote laboratory single server architecture for PBeL is shown in Figure 18.18. The test instruments and actuators are configured and controlled from the server using the GPIB interface and RS232-to-USB connectors. The experimental boards are accessed from the server also using the RS232-to-USB connection. The USB hub allows connection to a number experimental boards and measuring instruments.

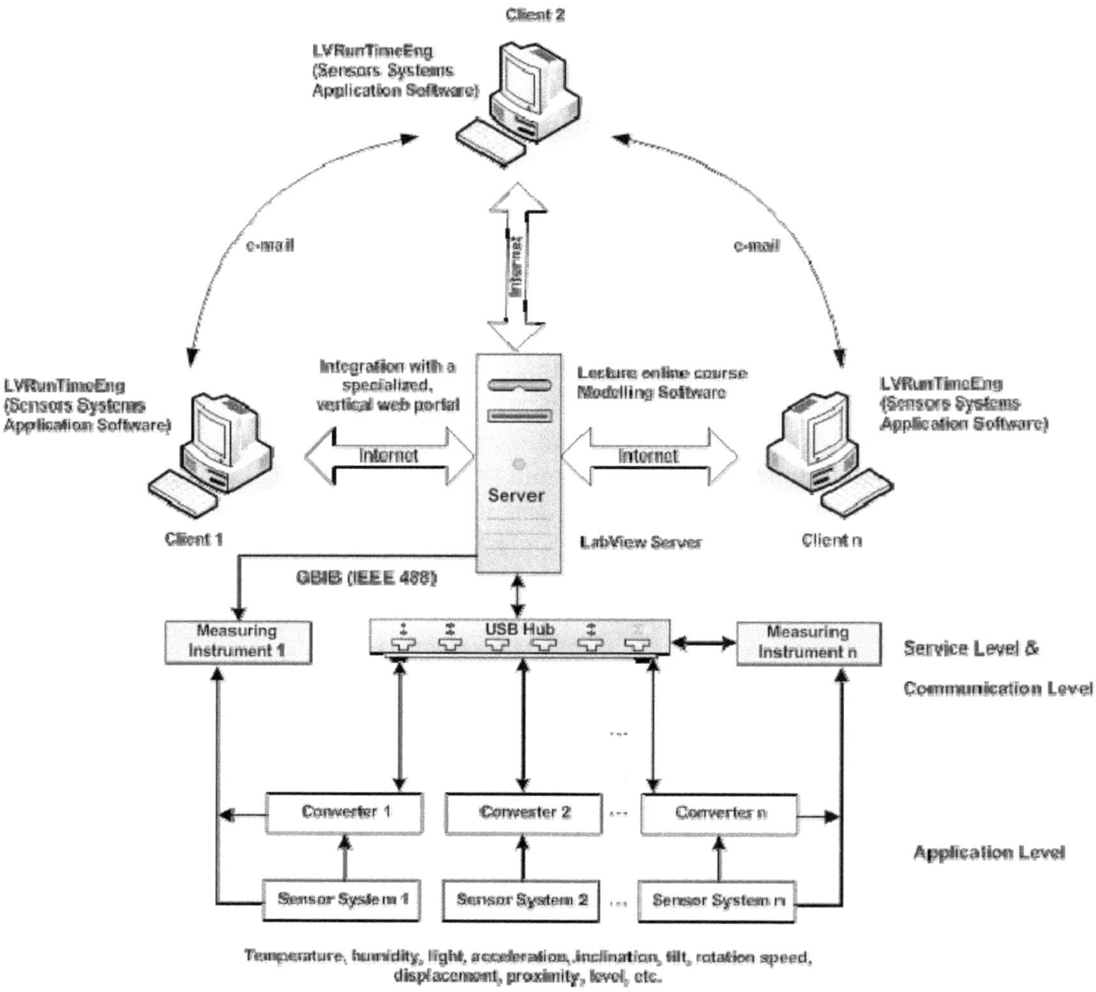

Fig. 18.18 Remote laboratory single server architecture for PBeL.

The remote laboratory's hardware should include standard equipments (Figure 18.18 in grey color) such as a server, Ethernet switch or USB-to-LAN Hub, measuring instruments (universal frequency counter, functional generators, oscilloscope, etc.) with LAN communication ports (Service Level); and non standard equipments (in white color) such as universal communication boards with SPI or RS232 to Ethernet bridges (Communication Level) for communication and actuation purposes; and universal sensors and transducers converter and interface boards with connected various sensors systems (temperature, humidity, light, acceleration, inclination, tilt, rotation speed, displacement, proximity, level, etc.) at the Application Level [44].

The possibility to control in a remote way the measuring instruments at the Service Level can be done also due to a free 'Agilent IO Control' application. This application, installed on the server allows sharing its instrumentation resources (GPIB, serial and USB ports, with the connected instruments) through the web with any computer using this same application configured as client. It is also expediently to use web cameras at sensors installations and actuators areas for additional visual control and verification of the correct operation of the lab equipment as a function of the control commands. The universal sensors and transducers converter and interface board contains the integrated circuit of novel, programmable Universal Frequency-to-Digital Converter (UFDC-1). It has 16 measuring modes of all frequency-time parameters of electric signals, one generating mode for calibration purpose and can work with all existing quasi-digital sensors and transducers. It means that many different laboratory activities and individual projects can be made on its basis. Only one difference between the hardware architecture of a hands-on laboratory and remote laboratory based on the proposal conception is that in the last one the Internet is the join between the end user and measuring devices. The rest of the architecture is the same. Using the remote laboratory the learners can work at home as if they were in the laboratory and working out an individual project according to the PBeL.

Due to numerous advantages, the standard LabVIEW software is used for the web server and clients. The LabVIEW Web Publishing Tool lets create a web server just in a few clicks. The .html files with front panels of virtual instruments embedded within will be easy generated. On the client side, the LabVIEW Run-Time Engine will be automatically instilled and run as a plug-in for free.

In contrast with previous versions of LabVIEW which allow only one user to interact with an application, the LabVIEW versions after v. 8.20 introduces simultaneous control of remote applications, consequently more students can retain simultaneous control of the same application remotely. However, it is obvious that multiple simultaneous accesses to the instrument server would not make sense because different settings from different students to one instrument could not be handled. Therefore it is necessary to do experiments sequent.

The PBeL includes all parts of PBL but taking into account its "distance" nature some activities should be done by appropriate specific ways. Main distinguishes between PBeL and PBL are showing in Table 18.4. In both cases students must learn how to apply the acquired, learned knowledge through problem solving and project design.

Table 18.4 Main distinguishes between PBeL and PBL.

Activities	PBL	PBeL
A short lecture course	In class	Online
Research & Development	In team	Individually
Project execution	In laboratory	In remote laboratory
Communication with other students and teacher	In class	In online chart, by e-mail, at forums
Presentation	In class	Online

Examples of measurement set up for online and on site laboratory activities are shown in Figures 18.19 and 18.20. A frequency counter (for example, Agilent 53132 A) with the relative error ± 0.0001 % or, at least, ± 0.0003 % should be used for a calibration procedure (Figure 18.19). The digital oscilloscope is used for pulse shape monitoring at the 8 MHz test output of UFDC-1. The calibration procedure is described in details in Chapter 2.

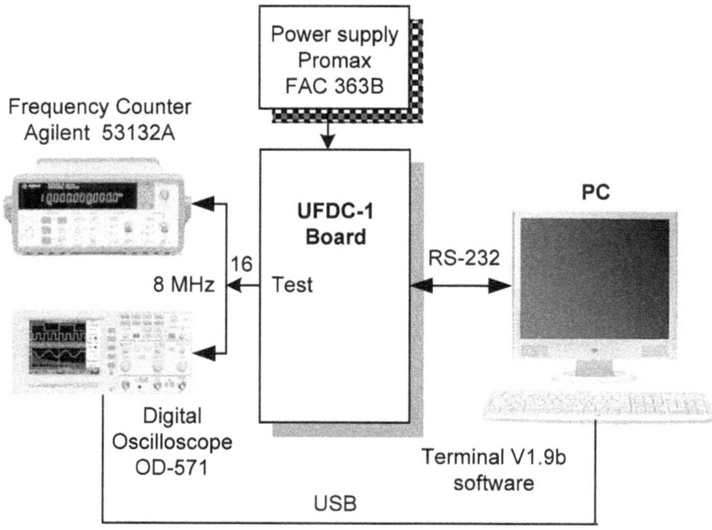

Fig. 18.19 'Quartz-Accurate Automated Calibration Technique' laboratory activities [57].

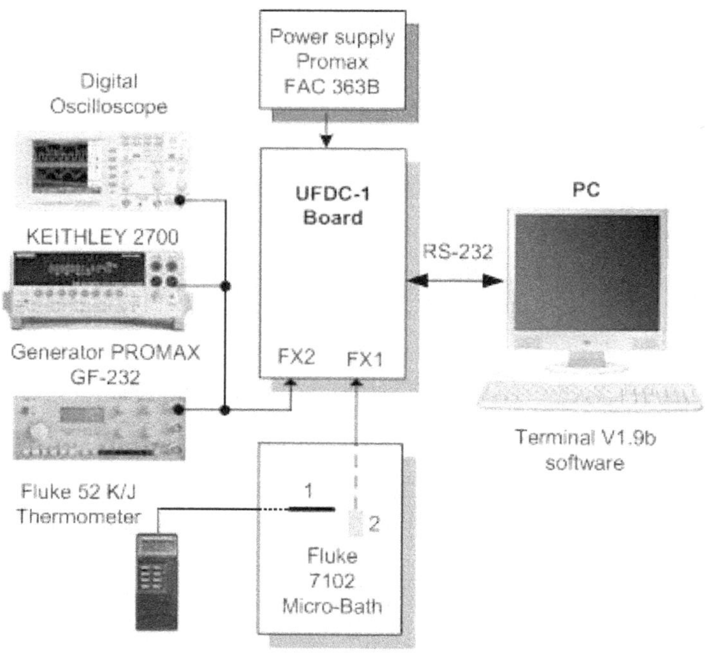

Fig. 18.20. Temperature Sensor Systems laboratory activities [57].

In the 'Temperature Sensor Systems; laboratory activity (Figure 18.20) students can make some metrological experiments with various quasi-digital temperature sensors, for example, semiconductor temperature sensors with frequency (MAX 6577), period (MAX 6576), duty-cycle (SMT-160-30) and pulse-width modulated (MAX 6673) outputs. During the metrological experiments a quasi-digital semiconductor temperature sensor should be connected to the first input (channel) of UFDC-1 boars and put into the Fluke 7102 Micro-bath with a temperature range – 5 ^{0}C to +125 ^{0}C. Appropriate relative error for the UFDC-1 and measuring mode should be chosen. The KEITHLEY 2700 digital multimeter with frequency (3 Hz to 500 kHz) and period (2 µs to 333 ms) measuring ranges and ± 0.01 % relative error is used for sensors output signals measurements [57].

The digital oscilloscope is used for signal shape monitoring at the sensors' outputs. The temperature in the Fluke 7102 micro-bath is monitored by the Fluke 52 K/J thermometer, with measurement error ± (0.1% of reading + 0.7 °C).

During the laboratory activities, any quasi-digital sensors or transducers with frequency, period, duty-cycle, PWM, pulse number or phase-shift output can be directly connected to the UFDC-1 board. In addition, the PROMAX GF-232 functional generator is used for different sensor output emulation and can be connected to the second UFDC-1 board input simultaneously.

During the project, students create both hardware and software for smart sensor systems and calculate their performance, including overall accuracy (resulting error). The calculations are based on an advanced engineering practical approach, which takes into account distribution laws for all components of error [59], see Appendix at the end of this book.

The Universal Sensors and Transducers Interface (USTI) IC [10] can be also used for various laboratory activities both: on site or online, and for PBL and PBeL due to numerous measuring modes (29 measuring modes and one generating mode, two channels and three serial interfaces and buses) including capacitance, resistance and resistive bridges measurements in any courses related to measurements, sensors and electronics.

Not any remote laboratory can be used for PBeL. Special requirements should be satisfied in order to create suitable remote laboratory for PBeL at the reasonable cost. It means that remote laboratory for global distance project based e-learning must be based on so called unified microelectronics component. In case of engineering project-based courses such unified components are the UFDC-1, UFDC-1M-16 and USTI ICs [8-10] and its evaluation boards. They can work with any frequency, period, duty-cycle, PWM, pulse number output sensors. In case of analog sensors a voltage-to-frequency converter should be used.

Finally, the remote laboratory is not limited to distance education. In research and industry, remote accesses also represent a useful and low cost opportunity to meet the growing need of scientists and engineers who wish to test unique or expensive, modern equipment or microelectronics components.

18.4 References

[1]. Sensors Web Portal: http://www.sensorsportal.com/HTML/Sensor.htm
[2]. Global Flow Meters Market to Reach US $ 4.6 Billion by 2015, According to New Report by Global Industry Analysts, Inc., Global Industry Analysts, Inc.,

9 February 2011 (http://www.prweb.com/pdfdownload/8121193.pdf).
[3]. David Wadlow, Turbine Flowmeters, Part 1: Details of the Basic Axial Turbine Flowmeter, *Sensors Magazine*, Vol.16, No.10, October 1999.
[4]. David Wadlow, Turbine Flowmeters, Part 2: Designs and Special Cases, *Sensors Magazine*, Vol.16, No.10, November 1999.
[5]. Jerry S. J. Chen, Paddlewheel Flow Sensors: The Overlooked Choice, *Sensors Magazine*, Vol.16, No.12, 1999.
[6]. Flow Sensors Manufacturers at Sensors Web Portal (http://www.sensorsportal.com): http://www.sensorsportal.com/HTML/SENSORS/FlowSens_Manuf.htm
[7]. Gabor Vass, Ultrasonic Flowmeter Basics, *Sensors Magazine*, October 1997.
[8]. Universal Frequency-to-Digital Converter (UFDC-1), Specification and Application Note, *Technology Assistance BCNA 2010, S. L.*, 2011.
[9]. High-Speed Universal Frequency-to-Digital Converter UFDC-1M-16, Data Sheet, *Technology Assistance BCNA 2010, S. L.*, 2011.
[10].Universal Sensors and Transducers Interface (USTI), Specification and Application Note, *Technology Assistance BCNA 2010, S. L.*, 2011.
[11].N. Viarani, N. Massari, M. Gottardi and A. Simoni, CMOS Interface of a Flow Sensor for Urodynamic Monitoring, in *Proceedings of the IEEE Instrumentation and Measurement Technology Conference (IMTC' 2004)*, Como, Italy, 18-20 May 2004, pp. 1574-1577.
[12].Level Sensors, A Global Strategic Business Report, *Global Industry Analysts, Inc.*, February 2009.
[13].Gabor Vass, The Principles of Level Measurement, *Sensors Magazine*, Vol. 17, no.10, October 2000.
[14].Manufacturers of Level Sensors at Sensors Web Portal (www.sensorsportal.com): http://www.sensorsportal.com/HTML/SENSORS/Level_Sens_Manuf.htm
[15].Load Cells Manufacturers at Sensors Web Portal (http://www.sensorsportal.com): http://www.sensorsportal.com/HTML/SENSORS/Load_Cells_Manufacturers.htm
[16].Digital Load Cells. A Comparative Review of Performance and Application, WP0803, *Institute of Measurement and Control*, London 2003.
[17].Jerome Johnston, Getting the Most out of Strain Gauge Load Cells, *Sensors Magazine*, Vol.17, no. 5, May 2000.
[18].Wilmar Hernandez, Improving the Response of a Load Cell by Using Optimal Filtering, *Sensors*, 6, 2006, pp.697-711.
[19].Miniature Load Cell - iLoad Mini at Loadstar Sensors web site: http://www.loadstarsensors.com/vmchk/lc004-iload-mini-series-miniature-load-cell.html
[20].T. Yan, B. E. Jones, J. Astrup, R. T. Rakowski, M. J. Tudor, S. P. Beeby and N. M. White, Stiff Load Cell With High Overload Capability and Direct Frequency Output, in Proceedings of the *18th European Conference on Solid-State Sensors (Eurosensors '2004)*, 12-15 September, 2004, Rome, Italy, pp. 442-443.
[21].Scott D. Welsby and Tim Hitz, True Position Measurement with Eddy Current Technology, *Sensors Magazine*, November 1997.
[22].Mark Howard, Measuring Position or Speed in Harsh Environments, *Control Engineering Europe*, 29 March 2011.
[23].Steven D. Roach, Designing and Building an Eddy Current Position Sensor, *Sensors Magazine*, September 1998.
[24].Gui Yun Tian, Sofiane Zairi, Ali Sophian, Craig A. Grimes, Frequency Output Sensors, in Encyclopedia of Sensors, Edited by C. A. Grimes, E. C. Dickey, and M. V. Pishko, Vol. 4, *American Scientific Publishers*, 2006, pp. 73-88.

[25].Temposonics® L Series LD 550645 E, Product Specifications, *MTS Systems Corporation*, USA, 2002.

[26].Manufacturers of Position Sensors at Sensors Web Portal (www.sensorsportal.com): http://www.sensorsportal.com/HTML/SENSORS/Position_SensManuf.htm

[27].Manufacturers of Proximity Sensors at Sensors Web Portal: http://www.sensorsportal.com/HTML/SENSORS/Proximity_Sens_Manuf.htm

[28].World Torque Sensors Market, *Frost & Sullivan*, 15 June 2011.

[29].Torque Sensors and Transducers at Sensors Web Portal: (www.sensorsportal.com): http://www.sensorsportal.com/HTML/SENSORS/TorqueSens_Manuf.htm

[30].David Schrand, The Basics of Torque Measurement, Technical Notes and Articles, *Sensor Developments, Inc.*

[31].Wilfried Krimmel, Evolution and Future of Torque Measurement Technology, *Lorenz Messtechnik GmbH.*

[32].T. Yan, B. E. Jones, R. T. Rakowski, M. J. Tudor, S. P. Beeby, N. M. White, Stiff Torque Transducer with High Overload Capability and Direct Frequency Output, in *Proceedings of the 19th IMEKO TC-3 International Conference on Force, Mass, Torque and Density Measurements*, Cairo, Egypt, 19-23 February, 2005.

[33].Rajender Thusu, Wireless Sensor Use is Expanding in Industrial Applications, Sensors Magazine, 1 June 2010.

[34].Comprehensive Analysis of Wireless Sensor Systems Market, *Research and Markets*, April 2006.

[35].A. Bayo, N. Medrano, B. Calvo, S. Celma, A Programmable Sensor Conditioning Interface for Low-Power Applications, in *Proceedings of the Eurosensors XXIV*, 5-8 September, 2010, Linz, Austria, *Procedia Engineering*, Vol. 5, 2010, pp. 53–56.

[36].Manufacturers of Voltage-to-Frequency Converters at Sensors Web Portal: http://www.sensorsportal.com/HTML/SENSORS/VFC_Manufacturers.htm

[37].Jim Williams, Designs for High Performance Voltage-to-Frequency Converters, Application Note 14, Linear Technology, March 1986.

[38].Wanghoon Lee, Yoshio Nishida, Kazuaki Sawada and Makoto Ishida, CMOS RF Transmitter using Pulse Width Modulation for Wireless Smart Sensors, Submitted for *IEEE Transaction for Components, Packaging and Manufacturing Technology*, 2011.

[39].Jose G. Villalobos, Zhen Xu, and Yi Jia, IDC Based Battery-free Wireless Pressure Sensor, *Sensors & Transducers*, Vol. 121, Issue 10, October 2010, pp. 121-132.

[40].J. M. Dias Pereira, O. Postolache, P. Silva Girão, A Low-Cost Tide Measurement System for Water Quality Assessment, in *Proceedings of the IEEE Instrumentation and Measurement Technology Conference (IMTC' 2006)*, Sorrento, Italy, 24-27 April 2006, pp.2226-2230.

[41].D. Ramírez Muñoz, D. Moro Pérez, J. Sánchez Moreno, S. Casans Berga, E. Castro Montero, Design and Experimental Verification of a Smart Sensor to Measure the Energy and Power Consumption in a One-phase AC Line, *Measurement*, Vol. 42, 2009, pp. 412-419.

[42].Single-Phase Multifunction Metering IC with di/dt Sensor Interface ADE7753, *Analog Devices*, USA, 2010.

[43].Papazian Petru, Băbăiţă Mircea and Popovici Adrian, Using the PIC16F84 Microcontroller in Intelligent Stepper Motor Control, *Journal of Electrical and Electronics Engineering*, Vol.1, Issue 1, 2008, pp.232-226.

[44].S. Y. Yurish, Project Based e-Learning: A New Education Technique for Distance Learning in Smart Sensors Systems, in *Proceedings of the 7th IEEE International Conference on Microelectronic Systems Education (MSE' 2009)*, San Francisco, CA, USA, July 2009, pp.93-96.

[45].D. G. Zutin and M.E. Auer, READ-Remote Analog ASIC Design System, International Journal of Online Engineering (iJOE), Vol. 2, No. 4, 2006.

[46].R. Hashemian and J. Riddley, FPGA e-Lab, a Technique to Remote Access a Laboratory to Design and Test, in *Proceedings of the 6th IEEE International Conference on Microelectronic Systems Education (MSE' 07)*, San Diego, CA, USA, June 2007, pp.139-140.

[47].S. Uran, D. Hercog and K. Jezernik, Remote Lab Experiment RC Oscillator for Learning of Control, *International Journal of Online Engineering (iJOE)*, Vol. 2, No. 4, 2006.

[48].Ilhyeon Moon, Saeron Han, et al, A Remote Laboratory for Electric Circuit using Passive Devices Controlled, in *Proceedings of the International Conference on Engineering Education (ICEE' 2008)*, Budapest, Hungary 2008.

[49].I. Gustavsson, J. Zackrisson and T. Olsson, Traditional Lab Session in a Remote Laboratory for Circuit Analysis, in *Proceedings of the 15th EAEEIE Annual Conference on Innovation In Education for Electrical and Information Engineering*, Sofia, Bulgaria 2004, pp.204-209.

[50].A. Mohtar, Z. Nedic and J. Machotka, A Remote Laboratory for Microelectronics Fabrication, in *Proceedings of the 38th ASEE/IEEE Frontiers in Education Conference*, Saratoga Springs, NY, USA, 2008, pp. S2F-7-S2F-12.

[51].B. Sánchez and R. Bragós, Modular Workbench for in-Situ Remote Laboratories, in *Proceedings of Instrumentation and Measurement Technology Conference (IMTC' 2007)*, Warsaw, Poland,), May 2007, pp.1-4.

[52].D. Anton, R. Bragós and P. J. Riu, Remotely Accessible Laboratory for Instrumentation and Sensors, in *Proceedings of Instrumentation and Measurement Technology Conference (IMTC' 2004)*, Vol.2, Como, Italy, May 2004, pp. 1272- 1276.

[53].Evaluation Board for Universal Frequency-to-Digital Converters UFDC-1 and UFDC-1M-16, EVAL-UFDC-1/UFDC-1M-16, *IFSA*, 2011.

[54].Yurish S. Y., Educational and Eurotraining Curriculum in Data Acquisition and Signal Processing for Smart Sensors, Microelectronics Education, in *Proceedings of the 5th European Workshop on Microelectronics Education (EWME' 2004)*, Lausanne, Switzerland, 15-16 April, 2004, ed. by Adrian M. Ionescu, Michel Declercq, Maher Kayal, Yusuf Leblebici, *Springer*, 2006, pp. 197-201.

[55].Yurish S. Y., Novel Curricula in Smart Sensors Systems Design in the Frame of Human Resources and Mobility Marie Curie Chairs (EXC) Action, in *Proceedings of 7th European Workshop on Microelectronics Education (EWME' 2008)*, Budapest, Hungary, 28-30 2008, pp. 132-133.

[56].Yurish S. Y., Integration of Marie Curie Chairs (EXC) Project Based Course on 'Smart Sensors Systems Design' into 'Sensors and Signal Conditioning' Curricula, in *Proceedings of 12th IMEKO TC1-TC7 joint Symposium on Man Science & Measurement*, Annecy, France, 3-5 September, 2008, pp.439-444.

[57].Yurish S. Y., Smart Sensors Systems Design: New Curricula Based on Marie Curie Chairs Excellence (EXC) Project's Results, *Measurement,* Vol.43, Issue 1, January 2010, pp. 21-30.

[58].Yurish S. Y., Smart Sensors Systems: Project Research Results Dissemination, Exploiting and Using in Teaching Activities, in *Proceedings of Marie Curie Conference, European Science Open Forum (ESOF' 2008)*, Barcelona, Spain, 17-18 July 2008, P-297.

[59].P. V. Novitsliy, I. A. Zograf, Errors Estimation for Measuring Results, *Energoatomizdat*, Leningrad, 1991 (in Russian).

Chapter 19

System-on-Chip (SoC), System-in-Package (SiP) and MEMS

It was mentioned in earlier predictions that "the use of IC compatible process lines for MEMS production is growing. There are clearly high benefits in using CMOS or BiCMOS-compatible manufacturing lines to target high-volume MEMS markets ... There is a general trend toward adoption of more integrated solutions." [1], and "the technological and cost factors were modified for the benefit of frequency output sensors and transducers" [2-4]. It also have been written some years ago: "Moving from the traditional analog (voltage and current) signal domain to the frequency-time signal domain (frequency, period, duty-cycle, time interval, phase-shift, etc.) lets achieve many benefits due to properties of frequency as informative parameters ..."[5]. Let see how modern microelectronics technologies make these predictions true.

19.1 System-on-Chip

Modern CMOS technologies allows manufacturers to greatly enhance silicon-based MEMS sensors by integrating signal conditioning, signal processing, digital output options, diagnostics and communications on a System-on-Chip (SoC). The integration of electro-mechanical device with IC circuit is interesting technology from low noise point of view and from low power point of view as well. Using frequency-time domain MEMS sensors in this design gives more precise control in comparison with tradition analog (voltage or current) sensor outputs [6]. Such approach delivers enhanced accuracy and reliability due to high precision sensors [7]. Interface circuit design has become a strategic technology for modern SoC including MEMS sensors, interfacing, processing and communication modules.

Achievements in the field of frequency-time measurements and frequency sensors are connected directly with the advent of new methods of measurement, and also perfection of modern microelectronic technologies [8]. And the first noticeably take the lead over perfection of microelectronic technologies in its development. So the first digital methods of measurement of frequency and time intervals have been patented in 40th years while digital frequency counters based on digital methods for frequency measurements have come in the stead of analog frequency meters only at the end of 60th. An occurrence of one-crystal, high-performance RISC-microcontrollers in a combination with advanced measuring methods for frequency-time parameters of electric signals has made today a possible creation of single-chip multipurpose frequency counters and high precision frequency-to-digital converters [9]. The further development of microelectronic technologies automatically leads to improvement of metrological characteristics and extension of functionalities for such devices without any changes in methods of measurements.

A possible sensor system-on-chip is shown in Figure 19.1. It should be based on CMOS digital ICs UFDC or USTI [10-12] as a SoC's core, and include two CMOS sensors, one of them can be a temperature sensor, for example. In order to working with analog (voltage or current output) sensor, the SoC can contain CMOS voltage-to-frequency converter. Additional sensing element such as capacitive, resistive and resistive bridge can be directly interfacing to SoC. The USTI IC can contain a TEDS table according to IEEE 1451 standard in its memory. Only one external component – quartz crystal oscillator (and reference components for capacitance, resistance and resistive bridges measurements) must be connected to the SoC.

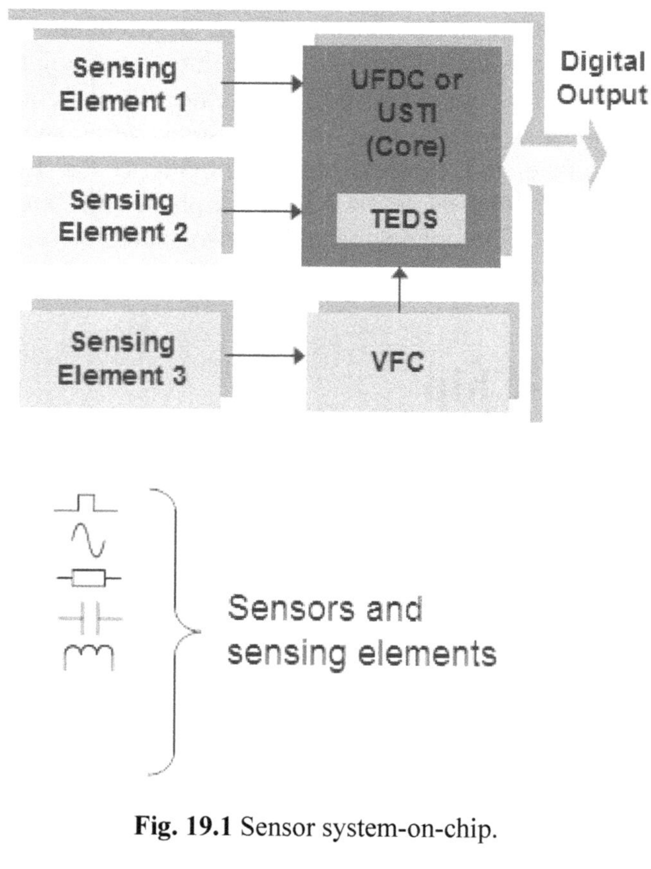

Fig. 19.1 Sensor system-on-chip.

19.2 MEMS-based Oscillators

The following milestone in the development of modern SoC and SiP with built in quasi-digital sensors was creation micro-electromechanical-system (MEMS) oscillators, which can be integrated onto CMOS chip [13]. Quartz crystal oscillators are widely used in different measuring instruments and devices such as frequency counters, tachometers, counter/timers and integrated frequency-to-digital converters, where an accurate frequency or time reference is required. They provide the integrated universal frequency-to-digital converter with stable, precision frequency base [14].

MEMS offer minimal packaging constraints, low power consumption, low damping, ease of parameter tuning, and relatively simple integration with electronics [15]. MEMS quartz oscillators promise smaller, higher-precision references that are unaffected by moisture, temperature or mechanical shock and vibrations [16]. The possible MEMS oscillator architectures are shown in Figure 19.2.

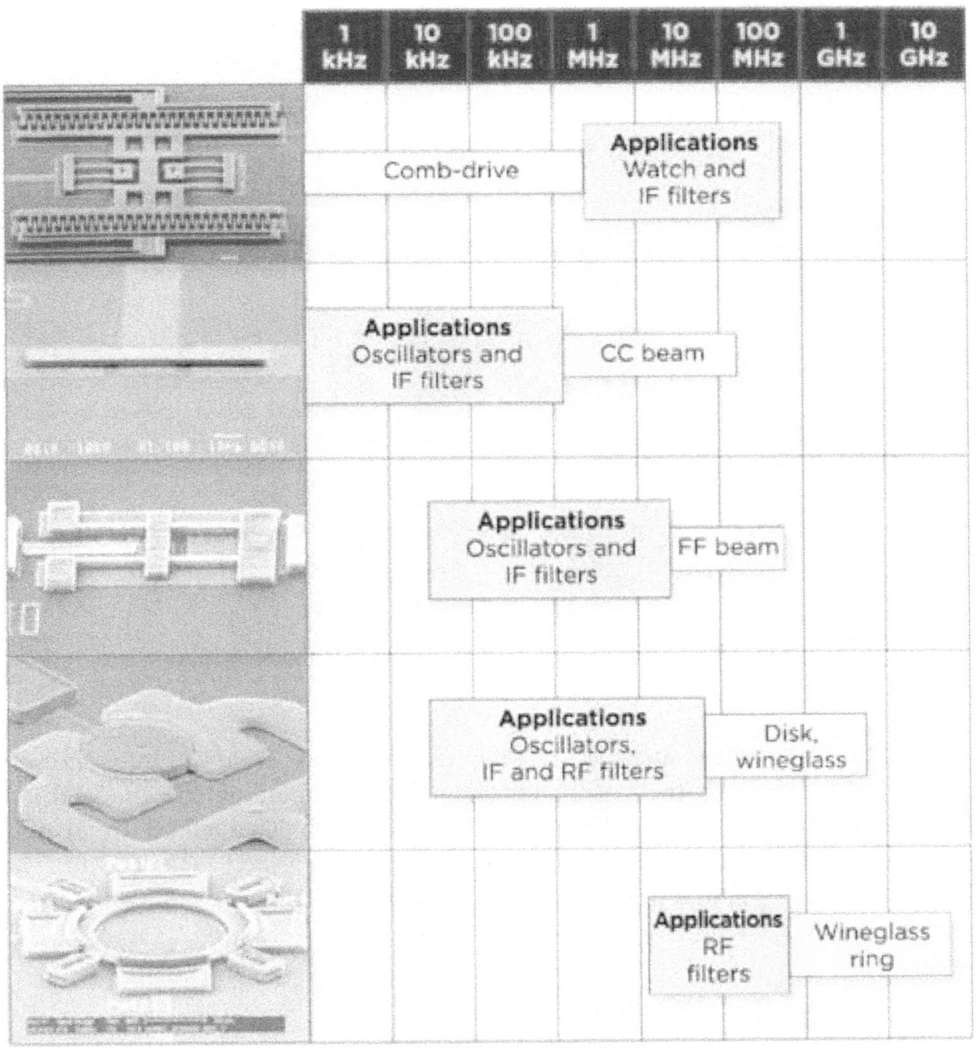

Fig. 19.2 MEMS oscillator architectures [16].

Taking into account the frequency range for reference oscillators for the UFDC and USTI ICs (16 MHz and 20 MHz respectively), and its future modification (up to 50-75 MHz), any of shown MEMS architectures can be used for this purpose.

According to R. Colin Johnson [16], "the total potential market for MEMS oscillators in all the place where crystal oscillators are used today exceeds US $ 3 billion, according to the MEMS chip vendors". Significant part of this market in the nearest future can belong to digital on-chip sensors systems based on MEMS frequency output sensors, advanced frequency-to-digital converter and MEMS quartz oscillators. The last one can output a signal ranging from 1 to 125 MHz, which is enough for all frequency measurements applications, frequency sensors and digital sensors and transducers on its basis.

The superior shock resistance comes courtesy of the MEMS oscillators' gravity-defying weight. Simulations predict the oscillators can survive shocks of 100,000 g's [17]. Without the conditioning circuitry on the CMOS chip, a micron-sized MEMS resonator is not as accurate as large quartz crystal resonator and can has 20 ppm frequency variations. However, when CMOS circuitry is added MEMS oscillators can be tuned more precisely than quartz crystal oscillators.

In 2009 *SiTime* introduced the world's first differential output MEMS oscillator SiT9102 with ultra-high frequency stability of ±10 ppm, 1 – 220 MHz frequency range and -40 ^{0}C to +85 ^{0}C operational temperature range (Figure 19.3) [18].

Fig. 19.3 Differential output MEMS oscillator SiT9102.

Today MEMS companies such as *SiTime, Discera* and *Sand 9* are continuing to commercialize and develop their oscillator technologies, including so called "all-silicon" oscillator technology. So, the *SiTime* proposes the SiT5000 standard frequency high performance Voltage Controlled, Temperature Compensated (VCTCXO) MEMS oscillator with ± 0.5 ppm to ± 5 ppm frequency stability [19].

19.3 System-in-Package

As it was written in [16] MEMS oscillators can be tuned for high-precision applications, where their features may justify the price premium. Form the other side, methods for lowering the price include combining ASIC and MEMS sensor chip in one package and decreasing the number of part/interconnections between separate chips [20]. MEMS can go in plastic packages, which costs less than half that for quartz. Taking into account these design approach, new applications and current MEMS technology achievements, the promised SoC or System in Package (also known as Multi-Chip Module) can have an architecture shown in Figure 19.4.

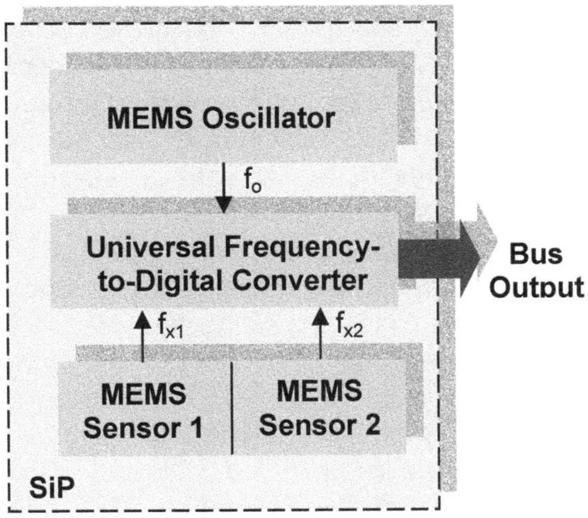

Fig. 19.4 System in Package with MEMS oscillator, USTI and MEMS sensors [6].

MEMS oscillator likewise behave like resonant RLC circuits, but they integrate all the function of millimeter-sized quartz crystal into a micron-sized "tuning fork" that is naturally immune to temperature and vibration. They are much more rugged than quartz crystal oscillators and keep on working reliably even under extreme temperatures. Designed sensors system based on such SoC or SiP does not need any external time or frequency references. For one's turn, the universal frequency-to-digital converter lets solve problems with the interface circuit design and additional circuitry for MEMS oscillators in order to increase its short frequency stability.

Today, different MEMS sensors and transducers, universal frequency-to-digital converters based on the most advanced methods and MEMS oscillators exist. All are based on CMOS technology. Many components exist as a standard library cells for various CAD tools or in IP form, for example, the UFDC and USTI ICs [21, 22], licensed through Technology Assistance BCNA 2010, S. L. [23]. It is necessary to make a little step ahead in a direction of creation essentially new SoC or Multi-Chip Module: to combine the mentioned above components in one crystal. The success and benefit of such devices is obvious.

19.4 References

[1]. Yurish S. Y., Kirianaki N. V., Myshkin I. L., World Sensors and MEMS Markets: Analysis and Trends, *Sensors & Transducers,* Vol. 62, Issue 12, December 2005, pp. 456-461.

[2]. Kirianaki N. V., Yurish S. Y., Shpak N. O., Deynega V. P., Data Acquisition and Signal Processing for Smart Sensors, *John Wiley & Sons*, Chichester, UK, 2002.

[3]. Yurish S. Y. Sensors and Transducers: Frequency Output Versus Voltage Output, *Sensors & Transducers,* Vol. 49, Issue 11, November 2004, pp.302-305.
[4]. Smart Sensors and MEMS, ed. by Sergey Y. Yurish and Maria Teresa S.R. Gomes, *Springer*, 2005.
[5]. Yurish S. Y. Forecasts and Modern Sensors Market: Today's Revolution Changes, *Sensors & Transducers*, Vol. 41, Issue 3, March 2004, pp.170-173.
[6]. Yurish S. Y., MEMS Based System-on-Chip and System in Package: New Perspectives, *Sensors & Transducers*, Vol.78, Issue 4, April 2007, pp.I-IV.
[7]. Yurish S. Y., Kirianaki N. V., Interface Circuit Design for Frequency-Time Domain MEMS Sensors, *Journal of Physics: Conference Series*, Vol. 34, 2006, pp. 101–105.
[8]. Yurish S. Y., Modern MEMS Technologies and Advanced Signal Processing: How to Obtain More Benefits in Smart Sensors Systems ? *Sensors & Transducers,* Vol. 36, Issue 10, October 2003, pp.I-III.
[9]. Yurish S. Y., Kirianaki N. V., Pallas-Areny R., Universal Frequency-to-Digital Converter for Quasi-Digital and Smart Sensors: Specifications and Applications, *Sensor Review*, Vol. 25, No.2, April 2005, pp.92-99.
[10].Universal Frequency-to-Digital Converter (UFDC-1), Specification and Application Note, *Technology Assistance BCNA 2010, S. L.*, 2011.
[11].High-Speed Universal Frequency-to-Digital Converter UFDC-1M-16, Data Sheet, *Technology Assistance BCNA 2010, S. L.*, 2011.
[12].Universal Sensors and Transducers Interface (USTI), Specification and Application Note, *Technology Assistance BCNA 2010, S. L.,* 2011.
[13].R. L. Kubena, F. P. Stratton, D. T. Chang, R. J. Joyce, T. Y. Hsu, M. K. Lim and R. T. M'Closkey, MEMS-Based Quartz Oscillators and Filters for on-Chip Integration, in *Proceedings of the Precise Time and Time Interval (PTTI) Systems and Applications*, 2005, pp.122-127.
[14].Yurish S. Y., Kirianaki N. V., Automated Calibration Technique for Programmable Universal Frequency-to-Digital Converter, in *Proceedings of IFAC Workshop on Programmable Devices and Embedded Systems (PDeS'2006)*, Czech Republic, Brno, 14-16 February 2006, pp.72-77.
[15].Steven W. Shaw, Jeffrey F. Rhoads, Kimberly L. Turner, Parametrically Excited MEMS Oscillators with Filtering Applications, in *Proceedings of 10th Conference on Nonlinear Vibrations, Stability, and Dynamics of Structures,* July 25-29, 2004.
[16].R. Colin Johnson, MEMS Eyes Quartz Crystal Apps, *EE Times*, 23 October 2006, pp.6 and 80.
[17].R. Colin Johnson, M/A-COM Qualifies MEMS Oscillator for Smart Munitions, *EE Times Online*, 9 April 2007.
[18].SiT9102 LVPECL / HCSL / LVDS / CML 1 to 220 MHz High Performance Oscillator, Data Sheet, Rev. 1.52, *SiTime Corporation*, 23 June 2010.
[19].SiT5000 Standard Frequency MEMS TCXO and VCTCXO, Advanced Information, Rev. 0.51, *SiTime Corporation,* 18 August 2011.
[20].Henne van Heeren, Jeremie Bouchaud, Richard Dixon, Patric Salomon, Rewards and Risk of Moving into New Applications: Case Study Accelerometers, *mstNews*, No.1, February 2007, pp.35-39.
[21].Worldwide Component Library: http://www.2wcl.com/
[22].Design & Reuse: http://www.design-reuse.com/
[23].Technology Assistance BCNA 2010, S. L. web site: http://www.techassist2010.com/

Appendix

Sensor System's Error Estimation: Engineering Approach

Modern sensor systems can consist of precision components such as accurate sensors; precision frequency-to-digital converters UFDC and USTI [1-3] with programmable relative error; quartz crystal oscillators or temperature compensated, voltage controlled generators. However, it is necessary to evaluate the overall (resultant) error of such sensor systems. There are many techniques for resulting error evaluation: from simple arithmetic summation, which is not recommended strongly for use, because of it gives overestimated result, to advanced modern computer-based techniques based on Lyapunov's characteristic functions and a piece-wise linear approximation of the respective probability distribution functions [4-7].

The advanced but simplified and easy to use an engineering technique for error estimation is based on a geometrical summation of n standard mean square non-correlated error's components [8]:

$$\sigma_{\Sigma}=\sqrt{\sigma_1^2+\sigma_2^2+\ldots+\sigma_n^2}=\sqrt{\sum_{i=1}^{n}\sigma_i^2} \qquad (1)$$

At errors summation, all error components must be presented by their standard mean square errors in order to eliminate an influence of distribution laws deformation. In order to use this technique in practice in is necessary to know (or to assume) distribution laws for each error components. Some useful considerations, which should be used in engineering practice, are described below.

Form one side the sensor system's accuracy is limited by sensor's error itself. Any further precision frequency-to-digital conversion can not improve this basic accuracy. From the other side, if suppose (in theory, of course) that we have a sensor with unlimited accuracy, in this case the sensor system's error will be limited by the frequency reference error – quartz crystal oscillator or generator.

For example, if you are using a sensor with the relative or full scale error $\delta_s = \pm 0.01$ %, you must select the UFDC's or USTI's quantization relative error in one order less (or in five times less, at least), than the sensor's error in order to be neglected in comparison with the sensor's error. You can also neglect the quartz crystal's (reference frequency) error if you are using a precision 10 ppm (± 0.001 % relative error) oscillator or generator, or calibrated UFDC or USTI ICs [9, 10]. In this case, the sensor system's overall (resultant) error is determined by sensor's error and equals to ± 0.01 %.

However, in practice, when you are using, for example, 30 ppm (± 0.003 %) quartz crystal oscillator or generator and non calibrated UFDC or USTI ICs, the sensor system's overall (resulting) error must be evaluated in details. First of all you need to know relative or full scale errors (δ_i) for all error components and its distribution laws in order to calculate appropriate standard mean square errors (σ_i). Let consider these errors and its distribution laws in details.

Sensor's error, (δ_s). Many various factors influence on the sensor's error. It contains many error's components distributed according to different distribution laws, as rule, more than five. Clear, it is not probably, that a manufacturer will indicate the distribution law in the data sheet, hence we can assume, that the sensor's error is distributed according to the Gaussian (normal) distribution law. In this case, for such distribution function, the standard mean square error should be calculated according to the following equation:

$$\sigma_s = \frac{\delta_s}{t_i}, \tag{2}$$

where t_i is the normalized quantile of normal distribution, which should be selected from the Table 1 [8], according to the required probability of estimation P:

Table 1 Normalized quintile of normal distribution.

P	0.8	0.9	0.95	0.98	0.99	0.995	0.998
t_i	1.28	1.64	1.96	2.33	2.58	2.81	3.09

For example, for probability P=0.95 the t_i = 1.96. Hence, in this case the standard mean square error should be calculated according to the following formula:

$$\sigma_s = \frac{\delta_s}{1.96} \tag{3}$$

Quantization error (δ_q). The possible programmable quantization relative error's discrete values for the UFDC and USTI ICs can be selected from the Table 2:

Table 2 Programmable relative errors for UFDC and USTI ICs.

	Programmable relative error (δ_q), %										
UFDC	1	0.5	0.25	0.1	0.05	0.025	0.01	0.005	0.0025	0.001	-
USTI	1	0.5	0.25	0.1	0.05	0.025	0.01	0.005	0.0025	0.001	0.0005

The quantization error is distributed according to symmetrical triangular (Simpson's) distribution law because of the quantization error consists of two components distributed according to the uniform distribution function each and the sum of two uniform distribution functions give the triangular distribution law [4, 11]. Hence, the standard mean square error should be calculated by the following way:

$$\sigma_q = \frac{\delta_q}{\sqrt{6}} \tag{4}$$

Reference frequency error (δ_{fo}). The reference quartz oscillator or generator (with the period of square wave T_0 and the pulse width $\tau << T_0$) is a source of the reference frequency error δ_{fo}. This error component is distributed according to the uniform distribution law, and the standard mean square error should be calculated according to the follow equation [4]:

$$\sigma_{fo} = \frac{\delta_{fo}}{\sqrt{3}} \tag{5}$$

The reference frequency error for calibrated UFDC-1, UFDC-1M-16 and USTI ICs [9, 10] is $\delta_{fo} = 0.0001$ %.

Other errors. There are some other errors, which can be taken into account at the total error determination. For example, a voltage-to-frequency converter's error (δ_{VFC}) at the use of analog sensors with voltage output instead of quasi-digital sensors. The assumed distribution law for VFC's error is Gaussian (normal). See consideration for the sensor's error above.

Other possible error is a calculation error by extended microcontroller or PC, if additional calculations are used to get measuring results. It is strongly recommended to select calculation error much less, then minimal component error in the sensor system:

$$\delta_{calc} << \delta_{i\min} \tag{6}$$

In opposite case, it should be taken into account at calculation of overall (resulting) error of sensor system. This error component is distributed according to the uniform distribution law (5).

Several additional components of error such as temperature error and errors due to other factors can be also taken in account if necessary.

After calculation of the standard mean square errors for all error components, the overall (resulting) standard mean square error for sensor system must be calculated according to the following formula, including the necessary components:

$$\sigma_\Sigma = \sqrt{\sigma_s^2 + \sigma_q^2 + \sigma_{fo}^2 + \sigma_{VFC}^2} \tag{7}$$

Now, it is necessary to make an assumption about distribution law of resulting error and calculate the overall sensor system error.

The following considerations will be useful for determination of resulting distribution function [8]:

1) If two error's components are distributed according to the Gaussian (normal) distribution law, its composition also will be distributed according to the normal distribution law;
2) If the number of error components have been summarized is $k > 5$, the resulting distribution law is normal;
3) However, in engineering practice at limited number of error components, it is also possible to assume, that the resulting distribution function will be close to the normal distribution law. For such case there is the following consideration:

$$\delta_\Sigma = 1.64 \times \sigma_\Sigma \text{ - for } n \leq 3 \text{ components at } P{=}0.90$$
$$\delta_\Sigma = 1.96 \times \sigma_\Sigma \text{ - for } n \geq 4 \text{ components at } P{=}0.95 \qquad (8)$$

Taking into account all considerations above, the overall reference error with the probability P=0.95 (see Table 1) should be calculated as:

$$\delta_{0.95\Sigma} = \sigma_\Sigma \times 1.96 \qquad (9)$$

Let consider a practical example of sensor system and determine its total error.

Example: let a sensor systems consist of an analog pressure sensor, for example, Type 740C Baratron from *MKS Instruments, Inc.* with a voltage output (0-10 V) and full scale error $\delta_s = \pm 0.1$ % [12]; a voltage-to-frequency converter (μ 2.0 Series from *Canopus Instruments*), with the relative error $\delta_{VFC} < \pm 0.1$ % [13], and non-calibrated UFDC-1 IC with the 30 ppm ($\delta_{fo} = \pm 0.003$ %) low cost quartz crystal oscillator form *Siward Crystal Technology Co., Ltd.* An example of such sensor system with four components of error is shown in Figure 1.

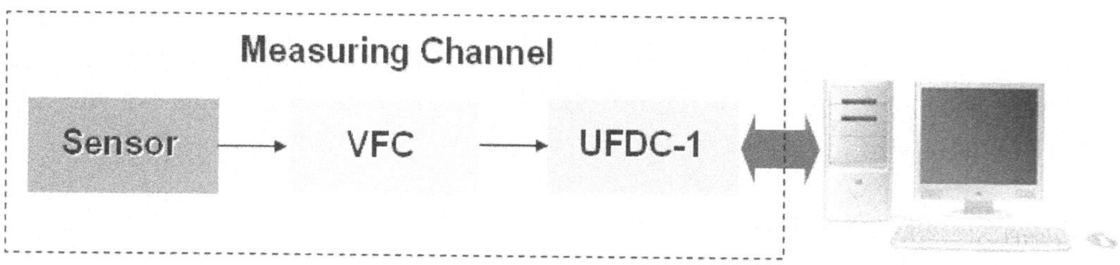

Fig. 1 An example of sensor system.

Taking into account that the UFDC and USTI ICs have two channels, the error estimation must be done for each of channel if two different sensors (measuring channels) are used.

The UFDC-1's programmable relative error, which is determinates mainly by the quantization error, should be selected $\delta_q = \pm 0.01$ % in one order less in comparison with the sensor and VFC's errors in order to be neglected. The frequency relative error

(δ_{fo} = ± 0.003 %) also can be neglected, but in this example we will take into account all components of total error for training purposes. Now we need to calculate standard mean square errors for each component taking into account of appropriate distribution law according to equations (3), (4) and (5). The calculation results are shown in Table 3.

Table 3 Components error calculation results.

Error's component	Relative error, %	Distribution law	Standard mean square error, %
Sensor's error	$\delta_s = \pm 0.1$	Gaussian	$\sigma_s = \dfrac{\delta_s}{1.96} = \pm 0.051$ %
VFC's error	$\delta_{VFC} = \pm 0.1$	Gaussian	$\sigma_s = \dfrac{\delta_s}{1.96} = \pm 0.051$ %
Quantization error	$\delta_q = \pm 0.01$	Triangular	$\sigma_q = \dfrac{\delta_q}{\sqrt{6}} = \pm 0.0041$ %
Reference error	$\delta_{fo} = \pm 0.003$	Uniform	$\sigma_{fo} = \dfrac{\delta_{fo}}{\sqrt{3}} = \pm 0.0017$ %

Let calculate now the overall (resulting) standard mean square error for sensor system according to equation (7):

$$\sigma_\Sigma = \sqrt{0.051^2 + 0.051^2 + 0.0041^2 + 0.0017^2} =$$

$$= \sqrt{0.0026 + 0.0026 + 0.000017 + 0.0000029} = \sqrt{0.00522} = 0.0723$$

Taking into account considerations (8) and equation (9) we will get the overall (resulting) relative error of designed sensor system:

$$\delta_{0.95\Sigma} = \sigma_\Sigma \times 1.96 = \pm 0.0723 \times 1.96 = \pm 0.142\,\%$$

In other words, the confidence interval for the total sensor system's error with probability P=0.95 is $\delta_{0.95\Sigma} \in$[- 0.142 % … + 0.142 %].

Let compare results obtained with the use of described above engineering technique for error estimation and results obtained with the advanced computer-based technique, described in [4-7].

The technique based on Lyapynov's characteristic functions (LCF) and a piece-wise linear approximation of the respective probability distribution functions lets precisely determine the resulting distribution law (for any number n of error components) and its characteristics. The initial data (distribution functions and errors' characteristic) are shown in Table 4. The resulting distribution function of designed sensor systems is very close to the Gaussian (normal) distribution law (Figure 2). Its characteristics are shown in Table 5.

Table 4 Error components and their characteristics (at probability $P = 0.95$).

Error component and their characteristics	**Distribution function**
Relative error: $\delta_s = \pm 0.1$ % Distribution function: Gaussian Coefficient of skewness, $K_{sk} = 0$ Coefficient of excess, $K_{exc} = -7.27\times10^{-13}$ Root mean square error: $\sigma_s = \pm 0.051$ % Dispersion: $D_s = 0.0026$ %	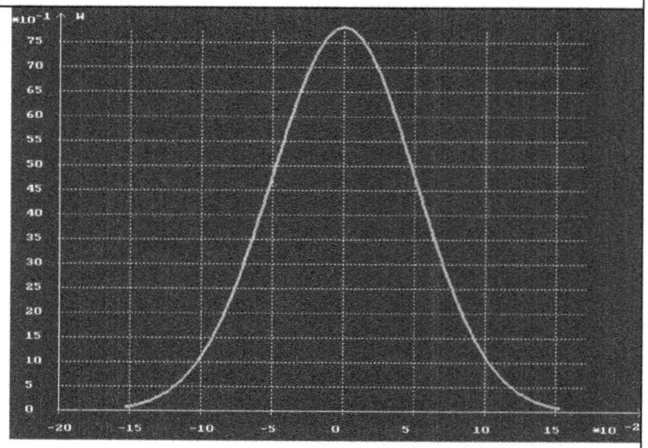
Relative error: $\delta_{VFC} = \pm 0.1$ % Distribution function: Gaussian Coefficient of skewness, $K_{sk} = 0$ Coefficient of excess, $K_{exc} = -7.27\times10^{-13}$ Root mean square error: $\sigma_{VFC} = \pm 0.051$ % Dispersion: $D_{VFC} = 0.0026$ %	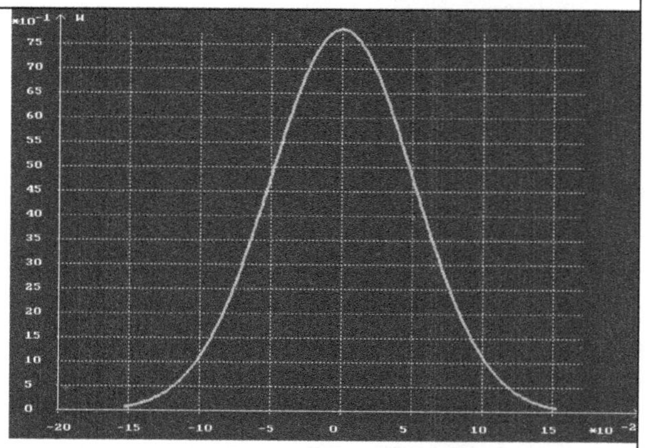
Relative error: $\delta_q = \pm 0.01$ % Distribution function: Triangular Root mean square error: $\sigma_q = \pm 0.0041$ %	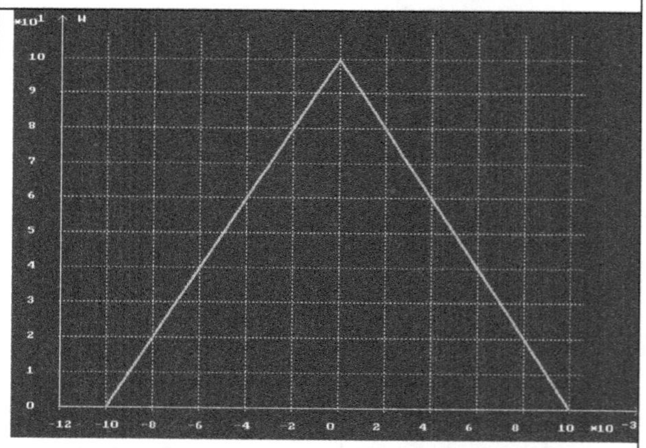
Relative error: $\delta_{f0} = \pm 0.003$ % Distribution function: Uniform Root mean square error: $\sigma_{fo} = \pm\, 0.0017$ %	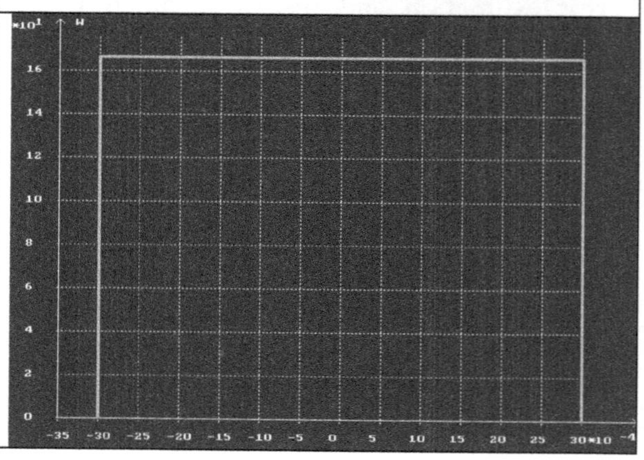

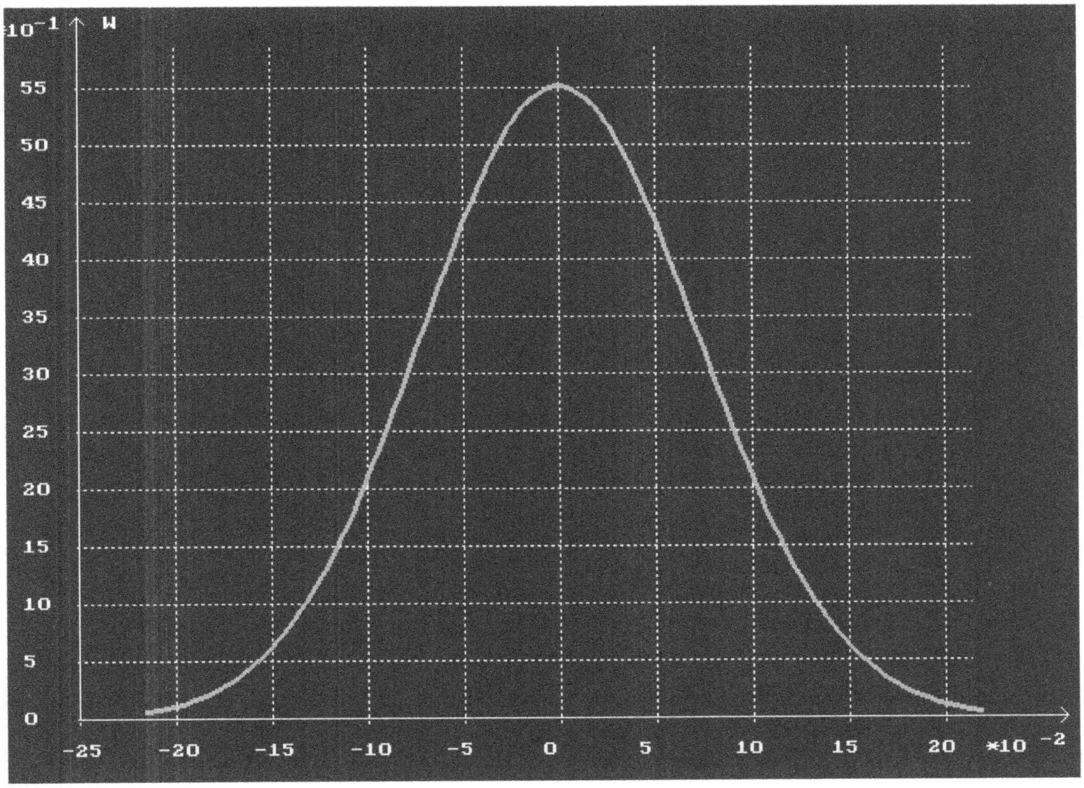

Fig. 2 Resulting distribution function of designed sensor system.

Table 5 Statistical characteristics of the total error.

Parameter	Value
Mathematical expectation, *M*	0
Dispersion, *D, %*	5.22×10^{-3}
Coefficient of skewness, K_{sk}	0
Coefficient of excess, K_{exc}	$- 6.51 \times 10^{-6}$
Confidence interval (at $P = 0.95$), %	- 0.142 … + 0.144

As it is visible form the Table 5 the confidence interval for sensor system's error at the same probability $P = 0.95$ is $\delta_{0.95\Sigma} \in$[- 0.142 % …+ 0.144 %], and difference of two estimations does not exceed 1.4 %. This difference is practically the same for other values of probability *P*: 0.98; 0.99; 0.995 and 0.998. It is clear that the difference between two techniques will be increased in cases when resulting distribution law is much differs from the Gaussian (normal) distribution law.

References A

[1]. Universal Frequency-to-Digital Converter (UFDC-1), Specification and Application Note, *Technology Assistance BCNA 2010*, S. L., 2011.

[2]. High-Speed Universal Frequency-to-Digital Converter UFDC-1M-16, Data Sheet, *Technology Assistance BCNA 2010, S. L.*, 2011.

[3]. Universal Sensors and Transducers Interface (USTI), Specification and Application Note, *Technology Assistance BCNA 2010, S. L.*, 2011.

[4]. Kirianaki N. V., Yurish S. Y., Shpak N. O., Deynega V. P., Data Acquisition and Signal Processing for Smart Sensors, *John Wiley & Sons*, 2002.

[5]. Kirianaki N. V., Shpak N. O., Yurish S. Y., Khoma V. V., Dzoba Y. S., New Method of Summation for Measurement Errors Based on Piece-Wise Linear Approximation of Probability Distribution, *Journal of Electrical Engineering*, 3-4, Vol. 51, 2000, pp.94-99.

[6]. Kirianaki N. V., Shpak N. O., Yurish S. Y., Khoma V. V., Dzoba Y. S., Novel Technique for Computer-based Estimation of Measurement Errors, in *Proceedings of the 10th International Symposium on Development in Digital Measuring Instrumentation*, September 17-18, 1998, Naples, ITALY, pp.196-200.

[7]. Yurish S. Y., Kirianaki N. V., Shpak N. O., Improved Analytical Approach to Evaluation of ADC's Error, in *Proceedings of the 4th IMEKO Workshop on ADC Modeling and Testing*, 9-10 September 1999, Bordeaux, France, pp.51-54.

[8]. Novitskiy P. V., Zograf I. A., Error Estimation of Measuring Results, *Energoatomizdat*, Leningrad, 1991 (in Russian).

[9]. Yurish S. Y., Kirianaki N. V., Automated Calibration Technique for Programmable Universal Frequency-to-Digital Converter, in *Proceedings of the IFAC Workshop on Programmable Devices and Embedded Systems (PDeS'2006)*, Czech Republic, Brno, 14-16 February 2006, pp.72-77.

[10]. Yurish S. Y., Advanced Automated Calibration Technique for Universal Sensors and Transducers Interface IC, in *Proceedings of the IEEE International Instrumentation and Measurement Technology Conference (I²MTC 2009)*, Singapore, May 5-7, 2009, pp.402-405.

[11]. Dietrich C.F., Uncertainty, Calibration and Probability, 2nd Edition, *Institute of Physics Publishing*, Bristol, UK, 1991.

[12]. Types 740C, 750C BARATRON® General Purpose Capacitance Manometer, *MKS Instruments, Inc.*, 2009.

[13]. Manufacturers of Voltage-to-Frequency Converters at Sensors Web Portal: http://www.sensorsportal.com/HTML/SENSORS/VFC_Manufacturers.htm

Index

www.ingramcontent.com/pod-product-compliance
Lightning Source LLC
Chambersburg PA
CBHW080934240926
51849CB00003BA/457

* 9 7 8 8 4 6 1 6 0 6 5 2 8 *